D0149475

# COMPUTATIONAL
# ASPECTS OF
# VLSI

# PRINCIPLES OF COMPUTER SCIENCE SERIES

**Series Editors**
**ALFRED V. AHO,** *Bell Telephone Laboratories, Murray Hill, New Jersey*
**JEFFREY D. ULLMAN,** *Stanford University, Stanford, California*

**David Maier**
*The Theory of Relational Databases*

**Theo Pavlidis**
*Algorithms for Graphics and Image Processing*

**Arto Salomaa**
*Jewels of Formal Language Theory*

**Jeffrey D. Ullman**
*Computational Aspects of VLSI*

## ANOTHER BOOK OF INTEREST

**Jeffrey D. Ullman**
*Principles of Database Systems, Second Edition*

# COMPUTATIONAL ASPECTS OF VLSI

## JEFFREY D. ULLMAN
*STANFORD UNIVERSITY*

COMPUTER SCIENCE PRESS

*Computer Science Press, Inc.*
*11 Taft Court*
*Rockville, Maryland 20850*

1  2  3  4  5  6                                              87 86 85 84


**Library of Congress Cataloging in Publication Data**

Ullman, Jeffrey D., 1942-
  VLSI.

  Bibliography: p.
  Includes index.
  1. Integrated circuits—Very large scale integration.
I. Title.  II. Title: V.L.S.I.
TK7874.U36  1983       621.381'73              83-7529
ISBN 0-914894-95-1

# TABLE OF CONTENTS

# PREFACE

I wish I could call this work the first fifth-generation Computer Science text, but unfortunately I have trouble accounting for about three of the generations. However, I do have the feeling that there is something special about the material in this book, because the subject matter builds on an unusually large number of different disciplines in Computer Science and Electrical Engineering. The book reminds me that Computer Science is coming of age, and concepts that were state-of-the-art in the 1960's and 1970's are now elementary background for a whole new layer of concepts and techniques.

## Prerequisites

While I have taught the material to first year graduate students, and it wouldn't surprise me if good Seniors could handle the material, there is no doubt in my mind that to use this book well, one needs a firm grounding in a diverse group of subjects, some of which are not generally perceived as related to one another. One needs to know something about compilers, about data structures, about automata theory, and about the theory of algorithms. From Electrical Engineering, one at least needs elementary switching theory and the Mead-Conway VLSI design methodology. I have provided a brief synopsis of the Mead-Conway theory in Chapter 1, and there is an appendix to introduce the reader to a few of the most central topics from Theoretical Computer Science, but I am counting on the reader having a good undergraduate education in Computer Science.

## Overview of the Book

The subject matter is divided into three parts. The first, lower bounds for VLSI-oriented computation, is a subject that belongs primarily to Theoretical Computer Science. In Chapters 2 and 3 we deal with the question of what does it cost to solve a problem in the VLSI environment. The second part is also theoretical, but more oriented to the design of algorithms, rather than the intellectually interesting but difficult-to-apply material on lower bounds that characterizes the first third of the book. In Chapters 4–6 we discuss different organizations for computers that are composed of many independent processors. These may be viewed either as many on one chip or as many chips or boards interconnected; many of the algorithms and issues are the same in either case. Since the key to the design of "fifth generation" computers with massive parallelism lies in the organization of processor networks and the algorithms such organizations support, we hope that these chapters will form a basis for the study of such issues.

The third part of the book concerns VLSI design tools and the algorithms that underlie them. Since algorithms in this area have immediate commercial application, I was faced with the curious phenomenon that although much thought has gone into the design of such systems, comparatively little concerning the algorithmic details has appeared in the open literature. Fortunately, there has been a recent surge of interest in the field from academic institutions and the industrial labs that do the best job of imitating an academic environment. Thus, I have managed to put together what seems a representative sample of the ideas in the area, although in selecting individual systems to cover I was evidently biased in favor of those whose authors either had published readable documents at a sufficiently detailed level, or were willing to talk to me about their ideas. I thank all of these people, and I hope that in the future there will be more public exchange of ideas on such subjects as algorithms for silicon compilation and design tools.

## Order of the Chapters

I put the most theoretical material, on lower bounds, first in the book; I think my motivation was to seduce theoretically oriented people into studying more pragmatic issues. On the other hand, I could just as well have started with the more applied material, in order to seduce designers into studying their subject with the point of view of the theoretician, asking questions like "how fast can this task be performed?" In fact, I want to do both, in accordance with my belief that the next "generation" of computer scientists will have to be broad enough in outlook to be at home in both the theoretical world and the world of design.

However, a course emphasizing the more pragmatic issues can do Chapter 1, proceed to Chapter 4 or even to Chapter 7, and then return to the earlier chapters. Chapters 2–6 are not a prerequisite for Chapters 7–9, although there are several places in Chapters 4–6 where the material on lower bounds in Chapters 2 and 3 is used to indicate why the algorithms discussed in Chapters 4–6 are good ones.

## Exercises

Each chapter includes exercises to test the basic concepts and, in some cases, to extend the ideas of the chapter. The most difficult exercises are doubly starred, while problems of intermediate difficulty have a single star.

## Acknowledgements

I am grateful for the comments and suggestions received from many people, including Al Aho, Peter Eichenberger, K. Gopinath, Leo Guibas, John Hennessy, Peter Hochschild, Kevin Karplus, Don Knuth, Charles Leiserson, Rob Mathews, Ernst Mayr, Alon Orlitsky, Ron Pinter, Arina Shainsky, Alan Siegel, Joe Skud-larek, Tom Szymanski, Howard Trickey, Peter Ullman, Alan VanGelder, Moshe Vardi, and Mihalis Yannakakis. The book was set using Don Knuth's TEX typesetting system. My son, Peter, designed the font used to set the stipple-pattern layouts in certain figures. Generation of the final manuscript was also Peter's doing.

J. D. U.
Stanford, CA

# 1

# VLSI MODELS

This book is divided into three parts, of three chapters each. Each part is on the interface between integrated circuit design and algorithm design. The first part is the most theoretical, being concerned with the inherent limitations on our ability to compute using VLSI (very-large-scale integration). The second group of chapters deals with general-purpose algorithms and how we can best use VLSI circuits to implement them. The last part of the book concerns design systems for VLSI circuits, their levels of abstraction, the algorithms used to compile design languages into languages at lower abstraction levels, and algorithms for implementation of the sorts of tools we need when we design circuits.

In this chapter we introduce the basics of very-large-scale integrated circuit design. We shall cover the rudiments of the Mead-Conway design rules for NMOS circuits and show how to relate logic circuits to the layouts that describe integrated circuits. A brief discussion of the underlying electronics is included. Then we shall develop an abstraction of integrated circuit layouts, the grid model, that will let us talk about the complexity of problems in a formal way. For instance, in later chapters we shall develop bounds on the area and the time, to within a constant factor, that certain problems require if they are to be solved with an integrated circuit. The grid model will allow us to discuss such issues formally, yet still remain faithful to the Mead-Conway design rules, neglecting only a constant factor.

## 1.1 INTEGRATED CIRCUITS AND THE MEAD-CONWAY RULES

Integrated circuits are built from conducting materials laid down on a wafer, usually of silicon. There are many different kinds of materials used to form such layers, and different collections of materials, together with the methods of depositing them, form differing *technologies*. One of the most common technologies, and the one we shall take as representative, is NMOS (negative-channel, metal, oxide, semiconductor).

Actually, for the purposes of this book, which is the investigation of algorithms and the ways they can be represented as integrated circuits, the particular technology matters little, since all known technologies allow us to build the same kinds of logic elements and wires to connect them. Generally, only

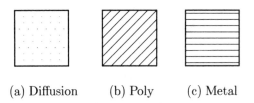

(a) Diffusion      (b) Poly      (c) Metal

**Fig. 1.1.** Stipple patterns representing layers.

constant factors of area and speed change from technology to technology.† Since
typically we cannot prove anything more precise than lower bounds to within
a constant factor anyway, nothing theoretical that we say here will depend
significantly on NMOS technology. Similarly, our investigation of VLSI design
tools and compilation of design languages in Chapters 7, 8, and 9 will use NMOS
for specifics, but the ideas carry over to other technologies well.

### Layers and Transistor Formation

NMOS offers three layers in which to run wires and build circuit elements.
These are

1.  *Diffusion*, represented by green or the stipple pattern of Fig. 1.1(a).
2.  *Polysilicon*, or "poly," represented by red or the stipple pattern of Fig.
    1.1(b).
3.  *Metal*, represented by blue or the stipple pattern of Fig. 1.1(c).

We also need a convention to distinguish wires in the three layers when wires
are to be represented as lines, without thickness. We use solid lines for metal,
dashed lines for poly, and dotted lines for diffusion. The reader should be
aware, however, that some diagrams use lines of those three types for purposes
unrelated to layers.

There are several differences among the layers. Physically, diffusion is
deposited first on the wafer, followed by poly, then metal. The most important
difference among layers is that when poly crosses over diffusion, as in Fig. 1.2,
it forms a *transistor*. The polysilicon cover is called the *gate*, and the section of
diffusion covered by the poly is the *channel*. The significant electrical property
of a transistor is that current can pass along the diffusion wire from $A$ to $B$ or
$B$ to $A$ only if the voltage on the poly wire $C$–$D$ is sufficiently high.

As we deal only with digital logic here, we shall refer to the voltages on wires
as "high" or "low," corresponding to logical 1 and 0, respectively. The highest
possible voltage, the voltage of the power source, is denoted $VDD$; typically

---

† However, there is another important issue that separates technologies: power dissipation.
There are technologies like CMOS that require much less power consumption than NMOS,
and in CMOS or similar technologies, certain observations about power consumption are
inapplicable.

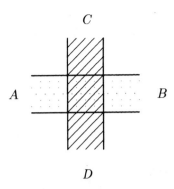

**Fig. 1.2.** A transistor.

$VDD$ is 5 volts. The lowest possible voltage is 0 volt, or $GND$ (ground). A *high* voltage is one that is "close to" $VDD$, perhaps 4 or more volts, while a *low* voltage is one that is "close to" ground, say 1 volt or less. A significant difference between the lowest "high" voltage and the highest "low" voltage is necessary to assure reliable operation of the circuit.

A low voltage on the poly does not allow current to pass, while a high voltage does. Thus, for example, if we attach $A$ to a power source, so that its voltage is always $VDD$, then if the voltage on the poly wire is high, $B$ becomes high. If the poly wire is low, the voltage at $B$ will not be affected by $A$.

The poly-diffusion interaction is the only interaction among the layers. Metal wires can cross over poly or diffusion wires without there being any interaction between the wires. The flow of current in the poly or diffusion wire is not affected by the voltage of the metal wire, and current flow between the wires is not possible unless a *contact cut*, or *via*, is made to connect them.† We shall discuss contact cuts shortly.

Another difference among the layers is in the ease with which they transmit signals. Diffusion wires have more capacitance than poly or metal. Also, poly and diffusion wires have much more resistance than metal. Without going into the physics involved, let us state an important fact: the time it takes a signal to propagate from one end of a wire to the other is proportional to the product of the resistance per unit length, the capacitance per unit length, and the square of the length. Thus, for a given supply of voltage and current, a signal will travel faster in metal than in poly, and faster in poly than in diffusion.‡ We

---

† There is the possibility that wires on different layers running one on top of the other for a long distance will slow down signals in the two wires, but we shall generally ignore such a phenomenon.

‡ We should note that there is another fundamental limit to the speed of signal propagation: the speed of light. However, in practice, the resistance and capacitance of the wires are what limit the propagation speed.

explore the subject more quantitatively in Section 1.3.

Of course, we can increase the supply of current (but not voltage) easily, so it is possible to drive signals quickly through diffusion, but the power consumption makes this undesirable. Therefore, wires of metal and poly are used almost exclusively, with diffusion used only to make transistors.

## The Size Parameter $\lambda$

A fundamental fact of life concerning the design of integrated circuits is progressive shrinkage of the minimum distance within which we can expect what we deposit on the wafer actually to appear. Thus, all designs are specified not in absolute sizes, but in terms of multiples of a size parameter, which is denoted by $\lambda$. This parameter is, approximately, the maximum amount of accidental displacement that we can expect when we deposit a feature on the wafer. In the early 1980's, $\lambda$ is usually taken to be about 2 microns (1 micron $= 10^{-6}$ meter).

The value of $\lambda$ limits, among other things, the width of wires. If a wire were less than $2\lambda$ wide, then since we cannot rely on either edge of the wire being closer than $\lambda$ from where we intend it, the edges could in fact turn out to be in the same place, i.e., the width of the wire would be zero.

## The Mead-Conway Design Rules

A step of great importance in simplifying the design of integrated circuits was the development by Mead and Conway [1980] of "universal" rules that if followed, assured the designer that his design could be fabricated by all, or at least most, of the manufacturers of NMOS circuits. Many of these rules follow from the principle alluded to above; i.e., the design must not change the connectivity of the wires even if all boundaries are moved, independently, by distances up to, but not including, $\lambda$. However, there are other factors concerning the process of fabricating circuits, which we shall not go into, that cause certain limits to be greater than what is implied by the general principle. Thus, we can summarize some of the Mead-Conway rules as follows.

1. *Minimum wire widths.* Poly and diffusion wires must be at least $2\lambda$ wide. Metal wires must be at least $3\lambda$ wide (this is an example of a "special case").

2. *Minimum wire separation.* Poly wires must be separated from one another by $2\lambda$. Diffusion wires must be separated from each other by $3\lambda$, and similarly for metal wires. Poly and diffusion wires that are not intended to form transistors must be separated by at least $\lambda$. If we allowed poly and diffusion wires closer, a shift of $\lambda$ by each could completely cover the diffusion by poly, forming a transistor and possibly cutting off current flow in the diffusion wire. Metal wires are not required to be separated from

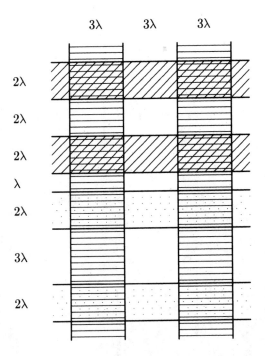

**Fig. 1.3.** Wire width and separation rules.

poly or diffusion wires. Fig. 1.3 summarizes the effects of these rules.

3.    *Transistor formation.* If a poly wire is to form a transistor with a diffusion wire, the poly must extend at least $2\lambda$ beyond the diffusion, as illustrated in Fig. 1.4. This rule follows from our basic principle. If the poly extended less, a shift of $\lambda$ in the edge of the diffusion and another $\lambda$ in the end of the poly could leave a small strip of diffusion not covered by poly, and the flow of current could never be cut off.

**Contacts**

In order to connect wires of different layers electrically, we must place *contacts* or *vias* where the wires touch. Contacts between metal and the other two layers are essentially the same. For example, a metal-poly via is shown in Fig. 1.5(a). It consists of a $4\lambda \times 4\lambda$ square with both metal and poly layers. Inside the large square is a $2\lambda \times 2\lambda$ *contact cut*, which is an area within which all three layers are burned through and make electrical contact. A metal-diffusion via is exactly the same, except that diffusion replaces the poly. Note that solid black denotes contact cuts, and in Fig. 1.5(a) there are metal and poly where the contact cut is, as well as around it.

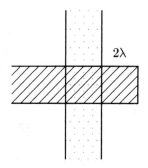

**Fig. 1.4.** Transistor rule.

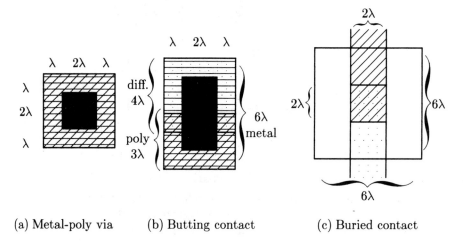

(a) Metal-poly via        (b) Butting contact          (c) Buried contact

**Fig. 1.5.**  Contacts.

To connect poly to diffusion directly is not so simple. One method, advocated by Mead and Conway [1980], is the *butting contact*, which is the arrangement shown in Fig. 1.5(b). It consists of a $4\lambda \times 6\lambda$ metal rectangle, with a $2\lambda \times 4\lambda$ contact cut in the middle. At one end is a $4\lambda \times 4\lambda$ diffusion region, and at the other is a $3\lambda \times 4\lambda$ poly region. The dimensions are chosen so that according to our basic principle, it is impossible that the diffusion or poly regions will fail to be connected, through the contact cut and metal, and it is impossible for the poly to completely cover the diffusion in the region of the cut and block current by forming a transistor.

There is some doubt that the butting contact is very universal, as its ability to work properly may depend upon the process by which the circuit is fabricated. Thus, other methods of connecting diffusion and poly have been developed, such as the *buried contact* shown in Fig. 1.5(c). A buried contact is formed by a $2\lambda \times 2\lambda$ region where poly and diffusion overlap, surrounded by a $6\lambda \times 6\lambda$ region of a layer called "buried contact." Another alternative is to use adjacent metal-poly and metal-diffusion vias (which form a $4\lambda \times 8\lambda$ piece of metal) to connect poly and diffusion wires.

The Mead-Conway rules for contacts are as implied by Fig. 1.5. In addition, there is a rule that contact cuts or buried contact regions may not come closer than $2\lambda$ to an unconnected object. There is also a general requirement that contact cuts should be small, usually no wider than $2\lambda$. In contrast, there is no upper limit to the widths of wires in any of the three layers. When designing circuits, we should remember that a contact has two or three layers in the same place, and the rules that apply to wires in those layers apply to contacts as well. For example, two metal-poly contacts that are not supposed to be connected to one another electrically must be at least $3\lambda$ apart at their closest because of the requirement that metal wires be separated by that distance.

### Pullup Transistors

There is another very important construct, the "pullup," that goes into making NMOS circuits. By way of motivation, observe in Fig. 1.6 how we can do logic, in particular the function $C = \neg(A \wedge B)$, with a resistor and two transistors formed from poly and diffusion. In that figure, connections among the output, the resistor, and the transistor channels is only suggested; we shall give the detailed picture shortly. We suppose that the resistance of the resistor is very much greater than the resistance of a transistor that is on† (gate is high) but very much less than the resistance of a transistor that is off, which is essentially "infinity." The top of the resistor is connected to $VDD$, the voltage of the power supply, and the bottom of the circuit in Fig. 1.6 is connected to ground.

When either wire $A$ or $B$ carries low voltage, one or both of the transistors formed by $A$ and $B$ will be off, and the resistance between point $C$ and the ground will be essentially infinity. Thus, the voltage at $C$ will be approximately $VDD$, that is, high. But when both $A$ and $B$ have high voltage, both transistors are on, and the resistance between $C$ and ground is much less than that of the resistor between $C$ and $VDD$. In this case, the voltage at $C$ will be close to ground, i.e., low. Thus, $C$ is high if and only if it is not true that both $A$ and $B$ are high, or put another way, the circuit of Fig. 1.6 performs the NAND (not-and) function. Other common logic functions can be performed by similar circuits, as we shall see in the next section.

It turns out that there is apparently no simple way to make a resistor in NMOS or similar technologies. What is done instead is to construct a transistor with two properties.

---

† We should be aware that the resistance of a transistor, even when it is on, is much greater than the resistance of a typical wire. We discuss the details in Section 1.3.

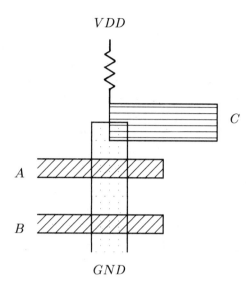

**Fig. 1.6.** Sketch of a NAND circuit.

1.   Its resistance is considerably greater than that of the transistors formed by simply crossing wires as by $A$ and $B$ in Fig. 1.6.

2.   The transistor is always on, independent of the voltage at $C$.

Condition (1) is satisfied by building a long, thin transistor, like that of Fig. 1.7(a). That is, the resistance of a rectangle of conducting substance (such as the three layers we have been discussing) depends only on the material and the ratio of the length in the direction of current flow to the width across the current flow. The reader can prove this with a little thought. A plausibility argument is given by the observation that the resistance of a square $\lambda$ on a side is the same as the resistance of a square $2\lambda$ on a side. If the resistance of the smaller square is $r$, then we can view the larger square as the series connection of two $\lambda \times 2\lambda$ rectangles, each of which is the parallel connection of two squares of resistance $r$. Thus, the $\lambda \times 2\lambda$ rectangles each have resistance $r/2$, and their series connection has resistance $r$. We therefore expect that the transistor of Fig. 1.7(a) will have high resistance, and the transistor of Fig. 1.7(b) will have low resistance.

Condition (2) is satisfied by a new layer, called *implant*, which covers the region of the transistor that we want always to be on. A transistor that is implanted is said to be a *depletion mode* transistor; ordinary, nonimplanted transistors like those shown in Fig. 1.2 are said to be in *enhancement mode*.

In addition to implanting the transistor, we tie the poly forming the transistor to the diffusion forming the same transistor, at the point $C$ of Fig. 1.6. The resulting transistor is called a *pullup*, and the transistors, such as $A$ and

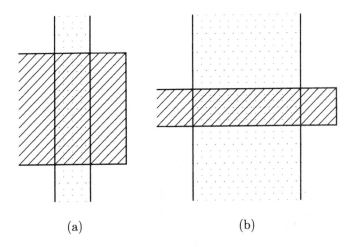

(a)            (b)

**Fig. 1.7.** High and low resistance transistors.

$B$ of Fig. 1.6, that form the inputs to the logic element are called *pulldowns*. Intuitively, the pullup tries to get the voltage at the output $C$ up (high), and the pulldowns, by switching on, try to get the output voltage down.

In order for the circuit to work properly, the resistance of the pullup must be at least eight times† the maximum resistance of the pulldowns whenever that resistance is not infinity. The purpose of this rule is to make the output voltage at $C$ be very close to ground whenever it is supposed to be low. While we showed a simple arrangement of pulldowns in Fig. 1.6, in fact any network of pulldowns could appear there, and finding the maximum, noninfinite resistance of the network is not trivial.

For example, we might draw Fig. 1.6 in more detail as Fig. 1.8. The implanted region, which we show as a "cloud," extends $2\lambda$ around the portion of the pullup where both poly and diffusion are present. There is no implant in the pulldowns, of course. The requirement for a $2\lambda$ border of implant for pullups, and also for a $2\lambda$ space between any implant and a transistor that is not to be implanted, are additional Mead-Conway rules.

Note the use of a butting contact to connect the poly and diffusion parts of the pullup transistor. Any of the methods we mentioned for connecting these two layers could be used instead. In Section 7.2 we shall illustrate a more compact form of pullup, where the butting contact is built directly into the channel.

If a square of transistor has unit resistance, then each of the pulldowns has resistance $1/3$ when on, since it is three times as wide, in the direction across current flow, as it is long, in the direction of flow. The only time the pulldowns

---

† There are circumstances where a 4:1 ratio will do.

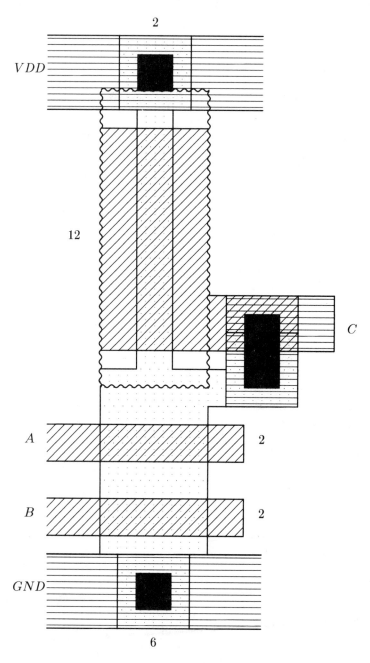

**Fig. 1.8.**  Detail of Fig. 1.6.

do not together have infinite resistance is when both are on, at which time the resistance is $2/3$, i.e., the sum of two $1/3$ unit resistances in series. The pullup, whose region where poly and diffusion overlap is six times as long as wide, will have a resistance of 6, so the ratio of pullup to pulldown resistance is 9, which exceeds the lower limit of 8 and is therefore more than sufficient.

### Input and Output

The last important ingredient of VLSI circuits is a way to get inputs onto the chip and take outputs from it. What is done is to create relatively enormous regions of metal, called *pads*, often around the border of the circuit, but in principle, anywhere. The pads are so large that it is possible to make contact to them with a thin wire (that is, a wire in the usual sense, not a strip of metal, poly, or diffusion). Typical pads in the early 1980's measure around $100\lambda$ on a side. Of course as $\lambda$ shrinks, pads must become relatively larger, unless the technology for bonding wires to pads improves as well.

## 1.2 VLSI IMPLEMENTATION OF LOGIC

Many logic functions can be implemented in NMOS with a network of simple transistors consisting of a piece of poly crossing a diffusion wire. Transistors used in this way are called *pass transistors*. The transistors we referred to as pulldowns in the last section are physically identical to pass transistors; we distinguish pass transistors from pulldowns only by the role they play in the circuit.

This form of logic is similar to the sorts of networks of relays that were used in telephone switching long before electronic switching became feasible. It is based on the principle that opening and closing a combination of gates can determine whether or not current can pass from one point to another. When gates are in series along a wire, current can pass only if all gates are open, so series connections realize logical "and." When gates are connected in parallel, they realize the logical "or" function, while more complex networks can realize complex logical expressions involving "and" and "or."

**Example 1.1:** The network of pass transistors shown in Fig. 1.9 expresses the exclusive or function on $A$ and $B$. That is, current can pass from $X$ to $Y$ if and only if exactly one of $A$ and $B$ is on (high voltage). We assume that the wires labeled $A$ and $\neg A$ are related in the obvious way; that is, $A$ is high if and only if $\neg A$ is low, and $B$ and $\neg B$ are related similarly. We shall discuss the construction of *inverters* below; these are circuits implementing the "not" function, which enables us to get $\neg A$ from $A$ and $\neg B$ from $B$. $\square$

While networks of pass transistors can be arbitrary, the Mead-Conway rules do place one important constraint upon us; the output of such a network cannot be input to another. For example, we could not connect point $Y$ in Fig.

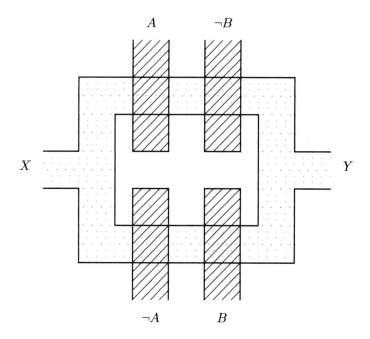

**Fig. 1.9.** A network of pass transistors implementing exclusive or.

1.9 to a poly wire, through a butting contact say, and make that poly wire be one of the wires that, like $A$ in Fig. 1.9, acts as a gate of another pass transistor. The reason for this rule is that the voltage at $Y$ will not be quite as high as at gates like $A$, and it is just possible that the $Y$ voltage will not be high enough to act as a "high" and open another pass transistor gate. The details of the problem are deferred to the next section. However, let us emphasize here that it is quite legal, and indeed frequently necessary, for a point like $Y$ to be the gate of a pulldown; it is only transistors playing the role of pass transistors that cannot be gated by $Y$.

**Switch Notation**

There is a convenient notation for displaying transistor diagrams such as Figs. 1.8 and 1.9. A pass transistor or pulldown is indicated by a detour for its channel, with the gate of the transistor forming a "T" with the top next to the detour. A depletion mode transistor is represented by a resistor, with a "T" adjacent to represent its gate. If the transistor is a pullup, the gate is connected to a point on the end of the resistor that is not connected to $VDD$.

**Example 1.2:** Figure 1.9 is redone in Fig. 1.10(a), and Fig. 1.8 is represented by the circuit of Fig. 1.10(b). □

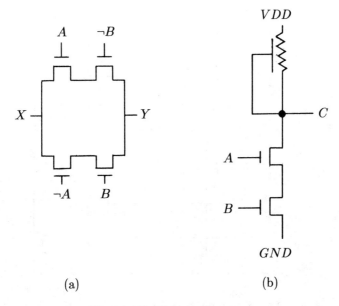

**Fig. 1.10.** Switch diagrams.

In general, we shall not indicate the layer in which wires run in switch diagrams such as Fig. 1.10. However, we have that option as a way of resolving possible ambiguities. Another useful convention we shall use is that crossing wires are assumed not to make electrical contact unless a dot appears at the point of intersection. On the other hand, wires meeting at a "T," such as a transistor gate, are assumed to be in the same layer and to make electrical contact.

## Logic with Pullups and Pulldowns

Another way to perform logic is to generalize Fig. 1.6, with a pullup that is pulled down by some network of pulldowns. The typical arrangement is for *power*, a wire that is connected to the pad at which the voltage $VDD$ is supplied, to run on top; below that are one or more pullups, and below each pullup is a network of pulldowns. The bottoms of the networks of pulldowns are attached to *ground*, which is a wire to a pad that is connected to the external ground, $GND$. Thus, any pass transistor network, such as Fig. 1.9, could become a network of pulldowns, if one end, say $X$, were connected to the bottom of a pullup, and the other end, $Y$, were connected to ground.

A pullup and its associated pulldowns form a logic element. The inputs are the gates of the pulldowns, and the output is at the point just below the pullup, that is, between the pullup and the network of pulldowns. The output is high

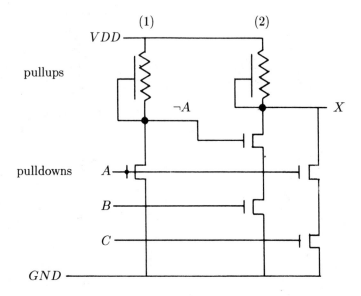

**Fig. 1.11.** Two pullups and pass transistor networks.

if there is no path through the pulldown network to ground. If there is a path from the bottom of the pullup to ground, then the output is low. In order for that to be the case, the ratio of the resistance of the pullup to the maximum noninfinite resistance of the pulldown network, considering all subsets of the pulldowns that, when on, make one or more paths to ground, must be at least 8:1 (or 4:1 in some special cases). This rule was discussed in Section 1.1, and we shall say more about this issue in Section 1.3.

**Example 1.3:** In Fig. 1.11 we see two pullups. The first has one path to ground, with only one pulldown along that path; the pulldown's gate is the poly wire labeled $A$. Note that the wires labeled $B$ and $C$, which we assume are metal below the first pullup, do not act as pulldowns; they interact in no way with the first pullup. The output of the first pullup is labeled $\neg A$, because this pullup functions as an *inverter*; its output is high if and only if the input ($A$) is low, which forbids current from reaching ground through the first pullup.

The second pullup has a much more complicated network below it. Two paths in parallel lead to ground. One, on the right, has transistors with gates connected to $A$ and $C$. The other has transistors whose gates are $\neg A$ and $B$. The network of pulldowns, viewed as pass transistors, therefore realizes the logic function

$$(A \wedge C) \vee (\neg A \wedge B) \tag{1.1}$$

in the sense that current can pass from point $X$, the point below the pullup, to ground if and only if $A$, $B$, and $C$ have values that make this function true, with high voltage identified with "true" and low voltage with "false." As with the inverter (1), the output $X$ of pullup (2) will be high if and only if the network below does not pass current. That is, the output $X$ of the second pullup is high if and only if (1.1) is false. Put another way, point $X$ realizes the logical expression $\neg((A \wedge C) \vee (\neg A \wedge B))$, i.e. $(\neg A \vee \neg C) \wedge (A \vee \neg B)$. $\square$

Notice in general that a pullup and network of pulldowns realizes the complement of the function that the network realizes. In essence, pass transistor networks are used to realize *unate* (or *monotone*) logic, combinations of "and" and "or," while the pullups are used to supply inversions, the "not" operator. However, because of the rule about not connecting the gates in pass transistor networks to the output of other pass transistor networks, often we have to use pullups and invert signals even when we would rather not.

Another difference between pass transistor logic and pullup-pulldown logic is that the former represents truth by the ability to pass current, while the latter represents truth by voltage. These are not quite the same. For example, the gate of a transistor will be high if charge is trapped on it, even if there is no current supply. For example, there may once have been a current supply to the gate through a pass transistor network, but that network closed off the supply, leaving the gate isolated, still holding the charge it had when current was available to flow into the gate. This charge will leak away in less than a second. That time is huge compared with the nanosecond ($10^{-9}$ second) or so that it would take a transistor to switch, but it does imply that if gates are to be operated by charge rather than by current, they will only work correctly if the circuit runs sufficiently fast. Such a circuit is called *dynamic*. In contrast, a circuit like Fig. 1.11, where current is constantly supplied to gates that are on,† will work correctly if run at any speed, because the current perpetually recharges the gate as needed. Such circuits are called *static*.

### Power and Ground Wires

Several rules about the power and ground wires that supply pullups must be observed. First, they are always metal, since neither poly nor diffusion can carry the heavy currents that these wires must carry. As a consequence, the power and ground wires never cross. This constraint makes routing power and ground to where they are needed somewhat tricky, but the reader may show that we can always arrange for power and ground to reach any assortment of points by placing the power and ground pads on opposite sides of the circuit.

Another observation about power and ground is that the more current they have to carry, the wider they must be, and it is common to see power and ground wires much wider than the minimum $3\lambda$. A rough rule of thumb is that

---

† We assume that inputs $A$, $B$, and $C$ represent current sources.

about 10 pullups of the usual size can be supplied by one $\lambda$ of width.

## Clocks

It is often convenient to design circuits that are operated by a *two-phase clock*, which is a pair of signals that alternate being on, while the other is off. It is important that the clock signals not overlap; i.e., one goes off before the other goes on. Conventionally, the clock signals are denoted $\phi_1$ and $\phi_2$, and called phases one and two.

Two-phase clocks provide a useful discipline when we design circuits, especially sequential circuits. In particular, they prevent an output of one logic element from feeding back to its own input, causing an error, or even a condition where the output oscillates indefinitely. One common way to use a two-phase clock is to compute the inverse of the desired outputs at phase one, in terms of the outputs computed by phase two; those outputs are trapped on the input gates to the logic elements used for phase one, by pass transistors that are gated by $\phi_2$. Then, at phase two, each output of the first phase, which is the inverse of the desired output, is trapped on the gate of the input to an inverter; the trap is formed by a pass transistor with gate $\phi_1$. The outputs of these inverters are fed back to become the inputs for phase one.

**Example 1.4:** *Registers* are arrays of circuit elements, each of which can store one bit, make it available to be read when necessary, and change the value stored whenever the register is *loaded*. One way to build a register cell is to use two pullups, as shown in Fig. 1.12. During phase one, the output of the first pullup is computed; this value will actually be the complement of the bit we regard as stored in the cell. Two wires, holding signals $L$ and its complement $\neg L$, indicate whether or not we wish to load the cell at present. If so, $L$ will be high, and the value to be loaded will be found on an input wire $A$. If not, $\neg L$ will be high, and we shall take as the value to be stored the old value of the cell, which will be the output of the second pullup, denoted $X$. Thus, the network of pulldowns below the first pullup realizes the function $(L \wedge A) \vee (\neg L \wedge X)$, and the output of the first pullup is the complement of this function, i.e., the complement of the bit we now desire to store in the cell.

The second pullup simply inverts the output of the first. However, the clock signal $\phi_1$ serves as the gate of a pass transistor along the wire connecting the output of the first pullup to the input of the second. Thus, during phase one, the input to the second pullup cannot change, and therefore neither can its output. As a result, during phase one, the output of the first pullup has a chance to adjust to the present values of the $A$ and $L$ signals, without complications that would be introduced if the output of the first pullup could influence the output of the second, which could change the input to the first immediately. Similarly, $\phi_2$ controls the passage of the output of the second pullup to the

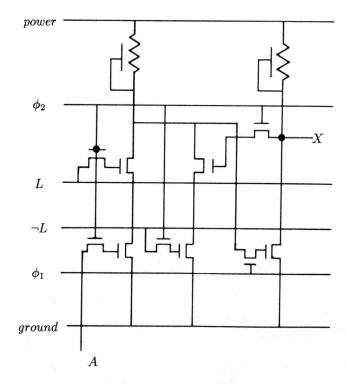

**Fig. 1.12.** A simple register cell.

input of the first, preventing feedback to the input of the second when that input changes during the second phase.

A switch diagram of such a circuit is shown in Fig. 1.12. This circuit is more complicated than is necessary, since it would probably be more economical to let signals $L$, $\neg L$, and perhaps even $A$, each go through one pass transistor gated by $\phi_2$, rather than have a clock gate for those signals appear in every register cell.

Note how the power, ground, $\phi_1$, $\phi_2$, $L$, and $\neg L$ signals all run from left to right. This arrangement allows us to string together as many of these cells as we wish, to make a 32-bit register, or whatever we desire. Also notice how the input $A$ comes in from the bottom, so an input to each cell could be made available without passing through the other cells of the register. Similarly, the output $X$ from each cell could be made available at the bottom or top. $\square$

Several remarks about clock signals should be made. First, signals $\phi_1$ and $\phi_2$ are supplied from pads, like power and ground. Unlike power and ground, the clock signals may be run in poly for some distance, since they normally do not draw much current. This is convenient, since clock signals often run to all parts of the circuit and must cross power or ground. It is also convenient because the principal use of clock signals is as gates of pass transistors, which must be poly. On the other hand, long wires normally should run in metal as much as possible, for speed. If we do not distribute the clock signals to all parts of the chip on fast wires, we face a situation known as *clock skew*, where the chip must be run with a long clock cycle because the clock gates open at different times in different parts of the chip.

Notice also that the use of clock phases often makes circuits dynamic, rather than static. For example, the clock signals in Fig. 1.12, when turned off, trap the output signal of each pullup on the gate of a pulldown belonging to the other pullup. If the clock rate were too slow, the trapped charge could leak away, changing a 1 output to an apparent 0.

## 1.3 ELECTRICAL PROPERTIES OF CIRCUITS

This section is devoted to a brief discussion of some of the electronic issues that are most important in understanding why the circuits we have been discussing behave the way they do. We shall discuss the rule regarding the use of pass transistor output, and what can be done when it is necessary to circumvent the rule. Another interesting electrical issue is the consequences of the nonlinearity of the resistance created by a transistor. A simplified view of the timing issues that come up when we design circuits will be given, and we mention some of the ways that it is possible to speed up circuits. In particular, we discuss "precharged" circuits, a third way (with pass transistor logic and pullup/pulldown logic) of performing logic. Precharged logic is applicable in only limited circumstances, but is very useful when the technique can be applied.

### The Pass Transistor Rule

As was mentioned in the previous section, one of the Mead-Conway design rules forbids the output of a pass transistor network from being the input to another such network. The reason is that for an ordinary, or enhancement mode transistor to be on, the voltage at the gate must be higher than the voltage in the channel. While the necessary difference varies, one volt is a reasonable value to take, and we shall work our examples assuming that value.

Thus, we see in Fig. 1.13 a string of pass transistors feeding another pass transistor (not a pulldown). We suppose that $VDD$ is 5 volts, and power is supplied at the left end. Further, we assume that all the gates are on. In order for current to flow, the first pass transistor must have a channel that is at least one volt lower than the gate voltage of 5, so the channel rises to 4 volts and no higher. Thus, the wire between the first and second pass transistors can be no higher than 4 volts.

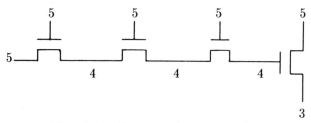

**Fig. 1.13.** Pass transistor networks.

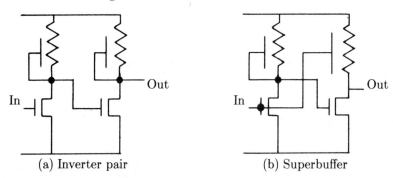

(a) Inverter pair                    (b) Superbuffer

**Fig. 1.14.** Methods of restoring signals.

Notice that as current passes through the second and subsequent pass transistors, there are no further drops in voltage, because the 4 volts on the wire is a volt below each of the gates. However, the drop in voltage will compound if the wire at 4 volts is the gate of another pass transistor, because then, as we see on the right end of Fig. 1.13, the wire through that pass transistor must drop to 3 volts in order to be a volt lower than the gate. Now 3 volts is still closer to $VDD$ than to $GND$, but it is sufficiently small that there is a chance of unreliable operation if we try to use it as a "high" voltage.

### Restorers

Frequently, after signals have passed through a pass transistor, such as a clocking gate, they become the input to a pullup/pulldown network, thereby restoring the signal to $VDD$, if it is high. However, if we wish to *restore* a signal that has fallen below $VDD$, so it can be the input to a pass transistor network, we need a special circuit to do so. The simplest restorer circuit is a pair of inverters, as shown in Fig. 1.14(a), since if the input to the first is high, the output of the second will be not only "high" but exactly at $VDD$, since the resistance of the second pulldown is infinite.

A restoring circuit called a *superbuffer* is recommended by Mead and Conway [1980] as superior to the inverter pair for speed of response. The

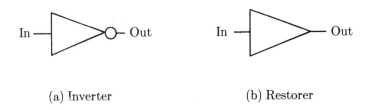

(a) Inverter                                   (b) Restorer

**Fig. 1.15.** Inverter and restorer notation.

superbuffer circuit is shown in Fig. 1.14(b). It differs from Fig. 1.14(a) in that the gate of the second inverter is connected to the circuit input, rather than to its own output.

The use of inverters and restorers is so common that a special notation for these elements is in use. Figure 1.15(a) shows the symbol for an inverter and Fig. 1.15(b) shows the symbol for a restorer. We can use these symbols in switch diagrams, although it should be realized that the new notation is not completely consistent with switch notation; e.g., there is no indication of power and ground supplied to inverters and restorers.

### Transistor Nonlinearities

We have pictured a transistor as a two-state device, one that is either off, in which case its resistance is infinite, or on, in which case its resistance is at some fixed finite value. In fact, the resistance does depend somewhat on the difference between the gate and channel voltages; the resistance is lowest when the gate is very much higher than the channel, and if the difference gets much below 1 volt (for an enhancement mode transistor), the resistance will be very large. However, the resistance, like essentially all physical phenomena, is a continuous function.

The fact that the resistance of an on transistor is not necessarily constant, together with the property of pass transistor voltages just discussed, justifies the rule about the ratio of pullup-to-pulldown resistances: a ratio of 8:1 is required if any input signal has gone through one or more pass transistors, but 4:1 suffices if the inputs have not done so, and are therefore at exactly $VDD$. If the 8:1 ratio is used, $VDD$ is 5 volts, and the minimum gate-to-channel voltage difference of an on transistor is 1 volt, then the voltage at the output of an inverter with a high input will be 5/9 volt. The voltages under our assumptions, for an inverter fed by a signal that goes through a pass transistor, such as a clocking gate, are shown in Fig. 1.16(a). The gate-to-channel difference of about

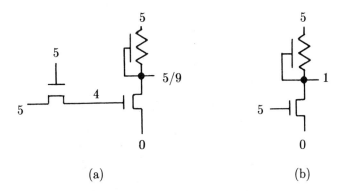

(a)

(b)

**Fig. 1.16.** Voltages on an inverter.

3.5 volts† is ample to keep the pulldown on, and at close to the lowest resistance it could ever be.

Fig. 1.16(b) shows an inverter whose input is at 5 volts rather than 4 and whose pullup/pulldown resistance ratio is 4:1 rather than 8:1. The output is now at 1 volt, but the gate-to-channel voltage difference is 4 volts, compared with only 3.5 in Fig. 1.16(a).‡ On the other hand, if we used a 4:1 resistance ratio in Fig. 1.16(a), the gate-to-channel voltage drop would be only 3 volts. That seems more than adequate to keep the pulldown on, but we should recall that it is essential that circuits be designed with a wide margin of error. For example, suppose that because of imprecision in the sizes of features, the pulldown, instead of having a square channel, had one twice as long as it was wide. Then its resistance would be twice what it should be, and instead of a 4:1 ratio of resistances, that ratio would be 2:1. In that case, the output voltage, given as 5/9 volt in Fig. 1.16(a) would really be 5/3 volt, and the gate-to-channel drop only 2.3 volts. That still seems adequate, but we must remember that the resistance of the pulldown is not constant as we increase the gate-to-channel voltage drop. Thus, its resistance may increase further, leading to an increase in the output voltage, a decrease in the gate-to-channel voltage, another increase in the resistance, and so on. It is therefore not clear that the final output voltage will converge to something sufficiently low to keep the pulldown on, and indeed for some fabrications of the circuit it might not.

---

† We shall, for simplicity, take the gate-to-channel voltage to be the difference between the gate and the higher of the voltages at either end of the channel. However, if the voltages at the two ends are different, then the voltage along the channel will vary with distance as we travel from one end to the other.
‡ This comparison should not be used to infer that a resistance ratio even less than 4:1 would suffice in Fig. 1.16(b), because all our voltage estimates are based on particular assumptions about gate-to-channel voltages, and these need not hold for circuits fabricated by any particular facility.

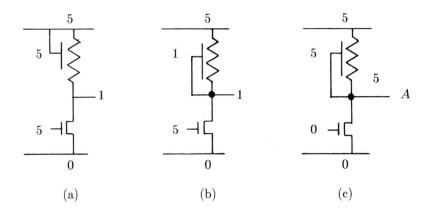

**Fig. 1.17.** Alternative inverter designs.

Another way that the variation in transistor resistance shows itself is in the way pullups are designed. One might expect that the problem of keeping a transistor perpetually on would be solved by a circuit such as Fig. 1.17(a), where a depletion mode transistor has its gate tied to $VDD$, which we continue to assume is 5 volts. Assuming that a 4:1 resistance ratio will do, the output of the inverter of Fig. 1.17(a) would be 1 volt, just as the output of the inverter of Fig. 1.17(b), which is the conventional type, would be.

The distinction between the inverters of Fig. 1.17(a) and (b) is that when the input to the inverter is high, the gate voltage is higher in Fig. 1.17(a). Although both pullups, being implanted, will stay on, the resistance per square will be greater in Fig. 1.17(b). Since we need a 4:1 ratio (say), we shall have to make the channel of Fig. 1.17(a)'s pullup longer than that of Fig. 1.17(b). That, in turn, requires more area for the circuit. Further, it increases the resistance of the pullup in the situation shown in Fig. 1.17(c), where the output is high and we want to send a signal to some point $A$ connected to the output. As we shall see, the higher the resistance of the pullup, the longer it takes to get the signal to $A$. Note, however, that the circuits of Fig. 1.17(a) or (b) will both cause the voltage of the output to rise all the way to $VDD$ when the inverter output is high. However, had we not used a depletion mode transistor in Fig. 1.17(a), the circuit would produce only a 4 volt output when high; it would in a sense act like a big pass transistor.

### Timing Calculations

Let us now consider briefly how to determine the speed with which signals will propagate in a VLSI circuit. There are two properties of the materials going into a chip that affect the speed of events: capacitance and resistance. The capacitance of a layer measures how much charge must be supplied to, or taken

| Substance | Resistance per Square | Capacitance per $\lambda^2$ |
|-----------|:---------------------:|:---------------------------:|
| Channel   | $10^4$                | $2 \times 10^{-15}$         |
| Metal     | .05                   | $4 \times 10^{-17}$         |
| Poly      | 50                    | $5 \times 10^{-17}$         |
| Diffusion | 10                    | $10^{-16}$                  |

**Fig. 1.18.** Typical resistance and capacitance figures.

away from, a $\lambda^2$ area of the layer in order to change its voltage from low to high or vice versa. The resistance measures the difficulty with which current can be supplied or taken away; the higher the resistance, the less charge that can pass through in a given time. Recall that while capacitance is measured in terms of absolute area, resistance depends only on shape and is measured in ohms per square.

It turns out that the resistance and capacitance of the layers depend not only on the value of $\lambda$, but on the particular process used to fabricate the chip. In Fig. 1.18 we give plausible values for resistance (in ohms) and capacitance (in farads), taken from the ranges given by Mead and Conway [1980] and particularized for $\lambda$ equal to 2 microns. Note that the resistance of a transistor channel is much greater than that of any layer, even when the transistor is on. The resistance of an off transistor is infinity, of course. The figure for channel capacitance is also greater than that of any layer. Also note that the figure for channel capacitance applies whether we are trying to change the voltage of the channel by supplying or removing charge, or trying to change the voltage of the gate by supplying or removing charge from it. However, channel resistance does not apply to the current moving in or out of the gate.

While exact analysis of timing involves a great deal of work, we can get rough estimates by taking the time to propagate a signal to be equal to the product of the resistance through which the supply of current must pass times the capacitance of the entire structure into which current is being supplied, or from which it is being removed. Strictly speaking, this estimate is inaccurate; for example, when filling a wire with current, the capacitance is distributed along the wire, and the effective resistance seen by each unit of change being brought to its position in the wire varies with how far down the wire the charge is brought.

**Example 1.5:** Consider the situation of Fig. 1.19, where an inverter sends its output through a $1000\lambda$ long polysilicon wire, into a small transistor gate. We assume that the poly wire is of minimum width, $2\lambda$, and the gate into which the current flows is $2\lambda \times 2\lambda$. The inverter pulldown has a channel with length-to-width ratio 1:2, while the pullup has ratio 4:1 and specifically is $8\lambda$ long and $2\lambda$ wide.

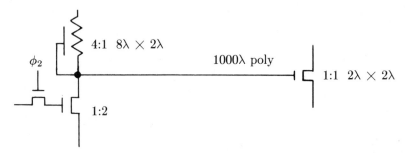

**Fig. 1.19.** Signal into a long wire.

Suppose that the pulldown is on, with charge trapped on its input by the clock gate $\phi_2$. At the second phase, the gate opens, and the charge drains away to ground, turning the pulldown off and forcing the output of the inverter to go high. We can see the changes as three events in sequence:

1. The pulldown turns off.
2. The output of the inverter becomes high.
3. The high output propagates down the wire to the gate at the right of Fig. 1.19.

The notion that these events can be separated is itself a major simplification; in reality they overlap in subtle ways. For example, as the output rises, it becomes possible for some charge to go down the wire, even though the output has not reached its final value. We further simplify by assuming that the pulldown gate drains through a small resistance, and that we may therefore neglect the time taken for the first event, the switching of the pulldown.

The second event, the switching of the pullup, requires that we calculate the capacitance of the channel, into which current must flow, and the resistance of the channel, which impedes that flow. The channel is four squares, so the total resistance of the channel is four times the resistance given in Fig. 1.18 for a square of channel, that is, $4 \times 10^4$ ohms. To compute the capacitance of the channel, we have only to observe that its area is $16\lambda^2$, so its capacitance is 16 times the capacitance of a $\lambda^2$ piece of channel given in Fig. 1.18, that is, $3.2 \times 10^{-14}$ farads. Our estimate of the switching time for the pullup is thus $4 \times 10^4 \times 3.2 \times 10^{-14} = 1.2 \times 10^{-9}$, or a little more than a nanosecond.

That estimate is high for two reasons.

1. Our estimate assumes that the gate-to-channel voltage switches from 0 volts to $VDD$, or vice versa. In reality, when the output was low, the gate was all low, but the channel voltage varied from $VDD$ down to almost 0. When the output goes high, the voltage of both gate and channel becomes $VDD$. Thus, the voltage differential, which varied along the channel from 0 to $VDD$, becomes uniformly 0. That change requires only about half the charge that would be necessary if the initial differential were $VDD$

throughout.

2. The charge being supplied to the channel does not all go through the entire channel; on the average it must go through about 2/3 of the channel, since the greatest differential between the gate and channel voltages occurs at the end of the channel away from $VDD$.

Thus, we expect that our estimate of $1.2 \times 10^{-9}$ seconds is too high, perhaps by a factor of 3. Probably half a nanosecond is a better estimate of the switching time of a typical pullup such as in Fig. 1.19. That time will shrink as $\lambda$ shrinks and will vary with the fabrication process in relatively random ways.

Now, let us consider the third event, as current fills the wire and propagates the output of the inverter to the gate on the right. The following table summarizes the resistances and capacitances relevant to the event.

|             | Pullup          | Wire                | Gate                 | Total               |
|-------------|-----------------|---------------------|----------------------|---------------------|
| Capacitance | —               | $10^{-13}$          | $8 \times 10^{-15}$  | $10^{-13}$          |
| Resistance  | $4 \times 10^4$ | $2.5 \times 10^4$   | —                    | $6.5 \times 10^4$   |

The time for the signal to propagate is approximately the product of the total resistance and the total capacitance, or 6.5 nanoseconds.

To see how we got the above figures, first note that we already calculated the resistance of the pullup at 40,000 ohms. The capacitance of the pullup is irrelevant, since we are not changing the relative voltage of the gate and channel of the pullup. The wire, being $1000\lambda \times 2\lambda$, has $2000\lambda^2$ worth of poly capacitance, which is $4 \times 10^{-17}$ per $\lambda^2$. The number of squares of resistance in the wire is 500, since it is 500 times as long as it is wide. Thus its resistance is $50 \times 500 = 2.5 \times 10^4$.

Finally, the gate resistance is not considered, because the gate of the transistor on the right is really part of the wire. Recall that channel resistance refers only to currents passing through the channel, while "channel capacitance" refers to both the channel and the gate. This capacitance for the transistor on the right is calculated by multiplying the capacitance per $\lambda^2$ by the area of the channel, which we assume to be $4\lambda^2$. $\square$

### Speeding Up Signal Propagation

There are a number of things we can do to make the propagation of the signal in Example 1.5 faster. First, we observe that the dominant time was the time to send the signal down the wire, not the time to switch the pullup, so we should concentrate primarily on improving the propagation time. However, some of the ways to speed the signal down the wire also speed the switching of the pullup.

To lower the time, we must lower either the resistance or capacitance, or both. If we examine the table of resistances and capacitances for the elements of Fig. 1.19, we see that the capacitance is dominated by the capacitance of the wire, while the resistance has large contributions from both the wire and the pullup. There is no good way to lower the capacitance, since poly is about as low in capacitance as anything.

Thus, we must consider ways to lower the resistance, and in order to have a major effect, both the pullup and wire resistance must be lowered substantially. Lowering the wire resistance is easy; just replace it by a metal wire. In fact, all long wires should run in metal for this reason. A metal wire 1000λ long and 3λ wide has 333 squares of resistance, but each square is only .05 ohms, from Fig. 1.18. Thus the 25,000 ohm poly wire may be replaced by a 15 ohm metal wire. The capacitance of the wire increases, since even though metal may have a little less capacitance than poly, the area of the wire is 50% greater. Thus, the capacitance increases to $1.2 \times 10^{-13}$ farads, but the resistance decreases to $4 \times 10^4$, and the propagation down the wires is reduced from 6.5 nanoseconds to under 5 nanoseconds.

To reduce the time further, we must decrease the resistance of the pullup, and the only way to do so is to decrease the length-to-width ratio of its channel. That, in turn, can be done only if we increase the width-to-length ratio of the pulldown, since we must maintain the 8:1 resistance ratio. In principle, there is little limitation to how low we can make the height-to-width ratio of the pullup, and therefore, how fast we can make signals propagate down the wire. The following are some practical limitations, however.

1.   Since the channel of the pullup must be at least 2λ high (or we violate the design rule that says a poly wire must be at least 2λ thick), as we make the channel wider, its capacitance increases. This does not affect the speed with which the signal propagates down the wire, but it does affect the switching speed of the pullup itself, possibly negating any improvements in the propagation speed.

2.   As we make the pulldown channel wider, we increase the capacitance on that channel, possibly to the point that the time to switch the pulldown, which we have neglected, becomes significant.

3.   As the channels get wider, they take more area.

4.   The speed of light limits the propagation time no matter what we do. In effect, if we speed up the circuit too much, our model that assumed the time could be measured by the product of the resistance and capacitance loses validity. Fortunately, this effect will not become significant until circuits become somewhat larger, or faster, or both. Light travels roughly a foot in a nanosecond. If λ is 2 microns, or $2 \times 10^{-6}$ meters, light traverses the 2 millimeters of a 1000λ wire in .006 nanosecond. Since the largest chips in the early 1980's are about 10 millimeters on a side, wires cannot get too much bigger than our example, and the speed of light generally can be ignored.

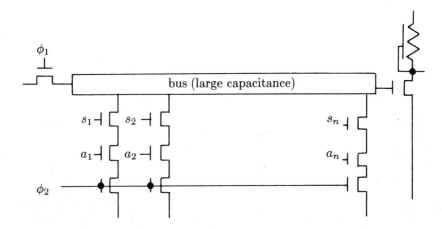

**Fig. 1.20.** One bit of a bus.

## Precharged Logic

Another way to speed up the propagation of signals through long wires is to use *precharging*, a logic implementation where the capacitance of the wire is used, in a sense, to substitute for the pullup. In phase one of a two-phase clock cycle, we allow charge to enter the wire, regardless of which signal we eventually wish to send through the wire. At phase two, we may or may not create a path to ground from the wire, depending on the setting of pulldowns, located between the wire and ground. If such a path is created, the wire drains rapidly, since current needs to go only through low-resistance pulldowns. Similarly, when we precharge the circuit at phase one, we charge through a small resistance, and the charging can be done much faster than through a typical pullup.

**Example 1.6:** Figure 1.20 shows an excellent application of precharged logic, one bit of a bus that connects $n$ registers. We presume that each register is, say, 32 bits, and the registers are connected by 32 wires like the one shown. For each register $i$ there is a *register select* signal $s_i$; that signal is on if and only if we wish to read the value of register $i$ onto the bus. Presumably only one $s_i$ will be on at any time.

For the particular bit position represented by Fig. 1.20, let us denote the value of register $i$ in that bit by $a_i$. Connecting the bus wire to ground are $n$ paths. Each has pulldowns gated by $s_i$ and $a_i$ for one $i$, and all paths include pulldowns gated by $\phi_2$. Thus, there cannot be a path from the bus to ground unless $\phi_2$ is on. Even in that event, the only path to ground possible is the one that goes through $s_i$, where $i$ is the register we have selected to be copied onto the bus. In fact, there will be a path to ground if and only if for this $i$, $a_i$ is high, i.e., the bit from the $i^{th}$ register is 1.

Thus, if the bit of the selected register is 1, the bus will drain, and the input to the inverter at the right of Fig. 1.20 will be 0, and its output 1. In this manner, the entire contents of the selected 32-bit register can be copied onto an array of 32 inverters. In practice, the contents of the bus would be fed to some destination, perhaps another register or an arithmetic unit.

To get an idea of the speed with which a circuit like Fig. 1.20 can operate, let us suppose that the bus wire is a $1000\lambda \times 3\lambda$ metal wire, and the pass transistor gated by $\phi_1$ has a channel ratio of 1:2. In practice, there is little limitation to how small we can make this channel ratio. The resistance of the channel is 5,000 ohms, since it offers half a square of resistance. The resistance of the wire is 15 ohms, as we calculated earlier, so the total resistance is in essence $5 \times 10^3$ ohms. The capacitance of the channel is that of $8\lambda^2$, or $1.6 \times 10^{-14}$ farads, while that of the wire is $3000 \times 4 \times 10^{-17} = 1.2 \times 10^{-13}$. Thus, the total capacitance is about $1.4 \times 10^{-13}$. Finally, we may estimate the time to precharge the circuit as the product of the resistance and capacitance, or about 0.7 nanosecond.

The time to discharge the circuit when $\phi_2$ goes on can be estimated similarly; it depends primarily on the channel ratios of the pulldowns. For example, if all pulldowns have a ratio of 1:2, then they each offer half a square of resistance, and three in series offer about 15,000 ohms. The discharge step will thus take about three times as long as the precharging, or 2 nanoseconds, but this time can be reduced if we give the pulldowns lower channel ratios. $\square$

Since it appears that we can make precharged logic that will send signals down long wires in virtually as little time as we wish, one might wonder why we would ever do anything else. Another argument in favor of precharging is that the use of current is limited to the amount used to precharge the wire at each cycle. In comparison, pullup/pulldown logic runs current perpetually through a pullup if there is a path to ground. That draws power, which heats up the chip and makes reliable operation harder.

The primary reason precharging cannot be used universally is that the method relies on the existence of a large capacitor, like the bus wire in Fig. 1.20, to work properly. That is, consider what would happen in Fig. 1.20 if the wire was small, and we wished to read, say, register 1, but the bit from register 1 was 0, i.e., $a_1$ was low. Then there would be no path to ground created, but charge could at least migrate into the channel of the pulldown for $s_1$. If a substantial fraction of the charge on the wire were able to go somewhere else, the voltage on the wire, and therefore the input to the inverter, would be significantly below $VDD$, and it might fall low enough to switch the inverter output to high, thus misreading the bit $a_1$.

The conclusion is that precharged logic is desirable when we have a large wire that must be charged anyway. However, to substitute precharged logic for pullup/pulldown logic everywhere would require the creation of too many area-consuming wires used only as capacitors, not to carry a signal anywhere.

## 1.4 ABSTRACTIONS OF VLSI CIRCUITS

A fundamental question in algorithm design is, given a problem, for example, the problem of multiplying two matrices, find the minimum area of a VLSI circuit needed to solve that problem. The reason the area of the circuit is so important is that the cost of fabricating a circuit is actually an exponential function of the area, at least if the circuit is large. In this sense, optimizing the area of a VLSI design is much more important than optimizing the speed of a program.

The reason the cost of a large circuit varies exponentially with the area is fairly subtle. When fabricating a chip, there is the opportunity for an imperfection, such as a minute piece of dust, to introduce a flaw somewhere in the circuit. To a first approximation, any circuit with a flaw will be useless.†

The largest circuits that it is feasible to fabricate have a *yield*, or probability of being produced unflawed, that is very low, say 10%. If we double the area of such a chip, the probability of both halves being unflawed is squared, i.e., it becomes 1%. In general, if we multiply the area by $c > 1$, the probability of no flaw is $(0.1)^c$. Since the cost of producing a good chip is inversely proportional to the yield, the cost per chip would grow as $10^c$, as $c$ grew past 1. Of course, if the area of the chip were small to begin with, so the yield was close to 1, the cost would not be a very sensitive function of the area at all. That is, if $y$ is the yield for chips of area $A$, the cost of chips of area $cA$ will be proportional to $(1/y)^c$, which grows rapidly if $y \ll 1$, but grows slowly if $y$ is near 1. We conclude, however, that at least for large chips, the area needed to perform a given operation is of great interest.

Another issue with which we shall want to deal is the speed of circuits. Often there is a tradeoff between the area needed and the time spent performing a function. For example, if we want to compute the product of $n \times n$ matrices as fast as possible, we must read them in at once, which means $2kn^2$ pads are necessary just to read the data, if the matrix elements are $k$-bit numbers. Thus, it seems $\Omega(n^2)$‡ area is needed just for the input pads.

However, we might consider the possibility of reading the matrices in one row or column at a time. We then might hope to make do with $O(n)$ pads and $O(n)$ area. However, if we did so, we could not produce the answer in $O(1)$ time, since it would take $O(n)$ time just to read the input.††

---

† Technology to repair or work around flawed pieces of circuit may or may not become available in the future. It is certainly an area of interest both for fabrication research and for circuit design research.

‡ See Appendix 1 for the definition of "big-omega" and "big-oh," if these notations for expressing the growth rate of functions are unfamiliar.

†† In fact, the reader should consider whether we can use only $O(n)$ area and $O(n)$ time under any circumstance. For example, is it possible to avoid using $O(n^2)$ area to store parts of one matrix that must later be used to multiply elements of the other that have not yet been read?

### The Role of Models in Algorithm Design

Let us digress for a moment and review the familiar process of discovering the best algorithm to solve a problem on the usual type of computer. When we tackle issues like finding the fastest program to perform a certain task, we must approach it from two directions. One part of our task is to find good algorithms, and when we find them, we express them in real terms; i.e, we write programs in one of the usual programming languages.

The second part of our task is to prove that, for a particular problem, we cannot do better than some time complexity $T(n)$. Here, $n$ represents the size of a problem instance; for example, we took $n$ to be the side of the matrices in the discussion of matrix multiplication above. For this proof, we usually cannot show a theorem about (say) all Pascal programs. Rather, we need an abstract model of computation, one that is sufficiently realistic that we believe it represents real computation, yet sufficiently formal and restricted that we can hope to prove a theorem of the form: "every program that solves problem $P$ takes at least time $T(n)$." A good example of this approach is the decision tree model applied to sorting (see Aho et al. [1974], e.g.), which is sufficiently precise that we can prove that any program that sorts a list of elements chosen from some arbitrary linearly ordered domain must take at least $O(n \log n)$ time to sort $n$ elements.

One problem with this approach is that we usually cannot claim that the lower bounds on running time obtained by our model are exact. Rather we claim that they are exact to within a constant factor, or in many cases, to within some polynomial. However, we can still get lower bounds that are useful, in that they tell us when we can expect large improvements in running time, say if the lower bound is $\Omega(n^2)$ and the best known algorithm has running time $O(n^3)$, and when we can expect only small improvements, say when the upper and lower bounds are the same to within a constant factor.

When dealing with the design of VLSI circuits, we similarly want to analyze problems in these two ways. We have been given, in Sections 1.1 and 1.2, enough of the rudiments of what circuits look like that we can hope to sketch the design of circuits much the way programs are sketched in a book about conventional algorithms. What we need are ways to prove lower bounds on the area and the time needed to solve certain problems in the VLSI environment, and that is the subject of this section.

### The Grid Model of Circuits

A variety of VLSI circuit models have been proposed, but most of them share many features in common with the one, called the *grid model*, that we shall use in this book. In the grid model, we postulate a rectangular grid, along whose lines wires run. There are one or more "layers," and on any grid line there can be at most one wire in each layer. It is important to note that, while we do not insist on any particular number of layers, the exact number of layers must be fixed once and for all, before we talk about a given problem. Otherwise, we might get fallacious results by assuming that the number of layers could grow as the size of the problem instance grows.

The spacing between grid lines we can view as (roughly) the minimum repetition rate at which wires on a certain layer can run. Unfortunately, this repetition rate, or *minimum pitch*, differs for different layers, being $4\lambda$, $5\lambda$, and $6\lambda$ for poly, diffusion, and metal, respectively. However, the reader should bear in mind that we are going to obtain only order of magnitude lower bounds with this or any model, so the difference between 4 and 6 is insignificant.

Our assumption that wires run only horizontally and vertically is again something that does not follow from the Mead-Conway model. However, many fabrication facilities do support only wires and other circuit elements whose edges are horizontal or vertical. Also, those that support more variety often limit edges to a small number of angles, e.g., horizontal, vertical, and 45° lines. If we wish to simulate a wire at some angle other than 0° or 90°, we can expand the grid by a factor of two and run wires in a staircase fashion. Thus, invoking again our assumption that constant factors will not influence lower bounds, we can assume for convenience that wires run along grid lines only.

Circuit elements, in particular input/output pads, contacts, and logic elements occur at the grid points. Wires running along grid lines that meet a grid point occupied by a logic element are assumed to be inputs or outputs of that element. It is not permitted to run a wire through a logic element without the wire being an input or output of that element. This rule is particularly relevant when the element is a contact.

As grid points have only four adjacent grid edges, we cannot bring an arbitrary number of wires to an element; we are limited to four times the number of layers permitted, and even that number may not be a realistic limit, say for a contact. If we need more wires impinging on a circuit element than can connect to a single grid point, we may represent the circuit element by a rectangle covering as many grid points as needed.

We assume for convenience that wires carry signals in one direction only, so it makes sense to speak of inputs and outputs of logic elements. In the rare case, such as a bus wire, that signals are carried in both directions at different times, we can replace the one wire in the real circuit by two wires in the abstract circuit.

We also assume that there is some limit to the *fan-in*, that is, the number of inputs one logic element can have. As with the number of layers, we do not demand any particular limit, but we assume the limit is fixed before we discuss a particular problem and its instances of different sizes. If we did not fix the fan-in beforehand, we would obtain silly results, since the solution to any problem instance could be expressed by one logic element with all the inputs. In contrast, we do not assume a limit on the *fan-out*, that is, the number of elements having inputs connected to the output of a logic element. While an output wire feeding many gates will have a large capacitance due to those gates, we still can deliver the signal in a reasonable amount of time if we drive the signal with a low-resistance pullup.

**Example 1.7:** Figure 1.21 shows a grid model abstraction of the register cell of Fig. 1.12. In Fig. 1.21 we assume there are only two layers, which we represent by solid and dashed lines. No correlation between these layers and the metal and poly layers of real circuits is necessarily implied. Pads are represented by open squares, and logic elements, such as the two pullup-pulldown combinations that form the register cell, are represented by shaded regions. Contacts are represented by squares with "X" in them.

Observe that between the two logic elements there is a pair of wires running in parallel on one grid line, in different layers, of course. These wires represent the fact that the output of each logic element is input to the other. We should imagine that the signals on these wires run in opposite directions. Notice also that we view all the logic, including the gating of the wires between logic elements by the clock signals, as taking place within the logic elements themselves. Thus, the topologies of the two circuits of Fig. 1.12 and 1.21 are only in rough correspondence. □

We might question the realism of the above assumptions for several reasons; some are easy to dismiss, while others require a certain amount of faith that circuits will be designed in ways that can be explained by logical elements. As an example of an easy-to-dismiss argument, one might question the wisdom of using a single grid point for an element as small as a contact or as large as a pad. The answer is that we are only concerned with area to within a constant factor, and there is "only" a factor of 625 between the area of a pad 100λ on a side and a via 4λ on a side.

A much harder issue is the question of whether an area for logic elements that is a constant per element with a fixed number of inputs is realistic. The exercises discuss a hypothetical case where real circuits of area $n$ can apparently implement logic that requires more than $O(n)$ logic elements to represent, although it is unclear how realistic this form of circuit is. More to the point is whether circuits such as the bus of Fig. 1.20, that can have arbitrary numbers of inputs, can be represented in the grid model in area proportional to that taken by the real circuit. It is our contention that we can always do so, and Fig. 1.22 illustrates how the bus can be represented abstractly in proportional area. There, the gates with + in them perform logical "or," while the other gates are "and" gates.

Where we need a certain faith is if we contemplate a logic element like the bus, taking $n$ inputs and using area $O(n)$, that does something much

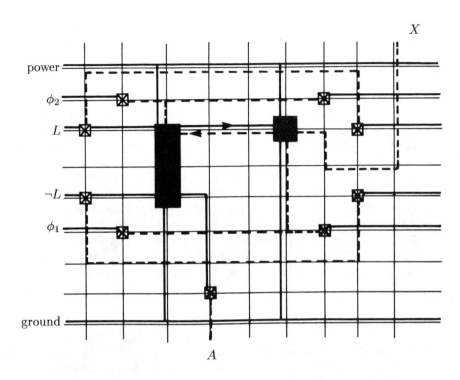

**Fig. 1.21.** Grid model representation of register cell.

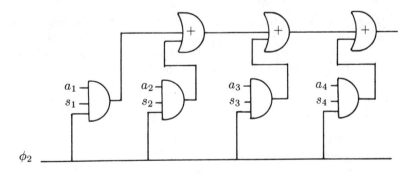

**Fig. 1.22.** Bus wire expressed in fixed-fan-in logic.

more complicated than the bus. For example, suppose there were an element that took inputs $a_1, \ldots, a_n$ and determined whether the integer with binary representation $a_1 \cdots a_n$ is a prime, producing output 1 if so. It is very unlikely that there exists a family of logic circuits, using $O(n)$ gates with a fixed fan-in, independent of $n$, that recognizes the primes. However, without demanding that there be a limit on the ways that logic can be performed in VLSI circuits, we could not hope to prove the nonexistence of such circuits in reality. Thus, while it is unlikely, we must keep open the possibility that there are ways to compute in real life that cannot be reflected in proportional area by our model.

Another way the fixed fan-in assumption is worrisome concerns networks of pass transistors or pulldowns of arbitrary size. The problem is that we cannot directly associate high and low outputs of these transistors with logical 0's and 1's. The trick to representing such networks by logic elements is not to associate logical values with voltages, but rather with the existence or nonexistence of paths to ground from the ends of each transistor. For example, a chain of pulldowns in series can each be represented by an "and" gate. The connections among the gates are such that the output of the gate associated with one transistor is the logical "and" of its (transistor) gate input and the output of the "and" gate next along the path to ground. This network of logic elements has area proportional to that of the chain of pulldowns. We leave as an exercise the general construction of a logic network of proportional area from any pulldown or pass transistor network.

## The Convexity Assumption

Another simplification of circuits that we make concerns how we measure the area of a circuit. Obviously we shall want to use the space between grid lines as the unit of distance and express areas in terms of the number of squares of the grid covered by a circuit.

However, we make one simplifying assumption that is not immediately justifiable; we assume that all circuits are in effect convex, by taking the *area of a circuit* to be the area of the smallest rectangle, with grid lines for sides, in which the circuit fits. If a circuit is convex and its major axis (longest diameter) is parallel to the $x$- or $y$-axis, then by taking its area to be that of the bounding rectangle, we overestimate the area by only a factor of two, a constant factor which we shall ignore as we ignore all other constant factors.

But what if the region used by a certain circuit were not convex, or its major axis were at the wrong angle. Recalling our discussion of why area decreases yield and increases cost, we see that a flaw within the bounding rectangle but outside the area actually used by the circuit will not make the circuit fail; therefore the area used by the elements and wires is the true cost, no matter what the shape of their region. Let us simply remark that a nonconvex circuit can be "bent" into a convex shape, and a circuit can be rotated, without increasing the total area used by more than a constant factor. Thus our assumption of convexity can be made without loss of generality, as long as constant factors of area are ignored.

## Timing

In addition to abstracting the area used by the circuit, we must model the time taken by the circuit. Again, we must make some presumptions about limits on the way logic can be implemented in VLSI circuits. We shall assume that there is a well defined notion of discrete computational steps, typified by the phases of a two-phase clock. In all common ways of performing logic on a chip, there is a notion of a minimum time in which an event occurs, e.g., the minimum time for a transistor to switch. As a case in point, we rule out analog computation, where outputs of transistors take values that are not only high or low but can be any real number, with precision that is in principle infinite. If such a circuit were perceived as performing logic, our hypothesis of discrete time units would not hold.

Another aspect to our assumption about the existence of a unit of time is the desire that the unit not be too large. In particular, we must be able to associate with each point in the circuit and with each time unit, a single value. At the beginning of the time unit, the signal at that point is driven toward that value if it is not already there; perhaps the switch does not occur at the very beginning of the time unit, but eventually the switch occurs. What we wish to avoid is a situation where in one nominal time unit, a significant amount of computation occurs at a point, with the point switching back and forth in response to signals computed at other points during the same time unit. On the other hand, we shall not rule out the possibility of "glitches," or temporary changes in a value that have no computational significance but are due to variations in the timing of events that take place at the same time unit.

Also because of our desire to keep the unit of time small, we shall assume that there is a limit on the number of levels of logic through which signals can propagate in one time step. Coupled with the assumption of fixed fan-in, this limit implies that if we wish to compute a logic function of $n$ variables, the time needed must grow at least as $\log n$. Such a growth in the time required can be observed in practice.

Having fixed on a notion of time unit that meets the above criteria, we can associate with any computation of a circuit the *time taken by the circuit* on a given input. This time is the number of time units that elapse from the first input signal until the last output signal. We need not fix on the notion of a time unit to within closer than a constant factor, because our results about the time required for solution will not be more precise than that. However, unless we specify otherwise, the time unit will be assumed to be a phase of a two-phase clock.

There are several points that need to be made about the restrictions that a discrete time model imposes on the operation of circuits. Logic elements are presumed not to store their values unless we explicitly design them to do so, as the register cell does. Thus, the time at which outputs of logic elements appear is important in our model, just as in "real life." Pads are incapable of storing their values, and we assume that their inputs and outputs appear only for one time unit each, so it is at least possible that one pad can be used for many input bits, e.g., to read in an entire row of a matrix, one bit at a time. Only one input or output may appear at any one pad at one time unit. If we wish to store the values that are input at a pad, we must connect the pad to a logic element capable of storing the value. Power and ground are obvious exceptions to this rule; they are available from their pads at every time unit.

It is also worth mentioning that we have made a reasonable, but potentially unrealistic assumption about the delays that wires introduce. We have assumed that in one time unit, signals have all the time they need to propagate down wires, as well as to switch transistors. In reality, as wires get longer, the time of propagation cannot be regarded as constant. For example, we mentioned the speed of light as a reason for regarding the time to send a signal down a wire as at least linear in the wire length. In fact, the delay due to resistance and capacitance implies an even faster growth of delay with length. Both the resistance and capacitance grow linearly with length, so their product, which is roughly the propagation time, grows quadratically with length.

We have adopted the constant propagation time model for two reasons.

1.    There seem to be few opportunities to prove stronger lower bounds on time by making the linear or quadratic assumption about the growth of delay with wire length.

2.    In practice, wire delays do not seem to dominate switching time, except for delays on a few long wires. In those cases, we often can reduce these wire delays by driving signals down them with large, low resistance pullups.

## The Number of Layers

We have assumed only that the number of layers is finite, not that it is exactly three. This assumption is reasonable because technologies other than NMOS are available, and they do not all have exactly three layers. For example, a second layer of metal, often used exclusively to supply power and ground, is becoming feasible, and there are proposed technologies where the number of layers is essentially as large as we like. Fortunately, the number of layers is irrelevant, as long as it is finite and we permit at least two. We shall make this statement precise with our first theorem.

**Theorem 1.1:** If $C$ is a circuit in the grid model that uses $k$ layers in which to run wires, then there is another circuit $C'$ that performs the same function as $C$ in the same time, uses only two layers for wires, and takes area $k^2$ times the area of $C$. Moreover, wires in one of the layers run only horizontally, and in the other layer they run only vertically.

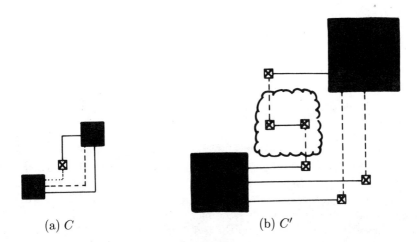

(a) $C$                                          (b) $C'$

**Fig. 1.23.** Many-layers-to-two construction.

**Proof:** Suppose for convenience that we have a coordinate system with the origin at the lower left-hand corner of $C$. We construct $C'$ on a grid with $k$ lines where one existed for $C$; line $i$ in some dimension becomes lines $ki$ through $k(i+1)-1$. Each circuit element of $C$ will be expanded by a factor of $k$ in each dimension, so if it formerly covered the grid from $i_{low}$ to $i_{high}$, it will now run in that dimension from $ki_{low}$ to $k(i_{high}+1)-1$. As a consequence, if grid line $i$ in $C$ touches a circuit element, in $C'$ the $k$ lines it becomes will also touch that element.

Let us number the layers of $C$ as $0, 1, \ldots, k-1$. A wire on layer $j$ along wire $i$ in $C$ will now run along wire $ki + j$ in the same dimension. In $C'$ we run all horizontal wires in the first layer, and all vertical wires in the second layer, with contacts connecting segments of the same wire when that wire changes direction. Note that contacts in $C$ become $k \times k$ regions in $C'$, so layer changes in $C$ can be reflected in $C'$ by having the wires connect within that region. Also note it is important that we have assumed nonconnecting wires do not go through a contact, or else we might not be able to make the connection within the allotted area. $\square$

**Example 1.8:** In Fig. 1.23(a) we see a simple circuit $C$ consisting of two logic elements and a contact where a wire changes layers. We assume that the three layers on $C$ represented by solid, dashed, and dotted lines are numbered 0, 1, and 2, respectively. Fig. 1.23(b) shows the corresponding $C'$, where the two logic elements have become $3 \times 3$ squares, and the contact of $C$ has been replaced by a $3 \times 3$ region, represented by a "cloud."

The solid line in $C$ runs along the lowest grid line in $C'$ until it reaches the seventh vertical line, whereupon it switches to the second layer at a contact and proceeds up that line. The dashed line in $C$, which follows the same path as the solid line, follows the second horizontal line and the eighth vertical line in $C'$. The dotted-solid line in $C$ follows the lines reserved for the dotted layer until it reaches the region of $C'$ representing the contact of $C$. The wire emerges from this region along the lines reserved for the solid layer. $\square$

We shall say a circuit is in *normal form* if it satisfies the conditions of Theorem 1.1; that is, there are only two layers and one layer is devoted to each direction.

Theorem 1.1 has a number of useful consequences. Evidently it allows us to use as many layers as we like, as long as that number is finite, independent of the size of the input to the problem that the circuit is intended to solve. As a more specific example, we shall hereafter feel free to ignore the need to run power, ground, and clocks to the logic elements of a circuit. Given a circuit without those wires, we could claim the use of four more layers for those wires, run them wherever we wished, then invoke Theorem 1.1 to reduce the number of layers to two.

We should be a bit careful here, since in reality both power and ground must run in one layer, metal, only. However, we can fix things up by the following construction, the details of which we leave as an exercise. Expand the scale of the grid by a factor of two in each direction, so wires now run on alternate lines. Let the power and ground lines, which we have previously argued need not cross each other, run exclusively in the metal layer, which we suppose corresponds to the horizontal direction, i.e., solid lines in Fig. 1.23(b). When power or ground, running vertically in metal, is to cross a horizontal metal wire, use the grid points between wires for contacts to change the other wire to the second layer.

## EXERCISES

1.1: A modulo 2 counter has a 1 output if it has seen an odd number of 1 inputs and a 0 output if it has seen an even number of 1's. Suppose we wish to build such a circuit that reads its input on phase one of a two-phase clock and produces its output at phase two.

    a)   Draw a switch diagram of such a circuit.

    b)   Draw a layout for your circuit from (a), that is a diagram indicating the layers and the sizes of all objects.

1.2: Draw a switch diagram of a circuit that realizes the majority function $xy + xz + yz$ of three logical variables, using

    a)   Pass transistor logic.

    b)   Pullup/pulldown logic.

    c)   Precharged logic.

1.3: Find all the design rule errors in Fig. 1.24. The grid may be assumed to be in units of $\lambda$.

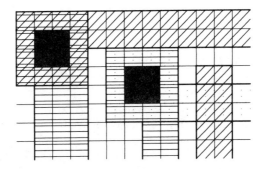

**Fig. 1.24.** Circuit with design rule errors.

1.4: In Section 1.3 we discussed the speed with which the output of an inverter rises. Suppose the pullup has a channel ratio of 4:1, and the pulldown has a ratio of 1:2. How will the speed with which the output falls to 0 when the input becomes 1 compare with the speed with which the output rises when the input goes to 0?

1.5: What would happen in Fig. 1.12 if there were no clock signals (i.e., $\phi_1$ and $\phi_2$ were always on)? What would happen if there were only one clock signal (i.e., one of $\phi_1$ and $\phi_2$ was always on, while the other alternated on and off)?

1.6: Suppose we have a circuit consisting of metal rectangles of areas $a_1, \ldots, a_n$, where each $a_i$ may be taken to be an integer in units of $\lambda^2$. At phase one of a clock, some subset $S$ of the rectangles are charged to $VDD$, while the others remain at ground. At phase two, gates open allowing the voltage of the collection of rectangles to reach the weighted average of the rectangles, that is,

$$VDD \frac{\sum_{i \in S} a_i}{\sum_{i=1}^{n} a_i}$$

This value is input to an inverter, which we may suppose will have output 1 if this average is at least $VDD/2$, and output 0 otherwise.

* a)  The area of this circuit is proportional to the sum of the $a_i$'s. Show that an equivalent logic circuit, using two-input logic elements, can be laid out in the grid model, in area proportional to the square of the area of the "real" circuit.

** b)  Show that there are values of the $a_i$'s that make quadratic area necessary.

c)  Is this sort of circuit likely to be useful in practice?

* 1.7: Show that any network of pass transistors or pulldowns, with distinguished source and sink nodes, can be represented in proportional area by a network

of two-input logic elements. The network of logic elements produces a 1 output if and only if the transistor network has a path of on transistors between the source and the sink.

∗ 1.8: Are there situations where the timing estimate rule of Section 1.3 (product of total resistance and total capacitance) can be wrong by more than a constant factor, e.g., at most a factor of three for any circuit?

1.9: In your pullup/pulldown network of Exercise 1.2(b), suppose that an 8:1 resistance ratio is necessary, because the values of $x$, $y$, and/or $z$ are supplied through pass transistors.

a)   Suggest reasonable channel ratios for the transistors in your circuit.

b)   If only $x$ and $y$ are on, and $VDD$ is 5 volts, what are the voltages at various points in your circuit?

c)   Suppose for sake of argument that the minimum gate-to-channel voltage drop of an on transistor is one volt. Suppose also that it is only $x$ that is supplied through a pass transistor; $y$ and $z$ are at $VDD$ when on. Further assume that we desire at least a 3.5 volt difference between the gate and the top of the channel of a pulldown, when that pulldown is to be on. How would we design the pulldown network so that the resistance ratio would be minimized?

1.10: We can show that it is always possible to wire power and ground so they do not cross, in the following formal way. Suppose we design a circuit in the grid model that is in normal form, but before doing the design we removed all power and ground wires. There are now some collection of circuit elements to which power and ground must be delivered. Prove that there is a fixed constant $c$ such that we can expand any such circuit by $c$ in both dimensions, then add two networks, exclusively in the same layer; one network connects all the elements that need power, the other connects all the elements that need ground. The resulting circuit is still in normal form.

∗ 1.11: Show that there is a constant $c$ such that, after expansion of each dimension by $c$, any circuit in the grid model can be rotated $d$ degrees about a point (in the sense that every grid point is mapped to the nearest grid point after the rotation), and the result will be a legal circuit in the grid model; i.e., wires in the same layer will not cross, two circuit elements will not share a grid point, and the same connections among elements exist.

## BIBLIOGRAPHIC NOTES

The fundamental reference on the design of VLSI circuits is Mead and Conway [1980]. Lyon [1981] gives another view of design rules, including some modifications that we have used. Recent design rules for use with a wide variety of fabrication processes can be found in MOSIS [1983].

Thompson [1979, 1980] and Brent and Kung [1981] are basic works on formal models for VLSI circuits, while other views, including issues regarding wire length, have been expressed in Bilardi, Pracchi, and Preparata [1981] and Chazelle and Monier [1981].

Recently, there have been some fundamental studies of how one can avoid the exponential growth of cost with area by using good modules of a circuit, while avoiding the failed parts. These works include Greene and El Gamal [1983], Leighton and Leiserson [1982], and Rosenberg [1982].

# 2

# LOWER BOUNDS ON
# AREA AND TIME

We shall explore the principal methods for proving lower bounds on the area and time, or some combination of these, that are required to solve problems in the VLSI domain. Frequently, these results are expressed as "$AT^2$ bounds"; the product of the area and the square of the time must be at least so much. The most common method of proof is a "fooling argument," where it is shown that in order to solve a certain problem, at least some minimum amount of information must be shipped from one part of the circuit to another, or else one half of the chip can be fooled into thinking the input to the other half was one thing, when in fact it was another. Since shipping large amounts of data requires either a long time or many parallel wires, we can obtain $AT^2$ lower bounds in this way.

We shall also consider several generalizations of the lower bound theory. Probabilistic chips contain random elements that can be used to solve certain problems efficiently; these solutions are not exact, but have arbitrarily high probability of being correct. A second generalization is to allow inputs to be repeated several times. This privilege presents an opportunity for improving the area or time of a circuit, since in a sense storage of the input is provided for free, external to the chip.

## 2.1 INTRODUCTION TO LOWER BOUND ARGUMENTS

In this section, we shall motivate the more difficult results to follow by presenting some simple arguments about what cannot be done with VLSI circuitry. To begin, we shall formalize the notion of a "problem."

### Problems

A *problem instance* consists of a list of input variables $x_1, \ldots, x_r$, a list of output variables $y_1, \ldots, y_s$, and functions $f_i(x_1, \ldots, x_r)$, for $1 \leq i \leq s$ that define the $y_i$'s in terms of the $x_j$'s. All these variables and functions are Boolean; that is, the values of variables are taken from { false, true }, or { 0, 1 }.

A *problem* is an infinite sequence of problem instances, one for each of an infinity of values of $n$, the *problem size parameter*; $n$ may or may not be the same as $r$. Normally, we expect some natural uniform interpretation for each problem instance in the sequence.

**Example 2.1:** For given $n$ and $k$, the problem instance of sorting $n$ $k$-bit numbers can be expressed as follows. Let $X_1 = x_1, \ldots, x_k$ represent the first number in binary, $X_2 = x_{k+1}, \ldots, x_{2k}$ represent the second, and in general, $X_i = x_{i(k-1)+1}, \ldots, x_{ik}$ be the $i^{th}$ number, in binary. The output bits $Y_1 = y_1, \ldots, y_k$ will equal, respectively, that $X_i$ which is the smallest number, $Y_2 = y_{k+1}, \ldots, y_{2k}$ will equal the second smallest, and so on. The formulas that express the output in terms of the input are exceedingly complicated to write, but it should be clear that such Boolean formulas could, in principle, be written. A "sorting" problem is some infinite sequence of these problem instances. For example, we could let $n$ be the problem size parameter, let $n = 2, 4, 8, \ldots$, and let $k = \log_2 n$.

As another example of a problem, consider the question of determining whether a graph is connected,† given its adjacency matrix. That is, let $n$ be the number of nodes in a graph. Then $x_{in+j}$ will be 1 if and only if there is an edge between nodes $i$ and $j$. Thus, the input variables $x_1, \ldots, x_{n^2}$ represent the adjacency matrix of a graph. The output is represented by a single variable $y_1$. This output is defined to have value 1 if the input graph is connected, and 0 otherwise. □

*Solutions* to problems are sequences of circuits, one for each instance of the problem. Note that we shall frequently talk of a particular circuit, designed in a way that depends on $n$, as if that one circuit were the solution to the problem, but in reality we are designing an infinite family of circuits, one for each $n$. This arrangement is somewhat at variance with the usual theory of algorithms, where we design one program that works for inputs of all sizes.

### Input/Output Schedules and Determinacy

In order to say that a VLSI circuit solves a certain problem, we must relate the variables of the problem to the inputs and outputs of the circuit. A *schedule* assigns to each variable a place (i.e., one of the pads) and a time at which the input is made available (if it is an input variable), or the output is produced by the circuit (if it is an output variable). It is normal for the schedule used by a chip to be independent of the actual input, although exceptions exist. For example, many systems make use of memory chips, because these are very carefully designed, and store information in significantly less area than we can under the Mead-Conway rules. A typical arrangement is for one chip, implementing an algorithm, to read data from one or more memory chips. The time at which the control chip reads data from the memory chip, or writes data

---

† A graph is *connected* if there is a path between any two nodes.

to the memory may well depend on what appears in the memory.

If the time at which inputs arrive and outputs are produced does not depend on the input, we say the circuit is *when-determinate*; otherwise it is *when-indeterminate*. If the pad at which inputs arrive and outputs are produced does not depend on the input, then the circuit is *where-determinate*; otherwise it is *where-indeterminate*.† Unless we state otherwise, a circuit claimed to solve a problem will be assumed to be when- and where-determinate.

**Example 2.2:** To see the difference that the two forms of determinacy make, let us consider the problem of sorting $n$ $k$-bit numbers. One circuit we might propose to "solve" this problem consists simply of $kn$ pads. The input schedule has the $kn$ input bits $x_1, \ldots, x_{kn}$ read in, one per pad, at the first time unit. At the next time unit, the output bits $y_1, \ldots, y_{kn}$ are also available at those pads. Of course, which pad holds $y_i$ depends rather heavily on the input! This "solution" to the sorting problem is when-determinate, since we know when each input bit arrives and when each output bit is available, but it is where-indeterminate, because the pad at which each output is produced depends on the input.

Another "solution" to the sorting problem is to have a circuit with $2k$ pads, $k$ for input and $k$ for output. At the first time unit, $X_1 = x_1, \ldots, x_k$ is read by the input pads, at the second time unit, $X_2 = x_{k+1}, \ldots, x_{2k}$ is read, and so on. At each time unit, the contents of each input pad is copied to a corresponding output pad. We can be sure that $y_1, y_{k+1}, \ldots, y_{(n-1)k+1}$ will be available at the first output pad, $y_2, y_{k+2}, \ldots, y_{(n-1)k+2}$ at the second, and so on, but when each of these bits appears is, of course, quite dependent on the input again. This circuit is where-determinate, but when-indeterminate.

Evidently, neither of these circuits really sorts in any useful sense, so we see the motivation for where- and when-determinacy. It should not surprise the reader that the area and time taken by these circuits, $A = kn$ and $T = 2$ in the first case, and $A = k$ and $T = n + 1$ in the second, are far better than it appears we can sort if we restrict ourselves to when- and where-determinate circuits. There is a restriction on when-indeterminate circuits that makes the lower bound results of this chapter apply. The ideas are given in the exercises, but intuitively, the requirement is that the circuit must not only make the output, but must be capable of indicating which $y_i$ has been emitted.

As examples of some "real" sorting algorithms, we could read the numbers to be sorted into a $k \times n$ array of cells, and perform a "bubble sort" in area $O(kn \log k)$ and time $O(n \log k)$. The "bubbling" is done in parallel, in $n$ stages.

---

† The term "oblivious" is frequently used in the literature for "determinate," but we avoid "oblivious" because it has somewhat the wrong connotation. One also sees in the literature the term "semilective," which means that the input or output appears at only one time (but perhaps at more than one place), and the term "semilocal," which means that the input or output appears at only one place (but perhaps several times).

During odd numbered stages, numbers in odd positions are compared with the number in the next higher position, and they are swapped if out of order. At even numbered stages, the same thing is done for numbers in the even positions. The factor $\log k$ in the time is needed to compare two $k$-bit numbers by a "divide-and-conquer" method, which the reader may discover as an exercise. This comparator requires a space $O(k)$ by $O(\log k)$, which accounts for the area expression. Alternatively, we could build a Batcher sorting network (see Knuth [1973]) that sorts in $O(\log^2 n \log k)$ time but that takes area $O(k^2 n^2)$. $\square$

Let us emphasize that in order to prove a lower bound on the area or time needed by a circuit that solves a given problem, we may assume that the circuit is when-determinate and where-determinate, if we wish, but we cannot in general dictate what the schedule is to be; or rather, if we do pick the schedule, then we have a weak result, one that does not rule out the existence of another schedule allowing a much faster or more compact circuit. Similarly, we cannot assume anything about the details of the circuit, such as where the pads are located, or how the pads correspond to the input and output bits.†

## The Three Basic Lower Bound Arguments

The history of the computation performed by a chip can be represented by the rectangular solid of Fig. 2.1. The vertical dimension represents time, while the other two dimensions represent the area of the chip. We have shown the history of a bit of information as well. It is read in at some time or place, it is possibly combined with other bits to form composite information. At some point, it may be forgotten completely, or it may be output at one or more points in the space-time solid.

We see three fundamentally different sorts of arguments that can be used to derive lower bounds on the area and/or time. First, we could argue about the volume of the solid. At most one bit can be input at any point in the solid, so the number of input bits cannot exceed the volume. The volume is the product of the area and time, so this method gives us a lower bound on the product $AT$, where $T$ is the time taken by the chip, and $A$ is the area.

The second possibility is to draw a boundary horizontally, as shown in Fig. 2.1, and argue that a certain amount of information must flow upward across that boundary, if the chip is to solve a particular problem. Since only one bit can flow through a unit area of the horizontal slice, we can sometimes obtain lower bounds on the area in this manner.

The third method is to make a slice vertically, cutting across the shorter dimension of area; the distance in that dimension is at most $\sqrt{A}$. If we can claim that a certain amount of information must flow across that boundary, then we can put a lower bound on the area of that slice, which is at most $T\sqrt{A}$

---

† However, sometimes we shall assume the pads are on the border of the chip, since this pattern is required if chips are to be packaged by standard equipment.

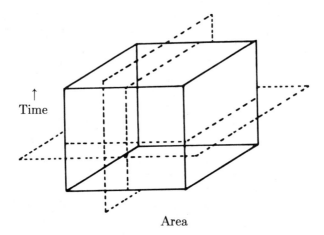

↑
Time

Area

**Fig. 2.1.** The space-time solid.

in terms of the area and time for the chip. If we square all quantities, we get the lower bound in its more familiar form, as a bound on $AT^2$.

In what follows, we shall informally discuss lower bounds of the $A$ type, the $AT$ type, and the $AT^2$ type, in that order.

### Memory-Based Lower Bounds

One of the simplest forms of lower bound arguments we can make concerns the memory requirement of a circuit and its relation to the area of that circuit. If a circuit has area $A$, it does not have more than $A$ circuit elements, and therefore cannot "remember" more than $A$ bits from one time unit to the next. To see this, note that each logic element and pad has a single output value at each time unit. Wires connected to the output of such an element have the value of that element at all times. Wires not connected to any input pad or output of a logic element do not have any well-defined value and can be removed without changing the output of the circuit. Thus the voltage at every point in the circuit is completely determined by the values of the input pads and the outputs of the logic elements, these being at most $A$ in number.

We can make the notion of remembering bits more precise by the following "fooling" argument. If up to time $t$, more than $A$ input bits have been read, then there are two assignments of values to the variables that have been read, such that after reading either of these assignments of values, the circuit is in the same *state*, meaning that the wires and circuit elements have the same (high or low) values in each case. Thus, if the subsequent inputs are the same, then the output made by the circuit in response to the two assignments of values will subsequently be the same (although prior outputs could have differed). If we can prove that there are $2^{f(n)}$ different assignments of values to the "early" inputs that require distinct outputs in response to the same subsequent inputs, then we have shown that any circuit to solve the problem in question, with input of size $n$, must have area at least $f(n)$. The reason is that if the area were less, then two of the $2^{f(n)}$ inputs would leave the circuit in the same state. Then the circuit output could not differ if these two "early" inputs were followed by the same "late" inputs.

**Example 2.3:** Let us use the above idea to get a lower bound on the area of any sorting circuit. First, let us observe that the minimum area required depends heavily on $k$, the number of bits in a word, at least when $k$ is small. As an exercise, the reader should be able to exhibit a circuit that sorts $n$ 1-bit words, yet uses only $O(\log n)$ area. However, when $k > \log_2 n$, we can prove that $\Omega(n)$ area is required, simply because that much storage is required at the time just before the first of the least significant bits is output.[†]

As a lemma, we first show that if any of the least significant bits,

$$y_k, y_{2k}, \ldots, y_{nk}$$

are output at or before the time unit when the last of the least significant input bits, $x_k, x_{2k}, \ldots, x_{nk}$, is read, then the circuit does not always sort properly. Suppose $y_{rk}$ is output before or when $x_{sk}$ is read. Then let $r - 1$ of the input words other than the $s^{th}$ word $X_s = x_{(s-1)k+1}, \ldots, x_{sk}$ be $00 \cdots 00$ and $n - r$ words other than the $s^{th}$ be $00 \cdots 01$. Let $x_{(s-1)k+1}, \ldots, x_{sk-1}$ be 0's. Then, since $y_{rk}$'s value cannot depend on $x_{sk}$, there is one value $y_{rk}$ assumes in response to whatever of the above values are read at time units before $y_{rk}$ is emitted. If that value is 1, then let $x_{sk}$ be 0, and we see that $y_{rk}$ is incorrect. Similarly, if the value chosen is 0, let $x_{sk} = 1$ and the output is wrong. We conclude that all least significant output bits follow in time the reading of all least significant input bits.

We now must construct a family of $2^n$ inputs that we know require different output values for $y_k, \ldots, y_{nk}$. All these input assignments agree on input bits other than the least significant ones, $x_k, \ldots, x_{nk}$. In those other bits, we count in binary, one number per word. That is, $x_1, \ldots, x_{k-1} = 00 \cdots 00$, $x_{k+1}, \ldots, x_{2k-1} = 00 \cdots 01$, and in general, $x_{ik+1}, \ldots, x_{(i+1)k-1}$ is $i$ in binary. The input assignment $\alpha_{a_1, a_2, \ldots, a_n}$, where each $a_i$ is 0 or 1, has $x_k, \ldots, x_{nk}$ equal to $a_1, \ldots, a_n$, respectively.

Let us focus on the time just prior to when the first of the least significant output bits is emitted. We have proved that all least significant input bits are input at or before this time. Since our $2^n$ input assignments differ only in the least significant bits, we know that henceforth all inputs will be the same.

---

[†] Remember that we are really discussing an infinite family of circuits, one for each $n$, and the big-oh and big-omega expressions refer to the growth rate of the area as $n$ gets large.

However, since in each of our input assignments, the inputs appear in sorted order, we know that for input $\alpha_{a_1,\ldots,a_n}$ the outputs will be $y_k = a_1$, $y_{2k} = a_2, \ldots$, and $y_{nk} = a_n$.

If there were fewer than $n$ circuit elements, then there would have to be two distinct input assignments, say $\alpha_{a_1,\ldots,a_n}$ and $\alpha_{b_1,\ldots,b_n}$, such that the state of the circuit, just before output of the first of the least significant bits, was the same for both inputs. Then, since subsequent inputs are the same in both cases, the values of $y_k, \ldots, y_{nk}$ output would be the same. But these values could not be both $a_1, \ldots, a_n$ and $b_1, \ldots, b_n$, a contradiction. We conclude that sorting requires $\Omega(n)$ area to sort $n$ words of $\log_2 n + 1$ or more bits. $\square$

### Input/Output-Based Lower Bounds

Often we can develop a simple lower bound not on the area or time needed by itself, but on the product of these. The principle is that if we must read $n$ bits or output $n$ bits, then a circuit of area $A$, which certainly has no more than $A$ pads, must take at least $n/A$ time units just to read or write all its data. Thus the following theorem is immediately obvious.

**Theorem 2.1:** If $D$ is the larger of the number of input bits and the number of output bits used by a problem $P$, then any circuit (even a when- or where-indeterminate circuit) that solves $P$ has area $A$ and time $T$ that satisfies $AT \geq D$. $\square$

**Example 2.4:** In Example 2.3 we learned that to sort $n$ ($\log_2 n+1$)-bit numbers requires $A = \Omega(n)$. Theorem 2.1 implies that if the circuit takes time $T$, then $AT = \Omega(n \log n)$. Thus, if, say, $T = O(1)$, at least $\Omega(n \log n)$ area is required. $\square$

### Introduction to Area-Time-Squared Lower Bounds

We have seen two different bound methods, one of which gives a lower bound on $A$, the area of a circuit, and the other of which gives lower bounds on the product $AT$, where $T$ is the time used by the circuit. In many cases these bounds are "weak," in the sense that there do not appear to be circuits as good as the bounds imply might be possible. Many "strong" lower bounds, that do match the best circuits we can construct, are lower bounds on the product $AT^2$. These bounds are based not on memory requirements or input/output rate, but on the requirements for information flow within the chip.

We shall give a powerful formulation of this result in the next section. Here we shall introduce some of the ideas with an example; we show informally that if a circuit sorts $n$ $k$-bit numbers, where $k \geq \log_2 n + 1$, and uses area $A$ and time $T$, then $AT^2 = \Omega(n^2)$. Notice that this result is stronger than either Example 2.3 or 2.4.

Suppose the circuit in question, in the grid model, looks like Fig. 2.2, a

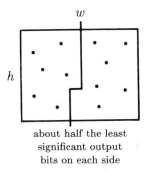

about half the least
significant output
bits on each side

**Fig. 2.2.** Hypothetical sorting circuit.

rectangle of height $h$ and width $w$. We assume without loss of generality that $h \leq w$. As a consequence of the normal form of Theorem 1.1, we may assume that along no grid line is there more than one wire.

Let us focus, as we did in Example 2.3, on the least significant output bits. These are emitted at pads located variously. However, we need not consider the case where as many as $1/3$ of them are emitted at the same pad. For if so, then $T \geq n/3$, so $AT^2 = \Omega(n^2)$ follows immediately.†

Thus, it is possible to divide the circuit by a line as shown in Fig. 2.2. This line runs between the grid lines and runs vertically, except possibly for a single jog of one grid unit. Most importantly, we can select the line so at least $n/3$ of the least significant output bits are emitted on each side. We select the line by sliding it from left to right, until the first point at which at least $n/3$ of the least significant bits are output to the left of the line.

If no more than $2n/3$ of these bits are output to the left, we are done. If not, start from the top, considering places to jog the line back one unit to the left. We know that if the line jogs at the very top, fewer than $2n/3$ of the least significant bits are emitted to the left, and if the line jogs at the very bottom, more than $n/3$ are. Thus, as no single grid point can be the place where as many as $n/3$ of the bits are emitted, we can find a suitable place in the middle to jog the line. There, we shall have between $n/3$ and $2n/3$ of the least significant output bits on each side.

Now assume without loss of generality that at least half of the least significant input bits are read on the left of the line, and let us, by renumbering bits, if necessary, assume that these are $x_k, x_{2k}, \ldots, x_{nk/2}$. Suppose also that least significant output bits $y_{i_1}, y_{i_2}, \ldots, y_{i_{n/3}}$ are output on the right. Using a trick similar to Example 2.3, we can pick values for bits other than the least significant, so that $y_{i_1} = x_k$, $y_{i_2} = x_{2k}$, and so on. Thus, information regarding what the first $n/3$ least significant input bits, $x_k, x_{2k}, \ldots, x_{kn/3}$, are must cross the line, or we can "fool" the chip by forcing it to give the same output for two

---

† Note that the number $1/3$ could be any $\epsilon > 0$.

different assignments of values to the least significant input bits.†

We may assume at most one wire or circuit element along any grid line, by Theorem 1.1, so intuitively, the number of bits crossing the line in one time unit is at most $h + 1$ ($h$ horizontal and one vertical, at the jog). Again, we shall make formal the notion of bits crossing a boundary in the next section. It follows that $(h + 1)T \geq n/3$, or else the required $n/3$ bits cannot cross the line in time. Since we assumed $w \geq h$, we have both $hT = \Omega(n)$ and $wT = \Omega(n)$. Since $wh = A$, we have $AT^2 = \Omega(n^2)$ by taking the product.

In comparison with this lower bound, consider the two methods discussed in connection with Example 2.2. Assuming $k = \log n$, the bubble sort has $A = O(n \log n \log \log n)$ and $T = O(n \log \log n)$. Thus, this method takes $AT^2 = O(n \log n (\log \log n)^3)$, which is not very close to the bound. The Batcher approach takes $A = O(n^2 \log^2 n)$ and $T = O(\log^2 n \log \log n)$. Here,

$$AT^2 = O(n^2 \log^6 n \log \log n)$$

which differs from the lower bound by only some powers of $\log n$. Note that any power of $\log n$ is asymptotically less than $n^\epsilon$, for any $\epsilon > 0$; therefore we are really quite close to the lower bound.

## 2.2 INFORMATION AND CROSSING SEQUENCES

In this section we shall define the information content of a problem formally and show how information content implies a lower bound on the $AT^2$ product. An important tool is the "crossing sequence," which is the sequence of events at an imaginary line through a circuit. A general development of techniques for proving lower bounds on $AT^2$ requires a number of concepts that must come together, each being rather strange in isolation. We shall thus attempt to give an overview of the methodology of lower bound proofs, and then to tackle the details one at a time.

In the world of lower bound proving, we may view the chip designer as our "adversary." He attempts to design chips that solve a given problem and have a low $AT^2$ product (or some other good property). We attempt to show that he cannot do so. An outline of the stages in this contest is as follows.

1. The adversary designs a chip, which he claims solves problem $P$ and does so with a small area and time.

2. We examine the chip and attempt to prove that it does not work. We do so by picking a partition of the input and output positions, as in Fig. 2.2. Then we exhibit a large set of input assignments $\mathcal{A}$ with the property that any two assignments in $\mathcal{A}$ require unique information flowing across the boundary. That is, if $\alpha$ and $\beta$ are in $\mathcal{A}$, and the chip does not distinguish, by the information crossing the boundary, between $\alpha$ and $\beta$, then the chip

---

† We make this fooling argument formal in the next section when we talk about "crossing sequences," the sequences of values that the elements and wires straddling the line can assume.

can be "fooled" by feeding it $\alpha$ on the left and $\beta$ on the right, or vice versa.

3. The chip designer is forced, because the chip is so small and fast, to admit that there are more input assignments in $\mathcal{A}$ than there are "crossing sequences," that is, histories of events at the boundary. Thus, he must give us two members of $\mathcal{A}$ that have the same crossing sequence.

4. We prove that the chip does not work, by showing that if we fed it one assignment on the left and the other on the right, one or more of the outputs would be wrong.

In this manner we show that chips which are too small and/or too fast cannot solve the problem at hand. Of course, the proof depends on our being able to select, for any chip, a partition and a large set of inputs that can "fool" the chip, or else we cannot get a large lower bound on the number of crossing sequences and, therefore, we cannot get a large lower bound on the $AT^2$ product. We proceed to formalize this argument with a series of definitions. In practice, we find that we cannot take advantage of our ability to pick the partition of inputs, so we in effect leave that to the adversary. Thus, in our formal development, we shall let the "adversary" pick the partition, subject only to the constraint that we focus on a subset $Z$ of the input variables and require him to divide those bits roughly equally on either side.

## Information

Let $P$ be a problem and $n$ a value of the problem size parameter, for which the input and output sets of variables are $X$ and $Y$. A *partition* for $P$ and $n$ is a division of $X$ into two disjoint sets $X_L$ and $X_R$, whose union is $X$, and a similar division of $Y$ into $Y_L$ and $Y_R$. An *input assignment* for $P$ and $n$ is a mapping from $X$ into $\{0, 1\}$, that is, an assignment of a value to each input bit.

Suppose $\pi = (X_L, X_R, Y_L, Y_R)$ is a partition, and $\alpha$ and $\beta$ are two input assignments. We use $\alpha_L$ for $\alpha$ restricted to $X_L$, and $\alpha_R$ for $\alpha$ restricted to $X_R$. Also, we use $\alpha_L\beta_R$ for the input assignment that agrees with $\alpha$ on $X_L$ and with $\beta$ on $X_R$.

A *fooling set* for a problem $P$, a size $n$, and a partition

$$\pi = (X_L, X_R, Y_L, Y_R)$$

is a set $\mathcal{A}$ of input assignments with the property that for any distinct $\alpha$ and $\beta$ in $\mathcal{A}$, one of the following four conditions must hold.

1. $\alpha_L\beta_R$ differs from $\alpha$ on some variable in $Y_L$.
2. $\alpha_L\beta_R$ differs from $\beta$ on some variable in $Y_R$.
3. $\beta_L\alpha_R$ differs from $\beta$ on some variable in $Y_L$.
4. $\beta_L\alpha_R$ differs from $\alpha$ on some variable in $Y_R$.

We should note how each of these four conditions implies that if $\alpha$ and $\beta$ caused the same data to cross the boundary between the left side (the variables

$X_L$ and $Y_L$) and the right side ($X_R$ and $Y_R$), then the chip could be "fooled."
For example, in case (1), feeding the chip $\alpha_L \beta_R$ looks the same to the left side
as $\alpha$, if the same data crosses the boundary. Thus, for both input assignments,
the outputs in $Y_L$ must be the same. But (1) implies that these outputs are not
the same, if the chip is to solve $P$. Note also that because we are given $\alpha$ and
$\beta$ as an ordered pair, we need all four of the above possibilities.

Next, we must discuss "acceptability" of partitions, a concept designed to
enforce the condition that certain essential input variables are divided relatively
evenly. Formally, let $Z$ be a subset of $X$, the set of input variables for a problem
$P$ and size $n$. We say a partition $\pi$ is *acceptable* for $P$, $n$, and $Z$, if between
one-third and two-thirds of $Z$ is in $X_L$ (and therefore between one-third and
two-thirds of $Z$ is in $X_R$).

Now we are ready to define the information content of a problem, which is
intuitively the amount of information that must cross a boundary in order to
solve the problem. Technically, we define three varieties of information, only
the last of which is relevant to lower bounds; the other two are used to define
the third. In what follows, we let $P$ be a problem, $n$ its size parameter, $\pi =$
$(X_L, X_R, Y_L, Y_R)$ a partition, and $Z$ a subset of $X = X_L \cup X_R$.

Define $I(P, n, \pi)$ to be the logarithm, base 2, of the size of the largest
fooling set for $P$, $n$, and $\pi$. Define $I(P, n, Z)$ to be the minimum over all
partitions $\pi$ acceptable for $P$, $n$, and $Z$, of $I(P, n, \pi)$. Finally, define $I(P, n)$,
the *information content of $P$*, to be the maximum over all $Z$ of $I(P, n, Z)$. We
shall, when no ambiguity results, delete $P$ and $n$ as arguments, referring to the
above quantities as $I(\pi)$, $I(Z)$, and $I$, respectively.

The above definitions make sense in the light of the adversary argument
that we gave in introduction to this section. We pointed out that after the
adversary designed the chip, we had the right to pick both the partition and
the fooling set of inputs. Clearly, the larger the fooling set, the larger or slower
we can prove the chip to be. We cannot simply pick that $\pi$ which maximizes
$I(P, n, \pi)$, because there might be no way to divide inputs and outputs according
to $\pi$, given the particular chip the adversary designs. However, we can use
an argument similar to that given in connection with Fig. 2.2 in the previous
section to claim that no matter what subset $Z$ of the inputs we pick, we can
find a simple boundary† that divides the given chip into parts that have roughly
equally sized subsets of $Z$. Thus, for each $Z$, we know there must be some
acceptable partition $\pi$ that can be achieved by making a simple cut of the chip.
Since we do not know what that partition is, we minimize over all acceptable
partitions for $Z$ to get $I(Z)$, but since we have the right to choose $Z$, we
maximize over $Z$, to get $I$.

---

† If we used a longer boundary than that of Fig. 2.2, we would find that more information
could cross the boundary in a given time; therefore we would not be any more successful at
proving lower bounds, and indeed we might do worse.

**Example 2.5:** We shall now give an example of how we can get a lower bound on $I(P, n)$ for a particular problem $P$. We do so by choosing a particular $Z$ and computing $I(P, n, Z)$, which is surely a lower bound on $I(P, n)$. In fact, our lower bound is, to within a constant factor, the best possible. However, in general we may not know the best $Z$ to pick, and therefore only obtain lower bounds on $I$, which is, fortuitously, exactly what we need.

The problem with which we shall deal is the *cyclic shift* or *barrel shift*. In the problem of size $n$, we are given inputs $x_0, \ldots, x_{n-1}$ and $c_1, \ldots, c_{\log_2 n}$, and we wish to cycle the $x_i$'s by $c$ places, where $c$ is the value of $c_1, \ldots, c_{\log_2 n}$ treated as a binary number. That is, there are $n$ output variables $y_0, \ldots, y_{n-1}$, and the functions defining the input/output relationship are $y_i = x_{i+c}$, where the sum $i + c$ is taken modulo $n$.

The particular value of $Z$ we use is $Z = \{ x_0, \ldots, x_{n-1} \}$, that is, we focus on the positions of the "data" variables and ignore the positions of the "control" variables, the $c_j$'s. In what follows, we refer to the sets $X_L$ and $Y_L$ as one *side* and $X_R$ and $Y_R$ as the other side. Intuitively, each $y_j$ is on the opposite side from at least one third of the $x_i$'s. If $c$ is chosen randomly, each $y_j$ is equally likely to take on any $x_i$ as value, and about one third the time or more this $x_i$ will be from the other side. There must be some $c$ that is at least as bad as average, and this is the one we want.

For a more formal proof, let $\delta_{ic}$ be 1 if $x_i$ and $y_{i+c}$ are on the opposite side, and 0 otherwise. No matter what $y_j$ we choose, there are at least $n/3$ $x_i$'s on the opposite side. Thus, for $0 \leq j < n$,

$$\sum_{c=0}^{n-1} \delta_{j-c,c} \geq n/3$$

Sum both sides from $j = 0$ to $n - 1$, let $i = j - c$, and interchange the order of summation on the left. Remembering that addition and subtraction in subscripts is modulo $n$, we get

$$\sum_{c=0}^{n-1} \sum_{i=0}^{n-1} \delta_{ic} \geq n^2/3$$

There must be some $c$ for which

$$\sum_{i=0}^{n-1} \delta_{ic} \geq n/3$$

or else the above double summation could not be as high as $n^2/3$. This is the $c$ for which at least $n/3$ $x_i$'s are on the opposite side from $y_{i+c}$. Let $W$ be this set of $x_i$'s.

We now construct $2^{n/3}$ assignments in our set $\mathcal{A}$ as follows. Let $\alpha_{a_1, \ldots, a_{n/3}}$

be defined as follows.

1. The bits $x_{i_1}, \ldots, x_{i_{n/3}}$ in $W$ are assigned $a_1, \ldots, a_{n/3}$, respectively.
2. All other $x_i$'s are assigned 0.
3. $c_1, \ldots, c_{\log_2 n}$ are assigned values to represent $c$ in binary.

Now let $\alpha$ and $\beta$ be different members of $\mathcal{A}$. Suppose these assignments differ in the value assigned to some variable on the right, that is, a variable in $W \cap X_R$, say $x_i$. Then $y_{i+c}$ is in $Y_L$, and $\alpha$ and $\alpha_L \beta_R$ differ in the output produced for a bit in $Y_L$. Thus, condition (1) in the definition of "fooling set" is satisfied. If $\alpha$ and $\beta$ agree on all the bits in $W \cap X_R$, then they must disagree in a bit of $W \cap X_L$, and we can similarly show that $\alpha$ and $\beta_L \alpha_R$ differ in a bit of $Y_R$ and therefore satisfy (4). As a consequence, we have $I(n) \geq n/3$, or $I = \Omega(n)$.

It should be apparent that if there are $r$ input bits, $I = \Omega(r)$ is the strongest result we can show, because there are no more than $2^r$ different assignments. In this example, $n$ is close to $r$ for large $n$, so we have in fact shown $I = \Omega(r)$, the strongest possible result. Thus, the cyclic permutations, although a very simple group of permutations, are as complex as can be proved. In fact, we can design a circuit that performs an arbitrary permutation of its inputs (as specified by about $n \log n$ control bits like the $c$'s) with little more area than needed for a barrel shifter. $\square$

We shall state the principal fact to be derived from the above sequence of definitions as the following lemma.

**Lemma 2.1:** Let $Z$ be any set of input variables, and $C$ a chip. If no more than one third of the $Z$'s are read at any one input pad of $C$, then we can draw a line with a single jog, as in Fig. 2.2, that divides $C$ into two halves, each with between one third and two thirds of the inputs in $Z$.

**Proof:** The ideas of the proof were given at the end of Section 2.1 and need not be repeated here. $\square$

### Crossing Sequences

Now, let us focus on a circuit that, like Fig. 2.2, has been divided left to right. We assume by Theorem 1.1 that the circuit is in normal form, and we also let $k$ be the limit on the number of inputs to any logic element. As there are $h+1$ grid lines that cross the boundary, there can be at most $h+1$ wires and logic elements crossing the boundary. Wires always carry a single value, while logic elements surely can be characterized by the values of their input variables. That is, if the inputs to a logic element are fixed, the output is determined. Thus, $k(h+1)$ is an upper bound on the number of values we need to determine completely the events at the boundary. There could be fewer independent values, either because some grid line is not part of any circuit element, or because two or more wires have values that must be identical. For example, a wire might cross the boundary twice.

Pick an order for the up to $k(h+1)$ voltages that characterize the boundary, and represent high and low voltage by 1 and 0, respectively. Then, at any time unit, the values of voltages across the boundary are completely characterized by a word of 0's and 1's of length at most $k(h+1)$. Given an input assignment $\alpha$ and a boundary line, the *crossing sequence at that line in response to* $\alpha$ is $W_0, \ldots, W_{T-1}$, where $T$ is the time taken by the circuit, and $W_t$ is the word that characterizes the voltages at the boundary after elapsed time $t = 0, 1, \ldots, T-1$.

The important property of crossing sequences is that the inputs on one side of the boundary and the crossing sequence at that boundary completely determine the activity of elements on that side of the chip. Thus, if we have two different input assignments that agree on the inputs on one side of the chip, say the left, and the crossing sequences in response to these assignments are the same, then the outputs (and all circuit activity) on the left will be the same for either assignment. Moreover, if we swap the assignments as far as input bits on the right are concerned, the crossing sequence remains the same, and the output on the left does not change. If the two assignments in question are among those used to show a high information content for the problem, then we shall be able to assert that some output must change, which contradicts the assumption that the circuit solved the problem and also associated the same crossing sequence with these two assignments.

We thus are able to claim that high information implies many different crossing sequences are possible. Since large numbers of crossing sequences can only be available if the product $hT$, which is essentially constant $1/k$ times the length of a crossing sequence, is high, we have a lower bound on $hT$. Since $h$, the smaller dimension, is no larger than $\sqrt{A}$, we have a bound on $\sqrt{A}T$ or, equivalently, $AT^2$. We proceed now to make these ideas formal.

**Lemma 2.2:** Let circuit $C$ have a boundary as in Fig. 2.2, creating a partition $\pi = (X_L, X_R, Y_L, Y_R)$ of the input and output bits. Let $\mathcal{A}$ be a fooling set for $\pi$ (for some problem $P$ and size $n$). Then if $C$ solves $P$, no two members of $\mathcal{A}$ can have the same crossing sequence at the boundary.

**Proof:** Suppose $\alpha$ and $\beta$ are two assignments in $\mathcal{A}$ with the same crossing sequence. Consider what happens when we run $C$ with input assignment $\alpha_L \beta_R$. An easy induction on the number of time steps shows that the crossing sequence remains the same, each element on the left behaves as if $\alpha$ were the assignment, and each element on the right behaves as if $\beta$ were the assignment. Thus, the outputs in $Y_L$ are as if $\alpha$ were the assignment, and the outputs in $Y_R$ are as if $\beta$ were the assignment. An analogous statement holds about input assignment $\beta_L \alpha_R$. If $C$ solves $P$, then what we have just said implies that none of the four conditions in the definition of "fooling set" can hold for $\alpha$ and $\beta$, contradicting the assumption that $\mathcal{A}$ is a fooling set. Thus, no members of $\mathcal{A}$ can share a crossing sequence. □

**Theorem 2.2:** Let $P$ be a problem with information content $I$, and let $C$ be a circuit of area $A$ and time $T$ that solves $P$. Then $AT^2 = \Omega(I^2)$.

**Proof:** Let $C$ have dimensions $h$ by $w$, with $h \leq w$ and $hw = A$, and assume that $C$ is in normal form. Let $k$ be the maximum number of inputs that a logic element in $C$ may have. Suppose that $Z$ is a subset of the input variables that maximizes $I$; that is, $I(P, n, Z) = I(P, n) = I$. Let $Z$ have $r$ members. Then it is easy to see that $I \leq r/3$, because the partition $\pi$ that has one third of the members of $Z$ on the left side and all other input and output variables on the right is acceptable for $Z$. If two assignments $\alpha$ and $\beta$ in a fooling set for $\pi$ had the same values for the $r/3$ variables on the left, then conditions (1) or (3) in the definition of fooling set could not hold, because $Y_L$ is empty, and conditions (2) and (4) could not hold because $\alpha_L \beta_R = \beta$ and $\beta_L \alpha_R = \alpha$. Thus, a fooling set for $Z$ has at most $2^{r/3}$ members, and $I \leq r/3$.

Now let us consider two cases, depending on whether $T$ is large or small. If $T \geq r/3$, then since $I \leq r/3$, we have $T = \Omega(I)$, and since $A = 0$ is impossible, we have $AT^2 = \Omega(I^2)$, and the theorem follows.

If $T < r/3$, then no more than $r/3$ inputs in $Z$ can be read at any one pad. Thus Lemma 2.1 applies, and we can partition the circuit as in Fig. 2.2, with the inputs of $Z$ divided with one third or more on either side. For this division of input and output bits, there is a fooling set $\mathcal{A}$ of size $2^I$. By Lemma 2.2, these assignments all have different crossing sequences associated, so there are at least $2^I$ crossing sequences at the boundary.

Since the number of bits in a crossing sequence is at most $k(h + 1)T$, we have $k(h + 1)T \geq I$. Since $w \geq h$, we have $k(w + 1)T \geq I$ and therefore $k^2(h+1)(w+1)T^2 \geq I^2$. Since $k^2(h+1)(w+1) = O(A)$, we have $AT^2 = \Omega(I^2)$, as was to be proved. □

**Example 2.6:** We know from Example 2.5 that for cyclic shifting, $I = \Omega(n)$, so by Theorem 2.2, any circuit to solve the problem must have $AT^2 = \Omega(n^2)$. The circuit in Fig. 2.3 uses three layers (solid, dashed, and dotted lines), and runs wires in different layers along the same grid lines.† In this circuit, we use $c_1$, the high order control bit, to decide first whether to cycle by $n/2$, then we use $c_2$ to decide whether to cycle by $n/4$, and so on. The circuit has height and width $O(n)$ and takes time $O(\log n)$, since there are $\log n$ stages that the inputs go through before being output. This circuit, with an $AT^2$ product of $n^2 \log^2 n$, comes very close to the lower bound. □

---

† We show wires running diagonally for clarity, but they could be replaced by wires running in a staircase fashion without increasing the dimensions by more than a factor of two. In fact, the circuit can be put in the normal form of Theorem 1.1 with only a constant factor expansion of size.

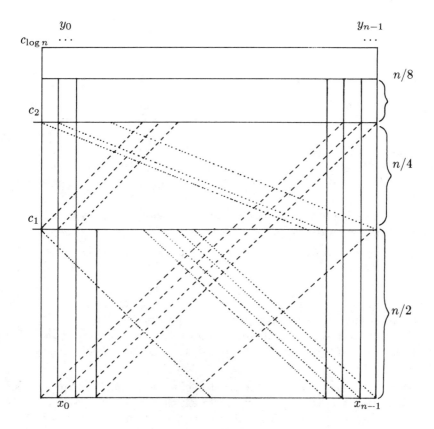

**Fig. 2.3.** Barrel shifter circuit.

## The Role of Aspect Ratio

The bound of Theorem 2.2 was proved assuming nothing about the shape of the circuit, except that it is rectangular. Sometimes, we can get a stronger bound if we make the *aspect ratio*, the ratio of the longer side to the shorter side, a parameter. The idea is summed up in the next theorem.

**Theorem 2.3:** Let problem $P$ and circuit $C$ be as in Theorem 2.2, but require that the aspect ratio of $C$ be at least $a$. Then $AT^2 = \Omega(aI^2)$.

**Proof:** The proof is like that of Theorem 2.2. However, we have $w \geq ah$, so in addition to $k(h+1)T \geq I$ we have $k(w/a+1)T \geq I$, from which $AT^2 = \Omega(aI^2)$ follows. $\square$

**Example 2.7:** Suppose we wanted a barrel shifter that, like Fig. 2.3, had area $O(n^2)$, but that had aspect ratio $n$, i.e., the circuit was $O(n^{3/2})$ by $O(n^{1/2})$. Then by Theorem 2.3, we would have $n^2T^2 = \Omega(n^3)$, since $a$ and $I$ would both be $\Omega(n)$. As a consequence, $T = \Omega(n^{1/2})$, which proves we could not have had a circuit that was as compact and as fast as Fig. 2.3, but was substantially less square in shape. $\square$

### Circuits with Pads on the Border

The typical chip has all of its pads on the border of the circuit, so it is significant to discover that we can often prove a stronger lower bound for circuits that obey this constraint than those that do not. In fact, there are problems for which the constraint of pads on the border really does affect the quality of the circuit we can expect. For example, Chapter 3 discusses tree layouts with and without pads on the border, and shows that forbidding interior pads hurts in regard to the area needed.

The following theorem is useful when we cannot prove a strong lower bound on the information content of a problem.

**Theorem 2.4:** Let $P$ and $C$ be as in Theorem 2.2, but suppose $C$ has all its pads on its border. Let $r$ be the number of input bits for a problem of size $n$. Then $AT^2 = \Omega(Ir)$.

**Proof:** Let $a$ be the aspect ratio of $C$. Then by Theorem 2.3, $\sqrt{A}T = \Omega(\sqrt{a}I)$. The longer side of $C$ has length $\sqrt{Aa}$, and therefore $4T\sqrt{Aa} \geq r$, or else we cannot read in all the input bits in time $T$. Thus $\sqrt{A}T = \Omega(r/\sqrt{a})$. If we multiply our two lower bounds, we find that $a$ drops out, and we get $AT^2 = \Omega(Ir)$. $\square$

**Example 2.8:** Let us consider the problem of sorting $n$ $(\log n + 1)$-bit numbers. We claim that $I = \Omega(n)$; this fact was essentially proved in Section 2.1. Also, $r = \Omega(n \log n)$ is apparent. Thus, if circuits have all their pads on the border, we have by Theorem 2.4 that $AT^2 = \Omega(n^2 \log n)$. $\square$

## 2.3 PROBABILISTIC CIRCUITS AND ALGORITHMS

There has been interest in probabilistic algorithms recently, because of the possibility that problems like factoring numbers are hard in the worst case, but can be solved efficiently with high probability by an algorithm that "flips coins." We might wonder if there is a similar phenomenon in the world of VLSI circuits, where problems that we can prove require, say, $AT^2 = \Omega(n^2)$ for a deterministic chip can be solved with a much smaller $AT^2$ product if we are willing to accept a chip that is correct with a high probability, but not with certainty.

We are thus motivated to consider the following model of a chip. A *probabilistic circuit* is an ordinary circuit with one or more random number generators that at each time unit emit a 0 or 1 with equal probability. Each time the circuit is run, the random number generators produce independent sequences of 0's and 1's.

A well-developed theory of how to design small circuits that behave almost as ideal random number generators is available (see Knuth [1969], e.g.), and we shall simply postulate the existence of single-grid-point elements that generate random numbers. Of course, real random number generators are much larger than, say, single pullups, and, in fact, as the time for which a circuit runs increases we need progressively larger random number generators to assure "random" behavior of the devices over suitably long periods of time. However, the growth in size is sufficiently slow that taking random number generators as unit area devices will not do serious violence to the validity of our lower and upper bounds.

We say a probabilistic circuit $C$ *solves* problem $P$ if on any run of $C$ with any input assignment $\alpha$, the probability that $C$ will produce the correct answer to $P$ is greater than $1/2$. Note that in our definition, success must be "odds-on" for every input, not just for the great majority of input assignments or even for all but a vanishingly small fraction of the input assignments, as $n$, the problem size, gets large.

Also note that with such a circuit $C$ we can for all intents solve $P$ exactly. Suppose that on any input, $C$ has probability at least $1/2 + \epsilon$ of success, and we wish probability $1 - \delta$ of success. Let us run $C$ $2k$ times and take as output that answer, if any, which is produced at least $k + 1$ times. The probability $\delta$ that we shall not thereby produce the correct answer is

$$\sum_{i=0}^{k} \binom{2k}{i} (\frac{1}{2} + \epsilon)^i (\frac{1}{2} - \epsilon)^{2k-i}$$

That is, we sum over all the insufficient numbers of successes, $i = 0$ through $i = k$, the probability of that number of successes occurring.

This sum is no more than the number of terms times the largest term. The factor $\binom{2k}{i}$ is no larger than $2^{2k} = 4^k$, and the remaining factors assume their largest values when $i = k$. Therefore an upper bound on the probability of an error is

$$(k + 1)4^k (\frac{1}{2} + \epsilon)^k (\frac{1}{2} - \epsilon)^k$$
$$= (k + 1)(1 - 4\epsilon^2)^k \tag{2.1}$$

If the probability of an error is not to exceed $\delta$, then the ratio of $\delta$ to expression (2.1) must not be less than 1, so the difference of the logarithms must be at least 0. That is,

$$\log_e \delta - \log_e(k + 1) - k \log_e(1 - 4\epsilon^2) \geq 0$$

If $\epsilon$ is small, then $\log_e(1 - 4\epsilon^2)$ is about $-4\epsilon^2$. Thus, (2.1) can be rewritten

$$4\epsilon^2 k - \log_e(k + 1) \geq \log_e(1/\delta) \tag{2.2}$$

If we let $k = \log_e(1/\delta)/\epsilon^3$, then (2.2) becomes

$$\frac{4\log_e(1/\delta)}{\epsilon} - \log_e(\log_e(1/\delta)) + 3\log_e \epsilon \geq \log_e(1/\delta) \qquad (2.3)$$

As long as $\epsilon$ and $\delta$ are small, the first term of (2.3) is positive and much bigger than the other terms. Thus, (2.3) is true, and so is (2.1). The conclusion is that as long as $\epsilon$ and $\delta$ are constants, independent of the problem size $n$, there is some constant number of trials $k$ we need to perform with a probabilistic chip in order to ensure the desired level of accuracy $1 - \delta$, no matter how close to zero we choose $\delta$.

It turns out that lower bounds on $AT^2$ for probabilistic circuits are frequently the same as for deterministic circuits. However, there is a difference in the definition of information that we need, and this difference leads to examples of problems that have significantly better probabilistic circuits than deterministic ones. Moreover, it is likely that there are large classes of problems for which the best probabilistic circuit for the problem is much superior to the best deterministic circuit. For example, there are seemingly intractable problems (see Garey and Johnson [1979], e.g.), for which efficient probabilistic algorithms exist, and these could be implemented as a chip, yet unless $P = NP$, there could be no family of deterministic chips that solved the problem and had a polynomial growth rate for their value of $AT^2$.

**One-Way Information**

It turns out that we can prove lower bounds on $AT^2$ for probabilistic chips, as long as we have a stronger kind of fooling set. The desired fooling set is one where all the assignments agree on one side; we call them "one-way" fooling sets because they imply that at least a certain amount of information must flow in one particular direction across a boundary.

Formally, a *left fooling set* for a problem $P$, a size $n$, and a partition $\pi = (X_L, X_R, Y_L, Y_R)$ is a set of assignments $\mathcal{A}$ such that
1.  All members of $\mathcal{A}$ agree on $X_L$.
2.  The outputs in response to two different members of $\mathcal{A}$ differ in at least one variable in $Y_L$.
A *right fooling set* is defined analogously, with "left" and "right" interchanged. A *one-way fooling set* is a left or right fooling set. Note that every one-way fooling set is a fooling set, but there are fooling sets in the original sense that are not one-way fooling sets.

Intuitively, the amount of information that must flow across the boundary, from right to left, in the case of a left fooling set, is at least what is necessary to determine the member of $\mathcal{A}$, for if not, then two members of $\mathcal{A}$ look identical to the left side of the chip, and the chip must err on a bit of $Y_L$. The analogous remark holds for right fooling sets.

Define $I_{1W}(P, n, \pi)$ to be the logarithm, base 2, of the size of the largest one-way fooling set for $P$, $n$, and $\pi$. Let $Z$ be a subset of the input variables, and define $I_{1W}(P, n, Z)$ to be the minimum, over all partitions $\pi$ acceptable for $Z$, of $I_{1W}(P, n, \pi)$. Then, define $I_{1W}(P, n)$, the *one-way information content* of $P$, to be the maximum over all $Z$, of $I_{1W}(P, n, Z)$. As with $I$, we shall drop $P$ and $n$ as arguments from $I_{1W}$ when no confusion regarding the problem referred to exists.

**Example 2.9:** Consider the cyclic shift problem from Example 2.5. As in that example, let us prove a lower bound on $I_{1W}$ by picking $Z$ to be the "data" inputs (the $x_i$'s) and evaluating $I_{1W}(Z)$. We already argued that we could find an assignment to the control variables $c_1, \ldots, c_{\log n}$ so that $n/3$ of the input variables were output on the opposite side. Thus, we can find $n/6$ inputs on one side, say the right, that become output variables on the other side, the left in this case. Let $\alpha$ be any assignment to the variables in $X_L$ that gives the control variables the proper values, and let $\beta_1, \ldots, \beta_{2n/6}$ be assignments to $X_R$ that also give the control variables the proper values and collectively give all possible values to the $n/6$ inputs that become outputs on the left. Then clearly all of $\alpha\beta_1, \ldots, \alpha\beta_{2n/6}$ differ in at least one bit of $Y_L$. We have thus exhibited a one-way fooling set of size $2^{n/6}$, and proved that $I_{1W} = \Omega(n)$.

**Theorem 2.5:** Let $P$ be a problem with one-way information content $I_{1W}$, and $C$ a circuit of area $A$ and time $T$ that solves $P$ in the sense that it has probability greater than $1/2$ of producing the correct answer. Then

a) $AT^2 = \Omega(I_{1W}^2)$.

b) If the aspect ratio of $C$ is at least $a$, then $AT^2 = \Omega(aI_{1W}^2)$.

c) If all the pads are on the border of $C$, then $AT^2 = \Omega(rI_{1W})$, where $r$ is the number of input variables.

**Proof:** The crux of the proof is in showing that the number of crossing sequences at some boundary of $C$ must be at least as great as $2^{I_{1W}}$. The rest of the proofs are analogous to Theorems 2.2, 2.3, and 2.4, and we shall omit them.

Assume without loss of generality that the largest one-way fooling set $\mathcal{A} = \{\alpha\beta_1, \ldots, \alpha\beta_{2^{I_{1W}}}\}$ is a left fooling set, and that there are at most $m$ different crossing sequences at the boundary corresponding to that partition for which $\mathcal{A}$ is a fooling set. Let $p_{ji}$ be the probability that when we run $C$ on input $\alpha\beta_i$, the $j^{th}$ crossing sequence is seen at the boundary. Let $q_{ij}$ be the probability that the output variables in $Y_L$ assume the proper values for $\alpha\beta_i$ when $\alpha$ is the assignment to $X_L$ and the $j^{th}$ crossing sequence appears on the boundary.

Note that

$$
\begin{bmatrix} >\frac{1}{2} \\ >\frac{1}{2} \\ \cdot \\ \cdot \\ >\frac{1}{2} \end{bmatrix} = \begin{bmatrix} q_{11} & q_{12} & \cdots & q_{1m} \\ q_{21} & q_{22} & \cdots & q_{2m} \\ \cdot & \cdot & & \cdot \\ \cdot & \cdot & & \cdot \\ q_{2^{I_1W},1} & q_{2^{I_1W},2} & \cdots & q_{2^{I_1W},m} \end{bmatrix} \begin{bmatrix} p_{11} & p_{12} & \cdots & p_{1,2^{I_1W}} \\ p_{21} & p_{22} & \cdots & p_{2,2^{I_1W}} \\ \cdot & \cdot & & \cdot \\ \cdot & \cdot & & \cdot \\ p_{m1} & p_{m2} & \cdots & p_{m,2^{I_1W}} \end{bmatrix}
$$

**Fig. 2.4.** Computation of output probabilities.

$$
\sum_{j=1}^{m} p_{ji} = 1 \qquad \text{and} \qquad \sum_{i=1}^{2^{I_1W}} q_{ij} \le 1 \tag{2.4}
$$

The first equality follows since in any run of $C$, exactly one crossing sequence occurs, and the second inequality follows because in any run, only one output assignment is made, but this assignment might not be the desired output for any of the inputs $\alpha\beta_i$.

Figure 2.4 shows the matrices represented by the $p_{ji}$'s and the $q_{ij}$'s, with their product also shown. The $i^{th}$ row and $j^{th}$ column of the product matrix gives the probability that when the input assignment is $\alpha\beta_j$, the output that is correct for $\alpha\beta_i$ will be made. Since we assume that $C$ works correctly in the probabilistic sense, the diagonal entries must each exceed $1/2$, as shown in Fig. 2.4.

Put another way, for each $i$ between 1 and $2^{I_1W}$, the sum

$$
\sum_{j=1}^{m} q_{ij}p_{ji} > \frac{1}{2}
$$

Since the $p_{ji}$'s sum to 1 as in (2.4), the above inequality could not hold for $i$ unless there were some $q_{ij} > 1/2$; that is, there is some crossing sequence $j$ that has probability greater than $1/2$ of producing the desired output when $\alpha\beta_i$ is input.

Now no crossing sequence can serve in this role for two different values of $i$, because if the $j^{th}$ crossing sequence did, then it would have probability greater than $1/2$ of producing each of two outputs known to be different, which would violate (2.4). Thus, $m \ge 2^{I_1W}$, as was to be shown. $\square$

### Comparison of Probabilistic and Deterministic Computation

We shall briefly sketch an example that exploits the difference between $I$ and $I_{1W}$ to exhibit a problem that has a much smaller and faster probabilistic chip than it has a deterministic one. The problem concerns the comparison of two numbers, but we must be somewhat careful how we phrase the problem. If we simply wanted a chip to read two binary numbers $w_1 \cdots w_n$ and $x_1 \cdots x_n$, and tell whether they were equal, we could arrange to read $w_i$ and $x_i$ next to one another and compare them. Then, a binary tree of "and" gates could determine whether all comparisons held, and the circuit would output 1 if so, 0 if not. This circuit is deterministic, has area $O(n \log n)$, and takes time $O(\log n)$.

In order to make the problem harder, so $I$ (but not $I_{1W}$) is $\Omega(n)$, we shall arrange to hide one of the numbers in a way we shall discuss in the next example. However, before proceeding to that example, let us discuss certain aspects of number theory that will prove important.

The way we shall test whether two numbers $w$ and $x$ are equal is to divide them by some number $v$ and check if their remainders are equal. If the remainders are equal, we say that $w = x$, and if they are not equal, we say that $w \neq x$. This method has a high probability of being correct; the probability, averaged over all $w$ and $x$, of saying "yes" when the answer is "no" is only $1/v$, and of course, if we say "no," that must be correct.

The trouble lies in a subtlety of our definition of probabilistic behavior. We (rightly so) want to guarantee better than 50% probability for every possible input. Unfortunately, if $w - x$ happens to be divisible by $v$, but $w \neq x$, we shall say "yes" when we should say "no." That in itself would not be fatal, because we could pick $v$ from some selection of small numbers. However, if the candidates for $v$ are too small, say in the range 2 to $n - 1$ when the numbers $x$ and $w$ are chosen from 0 to $2^n - 1$, then it is possible that $w \neq x$, but that $w - x$ is divisible by all the numbers that could be $v$, and for this input, we have 100% probability of being wrong.

Therefore, we shall have to pick our candidates for $v$ from a set of larger numbers. It turns out that we are safe if we choose $v$ to be between $n^2$ and $2n^2$. For the fraction of numbers around $m$ that are prime is $\Omega(1/\log m)$ (see Hardy and Wright [1938], e.g.), so in the range $n^2$ to $2n^2$ we may expect to find at least $\Omega(n^2/\log n)$ primes. As $\mid w - x \mid$ is no larger than $2^n$, it cannot be divisible by more than $O(n/\log n)$ primes in the range $n^2$ to $2n^2$. Thus, for every possible value of $w - x$, the probability that $w - x$ is divisible by a given prime in our range is $O(1/n)$.

Of course, we cannot tell easily which numbers in the range are prime. However, if, say, one out of $c \log_2 n$ numbers in the range $n^2$ to $2n^2$ is prime, then if we pick $c \log_2 n$ numbers from that range at random, the expected number of primes is 1, and the probability that at least one of the numbers is prime is very close to $1 - 1/e$, which is about .63. Thus, we can perform the following experiment. Pick $c \log_2 n$ numbers in the range $n^2$ to $2n^2$, check that for each number $v$ we get the same remainder when $w$ and $x$ are divided by $v$, and say "yes" if so, and "no" if not.

Whenever we say "no" we are correct, but we must check that when we say "yes," we have more than a 50% chance of being correct, independent of $w$ and $x$. Conservatively, assume that whenever we guess no primes, we are wrong. However, in the 63% of the cases where we do pick a prime, our chances are no more than $1/n$ of being incorrect, since that is the chance that the random prime divides $w - x$. As $n$ gets large, our chances of being correct are thus 63% as a lower bound, but in practice, for most inputs the probability will be very close to 1.

Thus $\epsilon$, where the probability of success is at least $1/2 + \epsilon$, is constant independent of $n$. It follows by the analysis earlier in this section that if we can design a small, fast chip based on the ideas we have sketched, then for any desired degree of accuracy $\delta$, say $10^{-10}$, there is a constant number of times the chip must be run to assure that accuracy. As a consequence, a probabilistic circuit for this comparison problem will realistically provide superior performance for large $n$, when compared to the best deterministic circuit. We now proceed to the example problem.

**Example 2.10:** The problem we consider is that of comparing a given number with a "hidden" one. The given number we represent in binary by inputs $w_1, \ldots, w_n$. The other number is hidden by supplying $2n$ input bits $x_1, \ldots, x_{2n}$ and $2n$ *mask* bits $c_1, \ldots, c_{2n}$, exactly $n$ of which are 1 (or the answer to the problem is "no").

The "hidden" number is formed by taking exactly those $x_i$'s for which $c_i = 1$. The resulting number, $x_{i_1}, \ldots, x_{i_n}$ is compared with $w_1, \ldots, w_n$, and we output 1 if $w_j = x_{i_j}$ for all $j$. The reader should note that we cannot solve this problem simply by reading the $w$'s and $x$'s next to one another, because depending on the $c$'s, we could require a large amount of motion of the $x$'s to get them next to the proper $w$'s. It turns out that the information content $I$ of this problem is $\Omega(n)$. The proof is not easy, but we leave it as an exercise.

Figure 2.5 outlines the efficient probabilistic circuit for comparison of numbers. In this circuit, the $w$'s are arranged at the leaves of a binary tree, and the $x$'s and $c$'s are arranged at the leaves of another. At each node is a processing element that can handle numbers $O(\log n)$ bits long, such as the guessed divisors $v$ (which, being in the range $n^2$ to $2n^2$, have at most $2\log n + 1$ bits), or such as the remainders when certain numbers are divided by $v$. Note that we use

$$(a_i, \ldots, a_j) \bmod v$$

to stand for the remainder when the binary number represented by $a_i, \ldots, a_j$, that is,

$$\sum_{\ell=0}^{j-i} a_{i+\ell} 2^{j-i-\ell}$$

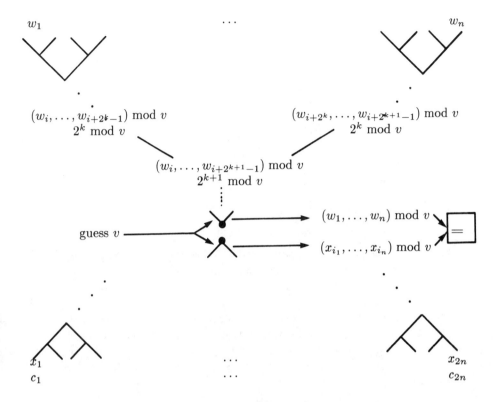

**Fig. 2.5.** Organization of probabilistic comparison circuit.

is divided by $v$.

We guess $\log n$ values of $v$, and for each guess the following computation is performed. The output of the circuit is 1 if and only if all guesses lead to equality of the remainders. We pass the guess down the binary trees, taking $O(\log n)$ time to do so, since we must pass it $\log n$ levels down. Then, we work up the trees, each node computing two values. For simplicity, let us focus on the $w$'s tree first.

A node at the $k^{th}$ level has $2^k$ descendant leaves, which we have supposed in Fig. 2.5 might have the values $w_i, \ldots, w_{i+2^k-1}$. At each node with descendant leaves $w_i, \ldots, w_j$, we compute $(w_i, \ldots, w_j) \bmod v$, and if that node is at level $k$ we also compute $2^k \bmod v$. The basis, $k = 0$, is trivial; $(w_i) \bmod v$ is $w_i$, and $2^k \bmod v$ is 1.

The induction step is illustrated in Fig. 2.5. $2^{k+1} \bmod v$ is the remainder when $2^k \bmod v$ is squared and divided by $v$. Note that if $v$ requires $2 \log n + 1$ bits, then these calculations can be performed with numbers of no more than $4 \log n + 2$ bits, and in time $O(\log^2 n)$. Similarly, $(w_i, \ldots, w_{i+2^{k+1}-1}) \bmod v$ is

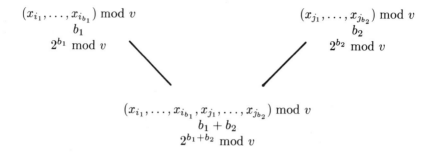

**Fig. 2.6.** Pattern of computation for masked bits.

$$[[2^{k+1} \bmod v] \times [(w_i, \dots, w_{2^k-1}) \bmod v]$$
$$+ [(w_{i+2^k}, \dots, w_{i+2^{k+1}-1}) \bmod v]] \bmod v$$

and this calculation can also be performed in $O(\log n)$ space and $O(\log^2 n)$ time.†

A modification of this technique also produces $(x_{i_1}, \dots, x_{i_n}) \bmod v$, the value modulo $v$ of the unmasked $x$'s. At each node we record the number $b$ of descendant leaves that are unmasked, as well as $2^b \bmod v$ and $(x_{i_1}, \dots, x_{i_b}) \bmod v$. The pattern of computations is shown in Fig. 2.6, which shows a node whose children have $b_1$ and $b_2$ unmasked descendant $x$'s, respectively. In $O(\log n)$ space and $O(\log^2 n)$ time we can compute the sum $b_1 + b_2$ of the number of descendant unmasked bits, we can compute $2^{b_1+b_2} \bmod v$ by an obvious formula, and we can compute

$$(x_{i_1}, \dots, x_{i_{b_1}}, x_{j_1}, \dots, x_{j_{b_2}}) \bmod v$$

by computing

$$[[2^{b_2} \bmod v] \times [(x_{i_1}, \dots, x_{i_{b_1}}) \bmod v] + [(x_{j_1}, \dots, x_{j_{b_2}}) \bmod v]] \bmod v$$

Our circuit thus takes $O(\log^4 n)$ time, there being $\log n$ iterations of a guess-and-compare operation, each iteration taking $O(\log^2 n)$ time at each of $\log n$ levels. The dimensions of the circuit in Fig. 2.5 are $O(n \log n)$ wide by $O(\log^2 n)$ high. To see this, note that we can arrange the arrays of cells that hold values at nodes vertically, thus taking $O(1)$ horizontal space per input bit. Each level of the trees requires height $O(\log n)$, since we must store numbers of $O(\log n)$ bits vertically. However, we must run $O(\log n)$ wires in parallel, both horizontally and vertically, from each node to the node at the next higher level.‡ Thus, the area is $O(n \log^3 n)$ and the $AT^2$ value for this circuit is $O(n \log^{11} n)$. We shall frequently represent large numbers of powers of $\log n$ by "$n^\epsilon$," recognizing

---

† There are other ways to perform the arithmetic that involve less time but more area; see Section 4.1.

‡ We can save a factor of $\log n$ in area if we can have inputs scattered throughout the chip, not just on the border, if we use the "H-tree" organization discussed in Section 3.1.

that for any $k$ and any $\epsilon > 0$, $\log^k n$ grows more slowly than $n^\epsilon$. Thus, our probabilistic circuit for comparing hidden numbers has $AT^2 = O(n^{1+\epsilon})$, which as $n$ grows to infinity is asymptotically less expensive of area and time than any deterministic circuit, which must have $AT^2 = \Omega(I^2) = \Omega(n^2)$. $\square$

## 2.4 CIRCUITS WITH REPETITION OF INPUTS

We now return the discussion to "ordinary" circuits, as opposed to probabilistic ones, and we consider what happens if we are allowed to read an input variable more than once. If we can receive the same input several times, we might save the circuit area needed to store its value, and thus make a great difference in the size of the circuit we need. We shall therefore generalize our notion of information content $I$ of a problem $P$, to a function $I(m)$. The parameter $m$ will be seen to represent both the number of times we can repeat an input, which for technical reasons is not $m$, but $m/2$, and also the number of regions into which we should divide chips in order to prove lower bounds on $AT^2$. Thus, $I(2)$ will be $I$, essentially, although there is a subtle difference due to our desire to simplify the definition of $I(m)$.

There are two possible viewpoints regarding such an opportunity. One aspect is that we are simply replacing memory on the chip with memory somewhere else, and are not being charged for that memory; therefore we are engaging in a "fraud" if we allow inputs to be available repeatedly. On the other hand, it is possible that the values are not really stored outside the circuit, but are regenerated as needed from a small amount of data or some other source, either in another portion of the chip or externally.

**Example 2.11:** In Example 2.1 we introduced the notion of a chip that solved a problem on graphs, connectivity to be specific, by reading the adjacency matrix of the graph, that is, input variable $x_{ij}$ is 1 or 0, depending on whether there is or is not an edge from node $i$ to node $j$. If the graph is undirected, we know that the matrix will be symmetric, and the entry $x_{ij}$ will be equal to the entry $x_{ji}$. Thus, if we read in the entire adjacency matrix, we are in a sense reading each datum twice, and we should consider the possibility that we are able to design a chip of small area only because we store the upper triangle of the matrix outside the chip to be read in a second time.

In many cases we are able to design circuits to solve undirected graph problems using only the upper triangle of the adjacency matrix as input. However, even if we needed the complete matrix, we cannot be certain that external storage is involved. For example, it is possible that the matrix is generated row by row on some other chip and passed to the chip we are designing. Alternatively, the values might be generated from a small amount of data. For example, we might have the coordinates of $n$ points in the plane and deem that there is an edge between two points if their Manhattan distance is below some threshold. That is, say there is an edge between nodes $i = (x_i, y_i)$ and $j = (x_j, y_j)$ if and only if $\mid x_i - x_j \mid\leq t$ and $\mid y_i - y_j \mid\leq t$, where $t$ is the threshold. Then only $O(nk)$ bits of storage are needed, where $k$ is the number of bits of precision used to represent positions, and the rows of the adjacency matrix can be generated in $O(k)$ time if we use an array of cells to hold the coordinates of the points and take turns distributing one point to all the cells. $\Box$

While repeating the inputs once does not seem to help much where adjacency matrices of undirected graphs are concerned, we might naturally ask whether there are problems for which repetition does help, and if so, how much. The following is an example where we get as much help as is theoretically possible.

**Example 2.12:** Let us reconsider the barrel shifting problem of Example 2.5. It turns out that the circuit of Fig. 2.3 cannot be improved much. Since it takes time only $O(\log n)$, repeating the inputs more than this amount would increase the time spent by the circuit, no matter what area savings could be made.

However, there is another interesting way to implement a barrel shifter that is suggested in Fig. 2.7. This circuit reads the $n$ bits to be cycled into a square array, $\sqrt{n}$ on a side, in the order shown. Then, for time $O(\sqrt{n})$ the bits are cycled around the horizontal lines, for as long as dictated by the high-order control bits. That is, each cycling step in this phase has the effect of cycling the bits as a group by distance $\sqrt{n}$. Thus, the values of the high order half of the control bits $c_1, \ldots, c_{(\log n)/2}$ determine how many times we should perform this "coarse adjustment" cycling operation.†

In the second phase, we shift bits up the columns, at most $\sqrt{n}$ times, depending on the values of the low order control bits $c_{(\log n)/2+1}, \ldots, c_{\log n}$. Bits reaching the top of one column are shifted to the bottom of the next column, and bits reaching the top of the last column are shifted to the bottom of the first. It is important to note that the order in which we shift in the two directions is unimportant. That is, horizontal and vertical shifts commute; the only important issue is how many of each we do.

The result is that the output $y_i$ will be at the position $x_i$ was at initially, and the where-determinate output can be made after $O(\sqrt{n})$ time. This circuit thus has an $AT^2$ value of $O(n^2)$, which is optimal if inputs are read only once.

However, if we are allowed to read inputs several times, we can do better. Suppose we divide the points of Fig. 2.7 into $n^{1/3}$ square regions $n^{1/3}$ on a side, as suggested by Fig. 2.8. The variables read into each of the smaller squares according to the pattern of Fig. 2.7 we shall term a *group* of variables. In

---

† For convenience we shall assume that $n$ is a power of 64, that is, $2^k$ for some $k$ divisible by 6; thus we get integer values here, and later in the construction, when needed. In practice, rounding figures up or down would produce a circuit of essentially the same performance if $n$ were not a power of 64, or even not a power of 2.

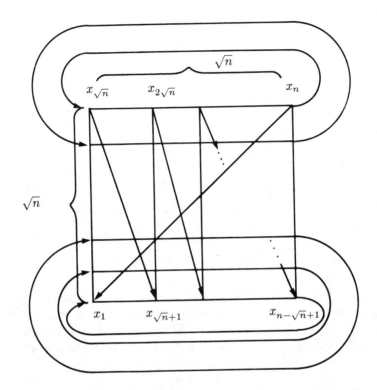

**Fig. 2.7.** Barrel shifter with area $O(n)$.

our new design, each group will be read into each square once, and the first and fourth sixths of the control bits will be used to determine which group is retained for each square.

By chosing the group retained in each square properly, we shall, in effect, have performed all of the horizontal cycle steps, except for the last $n^{1/3}$ or so (the exact number of which is determined by the second and third sixths of the control bits),† and except for the last roughly $n^{1/3}$ of the vertical cycle steps (the number of which depends on the fifth and sixth sixths of the control bits). Hence, we take $O(n^{1/3})$ time to read all the groups, and the same time to cycle in each direction. We thus have a circuit that takes area $O(n)$, time $O(n^{1/3})$, and reads its inputs $n^{1/3}$ times. This circuit has an $AT^2$ value of $n^{5/3}$, which we shall prove is optimal for the barrel shifting problem, given this amount of repetition of the input. □

---

† Note that this is where it is convenient to assume $n$ is a power of 64.

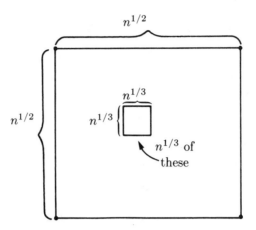

**Fig. 2.8.** Partition of square.

## Information with Respect to Overlapping Partitions

In order to prove lower bounds for circuits with input repetition, we must consider what happens when the area of the circuit is broken up into many small rectangles of limited perimeter. The perimeter is made small so that not much information can flow across at any one time, and the number of rectangles is determined by our desire that the probability that a given input variable is ever read into a pad within the boundary of the rectangle not be too high.

If a problem $P$ of size $n$ has input variables $X$ and output variables $Y$, we shall talk about *m-way overlapping partitions for $P$*, for even $m$, which are lists $\pi = (X_1, \ldots, X_m, Y_1, \ldots, Y_m)$ with the following properties.

1.   Each input variable $x_i$ occurs in at least one and at most $m/2$ of the $X_i$'s.
2.   For all $i$ and $j$, $Y_i$ and $Y_j$ are disjoint, and their union is $Y$.

Note that while the output variables are partitioned, an input variable appears in $m/2$ of the $X_i$'s, and therefore, for $m \geq 4$, the $X_i$'s are not disjoint.

Define a *fooling set* for $P$, $n$, and $m$-way overlapping partition $\pi$ to be a set $\mathcal{A}$ of input assignments such that there is an integer $k$, $1 \leq k \leq m$, such that for all $\alpha$ and $\beta$ in $\mathcal{A}$, the following two conditions hold.

1.   $\alpha$ and $\beta$ agree on $X_k$.
2.   $\alpha$ and $\beta$ disagree on one or more bits of $Y_k$.†

---

† The reader should be warned that this definition does not precisely generalize our notion of "fooling set" from Section 2.2. We could define for each assignment $\alpha$ the restriction $\alpha^{(k)}$ to those input variables that appear in $X_k$ and nowhere else. Then define $\alpha^{\neg(k)}$ to be the restriction of $\alpha$ to the other variables. Next, replace condition (2) by the requirement that either $\alpha^{(k)}\beta^{\neg(k)}$ differs from $\alpha$ on a bit of $Y_k$, or it differs from $\beta$ on an output bit not in $Y_k$. In this generalization, a fooling set for $m = 2$ is exactly the same as a fooling set in Section 2.2, as the reader may show as an exercise. We did not bother with this generalization because for large $m$, the chances of a bit being in $X_k$ and no other group gets small.

The intuitive motivation for this definition is that the sets of variables $X_i$ and $Y_i$ represent the variables read and written from a small region of the chip, representing roughly $1/m^{th}$ of the area of the chip. The two conditions above imply that if we do not get information into the $k^{th}$ region regarding exactly which member of $A$ is the input, then we shall make a mistake in one of the outputs that occurs in that region, and therefore occurs nowhere else. In order to have large fooling sets, there must be many input variables that are not read into at least one region, the $k^{th}$ in the definition above, that has many outputs to make.

Now, we define $I(P, n, \pi)$, where $\pi$ is an $m$-way overlapping partition, to be the logarithm, base 2, of the size of the largest fooling set for $P$, $n$, and $\pi$. Then, as before, we wish to concentrate on a particular set of input variables, $Z$. We define an $m$-way overlapping partition $\pi$ to be *acceptable* for $Z$ if for $1 \leq i \leq m$, the size of $X_i \cap Z$ is at most $2/3$ the size of $Z$; i.e., at least one out of every three members of $Z$ is not found in $X_i$. Then $I(P, n, m, Z)$ is the minimum, over all acceptable $m$-way overlapping partitions $\pi$ for $Z$ of $I(P, n, \pi)$.

Again, as in Section 2.2., we wish to take advantage of the fact that we can choose $Z$. Thus, we define $I(P, n, m)$ to be the maximum over all $Z$ of $I(P, n, m, Z)$. We shall use $I(m)$ for $I(P, n, m)$ when the problem and its size parameter are understood.

**Example 2.13:** Let us return to the barrel shifting example and compute a lower bound on $I(m)$. The argument is similar to that of Example 2.5, and we shall sketch it only. As in that example, let $Z = \{ x_1, \ldots, x_n \}$ be the "data" variables only, and let $\pi$ be an $m$-way overlapping partition acceptable for $Z$. In what follows, we refer to each pair of corresponding input and output sets $X_k$ and $Y_k$ as a *block*. Each output variable $y_i$ is in a set $Y_k$ such that at most $2/3$ of the input variables in $Z$ are in the same block, by the definition of $I(P, n, m, Z)$.

Thus, on a random setting of the control variables, the average $y_i$ will have a probability at least $1/3$ of being assigned an input $x_j$ that is not in the same block. Intuitively, there must be some setting of the control variables, and some value of $k$ such that the members of $Y_k$ do as well as average on that setting; that is, at least $n/3m$ members of $Y_k$ are assigned inputs not found in $X_k$. The formalization of this intuitive argument is similar to Example 2.5. Let the selected setting $c$ of the control variables map to these $n/3m$ outputs the input variables $x_{i_1-c}, \ldots, x_{i_{n/3m}-c}$, where subtraction in indices is modulo $n$.

We thus form our set $A$ by using all $2^{n/3m}$ possible assignments to

$$x_{i_1-c}, \ldots, x_{i_{n/3m}-c}$$

then setting the control variables to yield $c$, and assigning 0 to all other input variables. Then each $\alpha$ and $\beta$ in $A$ surely agree on $X_k$ but disagree on $Y_k$. We

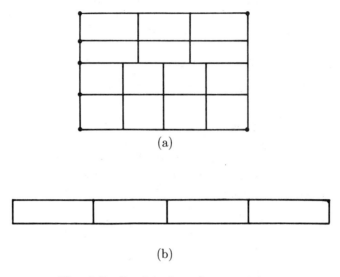

(a)

(b)

**Fig. 2.9.** Partitioning of a rectangle.

conclude that for barrel shifting, $I(m) = \Omega(n/m)$. $\square$

## Relating Information to Area and Time

We now prove that any circuit can be partitioned into regions that have small perimeter and not too many occurrences of inputs inside, and we use this fact to construct overlapping partitions of the input variables to satisfy the conditions in the definition of $I(m)$. We can then relate the existence of fooling sets to crossing sequences in the usual manner and get lower bounds on the $AT^2$ product by arguing that the small perimeter does not allow the needed information to flow too fast. We need two technical lemmas to begin.

For the purposes of this section, when we divide a rectangle into smaller rectangles, the *perimeter* of any of the smaller rectangles will be the sum of the lengths of the sides that are not sides of the large rectangle. Note that this definition of "perimeter" is nonstandard.

**Lemma 2.3:** It is possible to divide a rectangle of area $A$ into $d$ rectangles, each of perimeter no greater than $6\sqrt{A/d}$.

**Proof:** In the case that the aspect ratio of the large rectangle is not greater than $d$, we can divide it as suggested by Fig. 2.9(a), being sure that the small rectangles have equal area and aspect ratio no greater than 2. If $h$ and $w$ are the smaller and larger dimensions of one of the small rectangles, we have $hw = A/d$, and $w/2 \leq h \leq w$. It follows that $h \leq \sqrt{A/d}$, and the perimeter, which is at most $2(h + w)$, is bounded above by $6\sqrt{A/d}$.

The other case to consider is when the aspect ratio is so high that it is impossible to divide the large rectangle into smaller ones of aspect ratio as small as 2; Fig. 2.9(b) depicts this situation. In this case, the smaller dimension $h$ may be much smaller than the larger dimension $w$, but we still may assume that $hw = A/d$ and $h \leq w$. It follows that $h \leq \sqrt{A/d}$. But in this case, the perimeter is only $2h$, since the border of the large rectangle is excluded from the perimeter of small rectangles. The lemma is thus easily satisfied in this case. □

**Lemma 2.4:** Suppose we have a rectangular region in which there are $q$ "objects," each of which is at a point within the region. Suppose also that no more than $p/7$ of these objects are at any one point, and we wish to divide the region into smaller regions by making cuts, as in Fig. 2.2. Further, let us require that no region have more than $p$ objects. Then we can perform this task by selecting no more than $\frac{7q}{6p}$ cuts, each of which divides one of the current regions into two parts (and therefore the single region becomes no more than $1 + \frac{7q}{6p}$ regions).

**Proof:** Given such a rectangle to be partitioned, assume its width is at least as great as its height. We start from the left, sliding potential cuts with one jog, as in Fig. 2.2, to the right, until we get more than $p$ objects in one region. Then we give up the last point incorporated into the region, leaving at least $6p/7$ objects, since no one point has more than $p/7$ objects. In this manner we shall make at most $\frac{7q}{6p}$ cuts, and we shall successfully partition the rectangle. □

**Theorem 2.6:** Let $P$ be a problem with information content $I(m)$, and $C$ a circuit that solves $P$, has area $A$, and takes time $T$. Further, suppose that each input variable is read $m/2$ times, for some even $m$. Then $AT^2 = \Omega(mI^2(m))$.

**Proof:** Suppose that $I(m)$ is maximized when we consider set $Z$ of input variables, that is, $I(P, n, m, Z) = I(P, n, m) = I(m)$. Let $Z$ have $r$ members, so the total number of input occurrences for variables in $Z$ is $mr/2$. Let us begin by considering the case that as many as $2r/21$† of the input occurrences are at one pad. Then surely $T \geq 2r/21$. Also, in order to read all the input occurrences, we need $AT \geq mr/2$. Thus $AT^2 \geq mr^2/21$. We can prove $I(m) = O(r)$ by the method used to prove $I \leq r/3$ in Theorem 2.2. Thus, $AT^2 = \Omega(I^2(m))$ as desired.

Therefore, in what follows we assume that no more than $2r/21$ occurrences of inputs in $Z$ are located at any one point of $C$. We begin by breaking up the rectangle that represents $C$ into $m$ rectangles of perimeter‡ at most $12\sqrt{2}\sqrt{A/m}$. The construction is in two phases.

---

† The fraction 2/21 is, of course, somewhat arbitrary, but we shall see in a moment the role it plays.

‡ That is, perimeter in the sense that excludes the border of the large rectangle.

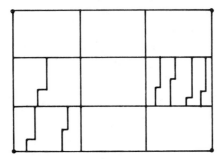

**Fig. 2.10.** Partition resulting from the two phases.

In phase 1, we break the circuit into $m/8$ rectangles of equal area. By Lemma 2.3, with $d = m/8$, we can be sure that these rectangles each have perimeter at most $12\sqrt{2}\sqrt{A/m}$. In phase 2, we look for rectangles that have more than $2r/3$ occurrences of inputs, and break them further, into strips, as suggested by Fig. 2.10. In case the original partition is like Fig. 2.9(b), we make further partitions vertically, so that in all cases we can be sure the perimeters of the rectangles created in phase 2 cannot exceed $12\sqrt{2}\sqrt{A/m}$.

Now we must show that a total of at most $m$ regions are made by the two phases. Apply Lemma 2.4, with $q = mr/2$ and $p = 2r/3$; the "objects" of Lemma 2.4 are the occurrences of inputs in $Z$. Since $2r/21 = p/7$, we know that not more than $p/7$ objects occur at any point. It also follows that we need to make no more than $7m/8$ additional cuts in phase 2, and with the $m/8$ regions of phase 1, we achieve our goal of $m$ regions, each with no more than $2r/3$ input occurrences and perimeter $O(\sqrt{A/m})$.

Now, we can relate this partition to the definition of $I(m)$ by defining $X_i$ to be the set of input variables that have an occurrence in the $i^{th}$ region, and $Y_i$ to be the output variables whose lone occurrence is in that region. This collection has a fooling set $\mathcal{A}$ of size $2^{I(m)}$, as in the definition of $I(m)$. Let $k$ be the integer such that members of $\mathcal{A}$ agree on $X_k$, but result in outputs that disagree on $Y_k$. If $12\sqrt{2}\sqrt{A/m}T\ell < I(m)$, where $\ell$ is the upper bound on the number of inputs to a logic element, then there is one crossing sequence at the border of the $k^{th}$ region that occurs in response to two members $\alpha$ and $\beta$ of $\mathcal{A}$. If the crossing sequences are the same, and $\alpha$ and $\beta$ agree on inputs in $X_k$, then the outputs in $Y_k$ will be the same when $C$ is run on $\alpha$ and $\beta$. But these input assignments produce outputs that differ in at least one bit of the $k^{th}$ region, by the definition of a fooling set. It follows that $\sqrt{A/m}T = \Omega(I(m))$, or equivalently, $AT^2 = \Omega(mI^2(m))$. $\square$

**Example 2.14:** For the barrel shifter, where $I(m) = \Omega(n/m)$, we have $AT^2 =$

$\Omega(n^2/m)$. If $m = n^{1/3}$, as in Example 2.13, then $AT^2 = \Omega(n^{5/3})$, which shows that the circuit of that example is optimal. $\square$

The reader should observe that if, as in Example 2.13, we have the relationship $I(m) = I/m$, which seems intuitively reasonable, then Theorem 2.6 gives us $AT^2 = \Omega(I^2(n)/m)$. That is, as the repetition rate increases, the $AT^2$ value decreases proportionally.

**EXERCISES**

2.1: Show that the following problems require $AT^2 = \Omega(n^2)$.
  a)  *Left shifting*: given data inputs $x_0, \ldots, x_{n-1}$ and control inputs

   $$c_1, \ldots, c_{\log n}$$

   representing integer $c$ in binary, set output $y_i$ to $x_{i+c}$ if $i + c < n$, and to 0 otherwise.
  * b)  *Integer multiplication*: Compute the product of two $n$-bit binary integers.
  c)  Given data and control inputs as in (a), set $y_i$ to $x_j$ if the $\log n$-bit binary representations of $i$ and $j$ differ in exactly those positions $k$ such that $c_k = 1$.
  * d)  The problem of Example 2.10 (comparing hidden strings of bits). *Hint*: One way to approach the problem is to find pairs of $c$'s that are on the same side of the boundary and consecutive or nearly so, and set one of their corresponding $x$'s to 0, the other to 1. In each input assignment of the fooling set, let exactly one of each pair be 1, and arrange that the $x$ "behind" that $c$ is matched with a $y$ on the opposite side.
* 2.2: Show that multiplication of $n \times n$ matrices over a field of two elements requires $AT^2 = \Omega(n^4)$.
2.3: Prove that sorting $n$ ($\log n$)-bit integers requires $AT^2 = \Omega(n^2)$.
2.4: Show that we can sort $n$ $k$-bit numbers in $AT^2 = O((n2^k)^{1+\epsilon})$.
2.5: A problem is said to be *transitive* if its inputs consist of "data" inputs $x_1, \ldots, x_n$ and "control" inputs $c_1, \ldots, c_m$, and for each setting of the control inputs, the data inputs are mapped to outputs $y_1, \ldots, y_n$ according to some permutation $\pi$ determined by $c_1, \ldots, c_m$; that is, $y_i$ is set to $x_{\pi(i)}$. Further, the set of permutations forms a group, and for each $i$ and $j$ there is some member $\pi$ of the group such that $\pi(i) = j$.
  a)  Show that the problems of Exercise 2.1(a, c) are transitive.
  ** b)  Prove that every transitive problem requires $AT^2 = \Omega(n^2)$.
2.6: A *nondeterministic chip* is a circuit with a hypothetical element called a *splitter*. At each time unit, each splitter on the chip may choose to be "active" or "inactive." Suppose $k$ splitters are active at some time unit. Then at the next time unit, there are $2^k$ copies of the chip, each having a

different combination of outputs (0 or 1) associated with the active splitters. The effect of splitting the chip is cumulative; each of the copies may split at the next time unit, and once split, the copies are not connected in any way. The output of such a device must be a single bit, and we take the output to be 1 if any of the copies that result from running the nondeterministic chip on a given input produce a 1 output. Only if none of the copies produces 1 is the output taken to be 0; so in a sense 1's "outvote" 0's. The area of the nondeterministic chip is the area in the ordinary sense, with splitters assumed to require area $O(1)$. Each copy of the chip is assumed to produce its output at the same time and place, and the time taken by the chip is the number of units that elapse from the time the process starts until all the copies produce their outputs. Show that the following graph problems can be "solved" by a nondeterministic chip with an $AT^2$ product of $n^{2+\epsilon}$, if the graphs are represented by $n \times n$ adjacency matrices. Following our convention, $n^\epsilon$ represents some power of $\log n$.

* a)   Produce output 1 if and only if the graph is connected. *Hint:* Use the splitters to guess a spanning tree for the graph, and check that the guess is correct, emitting 1 if so. Wrong guesses will result in output 0, even if the graph is connected, but that is unimportant because 1's "outvote" 0's in the output.

* b)   Produce output 1 if and only if the graph is not connected. Note that this problem cannot be solved simply by complementing the output to (a), since usually when the graph is connected there will be copies of the chip producing output 1 while others produce output 0. If we simply attached an inverter to the output, the new nondeterministic device would still produce both 1's and 0's from different copies, but because of our rule about how outputs are combined, the result would be 1, not 0.

* c)   Produce output 1 if and only if the graph is not planar.

** d)  Produce output 1 if and only if the graph is planar.

2.7: We can modify our notion of information so that it can be used to prove that certain problems with a single output variable have a strong lower bound on $AT^2$, even for nondeterministic chips. We require of a fooling set $\mathcal{A}$ that the output for each member of $\mathcal{A}$ be 1, but that for $\alpha$ and $\beta$ in $\mathcal{A}$, the output in response to at least one of $\alpha_L\beta_R$ and $\beta_L\alpha_R$ be 0. Prove that if $I$ is the logarithm base 2 of the size of the largest fooling set in the above sense, then $AT^2 = \Omega(I^2)$ holds even for nondeterministic chips.

2.8: Show that the problem of Example 2.10, comparison of hidden strings of bits, requires $AT^2 = \Omega(n^2)$ for a nondeterministic chip that is not probabilistic.

2.9: Show that a *predicate* (problem with one output bit) whose complement requires $AT^2 = \Omega(f(n))$ also requires $AT^2 = \Omega(f(n))$. What does this tell us about our ability to prove strong lower bounds for deterministic circuits for problems like connectivity and planarity mentioned in Exercise 2.8, if we use a method based on information content?

2.10: We can make many of the results of this chapter apply to circuits that are when-indeterminate (but still where-determinate), if we define the computational model correctly. In particular, imagine that when the circuit makes an output or requests an input, a label is sent to the pad indicating which variable is being written or read. We assume that the generation of labels takes no time or area, but that it is deterministic; a logic element generates labels as a function of its total history, but if we run the circuit twice and provide the same inputs to the logic element, it will generate the same label(s).

    a)   Does the where-determinate circuit of Example 2.2 meet this model?

    b)   Prove that the lower bounds of Section 2.2 apply to where-determinate circuits that are when-indeterminate but meet the above model.

2.11: The *union* of two problems is the problem that takes the inputs of both and produces the outputs of both, with no interaction.

    a)   Show that if $P$ and $Q$ are problems for which $AT^2 = \Omega(f(n))$, then the same is true of $P \cup Q$.

    ** b)   Show that the converse of (a) is false, that is, $P \cup Q$ can have $AT^2 = \Omega(n^2)$, yet for each of $P$ and $Q$, $AT^2$ can be $O(n)$.

** 2.12: Is Exercise 2.11 true for composition of functions; that is, is the composition of hard problems itself hard, and do there exist easy problems whose composition is hard?

2.13: Consider a "three-dimensional chip," where elements occupy a rectangular solid of volume $V$ and take time $T$ to solve some problem $P$ with information content $I$.† Prove that $V^2T^3 = \Omega(I^3)$.

2.14: Formalize the argument of Example 2.13, where we claimed that we could find a setting for the control bits and a set of output bits $Y_k$ such that at least an average number of the input bits mapped to these outputs come from the other side.

2.15: In Section 2.4 we mentioned that the definition of $I(2)$ could be modified to agree with that of $I$ from Section 2.2. Prove that our proposed modification is correct.

2.16: Theorem 2.6, giving a lower bound for circuits with input repetition, assumed that each input would be repeated the same number of times. Generalize the theorem to require only that the total number of repetitions of the $r$ inputs is $mr/2$.

---

† This problem is not as far-fetched as it looks. Not only is development of "high-rise" technology with many layers proceeding apace, but the lower bound we suggest in this problem applies to a room full of circuitry, where "elements," which are themselves chips or even circuit boards, are wired together by real wires.

2.17: Give a detailed proof that a rectangle of aspect ratio at most $k$ can be divided into $k$ rectangles of equal area and aspect ratio at most 2, as suggested in Fig. 2.9(a).

** 2.18: Show that there are problems for which
   a)   $I(m) \ll I/m$.
   b)   $I/m \ll I(m)$.

* 2.19: We can combine the barrel shifting technique of Fig. 2.7 with that of Fig. 2.3 to get a barrel shifter that runs, to within a constant factor, as fast as Fig. 2.3, and with essentially the same area, yet that processes many more elements. The basic idea is to replace each input $x_i$ in Fig. 2.3 with a column of $m$ inputs, and to connect the columns in a sawtooth fashion, as in Fig. 2.7, with the top of each column connected to the bottom of the next. We can then read $nm$ inputs, cycle them by up to $m$ places, and finally, send them through the network of Fig. 2.3 in $m$ waves.
   a)   What should the output schedule be to make this algorithm work?
   b)   What value of $m$, as a function of $n$, gives this circuit the best $AT^2$ value, as a function of the total number of inputs?

## BIBLIOGRAPHIC NOTES

Initial work on lower bounds involving area-time tradeoffs is found in Brent and Kung [1980a, 1981] and Abelson and Andreae [1980]. Thompson [1979, 1980] is a seminal work in this field. Historically, the next major step was that of Vuillemin [1980], who developed the theory of transitive functions (Exercise 2.5).

The treatment of the subject reflected by this chapter follows the work of Lipton and Sedgewick [1981] and Yao [1981], who independently developed the use of crossing sequences as a way to generalize Vuillemin's and Thompson's ideas. In particular, crossing sequences allow us to deal with predicates, which are problems with a single output bit. Our development follows Lipton and Sedgewick [1981] most closely; we take from there the notion of the fooling set, when- and where-determinacy, and also nondeterministic chips, as discussed in Exercise 2.6.

On the other hand, Yao [1981] has a somewhat more general view, replacing information content as defined by fooling sets by the notion of the minimum number of bits that two communicating devices must send to one another for them to determine their proper outputs. Fooling sets are the most widely used methodology for putting lower bounds on information transfer in the Yao sense, but in principle, one can prove lower bounds by crossing sequence arguments in many different ways. Yao [1979] was the original source for this notion of communication complexity.

The idea of one-way fooling sets combines one-way communication complexity in the works by Yao with the fooling sets of Lipton and Sedgewick

[1981]. The relationship between one-way information and probabilistic chips (Theorem 2.5) generalizes a proof from the latter paper.

The general adversary argument discussed at the beginning of Section 2.2 is from Aho, Ullman, and Yannakakis [1983]. Section 2.4, on circuits with repeated inputs, is based on ideas of Kedem and Zorat [1981], which has been generalized in an elegant way by Kedem [1982]. Recent progress concerning lower bounds for VLSI computation is found in the papers by Savage [1981a], Papadimitriou and Sipser [1982], Baudet [1981], and Brent and Goldschlager [1982].

Some specific problems that have been analyzed along the lines of this chapter are integer multiplication, in Brent and Kung [1980a, 1981] and Abelson and Andreae [1980], matrix multiplication in Savage [1981b] and Preparata and Vuillemin [1980], and Fourier transforms in Preparata and Vuillemin [1981].

Three dimensional circuitry, as discussed in Exercise 2.13, has been investigated by Rosenberg [1981] and Leighton and Rosenberg [1982].

Another interesting issue that we have not covered concerns the energy consumption of circuits. While NMOS circuits will consume energy because of current running through pullup/pulldown networks, there are other technologies, notably CMOS, where power dissipation comes primarily from the actual switching of wires. Lower bounds on power consumption have been studied by Lengauer and Mehlhorn [1981] and Kissin [1982].

Exercise 2.4 is by A. Siegel. Exercises 2.11 and 2.12 are from Lipton and Sedgewick [1981], and Exercise 2.18 is by P. Hochschild. Exercise 2.19 is by C. Leiserson.

# 3

# LAYOUT ALGORITHMS

Now we shall look with a somewhat different perspective at finding the area needed to solve problems on chips. In the previous chapter we assumed that the problem could be solved by any design at all. We therefore developed lower bounds on the area and/or time needed to solve the problem in any way. In this chapter we shall assume that we have a particular solution in mind, and that solution is expressed as a graph. For example, the nodes of the graph might be logic elements, and the edges might be wires between them. Alternatively, the nodes might be processing elements of considerable complexity, perhaps representing registers, memory, and arithmetic units.

In what follows, we need not represent in our graph every wire running between the circuit elements that correspond to nodes. Often, we can ignore a finite number of signals, such as power, ground, clocks, and even some inputs, that are distributed widely. After laying out the graph with these wires omitted, we run each of the omitted wires in a separate layer, and then invoke the normal form of Theorem 1.1 to argue that the resulting circuit area changes by only a constant factor if we reduce the number of layers to a realistic number.

In this chapter, we shall prove certain upper bounds on layout area by constructing layouts. We shall make these layouts have the normal form of Theorem 1.1, where only one wire runs along any grid line. As a result, we assume that all our graphs have degree at most four. Generally, straightforward modifications can be made to show that the results apply to graphs of degree greater than four, if the circuit model allows more than one wire along a grid line, e.g., we can replace nodes of degree greater than four by a network of nodes of degree four. When proving lower bounds on the required area, we are again entitled to assume the layout is in normal form; Theorem 1.1 assures us that our lower bounds will be correct to within a constant factor.

Even given the solution to some problem in the form of a graph, we may not be sure of the best layout to implement this graph. It is the purpose of this chapter to present constructions to map graphs into layouts in the most area-efficient way possible. First we discuss the layout of trees, and in particular the "H-tree" method of laying out complete binary trees. Section 3.2 discusses trees with leaves on the border of the chip and introduces the notion of wire length, i.e., how short can the longest wire in a tree layout be? Then we consider more general graphs, and relate the area a graph takes to the size of "separators" for that graph. We apply this idea to planar graphs in particular and show how to lay out any $n$ node planar graph in $O(n \log^2 n)$ area.

## 3.1 H-TREES

We shall show how to lay out a complete binary tree in $O(n)$ area by placing leaves throughout the chip. In contrast, we shall show in the next section that if the leaves are constrained to be on the border, then $\Omega(n \log n)$ area is required for complete binary trees. We shall see in the Section 3.3 that any tree can be laid out in $O(n)$ area, although the construction is not nearly as compact as the construction for complete trees.

### Binary Tree Layouts

A *complete binary tree* of height $h$ is a binary tree in which every interior node has exactly two children, and all paths from the root to a leaf have length $h$ (i.e., $h$ edges and $h+1$ nodes). There is a variety of problems for which networks of elements arranged in a complete binary tree offer a good solution. Thus, in Example 2.10 we computed a large number modulo a small one by constructing a tree of processing elements, distributing the bits of the dividend among the leaves, and letting the pieces of the answer propagate up the tree in $O(\log n)$ stages. The following is a somewhat nontrivial example where organizing the computation as a tree helps.

**Example 3.1:** Suppose we have a ring network and wish to pass data cyclicly around the ring of processors. However, as suggested by Fig. 3.1, some of the processors are not part of the cycle; either they are "down," or for some reason we do not want them to partake in the cycling process. The obvious solution is for each processor that is part of the cycle to pass its data counterclockwise, and if the processor receiving the data is not part of the cycle, then it must pass the data to the next, and so on, until a processor that is part of the cycle receives the data.

Two things are wrong with this approach. First, if the processors not selected to be part of the cycle are really broken in some sense, then they cannot pass data. But even if the unselected processors are working, and have simply been told by an input bit not to participate in the cycle, the process can take $\Omega(n)$ time if there are $n$ processors. For example, it is possible for all the processors on the left side of the circle to participate and all those on the right not participate, so that one datum travels half way around the circle.

A way to do the same task in $O(\log n)$ steps, and at the same time not depend upon processors that are out of the cycle, is to organize the processors

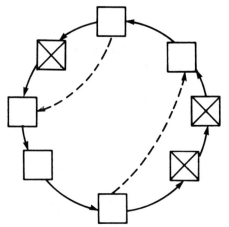

**Fig. 3.1.** Ring of processors.

as the leaves of a complete binary tree.† Each interior node is a processing element with the ability to

1. Store a datum of whatever particular size we are dealing with.
2. Compare the data items at its children, and move data between its children, its parent, and itself.
3. Store a direction indicator, telling a datum that is passed to the node from its parent which child to send that datum to.
4. Store information telling whether the datum it now holds was passed to it from above or below, and if the node is a leaf, whether the datum was placed their initially or was moved there as the result of cycling.

Intuitively, each datum $X$ to be cycled is passed up the tree, leaving a trail of direction indicators to tell the datum to its left how to get down to where $X$ started. When two items meet at a node, they cannot both be passed up. Fortunately, the one coming in from the left now knows where it should go, so it follows the trail deposited by the one on the right, back to the latter's place. The item coming in from the right proceeds up the tree. If the height of the binary tree is $h$, we proceed in $2h$ stages. At each stage, each interior node examines its parent and its children. The rules followed by an interior node $n$ are:

1. If the parent has a datum that it received from above (or its parent is the root), and the parent has its direction indicator set in the direction of $n$ (i.e., left if $n$ is a left child, right otherwise), then $n$ receives the datum from its parent.
2. If the left child of $n$ has a datum, either because we are at the first stage and the children of $n$ are leaves, or because the datum at the left child was

---

† Thus, there must be a power of two processors, although if we had fewer, we could pretend that there were enough additional "broken" processors to raise the number to the next power of two.

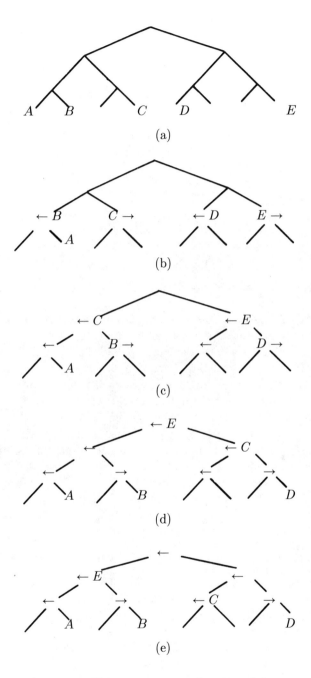

**Fig. 3.2.** Using a tree to cycle selected data.

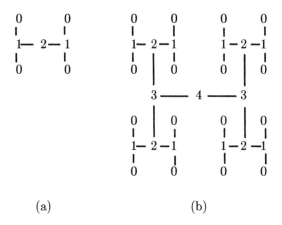

(a)                                         (b)

**Fig. 3.3.** The first two H-trees.

passed up to it at the previous stage, and the right child has no datum,†
then move the datum from the left child to $n$ and set the direction indicator
at $n$ to "left." Indicate at $n$ that the datum is moving up, except if $n$ is
the root, in which case the datum is now moving down.

3. If the right child has a datum moving up, and the left child has no datum,
   then do the same, but set the direction indicator to "right."
4. If both the left and right children of $n$ have data moving up, then
   a) move the datum at the right child to $n$,
   b) move the datum at the left child to the right child,
   c) set the direction indicator at $n$ to "left," and
   d) indicate that the datum at $n$ is moving up, except if $n$ is the root, in
      which case it is moving down.

Figure 3.2(a–e) shows five snapshots of what happens when five data items,
indicated $A, B, \ldots, E$, are subjected to this process. Two more stages are
needed, whereupon $E$ will move to where $A$ was initially, and $C$ will move
to where $D$ was, completing the cycle process. $\square$

## H-Trees

A convenient way to place the nodes of a complete binary tree on a grid is in
a recursive pattern that looks like the letter "H." Assume for the moment that
the nodes of the binary tree can be represented in a layout by grid points. Then
an *H-tree of order* $i$ represents a binary tree of height $2i$; it covers a square of
dimensions $(2^{i+1} - 1) \times (2^{i+1} - 1)$. Nodes are connected to their children and
parents by grid lines.

---

† It is easy to prove that a node cannot find data at its parent and one or more of its children
at one stage.

Figure 3.3(a) shows the H-tree of order 1. Nodes of the binary tree represented are indicated by numbers, and the numbers tell the height of the various nodes, i.e., 0 is a leaf, 1 the parent of leaves, and so on. To construct the H-tree of order $i + 1$ from that of order $i$, we do the following.

1.  Expand the scale by inserting grid lines between each pair of adjacent horizontal and vertical grid lines in the layout for the H-tree of order $i$. Also add single extra grid lines around the border of the layout. Thus the number of grid lines in each direction is doubled plus one.

2.  Replace each leaf in the expanded H-tree of order $i$ by an H-tree of order 1, attaching the root of the order 1 H-tree to the point where the leaf was. Figure 3.3(b) shows the H-tree of order 2 constructed by applying this construction to the H-tree of order 1.

The H-tree turns out to be, both theoretically and in practice, essentially the most area-efficient way to lay out a complete binary tree. In particular, note that the leaves are spread throughout the chip; as we shall see in the next section, this is essential for a small area layout of complete binary trees. The following theorem expresses the fact that, to within a constant factor, the H-tree is as good as possible, since surely $\Omega(n)$ area is needed to lay out any graph with $n$ nodes.

**Theorem 3.1:** The H-tree layout for a complete binary tree of $n$ leaves has area $O(n)$.

**Proof:** An easy induction on $i$ shows that the H-tree of order $i$ has area $O(4^i)$. But this H-tree is a layout for a complete binary tree of height $2i$, which has $4^i$ leaves. $\square$

## 3.2 LOWER BOUNDS ON TREE LAYOUTS

In this section we shall prove two facts. First, any layout of a complete binary tree with $n$ nodes, such that the leaves are on the border, requires $\Omega(n \log n)$ area. Second, we shall show that no matter how we lay out a complete binary tree, there must be some edge that has length $\Omega(\sqrt{n}/ \log n)$.

### Trees with Leaves on the Border

We shall now show that any $n$-node complete binary tree laid out with its leaves on the border requires area $\Omega(n \log n)$. In contrast, $O(n)$ suffices without the requirement about where the leaves are, as we discovered from the H-tree layout in the previous section. However, there may be good reason to put the leaves on the border; in particular, the leaves may each represent input/output ports, and as we mentioned, requirements for packaging chips imply that pads must be located on the border.

It is very easy to show that $O(n \log n)$ is an upper bound on the area needed. For example, the obvious tree layout, with all the leaves on a line and

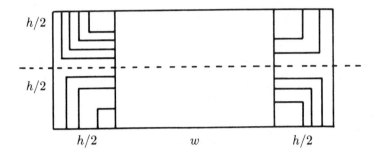

**Fig. 3.4.** Moving leaves onto two parallel lines.

the interior nodes piled above them in layers, meets this bound. The lower bound is not so simple, and we begin by reducing the problem to one of laying the leaves out on a straight line.

**Lemma 3.1:** Suppose a complete binary tree can be laid out in area $A$, with the leaves on the border of a rectangle, and with the layout in normal form. Then we can lay out the same tree with all leaves on the border in area $8A$.

**Proof:** Suppose the layout is $h \times w$, where $h \leq w$. First, we arrange that all the leaves will lie along two parallel lines, by opening up the shorter sides as shown in Fig. 3.4. Wires connect the original positions of the leaves on the shorter sides to corresponding positions on the extensions of the longer sides, whose lengths thus expand from $w$ to $w + h$. Since we assume $h \leq w$, we at most double the area of the layout.

Next, we change all wires above the vertical midpoint to two different layers, one for horizontal and one for vertical wires. We fold the circuit at that midpoint, so the lines at the top and bottom holding leaves coincide. If we expand the horizontal and vertical scales by a factor of two each, we can have wires in one horizontal layer running on even grid lines and the other on odd grid lines. Similarly, we can separate the two vertical layers, and we can at the same time be assured that no two nodes occupy the same grid point. Leaves that are one line too high can be moved down to the bottom grid line. We are now in a position to combine the two vertical layers into one and the two horizontal layers into one; the result is a normal form circuit of eight times the area of the original. $\square$

We now prove that any layout for a complete binary tree with leaves in a line must have area $\Omega(n \log n)$, if $n$ is the number of nodes. Our method is to prove a lower bound on the total length of the wires connecting the nodes. Since the circuit may assumed to be in normal form, the area must be at least

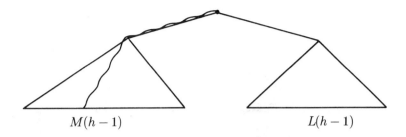

**Fig. 3.5.** Proof of Lemma 3.2.

as great as the sum of the lengths of the horizontal wire segments and also at least as great as the sum of the lengths of the vertical segments. At least one of these sums must be half or more of the total wire length, so the area is bounded below by the wire length divided by two.

We shall refer to the minimum total wire length for a complete binary tree of height $h$ as $L(h)$. We also define $M(h)$ to be the minimum sum of the lengths of the wires in any layout of a complete binary tree of height $h$, excluding the wires on the longest path in the layout from the root to some leaf. The relationships between $L$ and $M$ are exposed by the next lemmas.

**Lemma 3.2:** $M(h) \geq L(h-1) + M(h-1)$.

**Proof:** Consider the complete binary tree of height $h$ in Fig. 3.5, where the longest path from the root to a leaf for this layout is indicated by a wavy line. The subtree of the root that does not contain this path has wire length at least $L(h-1)$, while the other subtree has total path length at least $M(h-1)$. $\square$

**Lemma 3.3:** $L(n) \geq 2M(h-1) + 2^{h-1}$.

**Proof:** The leaves of the tree are laid out in some order, not necessarily related to the left-to-right order usually attributed to nodes in a tree. One leaf, $v$ in Fig. 3.6, is leftmost of all the nodes; we have suggested in that figure that $v$ is in the right subtree of the root, but which subtree $v$ is in does not matter. However, let us consider the rightmost leaf $u$ in the other subtree of the root. Between $u$ and $v$ horizontally are at least all the other nodes of the subtree that $u$ is in, a total of $2^{h-1} - 1$ nodes. Thus, the horizontal distance between $u$ and $v$ is at least $2^{h-1}$.

If we remove from the tree of Fig. 3.6 the path from $v$ to $u$, shown as a wavy line, we are left with all the edges of two subtrees of height $h - 1$, except for the removed path, which is certainly no larger than the two longest paths from root to leaf in this layout. Thus, the remaining edges have total length at least $2M(h-1)$. To this we add the lower bound $2^{h-1}$ on the length of the removed path, to get a lower bound on $L(h)$, the sum of the lengths of all the edges. $\square$

**Theorem 3.2:** The area of an $n$-node complete binary tree laid out with the

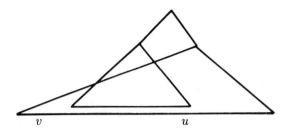

**Fig. 3.6.** Proof of Lemma 3.3.

leaves on the border of a rectangle is $\Omega(n \log n)$.

**Proof:** By Lemma 3.1 it suffices to consider layouts with all leaves on a line, and by what we have observed above, it suffices to compute $L(h)$, the sum of the lengths of the edges. By Lemma 3.3, we can get a lower bound on $L$ if we get a lower bound on $M$. Since $n = 2^{h+1} - 1$, we must get a lower bound on $M(h)$ that is $\Omega(h2^h)$.

If we substitute the result of Lemma 3.3 into the result of Lemma 3.2 to eliminate $L$, we get

$$M(h) \geq 2M(h-2) + M(h-1) + 2^{h-2} \tag{3.1}$$

Also, we observe that $M(0) = 0$ and $M(1) = 1$. We shall prove by induction on $h$ that $M(h) \geq h2^h/6$; the result surely holds for $h = 0$ and $h = 1$.

For the induction, assume that $M(i) \geq i2^i/6$ for $i < h$, and substitute that lower bound into (3.1). Then (3.1) becomes

$$M(h) \geq (h-2)2^h/12 + (h-1)2^h/12 + 2^h/4$$

from which $M(h) \geq h2^h/6$ follows by simplification. $\square$

### A Lower Bound on Wire Length

While we have generally ignored the length of wires, there are two factors that suggest we should strive to make wires as short as possible. In particular, the maximum wire length should be kept short, since it limits the speed at which the chip can run reliably, as discussed in Section 1.3. Thus, it would be nice to know that we could always lay out a graph so that nodes interconnected by wires could be placed close together. Unfortunately, that is not the case in general. For example, the H-tree has individual wires half as long as a side of the layout; in terms of the number of nodes of the tree, $n$, the longest wire has length $\Omega(\sqrt{n})$. We might hope to do better, perhaps by using a strategy different from the H-tree.

It turns out that we can do a little better by building a network of smaller H-trees, but not much better. The longest wire for an $n$-node tree still has length $\Omega(\sqrt{n}/\log n)$, and that is the best possible result. We shall show in the next theorem that $\Omega(\sqrt{n}/\log n)$ is a lower bound on wire length for complete binary trees, and then give the construction whereby that bound can be achieved.

**Theorem 3.3:** Every layout for a complete binary tree of height $h$ has a wire of length $\Omega(2^{h/2}/h)$. In terms of $n$, the number of nodes of the tree, the length is $\Omega(\sqrt{n}/\log n)$.

**Proof:** Suppose $d$ is an upper bound on the length of a wire in some layout. Then all the nodes of the tree are found within distance $dh$ of the root. The number of grid points within this circle is $\pi d^2 h^2$. But the number of nodes in the tree is $2^{h+1} - 1$. Since the number of nodes cannot exceed the number of available grid points, we have $\pi d^2 h^2 \geq 2^{h+1} - 1$, or

$$d \geq \sqrt{\frac{2^{h+1} - 1}{\pi h^2}}$$

from which $d = \Omega(2^{h/2}/h) = \Omega(\sqrt{n}/\log n)$ follows. $\square$

### An Upper Bound on Wire Length for Complete Binary Trees

We shall now show that the lower bound of Theorem 3.3 can be achieved. The plan of the layout is in Fig. 3.7. The root, at the lower right, fans out in $h/4$ levels horizontally, to $2^{h/4}$ leaves that are distributed uniformly on the upper surface of a rectangle of height $2^{h/4}$ and width $O(2^{h/2})$. Each of these leaves is the root of a tree looking exactly the same, but rotated to be vertical. Finally, each leaf of the vertical trees is the root of an H-tree of order $h/4$, representing the remaining $h/2$ levels of the complete binary tree. These H-trees each have sides of length $2^{h/4}$.

It is easy to check that the H-trees do not have edges that exceed our intended length limit of $O(2^{h/2}/h)$. The key inference to make is that the horizontal and vertical trees can be designed to use edges of that length; in particular, $16 \times 2^{h/2}/h$ is an upper bound on the needed edge length.

Figure 3.8 shows the basic idea behind the layout of the horizontal tree, which is within a rectangle $2 \times 2^{h/2}$ wide by $2^{h/4}$ high. Note the same idea works for the vertical trees. Also observe that the width of the horizontal tree, $2 \times 2^{h/2}$, is just enough for the $2^{h/4}$ columns of H-trees and the same number of vertical trees, each of width $2^{h/4}$.

What is really shown in Fig. 3.8 is what would happen if we wanted the $2^{h/4}$ leaves to be distributed uniformly on the right edge. The tree fans out gracefully from the left, with each level placed $8 \times 2^{h/2}/h$ to the right of the previous one, thus covering the width of $2 \times 2^{h/2}$ in the allotted $h/4$ levels. As $2^{h/4}$ grows much more slowly than $2^{h/2}/h$, allowing the edges to move vertically as desired does not more than double the length of edges, which is why we claim an upper bound of $16 \times 2^{h/2}/h$ for the length of an edge. For the same reason, there is enough width between the horizontal positions of the nodes at

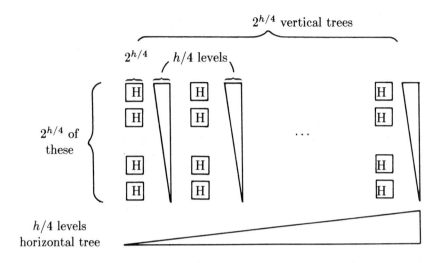

**Fig. 3.7.** Plan of binary tree layout with low maximum wire length.

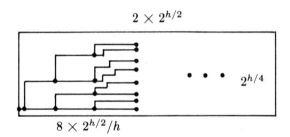

**Fig. 3.8.** Plan of horizontal tree.

consecutive levels for all the needed tracks to run vertically in parallel. Thus, the differences in vertical positions of nodes at different levels can be respected by wires that each jog once between levels.

However, the desired layout is not one in which the leaves are on the right border, but rather on the top, spaced $2 \times 2^{h/4}$ apart, so each can be connected to the root of a vertical tree. To make the necessary modification to Fig. 3.8, we do not allow any node to be placed to the right of the desired position of its leftmost node. In order to show that this rule is feasible, we must show that when a right child branches off from its parent, which was being held back by descendants of its left child, the right child still has enough levels available to allow its descendant leaves to reach their desired positions.

Suppose the child is at level $h/4 - i$, so each path from that child has $i$ edges to get leaves into the correct position. Its rightmost descendant is placed at a distance no more than $2^{i+1}2^{h/4+1}$ to the right of the parent, since there are $2^{i+1}$ descendants of the parent, and the separation between leaves in the horizontal tree must be $2 \times 2^{h/4}$. But the $i$ edges along paths can travel a distance $8i2^{h/2}/h$. We must verify that

$$8i2^{h/2}/h \geq 2^{i+1}2^{h/4+1}$$

or

$$\frac{2i}{h}2^{h/4-i} \geq 1$$

The above clearly holds if $i \leq h/4 - 2$. The largest possible value of $i$ is $h/4 - 1$, in which case the parent of the node in question is the root, and we need only to verify that there is enough room to get from the root to its rightmost leaf. As there are $h$ edges in paths from the root to a leaf, and each edge is of length $8 \times 2^{h/2}/h$, there is just enough length to travel all the way across the layout, a distance of $2 \times 2^{h/2}$. We have thus proved the following theorem.

**Theorem 3.4:** A complete binary tree of height $h$ can be laid out so that the longest edge length is $O(2^{h/2}/h)$. $\square$

## 3.3 A DIVIDE-AND-CONQUER LAYOUT ALGORITHM

There is a general-purpose algorithm that produces low area layouts for a wide variety of families of graphs, including all trees of degree four or less (not just complete binary trees) and planar graphs of degree four or less. Roughly, the algorithm works by finding ways to partition graphs recursively into pieces with few connecting edges; the pieces are laid out, and then the layout is modified to account for the edges between the pieces. In this section we give the general algorithm, and in the next sections we apply it to two interesting special cases.

### Separators

The driving force behind the algorithm is the ability to find small sets of edges that disconnect the graph; therefore the algorithm works well on classes of graphs that are easily disconnected but fails to perform well on graphs, such as complete graphs,† that cannot be disconnected with few edges.

Formally, we say a graph $G$ has an $S(n)$ *separator*, or is $S(n)$-*separable*, if either it has only one node, or the following two statements are true.
1.  Let $n_0$ be the number of nodes of $G$. Then we can find a set of at most $S(n_0)$ edges whose removal disconnects $G$ into two graphs $G_1$ and $G_2$, of $n_1$ and $n_2$ nodes each, such that $n_1 \geq n_0/3$ and $n_2 \geq n_0/3$. Thus, neither subgraph has more than twice the number of nodes of the other.

---

† A *complete* graph, having no relation to a "complete tree," is a graph with an edge between every two nodes.

2.   Both $G_1$ and $G_2$ are $S(n)$-separable.

More frequently, we shall talk about families of graphs. We say a family is $S(n)$-separable if every member of the family is $S(n)$-separable. Most frequently, graphs $G$ in the family will be decomposed into subgraphs $G_1$ and $G_2$ that are also in the family, in which case we have only to prove (1) for any $G$ in the family.

It is often useful to visualize the recursive partitioning of a graph in the form of a *partition tree*. If graph $G$ is a single node, then its partition tree is one leaf, representing $G$. If $G$ is split into $G_1$ and $G_2$, then the partition tree for $G$ consists of a root, representing $G$, with two children, which are the roots of the partition trees for $G_1$ and $G_2$. Note that the partition tree is a binary tree, and in fact each interior node has both left and right children present.

**Example 3.2:** The family of binary trees is 1-separable. In proof, we pick a root for the tree arbitrarily and let $n$ be the number of nodes in the whole tree. We shall travel down a path of the tree looking for a node that is the ancestor of between $n/3$ and $2n/3$ nodes. Then the edge from this node to its parent disconnects the tree into two pieces, each of which has at least $1/3$ of the nodes.

To find the desired node, we begin at the root. If one of the children of the root dominates between $n/3$ and $2n/3$ nodes, we are done; we pick that child. If there is no such child, there must be one that dominates more than $2n/3$ nodes.† We move to that child and repeat the argument recursively. That is, if we are at a node with $m > 2n/3$ descendants, either one of its two children dominates between $n/3$ and $2n/3$ nodes, in which case we are done, or one child dominates more than $2n/3$ nodes, in which case we repeat the argument at that child.

Consider the tree of Fig. 3.9, with $n = 20$ nodes. The children of the root dominate 18 nodes and 1 node, respectively, including themselves. The first of these, node 2, dominates more than $2n/3 = 13\frac{1}{3}$ nodes, so we move to node 2. Its children dominate 3 and 14 nodes, respectively, neither of which is in the desired range, so we move to the root of the larger tree, which is node 5. That node has a child, node 8, that dominates 12 nodes, which is between $n/3$ and $2n/3$. Thus, we can delete the edge from 5 to 8 and break the tree into two subtrees of sizes 8 and 12. The first, consisting of the set of nodes $\{1, 2, 3, 4, 5, 6, 7, 9\}$, can be broken recursively by removing the edge from 2 to 4, which divides this tree into $\{1, 2, 3, 5, 9\}$ and $\{4, 6, 7\}$. The former can be broken into $\{1, 3\}$ and $\{2, 5, 9\}$, and so on.

Figure 3.10 exhibits the partition tree for Fig. 3.9. Note that the partition tree would still be a tree even if the graph whose partition it represents were not a tree.

Since the scheme for finding a single edge that divides a binary tree into two

---

† Note we do not rule out the possibility that a node has only one child.

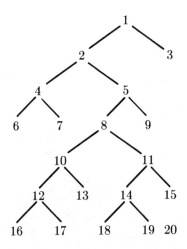

**Fig. 3.9.** Example tree.

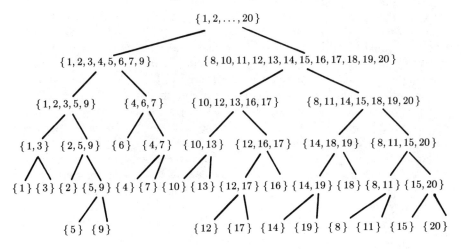

**Fig. 3.10.** Partition tree.

pieces whose sizes are in ratio no greater than 2:1 works on arbitrary trees of more than one node, without regard for the total number of nodes, we conclude that the family of binary trees have 1-separators. □

**Example 3.3:** The family of planar graphs is $\sqrt{8n}$-separable. We shall sketch the proof that such separators can always be found in Section 3.5. However, for the moment let us simply regard what is close to, but not quite, the worst case for planar graphs: the square grid. An array of $n$ nodes arranged into a $\sqrt{n} \times \sqrt{n}$ grid as in Fig. 3.11 cannot be separated by fewer than $\sqrt{n}$ edges.

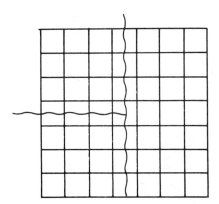

**Fig. 3.11.** Partition of square grid.

If we divide through the middle, as in Fig. 3.11, we are left with two subgraphs of $n/2$ nodes, each of which has a separator, through the middle the short way, consisting of $\sqrt{n}/2$ nodes, which is less than $\sqrt{n/2}$ nodes. The resulting subgraphs are squares that have side $\sqrt{n}/2$ and contain $n/4$ nodes. These have separators like the whole graph, of size $\sqrt{n}/2 = \sqrt{n/4}$. Since square grids are planar, and it should be clear that we cannot partition such graphs into roughly equal pieces by cutting fewer than $\Omega(\sqrt{n})$ edges, we cannot expect the family of planar graphs to have separators less than $O(\sqrt{n})$ in general.

Another interesting family is the set of rectangular grids, a simple generalization of the square grids and grids of aspect ratio 2 that we discussed above. An $h \times w$ grid with $h \leq w$ has $n = hw$ nodes. It can be separated by cutting the short way, in the middle; doing so removes $h$ edges, which is at most $\sqrt{n}$ edges. If $w$ is even, the resulting graphs are rectangular grids of dimension $h \times w/2$, while if $w$ is odd, the two graphs are rectangular grids of dimension $h \times (w-1)/2$ and $h \times (w+1)/2$. The reader can check easily that no matter what the value of $w > 1$, the smaller of the two graphs has at least $n/3$ nodes. If $w = 1$, then $h = 1$, and the "grid" was a single node, which we do not have to partition. $\square$

### Strong Separators

To make it easy to prove that the layout algorithm achieves the area bounds we claim for it, we shall present the algorithm in a version where graphs are divided exactly in half. Thus, we must consider under what conditions the 2:1 ratio of the pieces in the definition of a separator can be replaced by the ratio 1:1. We say a family of graphs has a *strong $S(n)$ separator*, or is *strongly $S(n)$-separable* if the conditions for an $S(n)$ separator hold, and in addition, we can always find subgraphs $G_1$ and $G_2$ that have at most $(n+1)/2$ nodes each. That is, if $n$ is even, the graph is divided exactly in half, and if $n$ is odd, the two pieces are as close to equal in size as possible.

**Example 3.4:** A simple corollary of Example 3.3 is that the family of rectangular grids whose sides $h$ and $w$ are each powers of two, is strongly $\sqrt{n}$-separable. That is actually no surprise, and does not even depend on the sides being powers of two, since we shall next show that as long as $S(n)$ grows at least as fast as $n^\epsilon$ for some $\epsilon > 0$, the existence of an $S(n)$ separator implies the existence of a strong $S(n)$ separator. $\square$

In the developments of this section, we shall introduce two functionals, $\Gamma$ and $\Delta$, as an aid to stating results. The first of these we define now. If $S(n)$ is any function, then the function $\Gamma_S(n)$ is defined by

$$\Gamma_S(n) = S(n) + S(\tfrac{2}{3}n) + S(\tfrac{4}{9}n) + \cdots \ (\lceil \log_{3/2} n \rceil \text{ terms})$$

$$= \sum_{i=0}^{\lceil \log_{3/2} n \rceil} S((\tfrac{2}{3})^i n)$$

**Example 3.5:** If $S(n) = n^\alpha$, then

$$\Gamma_S(n) = n^\alpha + (\tfrac{2}{3}n)^\alpha + (\tfrac{4}{9}n)^\alpha + \cdots \leq \frac{1}{1 - (2/3)^\alpha} n^\alpha = O(n^\alpha)$$

If $S(n)$ is a constant, $c$, we find $\Gamma_c(n) = c + c + \cdots + c$ ($\lceil \log_{3/2} n \rceil$ terms), so $\Gamma_c(n) = O(\log n)$. If $S(n) = \log_2 n$, we find

$$\Gamma_{\log_2} = \log_2 n + \log_2(\tfrac{2}{3}n) + \log_2(\tfrac{4}{9}n) + \cdots \ (\lceil \log_{3/2} n \rceil \text{ terms})$$

$$= \log_2 n \lceil \log_{3/2} n \rceil + \log_2(\tfrac{2}{3}) + 2\log_2(\tfrac{2}{3}) + \cdots$$

$$= \log_2 n \lceil \log_{3/2} n \rceil - \log_2(\tfrac{3}{2}) \lceil \log_{3/2} n \rceil (\lceil \log_{3/2} n \rceil + 1)/2$$

Since the second term is smaller than the first, we have shown that $\Gamma_{\log}(n) = O(\log^2 n)$. $\square$

We now show how to convert separators into strong separators while increasing the growth rate of the separators by only the functional $\Gamma$. We see from Example 3.5 that in most cases $\Gamma$ increases size by only a constant factor, and even when $S(n)$ is very slowly growing, like $\log n$, or even a constant, the size grows by only a factor of $\log n$. Thus we shall find it quite useful to assume strong separators for our families of graphs, rather than simply separators.

**Lemma 3.4:** If a family of graphs is $S(n)$-separable, then it is strongly $\Gamma_S(n)$-separable.

**Proof:** Consider the partition tree for a graph $G$. We shall actually show something stronger; for any target number of nodes $t$ less than all the nodes of $G$, we can find a path from the root of $G$ to some leaf, so that some set of siblings of the nodes on this path represent subgraphs with $t$ nodes of $G$. We call these the *selected* subgraphs. The proof is an induction on the height of

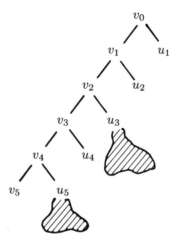

**Fig. 3.12.** Construction of path in Lemma 3.4.

the partition tree. The basis, height 0, is trivial, since such a partition tree represents a graph of one node, and therefore $t$ can only be 0. The basis thus holds vacuously.

For the induction, let the left and right children of the root have $n_1$ and $n_2$ nodes, respectively. If $n_1 \leq t$, begin the path by going to the right, and complete the path using target $t - n_1$ on the right subtree. Include the subgraph represented by the left child of the root in the selected set. Otherwise, if $n_1 > t$, but $n_2 \leq t$, then begin the path to the left with target $t - n_2$, and include the right child of the root in the selected subset. Finally, if $n_1$ and $n_2$ are both greater than $t$, pick one arbitrarily for the path, keep the target number of nodes at $t$, and include neither child in the selected set. Such a path is suggested by Fig. 3.12, where the path happens to extend to the left, along $v_0, v_1, \ldots$. Of the siblings of the path nodes, $u_1, u_2, \ldots$, we have selected the sets of nodes represented by $u_3$ and $u_5$, but not those represented by $u_1$, $u_2$, or $u_4$.

Now we must check that the number of edges connecting the selected subsets to the remainder of the nodes is not larger than $\Gamma_S(n)$. Let the path be $v_0, v_1, \ldots$, as in Fig. 3.12. If the nodes of the graph represented by $u_1$ are in the selected subset, then in the worst case, all $S(n)$ of the edges between $v_1$ and $u_1$† will connect to nodes of $v_1$ that are not in the selected set. If $u_1$ is not in the selected set, then in the worst case all $S(n)$ edges between $v_1$ and $u_1$ will connect to nodes of $v_1$ that are in the selected set.

Similarly, we can progress down the path and argue that the total number

---

† We shall use the name of a node in the partition tree to stand for the set of nodes of the graph that that tree node represents. The reader should bear in mind that the nodes of the graph represented by $u_i$ and $v_i$ are subsets of those represented by $v_j$, if $j < i$.

of edges between the $v_i$'s and the $u_i$'s is no more than

$$S(n) + S(\tfrac{2}{3}n) + S(\tfrac{4}{9}n) + \cdots$$

since the size of $v_i$ is no greater than $(\tfrac{2}{3})^i n$. This sum is $\Gamma_S(n)$, of course. But any edge between a node in a selected $u_i$ and an unselected node must be among the edges connecting one of its ancestor $v_j$'s to $u_j$, for some $j \leq i$. Thus the number of edges between the selected and unselected halves of the nodes of $G$ is bounded above by $\Gamma_S(n)$. $\square$

**Example 3.6:** We know by Examples 3.2 and 3.5 that the family of binary trees is strongly $\log n$-separable. Let us see how to divide the tree of Fig. 3.9 into two sets of 10 nodes each. We begin with the partition tree in Fig. 3.10, at the root with a target of 10. The left child represents 8 nodes, so we select it and work on the right child with a target of 2. Both children of the right child of the root are too large, so we pick one, say $\{10, 12, 13, 16, 17\}$ with a target of 2 again. The left child of this node represents exactly as many nodes as we need, so we select it and may end at this point.

The resulting selected set is $\{1, 2, 3, 4, 5, 6, 7, 9, 10, 13\}$. It is connected to the other nodes by the edges $(5, 8)$, $(8, 10)$, and $(10, 12)$. This number of edges, three, is less than the theoretical limit $\Gamma_1(20)$, which is $\lceil \log_{3/2}(20) \rceil = 8$. $\square$

**Channel Creation**

In order to understand the layout algorithm, we need to see first how to introduce certain edges back into a layout, once the layout for the graph without these edges has been constructed. Then, we can use the existence of a small separator for a graph to find its layout in a divide-and-conquer way. First the layouts for the pieces are selected and then the layout is modified to include the edges of the separator.

In what follows, we assume our graph has degree no more than four, and we assume that the layouts created are in normal form. As always, we shall assume that the nodes are mapped to single grid points, since if the elements represented by nodes were of any fixed finite size, we could expand the scale so that size was 1 and only change the constant factor in the results, which are order-of-magnitude results anyway.

The basic step in the process of inserting edges is *channel creation*. To create a channel between two adjacent vertical grid lines, $x$ and $x + 1$, we do the following.

1. Move all nodes and wires at position $x + 1$ or higher one position to the right.
2. Stretch all horizontal wires that formerly ran along a horizontal grid line from $x$ to $x + 1$, so they now span from $x$ to $x + 2$.

A similar definition holds for the creation of horizontal channels.

Now, suppose we wish to insert a wire between nodes $a$ and $b$ in a layout. We may assume that neither $a$ nor $b$ currently has four wires connecting, or else the graph would have degree five at least, which we have ruled out. Suppose for the moment that $a$ has an adjacent horizontal grid line unoccupied (i.e., there is no wire attached to $a$ from the left or none from the right), and $b$ has an adjacent vertical grid line unoccupied. We cut a vertical channel along the grid line adjacent to $a$ that is unoccupied, and we cut a horizontal channel along the unoccupied vertical grid line adjacent to $b$. These channels must cross. We therefore may run a wire

1.  from $a$, for unit distance along the adjacent unoccupied horizontal grid line, to the vertical channel,

2.  along the vertical channel's grid line (which must be completely unoccupied),

3.  along the horizontal channel's grid line (which is also unoccupied), and then

4.  for unit distance along the unoccupied vertical grid line adjacent to $b$, finally arriving at $b$.

**Example 3.7:** Figure 3.13(a) shows an example layout. Node $a$ has an unoccupied horizontal grid line to its left, and $b$ has an unoccupied vertical grid line above. The wavy lines indicate where we are going to cut channels to which these empty grid lines can feed a new wire. In Fig. 3.13(b), horizontal and vertical channels have been cut. Then in Fig. 3.13(c), the wire has been run. $\square$

If $a$ has a free adjacent vertical grid line and $b$ an adjacent horizontal grid line unoccupied, the same idea works. If both have adjacent horizontal grid lines free, then we cut three channels; two are vertical, on the sides of $a$ and $b$ that have the free grid lines, and the third channel is an arbitrary horizontal one to connect the vertical ones. The last case, where both $a$ and $b$ have unoccupied adjacent vertical grid lines, is solved similarly with two horizontal channels and one vertical channel.

**The Layout Algorithm**

Before describing the algorithm, we need to introduce our second functional, $\Delta$. We shall define $\Delta_S(n)$ assuming $n$ is a power of 4, for convenience, although the generalization should be obvious. If $S(n)$ is any function, then

$$\Delta_S(n) = S(n) + 2S(n/4) + 4S(n/16) + \cdots$$
$$= \sum_{i=0}^{\log_4 n} 2^i S(n/4^i)$$

**Example 3.8:** If $S(n) = n^\alpha$, then

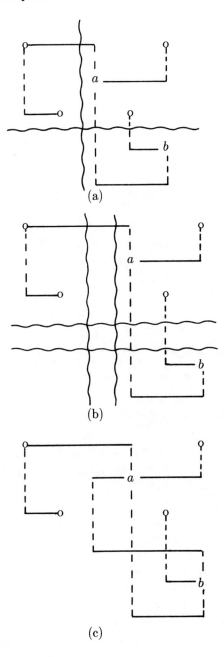

**Fig. 3.13.** The channel creation process.

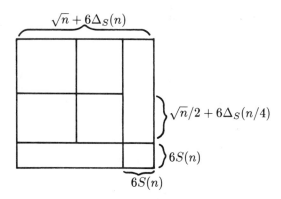

**Fig. 3.14.** Plan of recursive layout algorithm.

$$\Delta_S(n) = \sum_{i=0}^{\log_4 n} 2^i (\frac{n}{4^i})^\alpha = \sum_{i=0}^{\log_4 n} 2^{i(1-2\alpha)} n^\alpha \tag{3.2}$$

If $\alpha < 1/2$, then (3.2) is essentially equal to the largest term of the summation, i.e., the term for $i = \log_4 n = (\log_2 n)/2$, so $\Delta_S(n)$ is no more than some constant times $2^{(\log_2 n)(1-2\alpha)/2} n^\alpha = n^{\frac{1}{2}-\alpha} n^\alpha = \sqrt{n}$. Thus, $\Delta_S(n)$ is $O(\sqrt{n})$, independent of $\alpha$, when $S(n) = n^\alpha$, for some $\alpha < 1/2$.

If $\alpha > 1/2$, on the other hand, then (3.2) Is a decreasing geometric series in $i$, which is essentially the first term, where $i = 0$. Thus, $\Delta_S(n) = O(n^\alpha)$ when $S(n) = n^\alpha$ and $\alpha > 1/2$.

Finally, if $\alpha = 1/2$, then $S(n)$ is the square root function, $\sqrt{\ }$. All terms of (3.2) are $\sqrt{n}$, and there are $\log_4 n$ terms, so $\Delta_{\sqrt{\ }}(n) = O(\sqrt{n} \log n)$. $\square$

We shall now describe the layout algorithm in the proof of the next theorem. The algorithm is a recursive one; we take a partition tree representing the strong separators of a particular graph, and lay out the subgraphs represented by these nodes from the bottom up. If the graph has an $S(n)$ strong separator, then subgraphs of $n$ nodes are laid out in squares of side $\sqrt{n} + 6\Delta_S(n)$. Four of these are arranged in a larger square, and a total of at most $6S(n)$ horizontal channels and the same number of vertical channels are cut in the large square, to accommodate the wires representing edges between the four subgraphs.

The plan of the construction is illustrated in Fig. 3.14, although the channels, which are shown on the border, are in fact spread throughout the square and represent an expansion of at most $6S(n)$ in both the horizontal and vertical directions, beyond the size of the four smaller squares.

We should notice that if $S(n)$ grows more slowly than $\sqrt{n}$, then the side of the square is essentially $\sqrt{n}$, by our analysis of $\Delta$ in Example 3.8.† Thus, for slowly growing $S(n)$ we can lay out graphs in little more area than is needed to hold their nodes. When $S(n)$ grows as $\sqrt{n}$ or faster, the $6\Delta_S(n)$ term dominates, and we can in effect ignore the $\sqrt{n}$ term. Intuitively, it is when $S(n) \geq \sqrt{n}$ that the area required by the wiring suddenly takes over from the area required by the nodes as the deciding factor.

**Theorem 3.5:** Let $S(n)$ be any monotonically nondecreasing function. A graph of $n$ nodes with a strong $S(n)$ separator can be laid out in a square whose side is $O(max(\sqrt{n}, \Delta_S(n)))$.

**Proof:** We proceed by induction on $n$, assuming that $n$ is a power of 4, to show that a side of length $\sqrt{n}+6\Delta_S(n)$ suffices. If $n$ is not a power of 4, we can introduce dummy nodes to the graph without increasing the size of separators, thereby affecting the resulting area by only a constant factor of 4 at most. The basis, $n = 1$, says that a square of side $1 + 6\Delta_S(1)$ suffices for a graph of one node. Since such a graph requires only a single grid point for its node, the basis is clearly true.

For the induction, since $n$ is a positive power of 4, the partition tree for the graph in question has height at least 2, so we can divide the graph exactly in half and divide each of those halves exactly in half. The resulting four subgraphs of $n/4$ nodes each, have, by the inductive hypothesis, a square layout of side $\sqrt{n}/2 + 6\Delta_S(n/4)$. We put these four squares together, as shown in Fig. 3.14, and then cut the needed channels.

Between the two pairs of grandchildren of the root of the partition tree we must run at most $S(n/2)$ wires, which requires cutting at most $2S(n/2)$ channels for each pair, in each direction, a total of at most $4S(n/2)$ horizontal channels and the same number of vertical channels.‡ Then, to connect the children of the root in the partition tree requires at most $2S(n)$ channels in each direction. Since $S(n/2) \leq S(n)$ follows from the monotonicity of $S$, we require at most $6S(n)$ channels cut in each direction.

Now the side of the large square in Fig. 3.14 is $2(\sqrt{n}/2+6\Delta_S(n/4))+6S(n)$. However, it is easy to see that $\Delta_S(n) = S(n) + 2\Delta_S(n/4)$, so the bound $\sqrt{n} + 6\Delta_S(n)$ on the side of the square is immediate. $\square$

**Example 3.9:** We see from Example 3.6 that binary trees are strongly $\log n$-separable. Since $\log n$ grows more slowly than $\sqrt{n}$, Example 3.8 implies that $\Delta_{\log}(n) \leq \sqrt{n}$. Thus, in Theorem 3.5, the $\sqrt{n}$ term is the larger, and we may conclude that all $n$-node binary trees can be laid out in $O(n)$ area.

---

† It is tempting to suppose that we could drop the $\sqrt{n}$ term in general. However, in extreme cases, such as the family of graphs with no edges, where $\Delta_S(n)$ is 0 for all $n$, the $\sqrt{n}$ term is the dominant one.

‡ In fact, we could argue that the total number of channels cut at this step is not $8S(n/2)$ but at most $6S(n/2)$, but that observation affects the area by only a constant factor and makes the proof harder.

For another example, the family of planar graphs has, as we shall see, $O(\sqrt{n})$ separators. By Example 3.5 we know that $\Gamma_{\sqrt{}}(n) = \sqrt{n}$. Then by Lemma 3.4, we know that planar graphs have $O(\sqrt{n})$ strong separators. Finally, by Theorem 3.5 and Example 3.8, all $n$-node planar graphs can be laid out in $O(n \log^2 n)$ area. $\square$

## 3.4 LAYOUT OF REGULAR EXPRESSION RECOGNIZERS

This section is an extended example of the use to which the ideas of the previous section can be put. Here we are concerned with the layout of circuits that recognize the patterns defined by regular expressions.† In principle, we can express the action of any sequential circuit by giving a regular expression for each output bit of the circuit; that expression describes the sequences of inputs that cause this output bit to have value 1. In practice, regular expressions are adequately succinct for only a subset of all possible sequential processes. Regular expressions seem appropriate for describing control operations, where signals are made in response to patterns of events, but they are not appropriate for such tasks as counting, or sequential arithmetic circuits.

### Regular Expression Graphs

There are circuits that "recognize" regular expressions, in the sense that the output of the circuit is 1 if and only if the sequence of inputs is a member of $L(E)$, the set of strings generated by the regular expression $E$. Such circuits are really nondeterministic finite automata with $\epsilon$-transitions (see Appendix 3), and they can be represented by directed graphs whose nodes correspond to circuit elements of the following types.

1. A node labeled by an abstract symbol $a$ is a *recognizer* for that symbol. All recognizer nodes have one entering and one leaving edge. We assume all abstract symbols are defined in terms of the input wires of the circuit. For example, there could be one wire for each abstract symbol, in which case that wire must reach the recognizer node. Another possibility is that the abstract symbols are represented by combinations of input wires, e.g., there could be eight wires representing an ascii character, and the abstract symbol $a$ might be deemed present on the input when the inputs had the combination of 0's and 1's representing the letter "a" in ascii.

2. A node labeled $F$ is a *fork node*. This node has one edge entering and two edges leaving.

3. A node labeled $+$ is an *or-node*. This type of node has two edges entering and one edge leaving.

   The intuition behind these types of nodes is that together they define the sort of circuits we need to recognize any regular expression. The recog-

---

† The reader who is unfamiliar with regular expressions should consult Appendix 3.

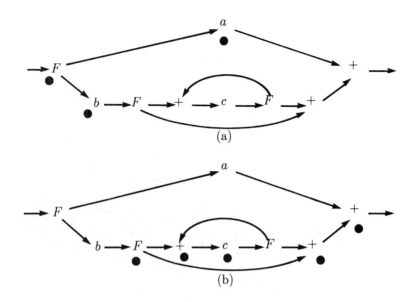

**Fig. 3.15.** Enabling in a regular expression network.

nizer nodes correspond to the operands of the expressions; when the current input could possibly correspond to a particular operand, then at phase 1 of a two-phase clock, the recognizer node corresponding to that operand will be "enabled." At phase 2 of the same clock cycle, the recognizer node checks to see whether its abstract symbol actually appears on the input. If so, then this node is still enabled at the end of phase 2. All other nodes, including the fork and or-nodes, are not enabled in phase 2.

At the next phase 1, the successors of the recognizer nodes that were "enabled" at phase 2 are themselves enabled. The nodes enabled at phase 2 are no longer enabled, but a rippling of "enabling" takes place during phase 1 that may turn some of these nodes back on. The rule followed to enable additional nodes is: if any fork or or-node is enabled, then enable its successor(s).

**Example 3.10:** In Fig. 3.15(a) is the graph that, as we shall later see, corresponds to the regular expression $a + bc^*$. Suppose that the fork node at the left were enabled at phase 1 of the first clock cycle. Then at that phase its successors, the nodes that recognize $a$ and $b$, would also be enabled. Note that the rippling of "enables" goes no further, since at phase 1 the successors of recognizer nodes are not enabled. We show the enabled nodes by placing dots at their left in Fig. 3.15.

Now suppose that at phase 2 the input is $b$. Then the node labeled $b$ stays enabled at phase 2, and all other nodes are turned off. For the next phase 1, the successor of $b$ is turned on, and it enables its two successors, the two or-nodes

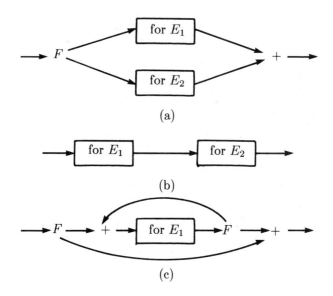

**Fig. 3.16.** Induction in the McNaughton-Yamada construction.

on the bottom row, and these enable the third or-node and the node recognizing
$c$. No further nodes are enabled; the enabled nodes at this time are indicated
in Fig. 3.15(b). □

## The McNaughton-Yamada Construction

There is a simple way to build a regular expression graph for any regular
expression $E$. This algorithm proceeds, by induction on the number of operators
in $E$, to construct a graph with the following properties.

1.  There is one *initial node* with no predecessors. The initial node is either a
    fork or recognizer node, so it has exactly one input; the predecessor feeding
    that input is not part of the graph.

2.  There is one *final node* with no successors. The final node is either a
    recognizer or or-node, so it has room for one successor, which is not part
    of the graph.

3.  Suppose that at phase 1 of some clock cycle, the initial node is enabled,
    and at successive phase 2's, the input symbols present are $a_1, a_2, \ldots, a_n$.
    Then for $1 \leq i \leq n$, at the phase 1 following the input of $a_i$, the final
    node will be enabled (and would enable any successor that we attached to
    it) if and only if $a_1, \ldots, a_i$ is in $L(E)$.

In the basis of the construction, for an expression that is a single operand
$a$, we "construct" a graph with a single node, which is a recognizer for $a$.

The induction is in three parts, depending on whether the outermost operator is +, *, or concatenation. If our expression $E$ is the union of $E_1$ and $E_2$, we take graphs for these two subexpressions, and add a fork node and an or-node. The fork node becomes the initial node, and its two successors are the initial nodes of the graphs for $E_1$ and $E_2$. The or-node becomes the final node, and its two predecessors are the final nodes of the two subgraphs. The construction is shown in Fig. 3.16(a).

To see why the union construction works, observe that if we enable the fork node in Fig. 3.16(a), it will immediately enable the initial nodes of the two subgraphs. By the inductive hypothesis, a sequence of inputs will cause the final nodes of these subgraphs to be enabled if and only if the input is in $L(E_1)$ or $L(E_2)$, respectively. But the final nodes for the two subgraphs each enable the or-node in Fig. 3.16(a), and that or-node cannot be enabled any other way. Thus, the final node for the whole graph is enabled after all and only the inputs in $L(E) = L(E_1) \cup L(E_2)$.

If $E = (E_1)(E_2)$, i.e., concatenation is the outermost operator, then we use the construction of Fig. 3.16(b), where the initial node of the graph for $E_2$ becomes the successor of the final node for $E_1$. The initial node for $E_1$ becomes the initial node for the whole graph, and the final node for $E_2$ becomes the final node for the whole graph. We leave it to the reader to observe why this construction works.

The third construction, depicted in Fig. 3.16(c), handles the case where $E = (E_1)^*$. We introduce two pairs of fork and or-nodes. The path from the first fork, which is the initial node, to the last or-node, which is the final node, is present so that zero occurrences of strings in $L(E_1)$, that is, $\epsilon$, will enable the final node. The feedback loop is so that one or more occurrences of strings in $L(E_1)$ will also enable the final node. Again, the reader may convince himself that this construction works.

**Example 3.11:** Figure 3.15(a) is an example of the McNaughton-Yamada construction applied to the expression $a + bc^*$, which is parsed $a + (b(c^*))$, following the usual convention that * takes precedence over concatenation, which takes precedence over +. The upper row consists of the recognizer for the expression $a$. The lower row is the concatenation of the graph for $b$ and the graph for $c^*$; the latter is constructed from a recognizer node for $c$ using the construction of Fig. 3.16(c). These rows are connected by the initial fork node and the final or-node, in accordance with the construction of Fig. 3.16(a). □

## A Separator Theorem for Regular Expression Graphs

We shall now show that the family of graphs constructed by Fig. 3.16 is 2-separable, so we can apply the results of the previous section. We actually find it easier to show the result for a somewhat larger family, the graphs formed

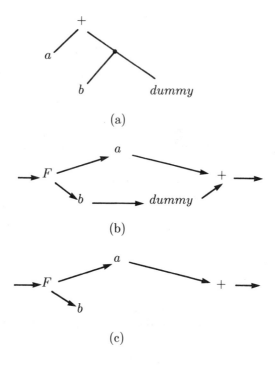

Fig. 3.17. Graph representing a regular expression with dummy operand.

from single nodes using the constructions of Fig. 3.16, but with the option of leaving out one or both of the subgraphs labeled "for $E_1$" or "for $E_2$" in that figure; the edges connecting a missing subgraph are also missing, of course. We call this collection of graphs the *extended M-Y family*.

The key to finding the right way to cut the graphs is to look not at the graphs themselves, but at the parse trees for the expressions that these graphs recognize. To account for the possibility of missing subgraphs in our extended family of graphs, we generalize the notion of regular expressions to include *dummy* operands, which are placeholders for the missing subgraphs.

**Example 3.12:** In Fig. 3.17(a) we see an expression with a dummy operand. To build the corresponding graph, we can treat *dummy* as if it were an ordinary operand, as shown in Fig. 3.17(b). Then, we delete *dummy* and the incoming and outgoing edges to get the correct graph for the expression of Fig. 3.17(a). This graph is shown in Fig. 3.17(c). □

In order that the parse tree's nodes each reflect the number of nodes of the regular expression graph that it represents, we define the *weight* of a tree node as follows.

1. A leaf has weight 1 if it is a real operand, and weight 0 if it is a dummy.

2.  An interior node labeled + has weight equal to 2 plus the sum of the weights of its children.

3.  An interior node corresponding to a concatenation has weight equal to the sum of the weights of its children.

4.  An interior node labeled * has weight equal to 4 plus the weight of its child.

It is straightforward to observe that the weight of any node is the number of nodes in the graph that would be generated for the tree of which it is the root, using the McNaughton-Yamada construction.

For convenience in what follows, we shall interpret the requirement for a separator that each piece contain "between 1/3 and 2/3 of the nodes" to allow rounding away fom 1/2. That is, we shall partition a graph of $n$ nodes into pieces with at least $\lfloor n/3 \rfloor$ and at most $\lceil 2n/3 \rceil$ nodes. It is easy to check that extending the definition of "separator" in this way only affects constant factors in the results of Section 3.3, and therefore does not negate the validity of any of those results. In fact, the constant 1/3 could have been any $\epsilon > 0$ without affecting the truth of the results in the previous section.

To find a pair of edges of the regular expression graph whose removal disconnects the graph into two parts, each with at least 1/3 of the nodes, we shall examine the weights of nodes of the parse tree, in a process almost, but not quite, identical to that of Example 3.2, where we partitioned binary trees themselves. By picking a node $n$ of the tree and disconnecting from the whole graph the subgraph corresponding to $n$, we guarantee that the two resulting subgraphs are in the extended M-Y family.

Suppose the root of some parse tree has weight $w$. We travel down the tree looking for a node of weight between $\lfloor w/3 \rfloor$ and $\lceil 2w/3 \rceil$. If we are at some node of weight greater than $\lceil 2w/3 \rceil$, and we find that one of its children also has weight greater than $\lceil 2w/3 \rceil$, we move to that child. If one of its children has weight between $\lfloor w/3 \rfloor$ and $\lceil 2w/3 \rceil$, then we take that tree node to be $n$.

We need to verify that it is not possible that a node has weight greater than $\lceil 2w/3 \rceil$, while each of its children has weight less than $\lfloor w/3 \rfloor$. If the node in question is a concatenation node, that is not possible because the weight of the node is the sum of the weights of its children. If the node is a +-node, we are also safe, since $2 + 2(\lfloor w/3 \rfloor - 1) \leq \lceil 2w/3 \rceil$, and thus a +-node of weight greater than $\lceil 2w/3 \rceil$ could not have two children both of whose weight was below $\lfloor w/3 \rfloor$.

A problem arises only if the node in question is a *-node. Such a node could have weight greater than $\lceil 2w/3 \rceil$, while its lone child has weight less than $\lfloor w/3 \rfloor$, provided there is an integer $i$ such that $i > \lceil 2w/3 \rceil$ and $i - 4 < \lfloor w/3 \rfloor$. We immediately observe that if there is such an $i$, then $2w/3 < w/3 - 4$, so $w \leq 12$. However, a moment's reflection tells us that, in fact, no such $i$ exists unless $w = 6$, when $i = 5$ satisfies the conditions. We leave it as an exercise for the reader to consider how to partition this small number of graphs; one

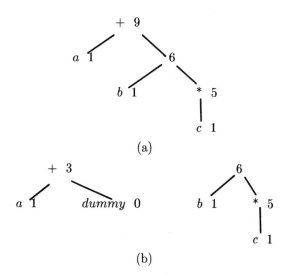

(a)

(b)

**Fig. 3.18.** Weighted parse trees.

case is covered in Example 3.13.†

We conclude that, except for a few small cases that are handled specially, given any graph in the M-Y family, we can examine its parse tree and find a node whose weight is between $1/3$ and $2/3$ of the total, and that therefore corresponds to a subgraph with between $1/3$ and $2/3$ of the nodes of the whole graph. Moreover, this subgraph is one of the boxes at some stage of the McNaughton-Yamada construction of Fig. 3.16.

An examination of Fig. 3.16 tells us that every such box can be disconnected from the rest of the graph by deleting only two edges. Of the resulting two subgraphs, the one "inside" the box surely belongs to the extended M-Y family. But the subgraph that is the original graph with the subgraph removed also belongs to the family, which was explicitly defined to allow the excision of subgraphs that correspond to boxes in Fig. 3.16. Thus, both subgraphs are within the family, and we may appeal to an inductive argument to say that each member of the family is 2-separable.

**Example 3.13:** In Fig. 3.18(a) we see the parse tree for $a + bc^*$, with weights on the nodes. The right child of the root has weight 6, which is $2/3$ of the total weight, so we may partition the graph by excising the subgraph that corresponds to the right child of the root. The whole graph was exhibited in

---

† Observe that for these 6-node graphs, we can partition in ways that take us outside the extended M-Y family. However, if we do so, we must continue the partition process until the pieces are single nodes, or we reach graphs that are in the family. We cannot appeal to the inductive argument that says all graphs in a family are separable if we can partition the graphs in the family into two pieces, both of which are in the family.

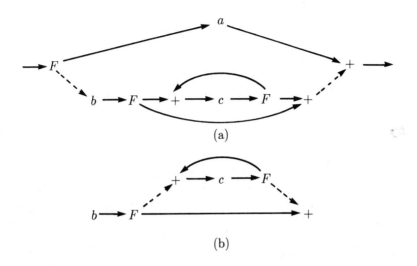

**Fig. 3.19.** First steps of graph partitioning.

Fig. 3.15(a), and the subgraph chosen is the six nodes on the bottom row of that figure. Figure 3.18(b) shows the two weighted parse trees that result from the split, including the dummy node of weight 0 that replaces the portion of the parse tree that has been removed. Figure 3.19(a) shows the graph itself split. Notice that only two edges were removed; they are shown by dashed lines.

The graph of three nodes is easily partitioned into two and one, and the two-node subgraph is partitioned into a pair of nodes. The six-node subgraph in the bottom row of Fig. 3.19(a) is one of the special cases where no descendant has between 2 and 4 nodes, that is, between 1/3 and 2/3 of the total. However, we can easily partition it into two subgraphs of three nodes each, by removing the two dashed edges in Fig. 3.19(b). Then these three node graphs are each partitioned in two obvious steps. $\square$

We can summarize what we have shown in this section, and its consequences according to Lemma 3.4 and Theorem 3.5 as follows.

**Theorem 3.6:**
a)    The extended M-Y graphs are 2-separable.
b)    The extended M-Y graphs are strongly $O(\log n)$-separable.
c)    Every regular expression of length $n$ ($n$ symbols including operators and operands) has a regular expression graph with $O(n)$ nodes, and this graph can be laid out in $O(n)$ area.

**Proof:** Part (a) is the consequence of the discussion of this section. Part (b) follows from Lemma 3.4, when we observe that $\Gamma_2(n) = O(\log n)$. For part (c), we note that no symbol of a regular expression introduces more than four nodes into the graph; the worst case is the closure construction of Fig. 3.16(c). Thus,

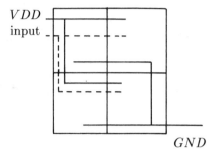

**Fig. 3.20.** Supplying utilities to the regular expression graph.

the number of nodes in the graph is $O(n)$, and (c) follows from Theorem 3.5 and part (b). □

**Some Practical Considerations**

While the graph of a regular expression can be laid out in $O(n)$ area, the edges of the graph represent only the signals that enable the various nodes. We must also supply power, ground, clock signals, and the input wires themselves to the nodes. Figure 3.20 illustrates one strategy to use. We assume that the graph has been laid out recursively, with large squares made up of four smaller ones as in Fig. 3.14. Suppose that each square at any level of the hierarchy has power entering at the upper left and ground leaving at the lower right. Then the arrangement of Fig. 3.20 allows power and ground to run in the same (metal) layer without crossing, and these utilities enter the smaller squares at the proper corners.

Figure 3.20 also shows a wire labeled "input," running in a different layer from power and ground. This wire could represent an input or clock signal. If we run the wires corresponding to the edges of the graph in poly, then we shall not have any problems when they cross power or ground. If input or clock signals cross the graph edges or each other, we can jump one over the other by changing layers from poly to metal.

As a final point, this overall strategy was actually used to compile regular expressions directly into circuits (Trickey and Ullman [1982]). It was found that in order to save space, it was not desirable to turn the regular expression into a graph whose nodes were recognizers for single operands of the expression. Rather, a regular structure, like a PLA† should be used to recognize subexpressions of 10–50 operands. The regular expression graph thus has nodes representing large subexpressions, and there are consequently many fewer nodes. Thus, although the PLA for, say, 50 operands will be much larger than a circuit that performs like a recognizer node, it will not be 50 times as large, and we

† See Section 7.7 for a discussion of PLA's.

shall achieve a considerable (although constant factor) savings in area.

## 3.5 LAYOUT OF PLANAR GRAPHS

In this section, we shall consider the family of planar graphs. First, we give the Lipton-Tarjan planar separator theorem, which in our terms shows that the planar graphs of degree 4 or less all have $O(\sqrt{n})$ separators. A consequence of this theorem is that all planar graphs of degree 4 with $n$ nodes can be laid out in $O(n \log^2 n)$ area, but it is not known whether so much area is ever needed.

To help resolve the question of how much area is really needed for planar graphs, we show how to relate the *crossing number* of a graph, that is, the minimum number of edge crossings needed to draw the graph in the plane (so the planar graphs are those graphs that have crossing number 0), to the area needed to lay out the graph. Then we discuss the *mesh of trees*, an interesting family of graphs that is $O(\sqrt{n})$-separable and yet can be shown to really require the most area that Theorem 3.5 implies might be needed: $\Omega(n \log^2 n)$. Further, we shall see that this family of graphs has a small crossing number, and therefore leads to a family of planar graphs that require $\Omega(n \log n)$ area. Thus, the best area result for the layout of planar graphs is known to lie somewhere in the range $n \log n$ to $n \log^2 n$.

### Concentric Presentation of Planar Graphs

An important preliminary to the construction of separators for planar graphs is a method for forming what we call the *concentric presentation* of a planar graph. We begin with any node as the origin, and group the nodes according to their graph-theoretic distance† from the origin. To form the concentric presentation of the graph, we add certain nodes, which we refer to as *dummies*, and which we do not count when we discuss the partition of a graph so that at most 2/3 of the nodes are in either group. That is, the term "nodes" will refer only to the original nodes of the graph, whenever there is a distinction to be made.

We also add certain edges to the graph. These edges form *rings* that connect all the nodes at a given distance from the origin. The set of nodes at distance $d$ from the origin is called the $d^{th}$ ring. It should be emphasized at the outset that the purpose of adding edges around the rings is only to make it clear that no edges of the graph cross certain boundaries. We are then able to partition the graph into two relatively equal pieces by finding a path in the graph that separates the pieces. The purpose of the dummy nodes is to enable us to introduce the necessary edges, while keeping the graph planar. There are several other properties that we require for a concentric presentation; these are:

1.   The concentric presentation is planar.

---

† The *distance* between nodes in the graph-theoretic sense is the number of edges on the shortest path between the nodes.

2.   Every node and every dummy on ring $r > 0$ has an edge connecting it to some node or dummy on ring $r - 1$. As a consequence, every node on ring $r$ has a path of length $r$ to the origin.

3.   All edges run between nodes and/or dummies whose rings differ by at most one. It is easy to show that this property holds for the nodes originally, since if $v$ is at distance $d$, and there is an edge $(v, u)$, then the shortest distance from $u$ to the origin cannot exceed $d + 1$. We shall see that the same property continues to hold when we introduce dummies.

4.   All edges between nodes and/or dummies on the same ring $r$ are outside the ring.

**Example 3.14:** Figure 3.21(a) shows a planar graph, and Fig. 3.21(b) shows a concentric presentation for this graph. The edges around the ring are shown by dashed lines. Edge $(b, c)$ was already present in the graph, but the other ring edges were added in the construction. There are six dummies introduced where we want the edges of the rings to cross edges of the original graph. Notice how the introduction of dummies allows us to keep the graph planar. $\square$

**Lemma 3.5:** For any planar graph, with any origin, we can add dummies and ring edges to form a concentric presentation of the graph.

**Proof:** The proof is a construction from the origin out. That is, we show by induction on the distance $d$ that we can put the first $d$ rings in the proper condition, where all edges that connect nodes or dummies on the ring are outside the ring, and each node and dummy on the ring has an edge to a node or dummy on the previous ring.

The basis, the ring for distance 1, is shown in Fig. 3.22(a). There, we show four nodes adjacent to the origin, although there could be any number. Each neighbor of the origin has edges leaving in various directions. However, there is surely room to connect the nodes of ring 1 inside any other edges leaving those nodes; the introduced edges are the ring edges, which are shown by dashed lines. If there already was an edge between two consecutive nodes, remove it, because it is important for the concentric presentation that the edge connecting these nodes be inside any other edges leaving either node. As a final touch, we may, if we wish, distort the plane so the dashed edges actually form a convex polygon.

For the induction, suppose that we have arranged for the nodes and dummies on ring $r$ to satisfy the conditions of a concentric presentation, as suggested by Fig. 3.22(b). Note that of the edges emanating from any node or dummy, the edges inside the ring go to nodes on ring $r - 1$, never to ring $r$. Edges outside the ring may go to nodes or dummies on rings $r$ or $r + 1$, but never to ring $r - 1$.

We order the nodes on ring $r + 1$ around the ring in the following way. Begin at any node on ring $r$, say node $v$. This node will have zero or more

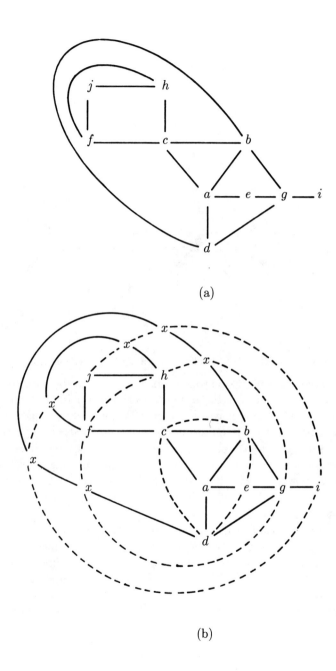

(a)

(b)

**Fig. 3.21.** Concentric presentation of a graph.

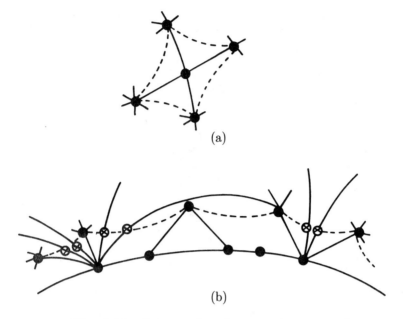

**Fig. 3.22.** Construction of concentric presentation.

edges emanating from it outside ring $r$. Some of these go to nodes on ring $r + 1$; the rest go to ring $r$. Order those nodes on ring $r + 1$ in the order they are connected to $v$. The plane may be distorted so the nodes in ring $r + 1$ form a circle, but we do not cause edges to cross as we slide nodes around the plane.

As for the basis, we connect the nodes on ring $r + 1$ by introduced edges that run inside any edges emanating from the nodes on ring $r + 1$. Also as in the basis, we remove edges of the original graph that are duplicated by ring edges. If these ring edges cut other edges emanating from $v$, we replace the points of intersection by dummy nodes.

We continue around ring $r$, adding new nodes to ring $r + 1$ as we find them adjacent to nodes on ring $r$. An example of the process is shown in Fig. 3.22(b). Notice how any edge from ring $r$ to ring $r$ that passes outside any node on ring $r + 1$ will be cut twice by that ring, introducing dummies at each crossing. Also, edges from ring $r$ to ring $r + 1$ that go around nodes on ring $r + 1$ are cut once. However, no other dummies are introduced. Since the dummies that are introduced on ring $r + 1$ clearly have edges to nodes on ring $r$, and the position of the introduced ring edges makes all other edges emanating from ring $r + 1$ to rings $r + 1$ and $r + 2$ lie outside ring $r + 1$, the condition for a concentric presentation is satisfied for ring $r + 1$. Thus, we may conclude the induction and the lemma. $\square$

## A Separator Theorem for Planar Graphs

We shall now prove a weak version of the Lipton-Tarjan planar separator theorem. The original version has a constant $2\sqrt{2}$ associated with it, whereas we use constant 4, and the original version is more general in allowing weights on the nodes, while we simply count the number of nodes in the partitions.

**Lemma 3.6:** Let $G$ be a planar graph of $n$ nodes. Then we can find a set $S$ of no more than $4\sqrt{n}$ nodes whose removal disconnects the graph into two parts, neither of which has more than $2n/3$ nodes.

**Proof:** We assume $G$ is connected; if not we add edges to connect $G$, keeping the result planar. Next, we pick an origin and perform the construction of Lemma 3.5 to put the graph in concentric form. In so doing, we introduce dummy nodes, but recall that the term "node" is reserved for the nodes of the original graph, and the number of "nodes," $n$, has not changed. In what follows, we attempt to find one or, in some cases, two closed paths that disconnect the graph into subgraphs with at most $2n/3$ of the original nodes.

Suppose there is some distance $d$ such that there are no more than $\sqrt{n}$ nodes at distance $d$. Then we shall call $d$ a *tight band*. Note that $d = 0$ is always a tight band. Also, the smallest distance at which there is no node is also a tight band. Thus there are at least two tight bands, and every node but the center is between tight bands.

If we can find two tight bands $d_\ell$ and $d_h$ such that at most $2n/3$ nodes are at distances between $d_\ell$ and $d_h$, and at most $2n/3$ nodes are at distances less than $d_\ell$ or greater than $d_h$, then we can select as our set $S$ the nodes at distances $d_\ell$ and $d_h$, which together total no more than $2\sqrt{n}$ nodes, and the lemma is proved.

Now consider the opposite case, where no such tight bands exist. Since all nodes are on or between tight bands, it must be that there is, at "middle distances," a group of consecutive distances $d_\ell + 1, \ldots, d_h - 1$, none of which are tight bands, and which include among them more than $2n/3$ nodes. We may assume that $d_\ell$ and $d_h$ are tight bands. It follows that $d_h - d_\ell$ cannot be greater than $\sqrt{n}$, because there are at most $n$ nodes between them, and each distance from $d_\ell + 1$ through $d_h - 1$ has at least $\sqrt{n} + 1$ nodes.

We shall attempt to find a set $S$ that divides the middle distances along two radii, as shown in Fig. 3.23. The two halves of the middle distances must each contain no more than $2n/3$ nodes. The radii each have no more than $\sqrt{n}$ nodes, since $d_h - d_\ell \le \sqrt{n}$. Then we separate the middle distances from the outer and inner distances, using the two surrounding tight bands, $d_\ell$ and $d_h$, for a total of no more than $4\sqrt{n}$ nodes. The inner and outer distances are grouped with the smaller of the two halves of the middle distances.

To find the desired radii, we begin by removing all nodes on rings other than those between distances $d_\ell + 1$ and $d_h - 1$. Since the remaining graph, like

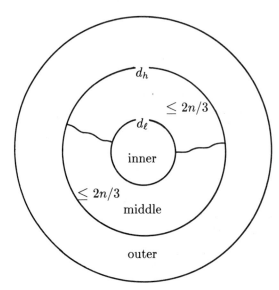

**Fig. 3.23.** Diagram for the proof of Lemma 3.6.

the original, is planar, and the number of nodes and dummies is finite, there must be some node or dummy that is exposed to the outside, in the sense that we can draw an edge from it to a point at infinite distance that does not cross any edge of the graph. A little thought is necessary to see this point; consider an edge running outside the outermost ring from node $v$ clockwise to $u$, say, where the clockwise distance from $v$ to $u$ around the ring is maximal. If $v$ (and $u$) were not exposed to the outside, there would be an edge between some $w$ and $x$ on the outer ring, where $w$ is clockwise of $v$, and $v$ is clockwise of $x$. If $w$ is counterclockwise of $u$, planarity is violated, because the edges $(w, x)$ and $(v, u)$ intersect. If $w$ is clockwise of $u$, or is $u$, we contradict the maximality of $(v, u)$.

We now cut the rings along the path from the selected node $v$ to the innermost ring; we know such a path exists because each node or dummy is connected to a node at the next lower ring by Lemma 3.5. The general situation is shown in Fig. 3.24. The rings, up to $\sqrt{n}$ of them, are shown as straight lines. No edges connect to any nodes other than the nodes at the sides. Initially, these two sides are the same nodes, the path from $v$ just found. However, in general, as we divide the structure, looking for a substructure of the same form having no more than $2n/3$ nodes and at least $n/3$, the two sides will consist of different nodes. It will also be the case, as we divide the structure, that while any ring may have edges connecting two nodes on that ring, running above the ring, as illustrated for the top ring in Fig. 3.24, no such edges run below the ring.

Now, assuming that the structure of Fig. 3.24 has more than $2n/3$ nodes, we must find a way to subdivide it into two pieces, each with the properties

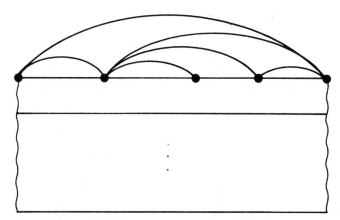

**Fig. 3.24.** General structure to be subdivided.

ascribed to that structure. Eventually, we shall find one substructure that has a number of nodes in the desired range. The first case we consider is shown in Fig. 3.25(a), where there is a node on the top ring, other than at the ends, that is either exposed upwards, or, as shown in the figure, on the same facet† as the edge connecting the nodes at the ends of the top ring. Then, we find a path, shown by wavy lines in Fig. 3.25(a), from the node in question down to the bottom ring.

We claim that either side satisfies the conditions on the structure of Fig. 3.24. The existence of the wavy path assures us that no edges running between the rings connect nodes on either side of that path. We are guaranteed that no such edge runs below the bottom ring. Finally, the selection of the node on the top ring guarantees us that no edge connects either side and runs above the top ring (except possibly the edge shown in Fig. 3.25 running between the ends of the top ring), or else the node in question could neither share a face with that edge nor be exposed upwards.

If there is no node like that shown in Fig. 3.25(a), then there can be only two nodes on the top ring, and these may even be the same node. We thus look to the next-to-top ring. If, as shown in Fig. 3.25(b), there is an edge between a node on the top ring and a node not at an end of the next ring, we can partition the structure along the wavy line shown, and again be guaranteed that no edges connect the two pieces. The proof is similar to that concerning Fig. 3.25(a), and we leave it as an exercise.

If Fig. 3.25(b) does not apply, then we must have the situation shown in Fig. 3.25(c), where there are one or two nodes on the top ring, and these do not connect with any nodes on the second ring, except the nodes at the ends. But then we can simply remove the top ring, sure that the structure of the remaining rings has no edge touching it from outside, except for the nodes at

---

† A *face* is a region of the plane not cut by any edges.

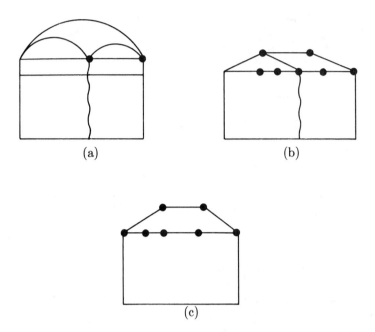

**Fig. 3.25.** Cases for the splitting of the structure of Fig. 3.24.

the ends of rings. The argument then proceeds with the smaller structure, which must be of one of the forms of Fig. 3.25(a–c).

By repeating the division process, we eventually get down to a structure that has at most $2n/3$ of the nodes, excluding dummies and the nodes at the ends of the rings, but has at least $n/3$ nodes, including the ends of the rings. If we delete the two tight bands, and we delete the radii represented by the ends of the structure (of the form of Fig. 3.24) finally arrived at, we split the graph into pieces: an inside, an outside, and two pieces of the donut at middle distances, as was suggested by Fig. 3.23. The tight bands and the radii are each no longer than $\sqrt{n}$, so the total number of nodes removed is at most $4\sqrt{n}$. Since none of these pieces are larger than $2n/3$, we may combine them into two groups of size no larger than $2n/3$.

One last detail remains. The graph so separated is not the original graph, but the graph with the middle distances replaced by a concentric presentation. The set of $4\sqrt{n}$ nodes selected for deletion may include some dummies. If that is the case, we replace each dummy by one of the real nodes at the end of the nonring edge whose crossing of a ring edge formed that dummy. Since only that edge crossed the selected radius at the point of the dummy, we continue to separate the graph into pieces; the number of deleted nodes does not increase. We have now selected only real nodes for deletion. □

**Theorem 3.7:** Every planar graph of degree 4 is $O(\sqrt{n})$-separable, and therefore can be laid out in $O(n \log^2 n)$ area.

**Proof:** By Lemma 3.6, we can find, for any degree 4 planar graph of $n$ nodes, a set $S$ of at most $4\sqrt{n}$ nodes whose removal disconnects the graph into two pieces, each with at most $2n/3$ nodes. Moreover, each of these pieces is planar of degree 4.† We pick as our separating set of edges the up to $16\sqrt{n}$ edges that are incident upon the nodes in $S$. The nodes in $S$ are included with the smaller set of nodes into which the removal of $S$ disconnects the graph, so neither set of nodes contains more than $2n/3$ nodes.

Since the pieces are in the family of planar graphs of degree 4, we have proved that all such graphs are $16\sqrt{n}$-separable. It is easy to check that $\Delta_{16\sqrt{}}(n) = O(\sqrt{n} \log n)$, so by Theorem 3.5, all graphs in the family can be laid out in $O(n \log^2 n)$ area. $\square$

### Crossing Numbers and Layout Area

Recall that the crossing number of a graph is the minimum number of pairs of edges that must cross when we draw the graph in the plane. For a family of graphs with relatively small crossing numbers, there is an alternative way to construct good layouts, besides the method of finding small separators. We take advantage of the fact that graphs with small crossing numbers can be made to look like planar graphs with not too many more nodes, and we then apply Theorem 3.7. The idea is expressed in the next theorem.

**Theorem 3.8:** Suppose all planar graphs of degree 4 with $n$ nodes can be laid out in area $A(n)$. Then every graph $G$ with degree 4, $n$ nodes, and crossing number $c$ can be laid out in area $O(A(n+c))$. In particular, $G$ requires no more than $O((n + c) \log^2(n + c))$ area.

**Proof:** Draw $G$ in the plane with $c$ edge crossings. At the point where each pair of edges cross, introduce a new node. The newly introduced nodes have degree 4, so the resulting graph $G'$ has degree 4, $n + c$ nodes, and is planar. Thus, $G'$ can be laid out in area $A(n+c)$, which is at most $O((n+c) \log^2(n+c))$ by Theorem 3.7.

To construct a layout for $G$ from the layout for $G'$, expand the scale of the latter layout by a factor of three, so each grid point becomes a $3 \times 3$ square. Consider some node of $G'$ that represents a crossing in $G$. Within the $3 \times 3$ square we can, if necessary, change layers of one wire to cross over the other wire. $\square$

---

† The reader should remember that the edges added in the concentric presentation are not really part of the graph or the pieces of the graph, so the property of having degree 4 is preserved for the pieces of the graph.

### The Mesh of Trees Family of Graphs

We shall now see an application of Theorem 3.8, but before we can, we must introduce an important family of graphs. The *mesh of trees of side* $m$, where $m$ is a power of 2, is defined as follows.

1.  Begin with $m^2$ nodes arranged in a square grid, but without any connecting edges.
2.  For each row of nodes, say $v_1, \ldots, v_m$, in order fom the left, build a complete binary tree with these nodes as leaves, adding the interior nodes, which are not among the original $m^2$ nodes.
3.  Do the same for each column of the grid.

There are thus $2m$ trees, each of which has $m - 1$ interior nodes, so the total number of nodes is $3m^2 - 2m$. Figure 3.26(a) shows the mesh of trees of side 2, and Fig. 3.26(b) shows the mesh of trees of side 4. In each graph, tree edges for the rows are solid and tree edges for the columns are dashed.

There are several interesting facts about the mesh of trees, whose proofs we shall sketch. First, the family of meshes of trees is $O(\sqrt{n})$-separable, where $n = 3m^2 - 2m$ is the number of nodes in the mesh. Second, layout of the mesh of trees requires $\Omega(n \log^2 n)$ area, and third, the crossing number of the mesh of trees is $O(n \log n)$. Let us consider the separability of this family first.

### Separators for Meshes of Trees

We can divide the mesh of trees of side $m$ evenly in half if we remove the roots of the row trees. Technically, we have defined separators to be sets of edges removed, so we shall find it convenient to remove more edges than we really have to. Specifically, let us remove the edges connecting the roots of the row trees to each of their children, a total of $2m = O(\sqrt{n})$ edges. The result is two rectangles of aspect ratio 2, and $2m$ isolated nodes.

We can divide the isolated nodes arbitrarily between the two rectangles, to get two identical subgraphs. Then, we divide the subgraphs by deleting the edges connecting the roots of the column trees to their children. The number of edges deleted in each subgraph is $m$, which is less than the square root of the number of nodes in each subgraph, which is $3m^2/2 - m$. Again the isolated roots are divided evenly between the subgraphs.

We have now four subgraphs, each of which is a mesh of trees of side $m/2$, together with $m$ isolated nodes. We can establish the fact that the meshes of trees are $O(\sqrt{n})$-separable most easily if we consider a slightly larger class, namely, the meshes of trees with additional isolated nodes, plus the rectangles of aspect ratio 2 that we get by removing the roots of the row or column trees from a mesh of trees, also with additional isolated nodes permitted. We have described the method for dividing each of these graphs exactly in half by deleting no more edges than the square root of the number of nodes. Thus, it

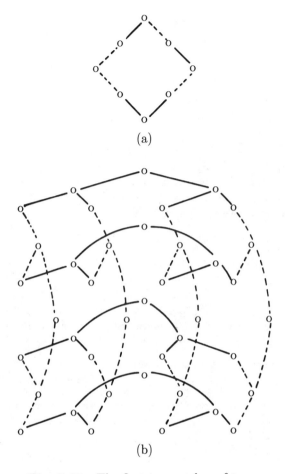

(a)

(b)

**Fig. 3.26.** The first two meshes of trees.

is easy to prove the following lemma by induction on the number of nodes.

**Lemma 3.7:** The meshes of trees are $O(\sqrt{n})$-separable. ☐

### Area Required by the Mesh of Trees

We shall now prove that the lower bound on area required for the mesh of trees matches the upper bound of $O(n \log^2 n)$ that we get from Lemma 3.7 and Theorem 3.5. Intuitively, the reason the meshes of trees need so much area is that the most efficient way to lay out the column trees is to lay them side by side, so the layout must be $\Omega(n \log n)$ wide. Similarly, the row trees force the layout to be $\Omega(n \log n)$ high. However, this observation is not a proof, and the proof is very difficult, since, for example, we must rule out the possibility that both row and column trees could be laid out in the same direction, giving a layout that was $O(n \log n)$ in one direction, but $O(n)$ in the other.

In outline, the proof is an induction on $m$, the side of the mesh. We pick a smaller side, $m2^{-i}$ for some $i$, $1 \leq i < \log_2 m$, and note that the mesh of trees of side $m$ is composed of $(2^i)^2 = 4^i$ of the smaller meshes, plus wires representing edges of the trees at higher levels.

Let us say an edge is at *depth* $i$ if it connects tree nodes at depths $i$ and $i-1$. Then the total area used for the wire in the layout for a mesh of trees of side $m$ must at least equal $4^i$ times the minimum amount needed for a mesh of trees of side $m2^{-i}$ plus the minimum amount needed for the edges of level $i$.†  If we let $A(n)$ be the minimum area needed for a mesh of trees of side $n$, and let $f(m, i)$ be some additional area that we can claim must be included in $A(m)$ because of edges at depth $i$, then we have

$$A(m) \geq 4^i A(m2^{-i}) + f(m, i)$$

Suppose we have proven that for all $p < m$, $A(p) \geq cp^2 \log^2 p$, where $c$ is a constant greater than 0. Let us, for convenience in what follows, also assume $c \leq 1/16$. Obviously if the statement is true for a larger $c$, it is true for $c = 1/16$, and since we are looking only for order-of-magnitude results, no harm is done by assuming $0 < c \leq 1/16$. But if the above-mentioned relation holds for $p = m2^{-i}$, then we must show

$$A(m) \geq cm^2((\log m) - i)^2 + f(m, i)$$

If we are to prove by induction that $A(m) \geq cm^2 \log^2 m$, then it suffices to prove $f(m, i) \geq 2cim^2 \log m$. We therefore prove in a series of lemmas that no matter what layout is used for the mesh of trees of side $m$, we can find some $i < m$ for which we can demonstrate that at least $2cim^2 \log m$ wire is used for the edges of depth $i$.

**Lemma 3.8:** Suppose that in some layout of the mesh of trees of side $m$ there are two sets of $k$ leaves (nodes in the grid on which the mesh is built) such that any leaf in one set is at least distance $d$ from any leaf in the other set. Let $L_i$ be the total length of all the edges of depth $i$ in this layout. Then there is some $i$ for which

$$L_i \geq \frac{k^2 d 4^i}{\beta m^3 i^2}$$

where $\beta$ stands for the limit of the sum

$$\frac{1}{1} + \frac{1}{4} + \frac{1}{9} + \cdots + \frac{1}{i^2} + \cdots$$

---

† In fact, we could consider the edges at all levels above $i$, but that seems not to help, and only complicates the proof.

which is easily seen to be finite.

**Proof:** Let us imagine that we draw a path from every leaf $p$ in the first set to every leaf $q$ in the second set, along a route of the following form. Assume without loss of generality that the column of $p$ is no higher in number (further right) than the column of $q$, or else interchange $p$ and $q$.

1.  Travel from $p$ to the leaf $r$ that is in the same row as $p$ and the same column as $q$. Do so by traveling up the row tree only as far as necessary, and then down the row tree to $r$.

2.  Travel from $r$ to $q$ in the same manner, but using the column tree shared by $r$ and $q$.

There are evidently $k^2$ such paths, and their total length is at least $k^2 d$. But to convert a lower bound on the lengths of these paths into a lower bound on the lengths of edges, we must consider how many paths may traverse any one edge. The answer depends on how high up its tree the edge is.

Consider an edge at depth $i$ that is in a row tree and connects some node to its left child. Then there are $m2^{-i}$ nodes in the row that could be $p$ and still require that the path traverse this edge; those nodes are the leaves descending from the left child. There are similarly $m2^{-i}$ nodes in the row that could be $r$, the descendant leaves of the right child. Each of these $r$'s is needed for paths to any of the $m$ leaves in its column. We conclude that this edge can be on $m2^{-i} \times m2^{-i} \times m = m^3 4^{-i}$ paths, but no more. It is easy to show that the same bound applies if the edge is at depth $i$ in a row tree, but connected to a right child, or if the edge is in a column tree.

Let us now introduce some notation. Let $\ell(e)$ be the length of edge $e$ in the layout in question, and let $P(e)$ be the number of paths, among the $k^2$ paths just defined, on which edge $e$ lies. We have just shown that $P(e) \leq m^3 4^{-i}$, if $e$ is at depth $i$.

Now, since each path from some leaf $p$ in one set to some leaf $q$ in the other set has length at least $d$, we have

$$k^2 d \leq \sum_{p,q} \sum_{\substack{e \text{ on path} \\ \text{from } p \text{ to } q}} \ell(e)$$

Interchanging the order of summations, we have

$$k^2 d \leq \sum_{e} \ell(e) P(e)$$

Then, we may organize the sum by depth of the edges. The sum of $\ell(e)$ over all edges at depth $i$ is $L_i$ by definition, and for all those edges, $P(e) \leq m^3 4^{-i}$. Thus

$$k^2 d \leq \sum_{i=1}^{\log m} L_i m^3 4^{-i} \tag{3.3}$$

Now, an obvious next step would be to say that there must be at least one $i$ for which $L_i 4^{-i}$ is at least average, that is, at least $k^2 d / m^3 \log m$. However, that turns out not to be good enough. The key step in the proof (which is due to F. T. Leighton) is to introduce some bias to the terms in the above sum, and claim that there must be some $i$ for which

$$L_i m^3 4^{-i} \geq \frac{k^2 d}{\beta i^2} \tag{3.4}$$

For if not, summing over $i$ gives us

$$\sum_{i=1}^{\log m} \frac{k^2 d}{\beta i^2} < \sum_{i=1}^{\log m} L_i m^3 4^{-i}$$

which, when we realize that

$$\sum_{i=1}^{\log m} \frac{1}{i^2} < \beta$$

by definition of $\beta$, provides an immediate contradiction of (3.3). But (3.4), solved for $L_i$, is exactly the relation claimed in the lemma. $\square$

Now we shall see how to establish that large sets of leaves separated by a large $d$ exist. Regard the diagram in Fig. 3.27, where we take an arbitrary layout for the mesh of trees of side $m$, and slice it vertically by two boundaries with a single jog each, so that $m^2/8$ leaves are to the left of the first boundary, and the same number are to the right of the second. Similarly, we divide the layout vertically by two boundaries, with $m^2/8$ leaves above and below the top and bottom boundaries. Thus, at least half the $m^2$ leaves are in the center rectangle in Fig. 3.27. We assume without loss of generality that $d$, the width of the center rectangle, is at least as great as the height; if not rotate the picture ninety degrees.

The next lemma tells how to get a lower bound on $d$ in Fig. 3.27, if we are performing an induction on $m$ and we know that for $p < m$, $cp^2 \log^2 p$ is a lower bound on the area required for the mesh of trees of side $p$, for constant $c$, $0 < c \leq 1/16$. In particular, we let $p = m^{1/4}$, so it appears $m$ must not only be a power of 2, but a power of 16. However, in the other cases, we can adjust $p$ up or down to the closest power of 2. The reader may check that the lower bound we claim is sufficiently conservative that the lemma to follow holds for any $m$ that is a power of 2. For $p = m^{1/4}$, our lower bound on the area needed by the mesh of trees of side $p$ is $\frac{c}{16} \sqrt{m} \log^2 m$.

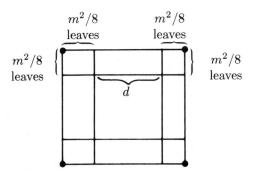

**Fig. 3.27.** Partition of layout for mesh of trees of side $m$.

**Lemma 3.9:** If the mesh of trees of side $m^{1/4}$ requires area $\frac{c}{16}\sqrt{m}\log^2 m$, for constant $c$, $0 < c \le 1/16$, then $d$ in Fig. 3.27 is lower bounded by $d \ge \frac{\sqrt{c}}{8}m\log m$.

**Proof:** We assume $m$ is a power of 16. The reader can easily modify the constants in the proof to verify the lemma for the cases in which $2m$, $4m$, or $8m$ is a power of 16. Divide the mesh of trees of side $m$ into $m^{3/2}$ meshes of trees of side $m^{1/4}$ by removing all edges of depth $\frac{3}{4}\log m$ or less. Since the center rectangle in Fig. 3.27 has half the $m^2$ leaves, at least half of the meshes of trees of side $m^{1/4}$ must have at least one leaf in the center, or there will not be enough leaves to fill the center. That is, at least $m^{3/2}/2$ of the small meshes must have a leaf in the center.

However, the perimeter of the center rectangle is no longer than $4d$, so no more than this number of the small meshes can have both a leaf inside and a leaf outside the center. Thus, at least $m^{3/2}/2 - 4d$ small meshes, with a total area of at least $(m^{3/2}/2 - 4d)(\frac{c}{16}\sqrt{m}\log^2 m)$, are wholly within the center. But the area of the center is no more than $d^2$, so

$$d^2 \ge \frac{c}{32}m^2\log^2 m - \frac{c}{4}d\sqrt{m}\log^2 m$$

If the lemma is false, i.e., $d < \frac{\sqrt{c}}{8}m\log m$, then

$$\frac{c}{64}m^2\log^2 m > \frac{c}{32}m^2\log^2 m - \frac{c^{3/2}}{32}m^{3/2}\log^3 m$$

or

$$\frac{1}{64} > \frac{1}{32} - \frac{\sqrt{c}\log m}{32\sqrt{m}} \tag{3.5}$$

But since we have assumed $c \le 1/16$, we have $\sqrt{c} \le 1/4$. Further, it is easy to check that $(\log m)/\sqrt{m}$ never gets as large as 2 (its largest value is $3/2\sqrt{2}$ when $m = 8$). Thus, (3.5) cannot be satisfied, and we have proved the lemma by contradiction. $\square$

We are now ready to prove the lower bound on area for meshes of trees.

**Theorem 3.9:** The mesh of trees of side $m$ requires area $\Omega(m^2 \log^2 m)$, or in terms of the number of nodes $n$ in the graph, $\Omega(n \log^2 n)$.

**Proof:** We proceed by induction on $m$ to show that for some constant $c$, $0 < c < 1/16$, $A(m) \geq cm^2 \log^2 m$. The basis, $m = 2$, is trivial.

For the induction, use Lemma 3.9 to claim that in the partition of an arbitrary layout for the mesh of trees as in Fig. 3.27, there are two sets of $k = m^2/8$ nodes separated by distance at least $d = \frac{\sqrt{c}}{8} m \log m$; these sets are the leaves in the left and right vertical strips. Then apply Lemma 3.8 with these values of $k$ and $d$, to establish that for this layout there is a value of $i$ for which

$$L_i \geq \frac{\sqrt{c}m^2 \log m 4^i}{512\beta i^2} \tag{3.6}$$

Since for no value of $i > 0$ is $4^i < i^3$, we may conclude from the above that

$$L_i \geq \frac{\sqrt{c}im^2 \log m}{512\beta}$$

Now we can complete the induction. We observed prior to Lemma 3.8 that $A(m)$ is at least as great as $4^i A(m2^{-i}) + L_i$, and by the inductive hypothesis, $A(m2^{-i}) \geq c4^{-i}m^2((\log m) - i)^2$. Thus

$$A(m) \geq cm^2 \log^2 m - 2cim^2 \log m + ci^2m^2 + L_i \tag{3.7}$$

The third term on the right side of (3.7) is positive, so we can ignore it. We must show that the sum of the second and fourth terms of (3.7) is nonnegative, that is, $L_i \geq 2cim^2 \log m$. But by (3.6) it suffices to claim that

$$\frac{\sqrt{c}im^2 \log m}{512\beta} \geq 2cim^2 \log m$$

which holds whenever $\frac{1}{512\beta} \geq 2\sqrt{c}$, i.e., $c \leq 1/2^{20}\beta^2$. Since we are free to choose this value of $c$ in the induction, we have proved the theorem. $\square$

### The Crossing Number for Meshes of Trees

We shall now show that the crossing number for the mesh of trees of side $m$ is $O(m^2 \log m)$. We then use this result to show that there are some planar graphs that require $\Omega(n \log n)$ area.

**Theorem 3.10:** The crossing number for the mesh of trees of side $m$ is $O(m^2 \log m)$, or in terms of the number of nodes $n$, $O(n \log n)$.

**Proof:** We show by induction on $m$ that we can lay out the mesh of trees in a rectangle, with $2m$ wires coming out of the top; the first $m$ wires are attached to the row roots, and the second $m$ lead to the column roots. Further, the number of crossings, $C(m)$, for this layout is at most $\frac{21}{8}m^2 \log m$. The basis, $m = 2$, is trivial.

For the induction, let us divide the mesh of trees of side $m$ into four meshes of side $m/2$, as indicated in Fig. 3.28(a). By the inductive hypothesis, we can lay out each of the meshes $A$, $B$, $C$, and $D$ with at most $\frac{21}{8}(\frac{m}{2})^2 \log(\frac{m}{2})$ crossings, and with the row and column roots accessible from the top. To complete the induction, we wire $A$, $B$, $C$, and $D$ together and add $2m$ roots, as shown in Fig. 3.28(b).

If we inspect Fig. 3.28(b), we see that there are seven places where $m/2$ lines cross $m/2$ lines (indicated by dashed squares), and another seven places where half that number of crossings occur (indicated by triangles). Thus the number of crossings in Fig. 3.28(b) is $4C(m/2) + \frac{21}{8}m^2$. If we assume by the inductive hypothesis that $C(m/2) \leq \frac{21}{8}(\frac{m}{2})^2 \log(\frac{m}{2})$, then to complete the induction we have to show only that

$$4\frac{21}{8}(\frac{m}{2})^2 \log(\frac{m}{2}) + \frac{21}{8}m^2 \geq \frac{21}{8}m^2 \log m$$

But the above follows by easy algebraic manipulation, and we have proved the theorem. $\square$

Now, suppose we could lay out any planar graph of $n$ nodes in $A(n)$ area, where $A$ is some monotonically nondecreasing function. By Theorems 3.8 and 3.10, we can deduce that a mesh of trees with $n$ nodes can be laid out in area $A(cn \log n)$, for some constant $c$. However, Theorem 3.9 then tells us that $A(cn \log n) = \Omega(n \log^2 n)$, whereupon we can show with some algebraic manipulation that $A(x) = \Omega(x \log x)$. That is, we have proved the following theorem.

**Theorem 3.11:** There is an infinite family of planar graphs that requires area $\Omega(n \log n)$, where $n$ is the number of nodes. $\square$

As a consequence of Theorems 3.7 and 3.11, the best result about the area needed to lay out $n$ node planar graphs must lie somewhere in the range $n \log n$ to $n \log^2 n$, but it is open exactly what the correct function is.

## EXERCISES

3.1: What happens in Example 3.1 if there is only one element being cycled?

** 3.2: What can you say about the minimum edge length when a complete binary tree is laid out with leaves on the border of a rectangle?

* 3.3: Suppose $S(n) = n^a \log^b n$. For what nonnegative values of $a$ and $b$ is $\Gamma_S(n) = O(S(n))$? What is $\Gamma_S(n)$ in the other cases?

3.4: Show that the family of rectangular grids is strongly $(\sqrt{n} + 1)$-separable.

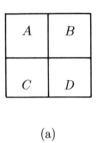

(a)

columns                                              rows

cols. rows        cols. rows        cols. rows        cols. rows

   A                 B                 C                 D

(b)

**Fig. 3.28.** Layout of mesh of trees with few crossings.

3.5: What is $\Delta_S(n)$ if $S(n)$ is
  a)   A constant?
  b)   Of the form mentioned in Exercise 3.3?

3.6: Construct a regular expression graph for the expression $a^*b^* + c^*$. How would this graph be partitioned by 2-separators?

3.7: A *hypercube of dimension* $d$ is a set of $n = 2^d$ nodes, connected in the form of a $d$-dimensional cube. That is, let the nodes be represented by binary numbers $a_1 a_2 \cdots a_d$, where each $a_i$ is 0 or 1. Then nodes $a_1 \cdots a_d$ and $b_1 \cdots b_d$ are connected by an edge if and only if there is exactly one value of $i$ for which $a_i \neq b_i$. Thus when $d = 2$ the graph is a square, and when $d = 3$ it is the ordinary cube.
  a)   Show that the family of hypercubes is strongly $n$-separable.
  b)   Technically, we cannot lay out hypercubes of dimension greater than $d = 4$, because the degree of nodes is $d$. However, we could replace nodes of the hypercube by binary trees of $d$ leaves, and for each neighbor $u$ of a node $v$, run an edge from some leaf in the tree for $u$ to some leaf in the tree for $v$. Show that this modified hypercube can be laid out in $O(n^2)$ area.
  ** c) Show that separators for the hypercube are of size $\Omega(n)$. *Hint:* The proof of Lemma 3.8 is instructive, in that it talks about a family of paths, some of which can traverse the same edge, yet there is a limit on how many of these paths can go through a single edge. Consider particular paths connecting pairs of nodes, one on each side of a partition, and show that together they determine $\Omega(n)$ edges that cross between the two sides.

3.8: An $n$-node *complete graph* has edges between every pair of nodes.
  ** a) Show that the complete graphs have crossing number $\Omega(n^4)$. *Hint:* Recall that the complete graph of five nodes is not planar, and therefore, any set of five nodes must have a pair of edges between them that cross. You must show that among the $\binom{n}{5} = O(n^5)$ sets of five nodes, there are $\Omega(n^4)$ different crossing points.
  b)   The complete graphs do not, of course, have bounded degree. However, we can use the trick of Exercise 3.7(b) to develop a family of graphs of degree three with essentially the same connectivity. Show that this family of graphs requires $\Omega(n^4)$ area for the layout of $n$-node graphs.
  c)   Show that the modified complete graphs from part (b) can be laid out in area $O(n^4)$.
  * d) Suppose that we instead lay out complete graphs directly, by replacing a node with a region of the grid having at least as many incident grid lines as the degree of the node. Edges not incident upon a node are not permitted to enter the region of the node. Show that under this model, $O(n^4)$ area is sufficient to lay out a complete graph of $n$ nodes

and $\Omega(n^4)$ is required.

&ast; e)   Prove that if under the model of part (d), edges are allowed to cross the region of a node without being incident upon that node, then $O(n^3)$ area suffices.

&ast;&ast; f)   Show that $\Omega(n^3)$ is necessary for the layout described in part (e).

&ast;&ast; 3.9: Show that if crossovers of edges are not permitted, then there are planar graphs of $n$ nodes that require $\Omega(n^2)$ area for their layout.

3.10: Complete the proof of Lemma 3.6 by considering the case of Fig. 3.25(b).

## BIBLIOGRAPHIC NOTES

The H-tree layout is discussed in Browning [1980]. Theorem 3.2 on the area needed for the layout of complete binary trees is from Brent and Kung [1980b]. Yao [1981] generalizes the result to noncomplete trees, and the general problem of laying out trees is also discussed in Bhatt and Leiserson [1982] and Fischer and Patterson [1980].

The discussion of the longest edges in tree layouts is from Patterson, Ruzzo, and Snyder [1981], as is the solution to Exercise 3.2. Further work on longest edge issues is found in Bhatt and Leiserson [1981] and Ruzzo and Snyder [1981].

The channel splitting algorithm for divide and conquer layout discussed in Section 3.3 is due independently to Valiant [1981], Leiserson [1980], and Floyd and Ullman [1982] (but really due to Floyd). The full generality of the result as expressed in this section is by Leiserson [1981], and Exercise 3.9 is from Valiant [1981]. Leighton [1982] gives a more general notion of separators, which he shows is both necessary and sufficient for the existence of layouts with the area given by Theorem 3.5. Storer [1980] discusses graph embedding from the point of view of the number of wire bends introduced.

The regular expression layout result of Theorem 3.6 is from Floyd and Ullman [1982], although Mukhopadhyay [1979] discussed a similar, but $O(n^2)$ layout algorithm. Foster and Kung [1981] discuss a different and promising way to lay out regular expression circuits.

The results of Section 3.5 are all due to Leighton [1981], with the exception of the planar separator theorem, which is from Lipton and Tarjan [1979]. See also Lipton and Tarjan [1980] for more on planar separators. Hong and Rosenberg [1981] and Dolev, Leighton, and Trickey [1983] investigate classes of planar graphs for which linear area layouts are known to exist.

# 4

# ALGORITHM DESIGN
# FOR VLSI

We now begin a study of the second area in which VLSI and computer science theory interact. VLSI technology makes possible the construction of "supercomputers," i.e., collections of many processors, each with capabilities similar to those of an ordinary computer. We may think of these processors as individual chips, or as several chips on a circuit board, with connections among the processors implemented by real wires running through a room full of such processors. Alternatively, we may envision a time in the not-too-distant future when many processors will be built on one chip, and interconnections among them implemented by wires on the chip.

Whether we take one or the other viewpoint or assume an intermediate view, where chips with several processors are interconnected to form networks larger than we can put on a single chip, we need to understand how best to use these supercomputers. Mechanically modifying algorithms that were designed for ordinary computers and expecting them to make good use of the enormous parallelism of a supercomputer has largely proved a failure.

In the next three chapters, we shall explore some of the ideas for highly parallel algorithms that have been developed recently, and also try to relate the algorithms to the processor interconnection patterns on which they are best run. It is important to note that the lower bounds on complexity developed in Chapter 2 apply to, and tell us many things about, supercomputer design. In particular, interconnection networks that have boundaries crossed by few wires will require a long time to solve a problem with a strong lower bound on $AT^2$. As we mentioned in Exercise 2.13, the same comment applies even if the interconnections are real wires in a room, rather than wires on a chip.

In this chapter, we consider some of the basic interconnection patterns, such as trees of processors and meshes of processors. The emphasis in this chapter is on layout patterns that are regular, and have limited interconnections between the processors, so they can be laid out in relatively little area. We begin by suggesting a possible processor organization and then discuss a language for describing the activity of a collection of processors.

In the next chapter, we shall consider "systolic arrays," which are designs that possess, in addition to area-efficient layouts, an efficient way to pass data from processor to processor, thus allowing good circuits for problems that require much more computation than input/output. Then, in Chapter 6, we consider some more interconnection patterns for processors, ones that cannot be laid out in an area-efficient way; that is, the wires connecting the processors take much more area than the processors themselves. These interconnection patterns sometimes allow faster solutions to problems than do the designs of this chapter, and they may also be suitable for general-purpose computations by a supercomputer whose processors are distributed over a large number of chips.

## 4.1 PROCESSORS AND PROCESSOR NETWORKS

Many of the organizations we discuss are based on the premise that there is a large number of identical, or almost identical, "processors," with each processor connected to a small number of other processors, in a regular pattern. For example, the processors might be the nodes of a complete binary tree, in which case a processor is connected to its two children and its parent; leaves are connected only to their parents and the root is connected only to its children, of course. As another example, to be discussed in Chapter 6, the processors might be the nodes of a $d$-dimensional hypercube, the connection pattern defined in Exercise 3.7.

A basic constraint that we try to meet is that processors should not know their positions in such a tree, hypercube, or other pattern, although some exceptions are necessary; for example, we shall normally expect the root processor in a tree to know it is the root. The reason for this constraint is so the processor we design for a certain application can be connected in arbitrarily large networks, such as trees of arbitrary depth, thus allowing us to solve progressively larger instances of the problem as the size of networks that are feasible to build grows. Fortunately, common interconnection patterns usually allow the processors to deduce where they are. For example, we can write a program that enables a tree network to count the number of nodes, whereupon the root can pass down enough information so that each leaf can discover how many leaves there are to its left.

Each processor is a small computer with limited storage capacity; we shall discuss the details of such a processor shortly. A large number of these processors may be brought to bear on a single problem, for example, sorting integers. Since the "sorting problem" is really a family of problems, parametrized by the number of elements $n$ to be sorted and by the number of bits $k$ used to represent each element, we actually are discussing a family of circuits, one for each $n$ and $k$.

Since the size of the circuit that can fit on one chip is limited, when we consider such a family of circuits, we should consider both designs for a single chip and designs for networks of chips, where the pattern of interconnections among processors includes not only the pattern on a chip but the pattern of interconnections among chips. The major problem with generalizing interconnection patterns from a single chip to many chips is that the number of pads that can appear on any one chip is limited by the number of (real) wires that can be attached physically to a chip (about 100 pads is the limit in the early 1980's).

Fortunately, good designs for a single chip often limit the number of wires leaving any particular rectangle on the chip.† Thus, breaking a design into squares or rectangles, making each a single chip, and replacing the wires of the layout that cross a boundary by pads and real wires, is often a successful way to generalize a design from a single chip to many chips.

## The Processor

A processor can be any circuit whatsoever, and in the following sections and chapters we shall often describe processors by what they do, confident that the reader could supply the details of a circuit that would perform the task in the area and time bounds that we claim. However, to convince oneself of our claims about circuit area, it helps to bear in mind the "normal" organization we expect for a processor.

A processor has some specific number of "registers," which are arrays of a specific number of cells, each of which holds one bit. Certain registers may have circuits associated with them to perform one or more arithmetic operations, e.g., addition; these registers are similar to arithmetic units. However, we shall not distinguish registers used for arithmetic from those used for storage, leaving open the issue of what operations are performed, and how many arithmetic units are actually available in the processor.

Often, the registers each hold the same number of bits, say $k$, and they are interconnected by one or more busses. We may view a bus as a collection of $k$ wires, one for each bit of the registers, as discussed in Section 1.3. The wires are used to store bits by being high or low. Of course, each bus wire stores only one bit at a time.

For each register, there are two "control wires." One causes the contents of the register to be placed on the bus; that is the wires of the bus will each be high or low depending on whether the corresponding bit of the register was 1 or 0. The other causes the contents of the bus to be stored in a register, or to have some operation performed on the data in the register and the data on

† The designs in Chapter 6 are an exception. There, we get high performance, but pay a price in terms of the wire area or volume, and we may face pad limitations as well. The ability of the networks in Chapter 6 to solve certain problems very quickly is one of the factors that go into selecting a design for a particular application.

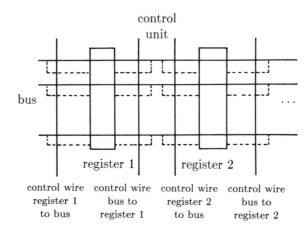

**Fig. 4.1.** Processor organization.

the bus. For example, we might add the contents of the bus to the contents of a particular register, using an adder circuit associated with that register, with the result appearing in the register. The arrangement is suggested in Fig. 4.1.

The control wires emanate from a "control unit," which is a logic circuit that implements a "program," thereby causing the processor, in cooperation with the other processors, to implement a particular algorithm. We assume that the control units for the processors are normally the same, although variations in the control units are permitted, and are frequently essential. For example, we might expect that the processors at the borders of a two-dimensional array of processors will "know" that they are at the edge, by having a slightly modified control to take account of the fact that certain of their neighbor processors are missing.

We shall leave undetermined the question of whether the control unit is executing a program that is built into it, e.g., a microcode program, or is executing a program in the ordinary sense, which has been loaded into a memory that stores a program. That is, the control unit is really nothing more than a sequential circuit, whose "states" correspond to places in a program. Each state executes the action required by that point in the program, e.g., moving data between a bus and a register, performing an operation on the data in a register, or making a comparison between data in a register and the bus, and then branching to one or another state depending on the outcome. Except for branch points, each state, on finishing its action, activates the state corresponding to the next step of the program being executed.

Communication among processors will also be performed via their busses. That is, when we say a pair of processors is "connected," we mean that the corresponding wires of their busses can pass values from one to the other. One or both of the processors have control wires that allow the contents of its bus to be copied onto the bus of the other processor. Thus, the single edge that we draw indicating the connection of two processors is really $k$ wires and logic gates used to connect the corresponding busses.†

Input and output to or from the chip is performed by an array of pads in a processor, used like the cells forming a register. Control wires allow data to move from a bus to the pads, or vice versa. In some designs, only a small subset of the processors, such as the processors on the borders of a two-dimensional array of processors, will have input or output pads.

**Example 4.1:** Recall from Example 2.2 our suggestion that we could "bubble-sort" an array of $n$ $k$-bit numbers. Let us investigate the details of such an algorithm here. Each cell of the array is a processor with several registers; each register is $k$ bits long. Clearly we need a register $A$ in each processor to hold the "current" number. We also need a register $B$ to hold a number passed by an adjacent processor when we wish to compare the contents of the $A$ register with the number currently held by that neighboring processor.

We can divide the operation of the sorting algorithm into "beats," in which half of the processors pass their current number to a neighbor for comparison, and the neighbor either returns the number it received or exchanges it for its own current number. Assuming $n$ is an even number, at odd beats, the odd numbered processors $p$ do the following.

1. Place the contents of register $A$ on the bus for processor $p$.
2. Send the contents of $p$'s bus to the bus of processor $p + 1$, which is $p$'s neighbor on the right.
3. Wait for the neighbor to the right to send a number to $p$'s bus.
4. Place the number received in register $A$.

At the same beat, the even numbered processors perform the following steps.

5. Wait for the next lower numbered processor, the neighbor to the left, to send a value to the bus of this processor.
6. Store the contents of the bus in register $B$.
7. Compare the contents of registers $A$ and $B$. If $A$ is larger, do steps (8a), (9), (10a), and (11a); otherwise do steps (8b), (9), (10b), and (11b).
8a. Copy register $B$ onto the bus.
8b. Copy register $A$ onto the bus.
9. Send the contents of the bus to the processor to the left.
10a. Wait (do nothing).
10b. Copy register $B$ onto the bus.

---

† While we shall not emphasize the point, we could also interconnect processors by single wires, no matter how many bits their busses had. Then, the contents of a bus is transmitted one bit at a time across the wire, an arrangement that is slower but saves the area needed to hold large numbers of wires running in parallel.

11a. Wait.

11b. Copy the bus into register $A$.

On even-numbered beats, we do the same steps, but with the roles of odd and even processors reversed. Also, the first and last processors do not participate in the even numbered beats.

In order that the above steps succeed in sorting the data, it is essential that the timing of data transfers is correct. We assume that the beats, which are not clock phases, but sequences of (possibly many) clock cycles, begin at the same times at all processors. However, within one beat, it may be necessary to time events so that when one processor sends data to the bus of another, the latter processor will be ready to use that data at the next cycle.

For example, each of the steps (1)–(11) can be executed in one clock cycle consisting of two phases. Steps (2) and (5) must be synchronized, as must (3) and (9). Thus, at the first cycle of any odd beat, the odd processors execute step (1), while the even processors do nothing. At the second cycle, steps (2) and (5) are performed. At the third through fifth cycles, the even processors do steps (6)–(8), while the odd ones wait. The sixth cycle sees steps (3) and (9) performed, while the seventh cycle is the time for step (4) by the odd processors and (10) by the even ones. Finally, at the eighth and last cycle, the odd processors wait and the even ones perform step (11).

In order that the algorithm terminate with the data in sorted order in the $n$ processors, we must arrange for the processors to know when they are done. The number of beats required is $n$. Thus, we may use a third register $C$ in each processor to count up to $n$. This register must be $\log n$ bits long, independent of $k$, and we must add to the above steps a step that increments $C$ by 1 and a step that tests if $C$ has reached $n$. If so, the sorting stops, and an output phase begins.

We prefer that the position of a processor in the array not be known to that processor, since if the number of a processor is built into it, arrays of processors will not be extendible in a simple fashion. If processors do not know their numbers, we can halt the sorting process by sending a signal from one end of the array to the other, and back again. The time taken by this signal will, naturally, be proportional to $n$, and we can arrange the timing so that on the back traversal, it can stop the activity of each processor as it passes. Ideas along these lines are discussed in Section 4.4 and Chapter 5.

It is unlikely that there could be enough pads for the data in all the processors to be output at once. A more likely arrangement is for only the first processor to have pads, and for the processors to pass their data to the left, with each processor's data appearing in turn at processor one, so the data is emitted in sorted order. If we halt the sorting by a signal running through the array, that signal can also be used to trigger the passing of data to the end at which the output occurs. □

## Time and Space Requirements for Processors

We can make some general observations about the time and space required for circuits that implement networks of processors such as the ones we have discussed. First, the space required for one bit of a register is a constant. Thus, a $k$-bit register requires $O(k)$ area, and $r$ registers require $O(rk)$ area. If we follow the pattern of Fig. 4.1, the bus wires and the gates whereby the control wires can move data to and from the bus will also take $O(rk)$ area.

The time required by a data movement operation—one that moves data between a register and the bus, between the bus and pads, or between the bus and a neighboring processor's bus—is also $O(1)$, since the data movement can be done in parallel, independent of how big $k$ is.

However, the need for operations like arithmetic, comparisons, and logical operations, like "and" and "or", introduce additional area into the processor represented by Fig. 4.1 and also may increase the time required for a single step. An operation like addition or comparison can be done in a variety of ways, but two methods stand out for each of these operations. The first is an $O(k)$ time, $O(k)$ area, straightforward method, and the second is an $O(\log k)$ time, $O(k \log k)$ area, divide-and-conquer-based method. We call an operation *normal* if it possesses implementations at least as good as these two combinations of area and time, i.e., the "straightforward" and "divide-and-conquer" methods.

For example, the ripple-carry method of addition is the straightforward way of adding; it requires only a full adder circuit to go with every register bit, taking a total of $O(k)$ area. But the time for the carries to ripple through the register is also $O(k)$, since we cannot compute any given bit position until we know the carry out of the previous position.

As an example of the divide-and-conquer approach, let us consider how to compare binary numbers $a_1 \cdots a_k$ and $b_1 \cdots b_k$. Assume $k$ is a power of 2, for convenience. We may imagine a complete binary tree of circuits, where the $i^{th}$ leaf holds $a_i$ and $b_i$, and each interior node computes the following two values. If some interior node $n$ has descendant leaves holding $A = a_i \cdots a_{i+2^j-1}$ and $B = b_i \cdots b_{i+2^j-1}$ (note $i$ will be an integer multiple of $2^j$), then we compute at node $n$ the Boolean value $x_{ij}$, which is 1 if and only if $A \geq B$, and $y_{ij}$, which is 1 if and only if $A > B$.

We can easily compute the $x$'s and $y$'s going up the tree. At the leaves, $x_{i0} = 1$ unless $a_i = 0$ and $b_i = 1$, while $y_{i0} = 0$ unless $a_i = 1$ and $b_i = 0$. For the induction, we may easily verify the following equations.

$$x_{i,j+1} = y_{ij} \vee (x_{ij} \wedge x_{i+2^j,j})$$
$$y_{i,j+1} = y_{ij} \vee (x_{ij} \wedge y_{i+2^j,j})$$

There are $\log_2 k$ levels in this tree. Each node requires $O(1)$ area, and as only two wires must pass from any node to its parent, the tree can be laid out, on its side to conform to Fig. 4.1, with height $O(k)$ and width $O(\log k)$. Further, as $O(1)$ time suffices to evaluate the bits for any node in terms of the bits at its children, the time for the bits at the root to be computed is proportional to the height of the tree, that is $O(\log k)$. As the bits at the root tell whether the first of two numbers, say the quantity in the register, is greater or at least as great as the second, say the quantity on the bus, we have the capability of making any comparison between $k$-bit numbers in $O(\log k)$ time and $O(k \log k)$ area, as claimed.

Of course, we cannot claim that what holds for addition or comparison holds for any operation, but these seem to be typical. Also, the construction of an $O(\log k)$ time, $O(k \log k)$ area adder follows a somewhat trickier line than the one we have given here for comparison, and we leave it as an interesting exercise for the reader who has not seen it.

Assuming the operations performed are normal, in the formal sense defined above, a processor with $r$ $k$-bit registers can be laid out in an area that is $O(k)$ high by $O(r \log k)$ wide, i.e., area $O(rk \log k)$; this divide-and-conquer implementation requires $O(\log k)$ time per operation performed. Alternatively, the straightforward implementation gives us area $O(rk)$ and time $O(k)$.

Another issue that may affect the area required for a circuit is the area taken by the wires connecting the busses of neighboring processors. If the processors are laid out in a line, the busses also line up, although they must be separated by arrays of gates, so that they do not unintentionally share data. Then the area used by $n$ processors is $O(nr)$ wide by $O(k)$ high if the straightforward implementation is used, or $O(nr \log k)$ wide and $O(k)$ high for the divide-and-conquer implementation.

However, some processor interconnection patterns require the wires connecting processors to run both horizontally and vertically; examples are a two-dimensional grid of processors, or processors arranged in an H-tree. In this case, the edges between processors, which must be $O(k)$ wide to carry all the bits of a bus, force the horizontal and vertical space between processors to be at least $k$. For example, an $n \times n$ array of processors, with connections in a grid pattern and each processor laid out as in Fig. 4.1, would, in the divide-and-conquer style of implementation, require height $O(nk)$ and width $O(max(nk, nr \log k))$.

**Example 4.2:** The processors designed in Example 4.1 have three registers, and the only operations performed are comparison, addition, and incrementation by one, each of which is normal. Thus, the $n$-processor bubble sorter has a straightforward implementation with area $O(nk)$. The time required, in this implementation, for the $O(n)$ beats required to sort is $O(nk)$.

Alternatively, we can use the divide-and-conquer implementation of the operations. The area goes up to $O(nk \log k)$, but the time goes down to $O(n \log k)$. $\square$

## 4.2 A PROGRAMMING LANGUAGE FOR PROCESSORS

We shall now describe informally a language based on Pascal that can be used to specify the algorithms performed by processors arranged in a network. Some of the ways in which our language differs from Pascal are the following.

1. Procedures are assumed to run on all processors at once. A procedure does not take any parameters except possibly the name of the processor. We prefer not to name processors, since that prevents us from extending networks to incorporate more processors. Instead, we prefer to limit differences in the executed programs to differences in the roles played by the processors, e.g., leaves and root, by the mechanism discussed in (4) below. If used, processor names may be of a variety of types. For example, if the processors were in a two-dimensional array, we would logically use pairs of integers to name processors, so processor $(i, j)$ would be the one in row $i$ and column $j$. If processors were arranged in a binary tree, then we might use path names from the root to name processors. For example, processor 110 would be the left child of the right child of the right child of the root.

2. We assume suitable names for the neighbors of a processor. For example, if the processors are in a linear array, we shall talk of the left and right neighbors. If the processors are in a binary tree, we talk of parents, left children, and right children.

3. Variable names refer to registers in each processor.

4. Conditional statements can use conditions that refer to classes of processors, e.g., **if** $p$ is a leaf **then**···. These statements are not conditionals in the usual sense; rather, they affect the program that each processor executes, like macros in a language, such as PL/I or c, that has a macro preprocessor.

5. The statement

   **send** $A$ **to** $p$

   places the value of variable (register) $A$ on the bus and then sends this value to the bus of processor $p$. The statement

   **receive** $A$ **from** $p$

   stores the value on the bus, which we assume will be placed there by processor $p$, into register $A$. In general, $p$ will be a designator for a neighbor of the processor at hand. For example, we might say "**send** $A$ **to** parent," or "**receive** $B$ **from** left neighbor."

### Synchronization

A critical aspect of the multiprocessor algorithms that we shall describe is the way actions of one processor synchronize with the actions of its neighbors. For instance, in Example 4.1, the odd processors sent a value to their neighbors on the right, and then had to wait until their neighbors sent a value in return. We shall introduce two primitives to help describe the desired synchronization.

The first idea is blocks called *beats*, designated by the keywords **beat begin** at the beginning and **beat end** at the end. The intent is that all processors execute the beats in synchronism; they all begin a beat together, and none can start the next beat until all have completed their beat. Processors that, because of conditional statements based either on their position in the network or on the values of their variables, have less to do, must wait for the other processors to finish.

We can implement beats by calculating the maximum number of clock cycles needed for any processor during a given beat, and by designing the control of the processors so that all processors consume this number of clock cycles in one beat. Arranging the steps within a beat so that all processors can execute their programs, and so that data passed from one processor to another will be available on schedule, is not trivial; Example 4.1 illustrated some of the considerations that go into designing the schedule and determining when processors have to wait for one or more cycles.

For example, if the program for an array of processors contains the two statements

> **receive** $A$ **from** left neighbor
> **send** $A$ **to** right neighbor

then within a single beat, the value of $A$ from the left end will propagate all the way to the right end. Thus, the number of cycles in a beat will be equal to the number of processors in the array, a situation that is intuitively unappealing, and which we shall not permit in designs. The reason to forbid beats whose length depends on the number of processors is that unless we forbid them, we cannot connect more than one copy of a chip to make a larger network than appears on a single chip. That is, since the network of chips will have more processors than any one chip, yet the time of a beat will presumably depend only on the number of processors on the chip itself, sufficiently large networks will not have time to complete their beats.

The second synchronization primitive is the labeling of **send** and **receive** statements. A signal sent by a statement with label $M$ is assumed to be received by a **receive** statement with label $M$. This feature is designed not so much to force synchronization as to make more transparent to the reader what is expected to happen.

**Example 4.3:** Let us write in our formal notation the bubblesort program of Example 4.1. The procedure *bubblesort(p)* uses variable $i$, which counts the number of beats, much as register $C$ did in Example 4.1, although for convenience, $i$ counts every other beat. Variables $A$ and $B$ play the same role

**procedure** *bubblesort(p)*;
(1)   **for** *i*:=1 **to** *n*/2 **do beat begin**
(2)       **if** *p* is odd **then** *M*1: **send** *A* **to** right neighbor;
(3)       **if** *p* is even **then begin**
(4)           *M*1: **receive** *B* **from** left neighbor;
(5)           **if** *A > B* **then** *M*2: **send** *B* **to** left neighbor
              **else** { *A ≤ B* } **begin**
(6)               *M*2: **send** *A* **to** left neighbor;
(7)               *A := B*
              **end**;
(8)       **if** *p* is odd **then** *M*2: **receive** *A* **from** right
          **beat end**;
          **beat begin**
              { similar steps where "odd" and "even" are replaced by
              "even and less than *n*" and "odd and greater than 1,"
              respectively }
          **beat end**
**end**;

**Fig. 4.2.** Bubblesort program.

here as they did in Example 4.1. The value *n*, assumed even, is a constant, the number of processors. We leave as an exercise the modification of the program in Fig. 4.2 that will make it independent of *n*, so it can run on arbitrary arrays of processors.

Figure 4.2 shows the program for bubblesorting. Note that it consists of a for-loop whose body consists of two beats, the first where the odd processors are compared with the even, and the second the opposite.

The reader should observe that line (2) in Fig. 4.2 corresponds to steps (1)–(2) in Example 4.1, and line (8) corresponds to steps (3)–(4). Line (4) corresponds to steps (5)–(6); line (5) is steps (7), (8a), and (9), while line (6) is steps (8b) and (9). Finally, line (7) is steps (10b) and (11b) of Example 4.1. □

**Example 4.4:** Let us consider a complete binary tree of processors that is designed to perform some operation on *n* values, each of which is stored initially at one of the leaves. In particular, we assume that we are to produce at the root the sum of values held in registers named *X* at each of the leaves. We assume for convenience that the number of leaves is a power of two.

In addition to variable *X*, which at each node will be made to hold the sum of the numbers held in all the descendant leaves of that node, all interior nodes use variables (registers) *Y* and *Z* to receive the values of *X* from their two children. Finally, our program uses variables *readyup* and *readydown*; these are one-bit values, true or false, and they are stored in one-bit registers at each processor.

The purpose of *readydown* is to allow the root to pass down a signal to the leaves that will get them to pass their $X$ values to their parent. Along with this $X$ value goes the signal *readyup*, which tells the parent that on this beat it must receive values from its two children, compute the sum, and pass that sum and the *readyup* signal to its parent at the next beat. A summary of the steps performed by each processor in a single beat follows.

1. Send the value of *readydown* to the two children, whether that value is true or false.

2. If the node is a leaf, set *readyup* equal to *readydown*. Thus, when the signal *readydown*=1 (true) reaches the leaves, it is turned upward and causes the addition to ripple up the tree.

3. Send the value of *readyup* to the parent. Only left children need to send the *readyup* signal, since this signal reaches all nodes at a given level of the complete binary tree at the same time. However, if nodes do not know they are left children, then all nodes can send *readyup*; the signal coming from the right child will be ignored by the parent.

4. Send the value of $X$ to the parent.

5. If the value of *readyup* just received is 1, then add the values of $X$ just received.

Figure 4.3 lists the steps of a single beat. Terms like "**if** nonleaf" mean "if the node at which the program is run is a nonleaf." We presume that the whole process is started by setting *readydown* to 1 at the root, and that this process is part of a larger program that, in effect, resides at the root. The completion of the process is indicated by *readyup* being set to 1 at the root, but the root processor could also count the number of beats necessary for the sum to reach it, provided either that the depth of the tree was built in to the root processor, or more desirably, the root calculated the depth.†

One point worth noting is the effort necessary to ensure that different types of processors, such as left and right children, are able to talk to the same parent in a correct way. For example, lines (1)–(5) simply send *readydown* to the two children, $\ell$ and $r$, of a given node $n$. If we allowed the value of *readydown* on the bus of $n$ to be gated onto the busses of $\ell$ and $r$ simultaneously, which in practice we could do, then there would be no problem. However, for simplicity, we have assumed that we can copy values from only one place to one other place at any time. Thus, we are required to undergo certain contortions in our language that may not be reflected by the actual circuits.

Thus, when in lines (1)–(2) we send *readydown* from $n$ to $\ell$, the latter cannot store the value in its own *readydown* immediately, because if $\ell$ is a nonleaf, it must later send its own value of *readydown* to its right child. To

---

† We leave the details of depth calculation as an exercise. However, the idea is simple, given the way the *readyup* and *readydown* signals are used in Fig. 4.3.

**beat begin**
(1)     **if** nonleaf **then** $M1$: **send** *readydown* **to** left child;
(2)     **if** left child **then** $M1$: **receive** *newreadydown* **from** parent;
(3)     **if** nonleaf **then** $M2$: **send** *readydown* **to** right child;
(4)     **if** right child **then** $M2$: **receive** *newreadydown* **from** parent;
(5)     *readydown* := *newreadydown*;
(6)     **if** leaf **then** *readyup* := *readydown*;
(7)     **if** left child **then** $M3$: **send** *readyup* **to** parent;
(8)     **if** nonleaf **then** $M3$: **receive** *readyup* **from** left child;
(9)     **if** left child **then** $M4$: **send** $X$ **to** parent;
(10)    **if** nonleaf and *readyup* **then** $M4$: **receive** $Y$ **from** left child;
(11)    **if** right child **then** $M5$: **send** $X$ **to** parent;
(12)    **if** nonleaf and *readyup* **then begin**
(13)         $M5$: **receive** $Z$ **from** right child;
(14)         $X := Y + Z$
        **end**
**beat end**

**Fig. 4.3.** One beat of procedure to add numbers at leaves of a binary tree.

avoid problems, we used a variable *newreadydown* to hold the value received until it is safe to store it in *readydown*.

Similarly, in lines (9)–(13), we pass the value of $X$ to the parent; each parent receives values first from its left child, then from its right. We need to store the values in temporary locations, since if the parent is a right child, it would receive a value before it sent its own value of $X$. We could, however, safely identify $Z$ with $X$ in Fig. 4.3. □

## 4.3 THE TREE-OF-PROCESSORS ORGANIZATION

We shall now consider one of several useful processor organizations, an arrangement in which the processors form a complete binary tree. Typically, the processors at the leaves, and the processors at the interior nodes perform somewhat different functions. We may view the leaves as processors that hold data and operate on that data in parallel. The interior nodes, on the other hand, are used for communication among the processors at the leaves. Thus, although we most likely would implement the tree of processors in the form of an H-tree, because it saves space, it often helps if we think of the organization as a one-dimensional array of processors, with a complete binary tree built upon them, like the trees of Fig. 2.5 or like one row or column of the mesh of trees discussed in Section 3.5.

Before looking at an example algorithm in detail, let us consider the sorts of operations this organization might be good for. First, let us observe that the single binary tree connecting the leaves can be a bottleneck. For example, we cannot implement an algorithm efficiently if that algorithm requires us to transmit much data from the left half of the tree to the right half, or more generally, from any subtree to any other. Thus, the tree organization will be suitable only for algorithms where most of the activity can be performed in parallel, with little communication needed.

In the following sections and also the next chapter, we discuss organizations based on two-dimensional grids of processors. The latter approach avoids the bottlenecks we find in trees, since a large amount of data can cross any boundary in unit time. However, the advantage of the tree is that global operations of certain types can be performed faster with the tree than the grid. For example, if we want to broadcast a value from one of $n$ processors to all the others, then $O(\log n)$ time suffices with the tree. In comparison, $O(\sqrt{n})$ time is required for the grid. In Chapter 6 we discuss organizations that provide both fast processor-to-processor communication and freedom from bottlenecks; the price we pay there is the large area needed for the layouts.

## Operations on a Tree of Processors

Let us enumerate some of the operations for which a tree of processors is especially suitable. These operations include the following.

1.   The select-and-rotate operation discussed in Example 3.1. The reader should consider how much more time-consuming this operation would be if the processors were arranged in a grid, with no tree connecting them.

2.   *Broadcasting.* A value at the root can be broadcasted to all leaf processors in $O(\log n)$ time. The steps that broadcast the signal *readydown* in Example 4.4 should serve as a prototype, although here we must not only broadcast the binary signal that says a value is being sent, but we must broadcast the value as well.

3.   *Census Functions.* A "census function" is a commutative, associative operation on $n$ values, each at one of the leaves. All these operations can be performed in time that is $O(\log n)$ times the time to perform the operation once. Some examples of census functions are:

   a)   Addition and multiplication. Example 4.4 shows how addition could be performed in $O(\log n)$ steps. Each step takes at most $O(\log k)$ time, if $k$ is the number of bits used to represent operands.

   b)   Logical "and" and "or". Again, Example 4.4 is a good prototype.

   c)   The minimum and maximum functions, which are also implemented as Example 4.4.

4.   The selection of an arbitrary number from among those offered by a subset of the leaves. To implement this operation, we might, for example, pick the value at the leftmost leaf that is selected. Following the manner in which values were passed up the tree in Example 4.4, each node, when it

is given the *readyup* value 1, passes to its parent a bit saying whether any of its descendant leaves were selected, and if so, also passes to the parent $p$ the value at the leftmost of its selected leaf descendants. Then, at the next beat, $p$ determines what it should pass up the tree. If either of its children indicated that there are selected leaves among their descendants, then so does $p$. In this case, if $p$'s left child has a value, $p$ passes that value to its parent. If only the right child has a value, then $p$ passes that value.

5. *Depth Computation.* We leave this as an exercise, but Fig. 4.3 gave essentially the method needed.

We shall discover a number of other operations that we can perform efficiently on trees; some are introduced when the need arises, and others are left as exercises.

### The Connected Components Problem

We shall now consider an extended example of an interesting algorithm that uses the power of the tree of processors organization. The problem we deal with is finding the *connected components* of a graph. We assume an undirected graph with $n$ nodes, numbered $1, 2, \ldots, n$, is represented by its adjacency matrix, with $e_{ij}$ equal to 1 if there is an edge between nodes $i$ and $j$, and 0 otherwise. Two nodes are in the same connected component if there is a path between them. We shall indicate the connected components by having the $i^{th}$ leaf hold the lowest number of any node in the connected component containing node $i$, when the algorithm is done. Then two nodes are in the same connected component if and only if the leaves for those nodes hold the same number.

The algorithm discussed here for finding connected components requires both time and space $O(n^{1+\epsilon})$, where following a convention introduced in Chapter 2, the factor $n^\epsilon$ represents some number of powers of $\log n$. This algorithm is unusual, in that it is when-indeterminate. We shall see in Section 4.5 that a when-determinate algorithm that even has a better $AT^2$ value exists. However, the latter algorithm uses much more space and requires that the entire adjacency matrix be available at once; it thus may be useful only in special circumstances.

**Algorithm 4.1:** *Lipton-Valdes Connected Components Algorithm.*

INPUT: The adjacency matrix of an $n$-node graph.

OUTPUT: A component number for each node.

METHOD: This algorithm operates in $n$ phases, if $n$ is the number of nodes in the graph. At the $i^{th}$ phase, the $i^{th}$ row of the adjacency matrix is read at the leaves, with the $j^{th}$ leaf from the left receiving the value of $e_{ij}$.† The current

---

† Strictly speaking, we only read $e_{ij}$ if $i < j$, so we do not read the edges in this undirected graph twice. For convenience, we assume that $e_{ij} = 0$ if $i > j$. We also assume that $e_{ii} = 1$, so each node is "adjacent" to itself.

component number of node $i$ is also kept at leaf $i$.

The basic idea of the algorithm is simple. When we read row $j$ of the adjacency matrix, we know all the nodes adjacent to node $j$. Node $j$, its adjacent nodes, and the nodes that are currently in the same component as one of these nodes, must be merged into one connected component. We do so by finding the smallest component number $c$ of $j$ and any of the nodes adjacent to $j$, and we broadcast messages to all the leaves saying that if the component of any node $k$ is one of those being merged, set the component of $k$ to $c$.

The reason that the algorithm is when-indeterminate is that the number of components being merged varies with $j$, but could be as high as $n$. The binary tree connection cannot broadcast more than one component number at a time. Thus, the time required to process the $j^{th}$ row of the adjacency matrix could be as high as $\Omega(n)$, and therefore the time of a when-determinate algorithm would be $\Omega(n^2)$ if this approach were used. However, we shall see that the total amount of time spent merging components is limited to $O(n^{1+\epsilon})$, so a much faster algorithm is possible if the processor network is allowed to prompt some external source for the next row of the matrix. Of course, that condition puts some constraints on how the adjacency matrix is to be provided as input.

At each leaf processor $i$, there is the integer variable $component[i]$, which is the current component number for the connected component containing node $i$. There is also the bit $edge[i]$, which is 1 if and only if there is an edge from node $i$ to the node $j$ whose row has just been read in, i.e., $edge[i] = e_{ij}$. Finally, there is a bit $selected[i]$, which is 1 if and only if node $i$ is currently "selected," meaning that it is one of the nodes adjacent to node $j$, and the current component of node $i$ still must be merged with the component containing $j$.

A sketch of the algorithm is given in Fig. 4.4. It helps if we view the control as residing at the root, with signals sent from the root to cause the actions indicated at each step. For example, to implement line (1), we can suppose that a signal to initialize is sent through the tree from the root to each leaf. If we do not wish to build the leaf number into each leaf processor, we may arrange for the leaves to be told their numbers before the algorithm commences. The details are left as an exercise, but the basic idea is for the root to find out the depth of the tree. Then, we can pass down to each node of the tree information telling it the number of its leftmost descendant and the number of descendants it has.

Steps that are done "**for** all leaves" are done in parallel at each leaf; these lines are (1), (3)–(4), and (10)–(13). Line (5) uses the binary tree to compute the minimum of the component numbers for the selected nodes; it is an example of a census function. The result of line (5) is broadcast to all the leaves at line (6), again using the binary tree. Likewise, lines (8) and (9) use the tree to make an arbitrary choice from among selected nodes (the fourth tree operation discussed at the beginning of this section), and broadcast the result. Finally,

**procedure** *connect*;
(1)  **for** all leaves $i$ **do** *component*$[i] := i$; { initialize }
(2)  **for** $j := 1$ **to** $n$ **do begin**
(3)      **for** all leaves $i$ **do** *edge*$[i] := e_{ij}$; { read row $j$ }
(4)      **for** all leaves $i$ **do** *selected*$[i] := edge[i]$;
             { $j$ and all nodes adjacent to $j$ are now selected }
(5)      compute $c := min\{$ *component*$[i] \mid$ *selected*$[i] = 1 \}$ at the root;
             { $c$ is now the lowest component number containing a node
             adjacent to $j$ }
(6)      broadcast $c$ to all leaves;
(7)      **while** selected processors remain **do begin**
(8)          choose at the root some $d = $ *component*$[i]$ such that
                 *selected*$[i] = 1$;
(9)          broadcast $d$ to all leaves;
(10)         **for** all leaves $i$ **do**
(11)             **if** *component*$[i] = d$ **then begin** { set component $d$ to $c$
                 and unselect nodes in that component }
(12)                 *component*$[i] := c$;
(13)                 *selected*$[i] := 0$;
                 **end**
             **end**
         **end**
**end**;

Fig. 4.4. When-indeterminate connected components algorithm.

the test of line (7) can be decided by computing another census function, the "or" of *selected*$[i]$ for all leaves $i$. $\square$

**Theorem 4.1:** Algorithm 4.1 correctly computes connected components and does so in $O(n^{1+\epsilon})$ area and time.

**Proof:** The correctness is easy to show. The inductive hypothesis is that after going around the loop of lines (2)–(13) for a fixed value of $j$, the component numbers of nodes $r$ and $s$ will be the same only if there is a path between $r$ and $s$, and will surely be the same if there is such a path that goes through no node numbered higher than $j$. This part of the proof is easy and left as an exercise.

Let us consider the time taken by the algorithm. In what follows, we assume that all operations are done in a divide-and-conquer way, taking $O(\log n)$ time on component numbers, which are in the range 1 to $n$. The "trick" in analyzing the algorithm is to account for different lines in different ways. First, we charge certain lines to the value of $j$; a total of no more than $O(n^{\epsilon})$ will be charged to any $j$. We also charge some time to the merger of particular components; again, no more than $O(n^\epsilon)$ will be charged to any merger. Since there are $n$ values of $j$, and at most $n - 1$ mergers, the total time is $O(n^{1+\epsilon})$.

Step (1), the initialization, takes $O(\log n)$ time, since we must send an initialization signal from the root. Once the signal is received, the copying of $i$ into the register $component[i]$ can be performed in $O(1)$ time, assuming that each processor is built with a register holding its own number. Preferably, so processors do not have to know their numbers, the root can signal each processor what its number is, by a method sketched above. However, then the calculation of numbers, as signals are sent up and down the tree, requires addition and subtraction operations at each level, so $O(\log^2 n)$ time will be required to send the initialization signal and the correct initial value to all the leaves, if the straightforward implementation of arithmetic is used.

Lines (3)–(6) are charged to $j$. The first two of these take $O(1)$ time, while line (5) takes $O(\log^2 n)$ time for a calculation to travel up the tree. Line (6) takes $O(\log n)$ time for broadcasting.

One time through the loop of lines (7)–(13), for each value of $j$, is charged to that value of $j$. In particular, we charge to $j$ that time when $d = c$, i.e., no change of component numbers occurs. It is easy to check that none of steps (7)–(13) takes more than $O(\log^2 n)$ time. Other times through that loop are charged to the merger of components $c$ and $d$. Since $d \neq c$ in this case, we can be sure that the merger of two distinct components is actually taking place, an event that can happen only $n - 1$ times.

We conclude that only $O(\log^2 n)$ time is charged to any $j$ or to any merger; therefore the total time is $O(n \log^2 n)$, which is less than $O(n^{1+\epsilon})$ for any $\epsilon > 0$.

Now consider the area required by this circuit. The $n$ processors each need registers of length $\log_2 n$ to store component numbers. Thus, busses must be $O(\log n)$ wide. Also, since arithmetic and comparison operations performed in a divide-and-conquer way require registers of width $O(\log \log n)$, each processor will be on the order of $\log n$ high and $\log \log n$ wide. If, say, we arrange the processors in an H-tree, the width of busses becomes the significant factor, and the area required is $O(\sqrt{n} \log n)$ wide and high both, for an area that is $O(n^{1+\epsilon})$. $\square$

Note that Algorithm 4.1 gives us an $AT^2$ value of $n^{3+\epsilon}$. That is not as good as possible, since we can almost achieve the lower bound of $\Omega(n^2)$; we leave the lower-bound result as an exercise. However, it appears that Algorithm 4.1 is quite good among those algorithms that use relatively little area. In fact, because of the difficulty of getting large amounts of data onto a chip quickly, approaches such as Algorithm 4.1, where data is read onto the chip a row at a time, are more likely to be practical than those that run faster in principle but require all the data to be on the chip before starting. For example, we shall give an $AT^2 = O(n^{2+\epsilon})$ connected components algorithm in Section 4.5 (Algorithm 4.3), but the area required for this algorithm is itself larger than $n^2$. If we had to read the adjacency matrix a row or column at a time, because we could not fit anything like $n^2$ pads onto the chip,† then Algorithm 4.3 would really take time $\Omega(n)$ and would be an $AT^2 = O(n^{4+\epsilon})$ algorithm.

## 4.4 THE MESH-OF-PROCESSORS ORGANIZATION

We shall now consider algorithms that are designed to operate on a square grid of processors, with each processor connected to its neighbors to the left, to the right, above, and below. Possibly, the processor at the left end of each row is connected to a processor at the right end, and each processor on the top row is connected to one at the bottom row. Fig. 2.7 was an example of such a processor array, but other connections among the processors on the border are possible; e.g., unlike Fig. 2.7, the processor at the top of column $i$ could be connected to the bottom of column $i$.

There are a variety of uses to which such an organization can be put. Example 2.12 was a study of one possibility, a barrel shifter with a near-optimal $AT^2$ value. In the next chapter, we shall consider "systolic" algorithms for fundamental matrix operations, such as matrix multiplication and transitive closure; these algorithms are based on a mesh organization. Here, we shall study a use of two-dimensional arrays of processors in a nonsystolic way; the particular algorithm we study is sorting.

### Odd-Even Merge Sorting

Before discussing the VLSI-oriented sorting algorithm we have in mind, let us review the sorting method known as odd-even merge sorting (see Knuth [1973], e.g.). The idea behind merge sorting in general is that to sort a list of $2n$ numbers, we divide the list into two sublists of length $n$, sort them recursively, and then merge the two sorted sublists. The odd-even merge technique is a recursive way to perform the merge.

Suppose now that we have two lists of numbers $a_1, \ldots, a_n$ and $b_1, \ldots, b_n$ that are sorted, i.e., $a_i \le a_{i+1}$ for all $i$ and similarly for the $b$'s. We wish to merge these numbers into one sorted list of length $2n$. One way to do so is a recursive algorithm called *odd-even merge*. As a basis, if $n = 1$, we simply compare $a_1$ with $b_1$ and order them lower first.

For the recursion, suppose $n$ is a power of 2. We first merge the odd positions, that is, we merge $a_1, a_3, \ldots, a_{n-1}$ with $b_1, b_3, \ldots, b_{n-1}$. These lists being of length $n/2$ each, the odd-even merge algorithm may be applied recursively. We must also merge the even positions, that is $a_2, a_4, \ldots, a_n$ with $b_2, b_4, \ldots, b_n$. In an environment such as a chip, it may well be possible to do the two recursive

---

† Recall that although our model uses a single grid point for a pad, in reality, pads are quite large compared with typical circuit elements, and moreover, pads generally are constrained to appear only on the border of the circuit.

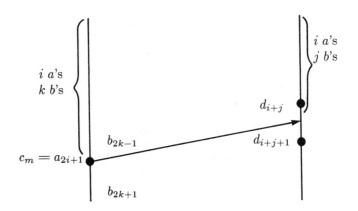

**Fig. 4.5.** Merger of odds and evens.

merge steps at the same time.

Whether the mergers are done sequentially or in parallel, we find that the resulting lists can almost be merged by shuffling (or interlacing) them. It turns out that the $m^{th}$ number on the odd list goes either before or after the $m-1^{st}$ element of the even list; therefore all we have to do is compare $n-1$ pairs of numbers, one from each list. We can formalize this observation in the next lemma.

**Lemma 4.1:** Suppose $a_1, \ldots, a_n$ and $b_1, \ldots, b_n$ are sorted lists, $c_1, \ldots, c_n$ is the result of merging $a_1, a_3, \ldots, a_{n-1}$ with $b_1, b_3, \ldots, b_{n-1}$, and $d_1, \ldots, d_n$ is the result of merging $a_2, a_4, \ldots, a_n$ with $b_2, b_4, \ldots, b_n$. Then the numbers $a_1, \ldots, a_n, b_1, \ldots, b_n$ can be found in sorted order by the following steps.

1.  Shuffle $c_1, \ldots, c_n$ with $d_1, \ldots, d_n$ to get $c_1, d_1, c_2, d_2, \ldots, c_n, d_n$.
2.  Compare each $c_i$ and $d_{i-1}$, and interchange them if they are out of order, i.e., if $d_{i-1} > c_i$.

**Proof:** For simplicity, assume that all elements are distinct. Surely, if the algorithm works when the elements are distinct, it will work when they are not. We shall show that each $c_m$ belongs either immediately before or immediately after $d_{m-1}$ in the sorted order. Suppose without loss of generality that $c_m$ is from the $a$ list, say $c_m = a_{2i+1}$, and among $c_1, \ldots, c_{m-1}$ are exactly $k$ elements from the $b$ list, which must be $b_1, b_3, \ldots, b_{2k-1}$.

Suppose that in the merged list, $c_m = a_{2i+1}$ belongs between $d_{i+j}$ and $d_{i+j+1}$, as shown in Fig. 4.5. Then among the first $i+j$ $d$'s there must be exactly $i$ $a$'s, namely, $a_2, a_4, \ldots, a_{2i}$, because those are exactly the even $a$'s that $a_{2i+1}$ follows.

We must deduce what value $j$ has, i.e., how many even $b$'s precede $a_{2i+1}$. Surely $a_{2i+1}$ follows at least $k-1$ even $b$'s, namely $b_2, b_4, \ldots, b_{2k-2}$, since $a_{2i+1}$

follows $b_{2k-1}$, which follows the latter $b$'s. However, $a_{2i+1}$ cannot follow more than $k$ even $b$'s, because $a_{2i+1}$ precedes $b_{2k+1}$, which precedes all even $b$'s except $b_2, b_4, \ldots, b_{2k}$.

Thus, $j$ is either $k - 1$ or $k$, and since $m = i + k + 1$, we now know that $c_m = a_{2i+1}$ goes either between $d_{m-2}$ and $d_{m-1}$ (if $j = k - 1$), or between $d_{m-1}$ and $d_m$ (if $j = k$). Thus, to resolve where $c_m$ goes, we need only to compare it with $d_{m-1}$, as we claimed. $\square$

**Example 4.5:** Suppose we start with the sorted lists

$$
\begin{array}{lcccc}
a: & 1 & 5 & 16 & 23 \\
b: & 2 & 3 & 4 & 6
\end{array}
$$

Let us then take the odd positions from each list, 1 and 16 from $a$, and 2 and 4 from $b$, and merge them, by a recursive application of the odd-even merge, to obtain the list $c$, and similarly merge the remaining even positions into the list $d$, given by

$$
\begin{array}{lcccc}
c: & 1 & 2 & 4 & 16 \\
d: & 3 & 5 & 6 & 23
\end{array}
$$

Now, we shuffle these lists, $c$ first, to obtain the list

$$
\begin{array}{cccccccc}
1 & 3 & 2 & 5 & 4 & 6 & 16 & 23
\end{array}
$$

Finally, we exchange even positions with the next odd position if they are out of order; this step is step (2) of Lemma 4.1, the comparison of $c$'s with their previous $d$'s. Thus, 2 is exchanged with 3 and 5 with 4, but 6 and 16 remain as they are. The resulting list is

$$
\begin{array}{cccccccc}
1 & 2 & 3 & 4 & 5 & 6 & 16 & 23
\end{array}
$$

$\square$

### Merge Sort and Odd-Even Merge on a Square Mesh

We shall now show how to implement the mergesort of $n$ numbers on a $\sqrt{n} \times \sqrt{n}$ grid in $O(\sqrt{n})$ time. We might comment that the original use of odd-even merge, by Batcher [1968], yields an $O(n^\epsilon)$ time sort in $O(n^{2+\epsilon})$ area. The latter, while the fastest known parallel sorting algorithm, may take too much area for large $n$. Further, the algorithm we propose allows us to read the data $\sqrt{n}$ numbers at a time, thereby filling the mesh from pads along one edge in $O(\sqrt{n})$ time without increasing the running time by more than a constant factor. Since reading large amounts of information at one time is not likely to be realistic, and Batcher's sort only works in time $O(n^\epsilon)$ if the data is all loaded at once, the issue of input availability is a second justification for why the two-dimensional mesh may be a useful structure for sorting.

We shall say that the mesh is sorted when the numbers appear in row major order. That is, the lowest-valued $\sqrt{n}$ elements appear in the first row, in sorted

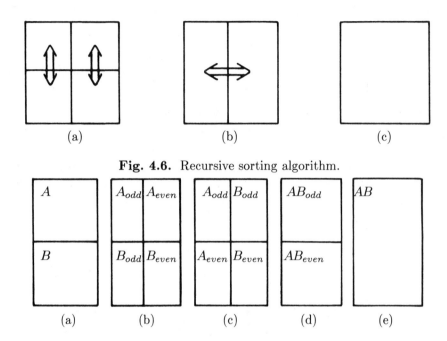

**Fig. 4.6.** Recursive sorting algorithm.

**Fig. 4.7.** Odd-even merge.

order, the next lowest $\sqrt{n}$ elements are in the second row, in sorted order, and so on. The recursive sorting algorithm is illustrated in Fig. 4.6. The square grid is divided into four quarter size squares, which are sorted recursively and simultaneously, as indicated in Fig. 4.6(a).

The two squares on the left are then merged by an implementation of odd-even merge, taking $O(\sqrt{n})$ time, that we shall describe in a moment. At the same time, the two squares on the right are merged in the same way. The result is indicated in Fig. 4.6(b). Finally, the two rectangular sorted lists are merged by a similar implementation of odd-even merge, taking $O(\sqrt{n})$ time.

The recursive odd-even merge step, assuming we start with lists $A$ and $B$ stored one above the other in squares, is shown in Fig. 4.7. The case where $A$ and $B$ start as rectangles twice as high as wide, as in going from Fig. 4.6(b) to 4.6(c), is similar and omitted. To go from Fig. 4.7(a), the initial condition, to Fig. 4.7(b), we divide the even and odd positions of the lists $A$ and $B$, with the odd positions sliding left and the even ones right.

This process is not difficult if we assume that $n$, the number of elements, is fixed, so that at any stage of the recursive algorithm each processor can determine by calculation within itself where it is in the rectangle of which it

$$1_0 \quad \_ \quad 3_1 \quad \_ \quad 5_2 \quad \_ \quad 16_3 \quad \_$$
$$\_ \quad 2_3 \quad \_ \quad 4_2 \quad \_ \quad 6_1 \quad \_ \quad 23_0$$

$$1_0 \quad 3_0 \quad \_ \quad 5_1 \quad \_ \quad 16_2 \quad \_ \quad \_$$
$$\_ \quad \_ \quad 2_2 \quad \_ \quad 4_1 \quad \_ \quad 6_0 \quad 23_0$$

$$1_0 \quad 3_0 \quad 5_0 \quad \_ \quad 16_1 \quad \_ \quad \_ \quad \_$$
$$\_ \quad \_ \quad \_ \quad 2_1 \quad \_ \quad 4_0 \quad 6_0 \quad 23_0$$

$$1_0 \quad 3_0 \quad 5_0 \quad 16_0 \quad \_ \quad \_ \quad \_ \quad \_$$
$$\_ \quad \_ \quad \_ \quad \_ \quad 2_0 \quad 4_0 \quad 6_0 \quad 23_0$$

**Fig. 4.8.** Separation of a row into odds and evens.

is currently a part, and what the dimensions of that rectangle are. Then, the division of a row into odds and evens could take place by using two registers in each processor, one for the odds and one for the evens. Each processor puts its current element into whichever register is appropriate for that processor. In a third register, the processor places a count of how far left or right the element must move, a distance that depends only on the position of the processor within its rectangle.† Finally, in a sequence of beats, the odds are shifted left and the evens shifted right, each as far as necessary.

**Example 4.6:** Figure 4.8 shows a row of length eight, first split into odds (upper track) and evens (lower track), with subscripts on the elements indicating the contents of the third register, that is, how far the element must travel. In three beats, also shown in Fig. 4.8, the elements all reach their destination. □

Going from Fig. 4.7(b) to 4.7(c) is straightforward. $B_{odd}$, the lower left rectangle in Fig. 4.7(b), slides as a body to the upper left, while at the same time $A_{even}$, the upper right, slides as a body to the lower right. Of course, extra registers are used to accommodate the moved rectangles, so they do not interfere with the data remaining in the upper left and lower right. Then the list $B_{odd}$, formerly in the lower left, slides in a body from the upper left to the upper right, at the same time that $A_{even}$, formerly in the upper right, siides from the lower right to the lower left.

The transformation from Fig. 4.7(c) to 4.7(d) involves the simultaneous application of the odd-even merge algorithm to merge the smaller lists $A_{odd}$

---

† Of course, we would prefer not to build into the processors the knowledge of their coordinates in the grid. It is an easy exercise to show how each processor can be told its position in $O(\sqrt{n})$ time. Moreover, after presenting the main algorithm, we shall discuss some of the ideas behind an implementation of the sorting algorithm that does not even assume there is room in a register to hold the coordinates.

with $B_{odd}$ and $A_{even}$ with $B_{even}$. Then, to reach Fig. 4.7(e) the resulting lists, which are $c$ and $d$ from Lemma 4.1, are shuffled. The $i^{th}$ row of $AB_{odd}$ is moved down $i - 1$ rows, and the $i^{th}$ row of $AB_{even}$ is moved up $m + 1 - i$ rows, if $2m$ is the height of the entire rectangle. Then the $i^{th}$ rows of the two sublists, which are now physically located in the same row of processors, are shuffled and distributed into that row and the row below it (rows $2i - 1$ and $2i$). The process of computing the distance to shift each element is similar to the way odd and even positions were split as in Fig. 4.8, and we leave this detail as an exercise.

**Theorem 4.2:** Odd-even merge sort takes $O(n^{1/2+\epsilon})$ time and $O(n^{1+\epsilon})$ area if the numbers being sorted can be represented by $O(\log n)$ bits.

**Proof:** If registers in processors are of length $\log n$, then the area of the grid of $n$ processors will be $O(\log^2 n)$ times the number of processors, that is, no more than $O(n^{1+\epsilon})$. The comparison and arithmetic operations will take $O(\log \log n)$ time, if we assume the divide-and-conquer implementation of arithmetic, whereas moving an element will take $O(1)$ time. Thus, we are conservative if we assume that each operation or data movement between neighboring processors takes unit time and then multiply the calculated time by $O(n^{\epsilon})$.

On that assumption, if we examine the merge operation of Fig. 4.7, we see that each step except going from Fig. 4.7(c) to 4.7(d), the recursive merge, takes $O(\sqrt{n})$ time units, if $n$ is the number of elements in the rectangle of Fig. 4.7. If $M(n)$ is the time to perform odd-even merge on $n$ elements, then going from Fig. 4.7(c) to 4.7(d) takes time $M(n/2)$, since the mergers may be done simultaneously. Thus

$$M(n) \le c\sqrt{n} + M(n/2) \tag{4.1}$$

for some constant $c$.

$M(1) = 0$, since a single element need not be merged. Thus, the solution to (4.1) is obtained by repeatedly substituting for $M$ on the right to get

$$M(n) \le c\sqrt{n} + c\sqrt{n/2} + c\sqrt{n/4} + \cdots$$

or, summing the geometric series, $M(n) \le c(2 + \sqrt{2})\sqrt{n}$.

Next, examine the recursive sorting step of Fig. 4.6. If $S(n)$ is the time to sort $n$ elements, then

$$S(n) \le d\sqrt{n} + S(n/4) \tag{4.2}$$

The term $d\sqrt{n}$ comes from the cost of the two odd-even merge steps, each of which takes $O(\sqrt{n})$ by (4.1), and the term $S(n/4)$ accounts for the recursive sort of the four quadrants, in parallel. The solution to (4.2) is easily seen to be $S(n) = O(\sqrt{n})$; thus when we multiply the time to sort by $n^{\epsilon}$ to account for the fact that what we considered to be unit time steps may in fact take slightly

```
0 0 0 0 0 0 0 0
0    1 1      0
0    1 1      0
0 2 2 1 1 2 2 0
0 2 2 1 1 2 2 0
0    1 1      0
0    1 1      0
0 0 0 0 0 0 0 0
```

**Fig. 4.9.** Border levels.

more, we have the theorem. □

## Implementation of Odd-Even Merge Sort Without Knowing $n$

As we mentioned, given a chip with, say, a $64 \times 64$ mesh of processors implementing the above algorithm, we cannot necessarily connect a mesh of these chips together at their borders and have them sort larger lists of numbers; the reason is that, as we have described it, the list length $n$ is built into the program that the processors execute, so they know when to perform each action. To make a chip that does not depend on $n$, we must arrange that the program executed by the processors can function correctly without knowing $n$.

There are a great many details to explain, and we shall only touch on some of the typical issues, enough, we hope, to convince the reader that the balance of the details could be filled in. The first important issue is how a processor is to relate to the "current" rectangle of which it is a member. The answer is that we cannot have each processor hold information telling where it is in the rectangle, because even to record, say, its distance to the left edge of the rectangle may in principle involve a number that is too large to fit in a register of the processor. Thus, operations like that of Fig. 4.8, where each number is told by its processor how far to go, cannot be performed.

The approach we take is to have processors on the border of a rectangle know that they are on the border, and also know the *level* of the border, the number of times we divided the whole square into halves before we introduced that border. This approach puts a limit on the size of arrays, just as fixing $n$ does, but it is a much weaker limit since the number of levels will be the logarithm of $n$, the number of elements to be sorted. Thus, if a register of 32 bits is available to store the level of a border, we could sort up to $2^{2^{32}}$ elements, probably more than enough.

**Example 4.7:** Figure 4.9 shows the first three levels of border that we would obtain if we divided a square first vertically, then horizontally. □

It helps if we view control as residing in the processor at the upper left-hand corner of each of the innermost rectangles. For example, in Fig. 4.9, the processors with coordinates $(1, 1)$, $(1, 5)$, $(5, 1)$, and $(5, 5)$ each control a $4 \times 4$ rectangle. To create a division of rectangles into two, as during the recursive sorting step where we divide squares into four quadrants, or during the merge in going from Fig. 4.7(a) to 4.7(b), we "find the middle," by a process to be described, and create a new wall at the level one higher than the highest level of any wall of the present rectangle.

To break down a wall and combine two rectangles into one, as in the two steps of Fig. 4.6 or when going from Fig. 4.7(c) to 4.7(d), each controlling processor destroys the highest numbered of its two walls that run in the proper direction (horizontal or vertical). For example, if we were going from Fig. 4.7(c) to 4.7(d), and wanted to break down alternate vertical walls, the processor at $(1, 1)$ in Fig. 4.9 would send a signal right, along its second row, until it met a vertical wall. The signal returns to $(1, 1)$ with information telling the level of the wall, whereupon the controlling processor at $(1, 1)$ knows that its right wall must be removed. Similarly, the processor at $(5, 1)$ sends and receives a signal which tells it that its left wall is at a higher level than its right, and therefore the left must be destroyed. Signals sent from the controlling processors reach the sides to be destroyed and tell all the processors along those sides to set themselves to "not a wall," or in the case of the corners, to become part of the crossing wall. Thus, the central four 1's in Fig. 4.9 become 2's, while the other 1's become blank.

Now let us consider three example operations and sketch how they are done in an environment that knows only levels of walls, not sizes of rectangles.

1. *Find the middle of a rectangle.* Assume we want to bisect a rectangle vertically; horizontal bisection is analogous. The processor in the upper left sends out two signals; both travel horizontally along the top row of the rectangle. For convenience let us assume that all corners are marked by a single bit, so the right wall can be detected when it is reached. The first moves one position at every beat, while the other moves once every third beat. When the first hits the right wall it bounces off and returns at its same speed. The two signals will therefore meet in the middle.

2. *Move the contents of one rectangle up to occupy the rectangle above.* (An example is the movement of $B_{odd}$ in Fig. 4.7(b) to the upper left quadrant.) First, note that not every rectangle performs this operation, only those that at the most recent horizontal partition were made lower halves, such as $B$ in Fig. 4.7(a). A processor can tell whether its rectangle is a lower half by checking that its top wall has a higher level number than its lower wall. Assuming the rectangle qualifies, its controlling processor sends out a signal down its left edge. At each row, that signal sets another signal heading right, which tells each number it encounters to start moving upwards, until that signal reaches the right wall, where it "dies." The numbers set in upward motion travel in diagonal waves, as illustrated in Fig. 4.10.

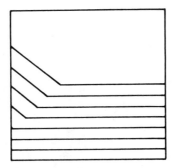

**Fig. 4.10.** Data traveling upward in waves.

If a number reaches the second wall, i.e., the top of the rectangle above the one from which it started, it stops and sends to the processor below it a signal that it is to halt the motion of the number that reaches that processor. When a number reaches a processor that is holding a halt signal, the processor below is passed a halt signal. Note that the diagonal portions of the waves in Fig. 4.10 are distance 2 apart, so there is time for a halting element to send a signal and halt the element below it.

3. *Slide the elements in even positions left to close up gaps.* (This operation is part of what was done in Fig. 4.8.) The controlling processor can send a signal down its left edge, and from there send signals right, along each row of the rectangle, to tell each processor whether it is in an odd or even column†. Each processor then places its number in the appropriate register, as in Fig. 4.8. Now, from the left edge of each row, two signals are sent to the right. One, $x$, moves right at each beat, telling all elements in even positions to move left, until they meet the other signal, $y$. Signal $y$ behaves like the "halt" signal discussed in (2), since it stays where it is until an element reaches its processor, whereupon it moves one position right and waits for the next element. The initial portion of this process is shown in the series of snapshots in Fig. 4.11.

All of the operations described above take time that is on the order of the length of the height or width of the rectangle in question. Since the odd-even merge sort never uses rectangles with an aspect ratio greater than 2, the time required for each operation is $O(\sqrt{n})$ if the rectangle holds $n$ elements. Since Equations (4.1) and (4.2) each have $O(\sqrt{n})$ terms on the left anyway, the new operations that we are required to perform to compensate for not knowing $n$

---

† Actually, since all rectangles are of even height and width, except for those with a dimension of 1, for which the operation being described is never needed, every processor can know whether it is odd or even independent of the size of its current rectangle.

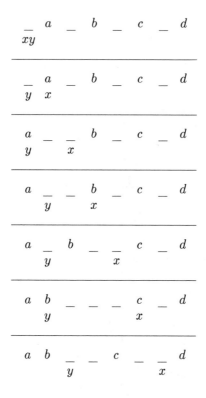

**Fig. 4.11.** Closing up gaps by shifting left.

affect only the constant factor in the $\sqrt{n}$ terms and, thus, do not change the growth rate of the solutions to (4.1) and (4.2). Therefore, Theorem 4.2 continues to hold.

## 4.5 THE MESH-OF-TREES ORGANIZATION

We met the mesh of trees in Section 3.5, where it was used to prove facts about layout algorithms. It turns out that this graph has some interesting capabilities as an organization for processors. We already know from Section 3.5 that $n$ processors organized as a mesh of trees and using normal operations on registers of length $O(\log n)$ can be laid out in area $O(n^{1+\epsilon})$, where $\epsilon$ represents four powers of $\log n$. That is, $O(n \log^2 n)$ is the area assuming wire widths of 1 and processors of size 1, while the other two factors of $\log n$ represent the fact that processors and wires are really $O(\log n)$ units wide and high.

   Thus, for the mesh of trees, a large fraction of the area of the chip is devoted to processors, unlike some of the efficient organizations we shall study in Chapter 6, where wire area dominates the processor area. Like these wire-dominated organizations, the mesh of trees has some very fast algorithms. The negative side of the coin regarding the mesh of trees is that in some cases the number of processors it uses is just too large to make the algorithms practical. For example, it uses $n^2$ processors to sort $n$ numbers.

In this section we shall discuss algorithms for sorting and connected components based on the mesh of trees. We shall then discuss another connected components algorithm that uses a variant of the mesh of trees, a $(\log n) \times n$ array, with columns connected by column trees, and a single row tree connecting the roots of the column trees. This algorithm is not as fast as the one using the normal mesh of trees, but it allows the adjacency matrix to be read a row at a time. Like Algorithm 4.1, it is when-determinate, takes much less area than the mesh of trees, and it is almost as fast as Algorithm 4.1.

## Sorting

Let us begin by giving some general observations regarding programs that run on the mesh of trees. First, any of the census functions or other tree operations discussed in Section 4.3 can be performed on either all the row trees at once, all the column trees at once, or any combination or subset of these.

Second, it is often useful to imagine that the root of the $i^{th}$ row tree and the root of the $i^{th}$ column tree are the same processor, which in fact they could be if we added connections between them. We shall, therefore, refer to this pair of nodes as *the $i^{th}$ controller*, for each $i$. We may invoke Theorem 1.1 to claim that the layout area of the mesh of trees with these extra connections increases only by a constant factor. Alternatively, we may use no extra wire, but instead use the leaf processor in position $(i, i)$ as the $i^{th}$ controller, understanding that the roots of the $i^{th}$ row and column trees both pass any information they obtain to that leaf. Since the time to pass the data to the leaf is $O(\log n)$, and it is our intent that factors of $n^{\epsilon}$ will stand for any number of powers of $\log n$, we shall ignore the extra time that may be required to allow leaves to serve as controllers.

Before giving the algorithm for sorting, let us mention one new operation that we can perform on trees. Given a datum $X$ and an integer $i$, we can store $X$ in the $i^{th}$ leaf from the left. To do so, let $n$, the number of leaves, be $2^k$, and express $i$ as a $k$-bit binary number $a_1 \cdots a_k$. If $a_1 = 0$, pass $X$ and $a_2 \cdots a_k$ to the left child of the root; if $a_1 = 1$, pass $X$ and $a_2 \cdots a_k$ to the right child. Repeat the same steps at each level, until a leaf is reached.

**Algorithm 4.2:** *Mesh-of-Trees Sorting Algorithm.*

INPUT: The elements to be sorted, $a_1, \ldots, a_n$, are initially read into the controllers, with $a_i$ read into the $i^{th}$ controller.

OUTPUT: At the end of the algorithm, the $i^{th}$ controller holds the $i^{th}$ element in sorted order.

METHOD: The processors perform the following steps.

1. For all $i$, broadcast $a_i$ from the $i^{th}$ controller to all processors of the $i^{th}$ row, using the row trees.

2. For all $j$, broadcast $a_j$ from the $j^{th}$ controller to all processors of the $j^{th}$ column, using the column trees. Now the processor $P_{ij}$ in position $(i, j)$ of the mesh has both $a_i$ and $a_j$ available.

3. At the processor $P_{ij}$, set variable $X$ to 1 if either $a_i < a_j$, or $a_i = a_j$ and $i < j$. Otherwise, set $X = 0$. The result of this step is that in the $j^{th}$ column, the number of $X$'s that are 1 is exactly the number of elements that precede $a_j$ in a sorted order, where ties are broken in favor of the element that initially preceded the other (that is, the sort is stable).

4. For all $j$, compute at the $j^{th}$ controller the sum of the $X$'s in the $j^{th}$ column; let this sum be $s_j$. Thus, $s_j + 1$ is the correct position for $a_j$ in the sorted order.

5. For all $j$, send the signal $flag = 1$ to the $s_j + 1^{st}$ processor in the $j^{th}$ column, that is, to the processor in row $s_j + 1$ and column $j$. Send $flag = 0$ to the other processors.

6. Each processor $P_{ij}$ with $flag = 1$ sends the value of $a_j$, which it received in step (2), to the $i^{th}$ controller, through the $i^{th}$ row tree. Thus, the $i^{th}$ controller receives $a_j$ if and only if that element belongs $i^{th}$ in the order. The elements now appear in sorted order in the controllers. $\square$

It is easy to analyze the above algorithm. Steps (1) and (2) are broadcast steps, taking $O(\log n)$ time. Step (3) is a comparison of $O(\log n)$ bit numbers, taking $O(\log \log n)$ time assuming an efficient implementation. Step (4) is a census function, taking $O(\log n \log \log n)$ time, while step (5) is the directed broadcast to a single leaf; it takes $O(\log n)$ time. Finally, step (6) can be viewed as a census function, addition, if we suppose that each processor with $flag = 0$ contributes 0 and the processor with $flag = 1$ contributes $a_j$. Thus it takes the same time as step (4). Therefore, we have the following theorem.

**Theorem 4.3:** $n$ elements of $O(\log n)$ bits can be sorted in area $O(n^{2+\epsilon})$ and time $O(n^\epsilon)$ on the mesh of trees. $\square$

Note that both Algorithm 4.2 and the sorting algorithm of Section 4.4 have an $AT^2$ value of $n^{2+\epsilon}$. This is about the best we can do (Exercise 2.3).

## A Connected Components Algorithm

We shall now consider a very fast algorithm for computing connected components, using the mesh-of-trees organization. Later, we shall consider another algorithm for the same problem that is slower, but uses a modified mesh of trees with considerably less area.

**Algorithm 4.3:** *Mesh-of-Trees Connected Components Algorithm.*

INPUT: The adjacency matrix of the graph is initially read into the array of processors, with $e_{ij}$, the bit that indicates whether there is an edge between nodes $i$ and $j$, read into processor $P_{ij}$, the processor in row $i$ and column $j$.

OUTPUT: At the end, the $i^{th}$ controller holds the group to which node $i$ belongs, i.e., the lowest-numbered node to which node $i$ is connected.

METHOD: The overall outline of the algorithm is that we start with each node in a group by itself. Repeatedly, each group $g$ finds the lowest-numbered group $h$ to which it is adjacent, in the sense that there is an edge from some node in $g$ to some node of $h$. If $h$ is lower than $g$, we merge $g$ into $h$, by giving the nodes of $g$ the same number as $h$ (which may itself change because $h$ is adjacent to a lower group, which is adjacent to a lower group, and so on).

We claim that within two repetitions of this process, every group will be merged at least once with a neighbor group, if it has any neighbor groups at all. To see why, suppose at one pass, group $g$ has one or more neighbor groups, but all its neighbors are higher numbered than $g$. If $g$ has a neighboring group $h$, and $g$ is $h$'s lowest neighbor, then $h$ and $g$ will be merged on this pass. If $h$ has a neighbor lower than $g$, then all the nodes in $h$ will be given a number lower than $g$, guaranteeing that on the next pass, $g$ will merge with some group. Thus, the number of groups that are not complete connected components is at least halved in two iterations, and after $2 \log n$ iterations, every group is a connected component by itself. Hence, within $2 \log n$ passes, all mergers of groups that will ever take place have been done, and the groups that remain are the connected components.

This process is sketched in Fig. 4.12. The algorithm uses the variables $e_{ij}$, at each processor $P_{ij}$, to represent the edge between nodes $i$ and $j$; $e_{ij} = 1$ means the edge is present and $e_{ij} = 0$ means it is not. Variable $g_i$ represents the current group of node $i$ at the $i^{th}$ controller, and $m_i$ represents the minimum group of a neighbor of node $i$. Then, for each group $g$, the $g^{th}$ controller computes variable $n_g$, which is the minimum group of a node neighboring group $g$, and $h_g$, which is the lowest group to which group $g$ is connected by a path that follows a monotonically decreasing sequence of groups.

Step (1) of Fig. 4.12 can be implemented using the following two tree operations.

1a) For each row $i$ broadcast $g_i$ to all $P_{ij}$'s, using the row trees.

1b) For each column $j$, set $m_j$ equal to the minimum of all $g_i$'s at those $P_{ij}$'s for which $e_{ij} = 1$. The effect is that $m_j$ is the minimum group of $j$'s neighbors, which are the nodes $i$ for which there is an edge from $j$ to $i$.

We can implement line (2) of Fig. 4.12 (the calculation of $n_g$, which is the minimum group that borders on group $g$), as follows.

2a) For each row $i$, broadcast $g_i$ and $m_i$ to the $P_{ij}$'s, using the row trees.

2b) For each column $j$, broadcast $j$ to the $P_{ij}$'s, using the column trees.

2c) For each processor $P_{ij}$, set $flag = 1$ if $g_i = j$, and $flag = 0$ otherwise.

Initialize by reading the edge $e_{ij}$ at processor $P_{ij}$
   and setting $g_i = i$ at the $i^{th}$ controller;
**repeat** $2 \log n$ times
(1)    **for** each node $i$ in parallel **do**
        compute $m_i = $ minimum group of a neighbor of $i$ at
          the $i^{th}$ controller;
(2)    **for** each group $g$ in parallel **do**
        find $n_g = $ minimum of $g$ and the $m_i$'s for $i$ in group
          $g$, at the $g^{th}$ controller;
(3)    **for** each group $g$ in parallel **do**
        find $h_g$, the limit of the sequence $g, n_g, n_{n_g}, \ldots$,
          at the $g^{th}$ controller;
(4)    **for** each node $i$ in parallel **do**
        $g_i := h_{g_i}$;
**end**

**Fig. 4.12.** Fast connected components algorithm.

2d) For each column $j$, compute at the $j^{th}$ controller the minimum of those $m_i$'s such that $flag = 1$ at $P_{ij}$, i.e., we compute the minimum group of any neighbor of those nodes $i$ that are in the group $g_j$; these are the nodes for which $flag$ was set to 1 in $P_{ij}$ in step (2c).

2e) At the $g^{th}$ controller, set $n_g$ equal to the smaller of $g$ and the minimum of the $m_i$'s computed in (2d).

Now, $n_g$ is the lowest-numbered neighboring group of $g$. Imagine a directed graph (*digraph*) whose nodes correspond to the groups, and for which there is an arc from $g$ to $n_g$ for each group $g$. Remembering that $n_g$ could be $g$, and that each node has only one arc out, we see that this digraph must look like Fig. 4.13. We wish to compute for each group $g$ the group we reach by following the unique path from $g$ until we can go no further. We do so by first replacing each $n_g$ by $n_{n_g}$; this group is the one we reach by following two arcs from $g$. If we make this replacement a second time, for all groups, $n_g$ will then be the group reached by following four arcs. After a third round of replacements, we get the group reached by eight arcs, and so on.

The conclusion is that after repeating the following steps $\log_2 n$ times, $n_g$ will be equal to the smallest group reachable from group $g$ along a path that travels through a decreasing sequence of groups.

3a) **repeat** $\log n$ times
   $i$    For each row $g$ broadcast $n_g$ to the $P_{gj}$'s, using the row trees.
   $ii$   For each column $j$, set $flag = 1$ in the processor $P_{n_j,j}$; set $flag = 0$ otherwise. The processor $P_{n_j,j}$, like all in row $j$, holds $n_{n_j}$, because of step ($i$).

**Fig. 4.13.** Digraph representing $n_g$.

*iii*  In each column $j$, send to the $j^{th}$ controller the value of $n_i$ in the processor $P_{ij}$ that has $flag = 1$. As a result, the value sent is $n_{n_j}$.

*iv*  For all $g$, set $n_g$ to the value, $n_{n_g}$, computed in (*iii*).

**end**

3b)  For all $g$, set $h_g$ to the final value of $n_g$ computed in (3a). Now, $h_g$ is the new number for the members of group $g$.

Finally, step (4) of Fig. 4.12 can be implemented by a sequence of steps similar to the four steps of (3a) above.

4a)  For each row $i$ broadcast $h_i$ to the $P_{ij}$'s, using the row trees.

4b)  For each column $j$, set $flag = 1$ in the processor $P_{g_j,j}$, and set $flag = 0$ at the others.

4c)  In each column $j$, send to the $j^{th}$ controller the value of $h_i$ in the processor $P_{ij}$ that has $flag = 1$. As a consequence, $h_{g_j}$ is sent to the $j^{th}$ controller.

4d)  For all nodes $j$, set $g_j$ equal to the value $h_{g_j}$ received in (4c). □

**Example 4.8:** Consider the graph of Fig. 4.14(a). Initially, each node is in a group by itself, and the number of the node and the number of its group are the same. The first time through the loop of Fig. 4.12, we compute in step (2) the values of $n_g$ represented by the digraph of Fig. 4.14(b). That is, each node $g$ in Fig. 4.14(b) points to $n_g$. Notice that because all groups are singletons, a node simply points to its lowest neighbor, or to itself if it has no lower numbered neighbor. In step (3), we chase paths in Fig. 4.14(b), until each node reaches its ultimate destination. At that time, group 1 consists of $\{1, 4, 5\}$, group 2 consists of $\{2, 6, 7, 9\}$, and groups 3 and 8 consist of the single nodes 3 and 8, respectively. There are no other groups.

Fig. 4.14(c) shows the arcs representing $n_g$ for the surviving groups during the second pass around the loop of Fig. 4.12. We merge group 3 into 1, so group 1 now consists of $\{1, 3, 4, 5\}$, and we merge 8 into 2, so the remaining nodes are in group 2. The third time around the loop, we compute the $n_g$'s represented by the digraph of Fig. 4.14(d), and in this pass we merge all nodes into group 1, meaning that there is but one connected component. □

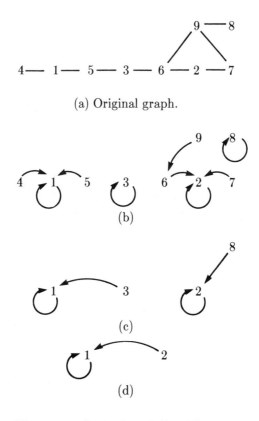

(a) Original graph.

(b)

(c)

(d)

**Fig. 4.14.** Operation of Algorithm 4.3.

**Theorem 4.4:** Algorithm 4.3 correctly computes connected components and does so in area $O(n^{2+\epsilon})$ and time $O(n^\epsilon)$ on a graph of $n$ nodes.

**Proof:** As we presented the algorithm we explained why $\log n$ iterations of step (3a) were sufficient; the length of path traversed doubles each time. We also explained why $2 \log n$ iterations of the outer loop of Fig. 4.12 were sufficient; each group that will ever be merged is merged at least every other iteration.

The area of the mesh of trees with processors that require registers of length $\log n$ has already been established to be $O(n^{2+\epsilon})$ from our analysis of Algorithm 4.2. To see the time bound, we can check that each individual step of the algorithm requires no more than $O(\log n \log \log n)$ time. There are two nested loops, each of $O(\log n)$ iterations; thus the total time is less than $O(\log^4 n)$ and, therefore, $O(n^\epsilon)$. $\square$

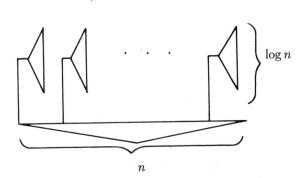

**Fig. 4.15.** Layout for connected components algorithm.

## An $AT^2 = n^{3+\epsilon}$ When-Determinate Connected Components Algorithm

We shall now consider a connected components algorithm that uses the processor organization suggested in Fig. 4.15, that is, a one-dimensional array of $n$ processors connected by a complete binary tree; each processor is the root of another tree of depth $\log \log n$, having $\log n$ nodes, which we show conceptually as a column tree. In fact, the whole arrangement is one tree of depth

$$\log n + \log \log n$$

but it is useful to view the function of the row tree and column trees differently.

The algorithm we now describe consists of two phases. First, the adjacency matrix is read in, row by row, and we compute from it a new graph with different edges but the same connected components. There is a strong limit on the amount of time needed to process a row, so the input phase can be made when-determinate without slowing up the algorithm by more than a constant factor. In the second phase, we run Algorithm 4.1 on the new graph, generating the rows of its adjacency matrix from the edge information that we have stored in the processors.

The new graph has a special set of edges. First, we record an edge at most once; an edge $(i, j)$ may be recorded in column $i$ if $i < j$ and in column $j$ if $j < i$. Only $\log_2 n$ edges can be recorded for any node. We use the $i^{th}$ column tree to record, one to a leaf, certain edges $(i, j)$, for $j > i$. However, we do not even have free choice of the $\log n$ edges. Rather, the edges that might emanate from node $i$ are divided into *ranges*. Node $j$ is in range $r$ (with respect to $i$) if $2^{r-1} \leq j - i < 2^r$. Thus, only node $i + 1$ is in range 1, nodes $i + 2$ and $i + 3$ are in range 2, $i + 4$ through $i + 7$ are in range 3, and so on.

We shall use the notation $edge(i, r)$ to refer to the edge in range $r$ stored in the column tree for node $i$; $edge(i, r) = 0$ if there is no such edge. We also

```
        procedure tell(i, j);
        begin
(1)         r := range(j - i);
(2)         if edge(i, r) = 0 then edge(i, r) := j
(3)         else if edge(i, r) < j then tell(edge(i, r), j)
(4)         else if edge(i, r) > j then tell(j, edge(i, r))
            { if edge(i, r) = j do nothing }
        end
```

**Fig. 4.16.** Recursive edge insertion algorithm.

use $range(k)$ for the range of $k$, that is, $\lfloor \log_2 k \rfloor + 1$.

The heart of the algorithm is a procedure $tell(i, j)$ that is run when we must take account of the fact that node $i$ is connected to node $j > i$. If possible, we would like to "tell" node $i$ about $j$ by setting $edge(i, range(j - i))$ to $j$. However, there may already be a value for that edge, in which case $tell$ must make a recursive call to insert another edge that will keep the desired connectivity. A key observation is that the range involved in the call is strictly less than $range(j - i)$, so after $\log n$ calls the process must stop. The procedure is defined formally in Fig. 4.16. The following lemmas about $tell$ will prove useful.

**Lemma 4.2:** If there are $n$ nodes, a call to $tell(i, j)$ cannot result in more than $\log_2 n$ recursive calls to $tell$.

**Proof:** Let $k = edge(i, r)$. Then, as both $j$ and $k$ are in range $r$ with respect to $i$, we have $2^{r-1} \leq k - i < 2^r$ and $2^{r-1} \leq j - i < 2^r$. It follows that $| k - j | < 2^{r-1}$. Thus, whether we call $tell(k, j)$ at line (3) or $tell(j, k)$ at line (4), the range computed in line (1) of the recursive call to $tell$ will be at most $r - 1$. Since we start with a range no greater than $\log_2 n$, and the range decreases with each call, at most $\log_2 n$ calls can be made. $\square$

**Lemma 4.3:** If we call $tell(i, j)$, then the edges actually stored in the column trees will represent a graph with the same connectivity as the graph before the call, plus the edge $(i, j)$.

**Proof:** The result is obvious in the case $edge(i, r) = 0$, where we simply add the edge $(i, j)$ to those in the graph, and in the case $edge(i, r) = j$, where no change to the graph is needed. Otherwise, let $k = edge(i, r)$. In lines (3) and (4) we add the edge $(k, j)$ instead of adding $(i, j)$. However, since the edge $(i, k)$ is recorded, the edge $(k, j)$ also allows us to go from $i$ to $j$, and does not allow us to go between two nodes that we could not travel between using the edges $(i, j)$ and $(i, k)$. While the recursive call to $tell$ may not result in adding $(k, j)$, but may instead add some other edge, we need a simple induction on the range $r$ that the resulting graph will have a path between $k$ and $j$, and, therefore, the

graph will have a path between $i$ and $j$. This induction is left as an exercise.
□

Before giving the algorithm formally, we must discuss the way the row tree is used to communicate among the nodes. That is, when *tell* is called recursively in line (3) or (4) of Fig. 4.16, the recursive call must be implemented at a column other than $i$. In line (3) the call is at column $edge(i, r)$, and in line (4) the call is at column $j$. We say that column $i$ sends a *packet* to whichever of these columns is appropriate. The packet consists of the destination and the second argument of the call to *tell*. It is sent along the unique route in the row tree, from the leaf of the row tree for column $i$ to the leaf for the destination column.

We shall assume that each interior node of the row tree has room to store $\log_2 n$ packets, although in any one time unit, at most one packet may leave any node, and at most three may arrive, one from each child and one from the parent. As we make recursive calls to *tell* from various columns, packets may follow routes that cross each other, causing delays. However, we shall prove that no packet ever meets as many as $2 \log_2 n$ other packets at any one node, so while it may have to wait $O(\log n)$ time to move from an interior node of the row tree, it cannot be delayed more.

**Algorithm 4.4:** *Mayr-Siegel Connected Components Algorithm.*

INPUT: An adjacency matrix of an $n$-node graph, read row by row into the leaves of the row tree in the structure of Fig. 4.15. We assume that in row $i$, bit $i$ is 1, and the only other bits that are 1 are in positions above $i$. That is, we read the symmetric matrix in with its lower triangle zeroed and the diagonal set to 1's.

OUTPUT: Ultimately, the $i^{th}$ leaf of the row tree will contain the connected component of $i$, that is, the lowest numbered node connected to $i$.

METHOD: We assume that each leaf of the row tree knows its number; that number can easily be computed by fast operations on the row tree. The value of $edge(i, r)$ is set to 0 for all $i$ and $r$, and we then do the following steps for each row that is read in.

1.  Each leaf of the row tree that reads input 1 sends its number to the closest leaf to its left that has also read a 1. The method used is essentially that of Example 3.1, although there messages were passed to the right and the last element was cycled to the place of the first.

2.  Each node $i$ that receives a message from node $j > i$ in step (1) executes $tell(i, j)$. If that call to *tell* results in a recursive call to $tell(k, \ell)$, a packet is sent from column $i$ to column $k$, and $tell(k, \ell)$ is executed at column $k$. Similarly, all recursive calls to *tell* are executed at the column of the first argument of the call.

    After executing the above two steps for each row, we are left with a

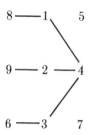

**Fig. 4.17.** Input graph for Example 4.9.

collection of edges whose connectivity is the same as that of the original graph, by Lemma 4.3. We therefore may construct from the stored edges a new adjacency matrix that is fed row by row into Algorithm 4.1, which runs on the row tree only. Row $i$ of this new adjacency matrix needs to represent only the edges $(i, j)$ for $j > i$; these are exactly the edges stored in the $i^{th}$ column tree. The information corresponding to these edges can be read out of the $i^{th}$ column tree and distributed to the desired leaves of the row tree in $O(\log^2 n)$ time, in an obvious fashion. Algorithm 4.1 may call for new rows whenever it wants, since the input schedule is not dependent on when new rows are to be read in; the rows are already in the data structure. Thus, the whole algorithm is when-determinate. $\square$

**Example 4.9:** Consider the graph of Fig. 4.17. When we read row 1 into Algorithm 4.4, the leaves for 1, 4, and 8 hold the value 1; the others hold 0. Therefore, leaf 1 receives a message from 4 and calls $tell(1, 4)$, while leaf 4 receives a message from 8 and calls $tell(4, 8)$. Since $edge(i, r)$ is set to 0 initially, both of these calls to $tell$ are implemented directly, by setting $edge(1, 2) = 4$ and $edge(4, 3) = 8$. Note that with respect to 1, 4 is in range 2, since

$$2^1 \leq 4 - 1 < 2^2$$

and 8 is in range 3 with respect to 4.

When we read the second row, 2 is told about 4, and 4 is told about 9. The call to $tell(2, 4)$ at column 2 simply results in setting $edge(2, 2) = 4$. However, the call $tell(4, 9)$ at column 4 cannot be executed directly, because 9 is in range 3 with respect to 4, and $edge(4, 3)$ is already set to 8. As $8 < 9$, line (3) of $tell$ applies, and we call $tell(8, 9)$. That results in a packet being sent from column 4 to column 8. The call to $tell$ there results in $edge(8, 1)$ being set to 9.

Next, consider row 3. Columns 3, 4, and 6 receive 1's, so 3 is told about 4 and 4 is told about 6. The calls to $tell(3, 4)$ and $tell(4, 6)$ result in no further calls, but the new values $edge(3, 1) = 4$ and $edge(4, 2) = 6$ are entered. Note that the fact $edge(4, 2)$ is set to 6 does not affect the value of $edge(4, 3)$, which is 8.

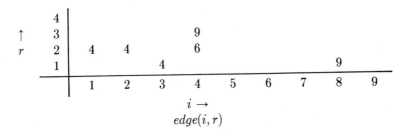

$i \rightarrow$

$edge(i, r)$

**Fig. 4.18.** Edges constructed by Algorithm 4.4.

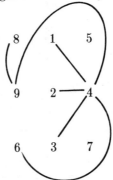

**Fig. 4.19.** Output graph for Example 4.9.

Row 4 results in no new edges, since 4 is only connected to nodes with lower numbers. In fact we have now accounted for all the edges, since all of nodes 4 through 9 are connected only to lower numbered nodes. The resulting table of *edge* values is shown in Fig. 4.18, and the graph with these edges is shown in Fig. 4.19. □

**Running Time of Algorithm 4.4**

We shall now prove that Algorithm 4.4 takes time $O(n^{1+\epsilon})$. Since we have already that result for Algorithm 4.1, we need to prove only that the time spent by Algorithm 4.4 on steps (1) and (2) is limited by $O(n^{1+\epsilon})$. For that, we need to show that the time spent for any individual row is some power of $\log n$. That clearly holds for step (1) by the analysis that accompanied Example 3.1.

The analysis of step (2) is far more subtle. Although the initial calls to *tell* can surely be carried out in parallel at separate leaves of the row tree, packets sent by recursive calls from these can sometimes meet one another.

**Example 4.10:** Suppose nodes 20, 36, and 40 are three consecutive nodes whose leaves receive 1 in the current row, so we call *tell*(20, 36) and *tell*(36, 40). Suppose also that *edge*(20, 5) = 50, so from *tell*(20, 36) we call *tell*(36, 50). Also, let $edge(36,3) = 43$, so from $tell(36,40)$ we call $tell(40,43)$. For one interference, the packet from 20 to 36 telling 36 to call $tell(36,50)$ could meet the packet from 36 to 40 announcing $tell(40,43)$. However, much more complex interactions are possible. For example, if $edge(36,4) = 45$, a packet from 36 to 45 will announce $tell(45,50)$. If $edge(40,2) = 42$, that packet could meet the packet announcing $tell(42,43)$ that travels from 40 to 42. In particular, each of these packets will have to go through the node of the row tree that is an ancestor of all of nodes 33 through 48 (assuming the nodes are numbered starting at 1 and the tree is a complete binary tree). $\square$

We need several lemmas that put a limit on how much interaction there can be. The results we give are not as tight as they could be, but they serve to prove the order-of-magnitude result that we want. The proof of the tightest possible result is left as an exercise.

**Lemma 4.4:** Suppose a call to $tell(i,j)$ results in a call to $tell(x,y)$. Then $i < x, y \leq i + 4(j - i)$. That is, $tell(i,j)$ can have no effect further to the right of $j$ than three times the distance between $i$ and $j$.

**Proof:** Let $range(j - i) = r$. We prove by induction on $r$ that $x$ and $y$ are at most $i + 2^{r+1}$. Since $j - i \geq 2^{r-1}$, the lemma then follows. The basis, $r = 1$ is trivial. For the induction, suppose $tell(i,j)$ results in a call $tell(j,k)$, where $k = edge(i,r) > j$. The case where $k < j$ is similar and left as an exercise. First, we observe that $k < i + 2^r$ and $j - i \geq 2^{r-1}$, so the induction holds for $x = j$ and $y = k$.

Suppose that $tell(j,k)$ results in a call to $tell(x,y)$. By Lemma 4.2,

$$range(k - j) \leq r - 1$$

Thus we may apply the inductive hypothesis and claim that $x, y \leq j + 2^r$. Since $j \leq i + 2^r$, we have $x, y \leq 2^{r+1}$, and the induction is proved. $\square$

Next, we must develop a bound on how many packets can be at an interior node of the row tree at any time. Our approach is to consider how many packets can cross a particular boundary. That is, we assume the leaves of the row tree are numbered $1, 2, \ldots, n$, and we ask how many packets can be sent at one time from some node $i < b$ to some node $j \geq b$. These are the packets that *cross boundary b*.

One useful way to divide packets is by *family*. That is, each of the original calls to $tell(i,j)$ in step (2) of Algorithm 4.4 results in zero or more recursive calls to *tell*. The packets announcing these calls are in the same family; otherwise, packets are in different families. Lemma 4.4 tells us that all the packets of the family of $tell(i,j)$ cross only boundaries that lie between $i$ and $i + 4(j - i)$. Moreover, since only one call from a family is active at any time, no two packets in the same family can ever be at the same place at the same time. We use these facts in the next lemma.

**Lemma 4.5:** There is a constant $c$ such that at most $c \log_2 n$ families have packets that cross boundary $b$, when step (2) of Algorithm 4.4 is run for a particular row of the adjacency matrix.

**Proof:** Let $\ldots, i_3, i_2, i_1$ be the sequence of leaves that

1. Received input 1 for some row of the adjacency matrix,
2. Lie to the left of the chosen boundary $b$, and
3. Send packets across boundary $b$.

By Lemma 4.4,

$$i_{j-1} - i_j \geq \frac{1}{3}(b - i_{j-1}) \tag{4.3}$$

Note that the lemma assumes $i_{j-1}$ and $i_j$ are consecutive nodes that were given input 1, but if there are such nodes intervening, whose families do not send packets across boundary $b$, it is easy to show that (4.3) must still hold.

We show by induction on $j$, that $i_j \leq b - (\frac{4}{3})^{j-1}$. The basis, $j = 1$, is trivial, since $i_1 < b$, i.e., $i_1$ is to the left of the boundary. For the induction, (4.3) gives us

$$i_j \leq \frac{4}{3} i_{j-1} - \frac{1}{3} b$$

By the inductive hypothesis, this becomes

$$i_j \leq \frac{4}{3}(b - (\frac{4}{3})^{j-2}) - \frac{1}{3} b = b - (\frac{4}{3})^{j-1}$$

proving the induction. It follows that since $b \leq n$, and $i_j$ cannot be negative, that $(\frac{4}{3})^{j-1} \leq n$. Therefore, $j \leq 1 + \log_{4/3} n$, which means that the number of different $i_j$'s whose families can send packets across boundary $b$ during the processing of any one row of the adjacency matrix is $O(\log n)$. $\square$

**Theorem 4.5:** Algorithm 4.4 takes area and time $O(n^{1+\epsilon})$, correctly computes connected components, and can be made when-determinate.

**Proof:** The area of the structure of Fig. 4.15 is easily seen to be $O(n^{1+\epsilon})$, since there are $O(n \log n)$ nodes in the row and column trees, and each node only needs to store a number of size $n$, i.e., $O(\log n)$ bits.

The correctness follows from Lemma 4.3 and the correctness of Algorithm 4.1. The fact that the input schedule can be made when-determinate, while keeping the time bound claimed will follow when we prove that the time to process each row is upper bounded by $O(\log^3 n)$.

Therefore, let us consider how long it could take all the recursive calls to *tell* to be completed, for one row of the matrix. First, a call to *tell* requires that the lookup of an *edge* value in a column tree be accomplished. However, this task is a binary search in a tree of depth $\log \log n$; thus this time will be seen to be much less than the maximum time needed to send a packet,

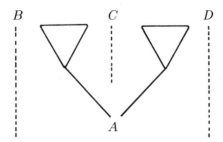

**Fig. 4.20.** Boundaries whose crossing can result in a packet at node $A$.

and we can ignore it. The comparison to determine which case of *tell* applies also can be accomplished in time $O(\log \log n)$ if we use the divide-and-conquer implementation of arithmetic; this time will be negligible as well.

Now, let us consider the time that a packet could spend on its journey. First, its route in the row tree takes it to at most $2 \log n$ nodes. We claim that at any one of these nodes, such as $A$ in Fig. 4.20, it can wait for at most $3c \log_2 n$ packets, where $c$ is the constant from Lemma 4.5, if a first-come-first-served strategy is used to determine which packet is forwarded at any time. To see why, we first observe that the families that send packets through node $A$ in Fig. 4.20 are limited to those that cross boundaries $B$, $C$, or $D$ in that figure. For any packet that does not either

1.  Have a source and destination exactly one of which is between $B$ and $D$, or

2.  Have a source between $B$ and $C$ and a destination between $C$ and $D$, will not pass through $A$.

The number of families that can send a packet across one of the three boundaries shown in Fig. 4.20 is limited to $3c \log_2 n$ by Lemma 4.5. If we use a first-come-first-served strategy to handle conflicts at a node, no packet can wait for two packets from the same family. Thus, we have an $O(\log n)$ upper bound on the delay of a packet at any one node. The fact that a packet travels a route of length $O(\log n)$ tells us that $O(\log^2 n)$ is an upper bound on the length of time it takes any one packet to travel. Then, by Lemma 4.2, a family of packets consists of at most $\log_2 n$ packets. By the analysis of the time taken by a single call to *tell*, one packet of the family is launched within time $O(\log \log n)$ of the arrival of the previous packet in the family, so each family completes its activity in time $O(\log^3 n)$.

Since families of calls to *tell* operate in parallel, we have shown an $O(\log^3 n)$ upper bound on the time it takes Algorithm 4.4 to process a row of the adjacency matrix. Thus, we may adapt the input schedule to match this upper bound, creating a when-determinate algorithm running in time $O(n \log^3 n)$, which is surely $O(n^{1+\epsilon})$. $\square$

**EXERCISES**

4.1: Design algorithms, which may be when-indeterminate, to run on the tree of processors organization and have area and time $O(n^{1+\epsilon})$, to solve the following problems.

    a)   Determine whether the graph is bipartite, i.e., the nodes can be divided into two groups, such that no edge connects two nodes from the same group.

    b)   Determine whether the graph is acyclic.

    c)   Generate the adjacency matrix of a spanning forest for the graph. A *spanning forest* is a subset of the edges that connects all nodes that are in the same connected component, yet contains no cycles.

4.2: Solve the same problems as Exercise 4.1 on the mesh of trees, with $AT^2 = O(n^{2+\epsilon})$.

\* 4.3: Give an $AT^2 = O(n^{2+\epsilon})$ algorithm on the mesh of trees to find a minimal spanning forest. That is, let processor $P_{ij}$ start out with the "weight" of edge $(i, j)$, or infinity if there is no such edge. At the end, let $P_{ij}$ hold 1 if the edge $(i, j)$ is part of a particular spanning forest, the sum of whose edge weights is as small as possible, and hold 0 if not.

\*\* 4.4: Implement the problems of Exercise 4.1 on the structure of Algorithm 4.4. Make your algorithms when-determinate, with $AT^2 = O(n^{3+\epsilon})$.

4.5: Give "divide-and-conquer" circuits taking area $O(n \log n)$ and time $O(\log n)$ to perform the following operations on $n$-bit words.

    a)   Logical "and".

    \* b)   Addition.

4.6: Rewrite the bubblesort program of Fig. 4.2 to make its operation independent of $n$, the number of elements to be sorted.

4.7: Show how to calculate the depth of a complete binary tree and let each leaf know its position from the left, in time $O(n^\epsilon)$.

4.8: Complete the correctness proof of Algorithm 4.1 by performing the induction indicated at the beginning of the proof of Theorem 4.1.

4.9: Show the steps performed by the odd-even merge algorithm when it merges the lists $1, 5, 9, 12$ and $2, 3, 6, 8$.

4.10: Show how it is possible, in $O(\sqrt{n})$ time, to have each node in an $n$-node rectangular grid determine

    a)   Its coordinates.

    b)   After recursively dividing the grid in half $k$ times vertically and $\ell$ times horizontally, how far from the left and top edges is the node, in the rectangle to which it belongs?

4.11: Give the details of how to go from Fig. 4.7(d) to Fig. 4.7(e).

4.12: Algorithm 4.2, the mesh of trees sorting algorithm, requires that processor $P_{ij}$ know if $i < j$. Show how this information can be determined for all the processors, in time $O(n^\epsilon)$.

* 4.13: Show that storage of packets in row tree nodes by Algorithm 4.4 is not really necessary. That is, in the same order of magnitude time, we can arrange a schedule of the packets so that no interference is possible.

4.14: Complete the proof of Lemma 4.4 by considering the case $k < j$.

4.15: Show how it is possible, in Algorithm 4.4, to distribute the contents of a single column tree to the relevant leaves of the row tree in time $O(\log^2 n)$, thereby simulating the input, to Algorithm 4.1, of a row of the adjacency matrix.

4.16: Show how the lookup of $edge(i, r)$ in the $i^{th}$ column tree can be accomplished in $O(\log \log n)$ time.

## BIBLIOGRAPHIC NOTES

Algorithm 4.1 and the notion of a census function are from Lipton and Valdes [1981]. Exercise 4.1 is also from there. The sorting algorithm of Section 4.4 is from Thompson and Kung [1977]. Algorithm 4.4, as well as the results of Exercise 4.4, are based on an oral communication from Ernst Mayr and Alan Siegel.

Solution of the connected components problem in the VLSI or parallel computing environment has received recent attention by Hirschberg, Chandra, and Sarwate [1979], Nassimi and Sahni [1980], Hambrusch [1981], and Hambrusch and Simon [1982]. Parallel algorithms for other graph-theoretic problems have been examined recently by Savage [1977], Dekel, Nassimi, and Sahni [1979], Kosaraju [1979], Chin, Lam, and Chen [1982], and Hirschberg and Wolper [1982]. The reader should note that not all these papers assume the most realistic model of VLSI computation. Chapter 6 discusses how to convert "parallel" algorithms, where processors are each allowed to access any datum, regardless of the processor interconnection pattern, into realistic VLSI algorithms, with relatively little slowdown.

A number of other papers deal with the solution of other familiar problems in the VLSI model of computation. For example, see Luk [1981], Preparata [1981], and Preparata and Vuillemin [1981] on integer multiplication, Preparata and Vuillemin [1980] on matrix multiplication. Also, see Shiloach and Vishkin [1981] and Thompson [1981] for more on sorting.

# 5

# SYSTOLIC ALGORITHMS

This chapter introduces the idea of a systolic array of processors and algorithms based on arrays of this type. Systolic arrays pass data from one processor to neighboring ones in a regular, rhythmic pattern. Frequently, although not necessarily, the processors are also laid out in a regular pattern, such as a one- or two-dimensional array. The merit of this approach lies principally in the ability of systolic algorithms to use a limited number of input/output pads, while operating at roughly at the same speed as more obvious algorithms that require many more input pads. Further, if the processors are arranged in a grid-like way, systolic algorithms will minimize propagation delay among processors, allowing the clock cycle to be relatively short.

After an introduction showing how one might go about designing a systolic algorithm, we study mechanical ways of converting nonsystolic algorithms to systolic ones. We then consider several important examples of systolic algorithms: matrix multiplication, transitive closure, and applications of these.

## 5.1 INTRODUCTION: SYSTOLIC CONVOLUTION

In this section we shall go through a possible design process resulting in a systolic algorithm. The particular problem with which we deal is *convolution*, where we are given two sequences of numbers $a_0, \ldots, a_{n-1}$ and $b_0, \ldots, b_{n-1}$ and we produce the sequence of double the length $c_0, \ldots, c_{2n-1}$, where

$$c_i = \sum_{j=0}^{n-1} a_j b_{i-j}$$

For example, $c_0 = a_0 b_0$, $c_1 = a_1 b_0 + a_0 b_1$, and $c_{2n-1}$ is always 0. Another way to look at convolution is that $c_i$ is the coefficient of $x^i$ in the product of polynomials

$$\left( \sum_{i=0}^{n-1} a_i x^i \right) \times \left( \sum_{j=0}^{n-1} b_j x^j \right)$$

The convolution operation is important in digital signal processing, and it is also closely related to integer multiplication, in the case where the $a$'s and $b$'s are each 0 or 1.

Let us imagine that we have a linear array of processors, which might actually be in a row, or they might be snaked around a chip to make the aspect ratio reasonably low. However, to limit the number of pads, let us suppose that only the processors on the ends have pads; the number of pads at each end will be the number of bits needed to represent the $a$'s, $b$'s, or $c$'s.

Since we cannot read all the $a$'s or all the $b$'s at once, let us do the next best thing and read the $a$'s one at a time, from the pads at the left. We shall have to do something similar with the $b$'s, and we must write the $c$'s at one end, one at a time, as well. There are many ways to approach the design after we have agreed to pipe the $a$'s in from the left; we shall pick one way here. Let us suppose that the $c$'s travel from the right. Initially they are each zero, but as they travel, they meet pairs of $a$'s and $b$'s, and the products of these $a$'s and $b$'s are added to the $c$. In order that the algorithm work properly, it is required that every $a$–$b$ pair that $c_i$ meets must be part of the sum for $c_i$; that is, $c_i$ must meet only pairs $a_j$–$b_{i-j}$.

One other essential observation is that the streams of $a$'s and $c$'s must travel in alternate processors. The reason is that if they were in consecutive processors, and each moved at each beat, half the $a$'s would skip over a given $c$, and there would be no hope that the $c$ could ever collect a term involving one of the $a$'s that it missed.

The pattern we wish to achieve is shown in Fig. 5.1. There we see the first five beats from the time the $a$ stream meets the $c$ stream. We have indicated the desired member of the $b$ stream that should appear at each point where an $a$ meets a $c$; the proper value is $b_{i-j}$ when $a_j$ meets $c_i$, provided $i \geq j$; if $i < j$, then no $b$ value should be used.

The pattern of Fig. 5.1 should be clear. The value $b_0$ is used by the processor in the middle column of Fig. 5.1. No other processor needs that value. Similarly, $b_1$ is needed only at the next processor to the right; $b_2$ is needed only at the processor to the right of that, and so on.

We see two ways to provide the $b$'s where needed. In each case, the actual array of processors extends only from the middle of Fig. 5.1 to the right. That is, the values $a_2$ and $a_1$ shown in the first beat of that figure represent values that will be read in at future beats. Similarly, the $c$'s to the left of center represent values that have already been written out. The approaches to be considered are the following.

1.  Pipe the $b$'s in from the right, with $b_i$ traveling along with $c_i$ (which is not read in, but is initialized to 0). When any $b$ meets $a_0$, which is identified by a special flag traveling with it, the $b$ freezes where it is.

2.  The $b$'s are read in from the left end along with the $a$'s. However, the

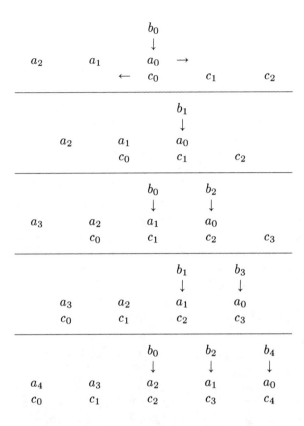

**Fig. 5.1.** Beats of a systolic convolution algorithm.

$b$'s travel twice as fast as the $a$'s, i.e., the beats shown in Fig. 5.1 really represent two beats. When a $b$ catches up with $a_0$, again identified by a flag, the $b$ freezes.

Method (1) has the disadvantage that the $b$'s must travel from right to left through $n$ processors before they begin to be used. In comparison, with method (2), the $c$'s can begin coming out at the second beat. The fact that we must wait for the output to begin in method (1) does not necessarily mean that it finishes after method (2), because the latter, as we shall see, requires more beats than (1). The most significant advantage of method (2) is that its output can be piped to another systolic process with only a one-beat delay, rather than an $n$ beat delay. That might be significant if, say, we were multiplying together many polynomials, and the $c$'s from one product became the $a$'s for the next product. Thus, we choose to consider method (2) in detail.

The algorithm is sketched in Fig. 5.2. There, we see two consecutive beats of the process. The first shifts all the variables; the second shifts only $b$'s, which we call *traveling_b* when they are shifted and *b* when they are finally frozen. These two beats are repeated $2n$ times, so all the data can be read in and the answers written out, all at the left end. In addition to the variables $a$, $b$, $c$, and *traveling_b*, the flag *leading_edge* is used to mark the position of $a_0$. We assume that $a$, $b$, $c$, *leading_edge*, and *traveling_b* are initialized to 0 in every processor.

The total number of processors is $n$. We can see why from the fact that as $a_0$ moves right $n$ times, it meets $c_0, \ldots, c_{n-1}$, which are all the $c$'s it needs to meet. Similarly, each $a$-value needs to meet $n$ $c$-values, which it does in the first $n$ processors it reaches. Finally, the $n$ $b$-values also require exactly $n$ processors for their storage.

**Example 5.1:** Figure 5.3 shows the first five beats of the algorithm of Fig. 5.2. They correspond to the first three beats in Fig. 5.1. In each beat the four rows represent $a$, *traveling_b*, $b$, and $c$. The center column represents the leftmost processor; everything to the left of center represents $a$'s and $b$'s that we shall read in in the future. We allow $c$'s to "fall off the left edge," representing the fact that they are written out. $\square$

### Timing Analysis of Systolic Convolution

It is easy to deduce that if the $a$'s and $b$'s are $O(\log n)$ bits long, then the convolution of $n$-bit streams uses both time and area $O(n^{1+\epsilon})$, for an $AT^2$ product of $O(n^{3+\epsilon})$.

On the other hand, it is not hard to show that the information content of convolution is $\Omega(n)$. Let the $b$'s each be 0 or 1, and let all of the $a$'s but one be 0; the remaining one is 1. Then the effect of the convolution is to shift the $b$'s by some amount, and the proof that the information content is $\Omega(n)$ is similar to Example 2.5 on barrel shifting. Thus, a lower bound of $\Omega(n^2)$ on $AT^2$ is immediate.

The discrepancy between the upper and lower bounds lies primarily in the fact that systolic algorithms are appropriate only when the input/output rate must be limited. If we consider only algorithms for convolution that read some limited number of bits at any time unit, independent of $n$, as the systolic algorithm does, then we see that $T = \Omega(n)$ is necessary. We shall leave as an exercise the fact that if a constant number of bits are read at a time, then after reading half of the bits, $\Omega(n)$ bits must be remembered, or we shall not be able to compute the convolution correctly. Thus, an $AT^2 = \Omega(n^3)$ bound is provable, given this constraint on the input/output rate, and among algorithms with limited input/output, the algorithm of this section is close to optimal.

### 5.2 TRANSFORMATION RULES FOR SYSTOLIC ALGORITHMS

In this section we shall give a few simple rules that enable us to change the

**beat begin**
 { shift $a$'s }
 **if** not right end **then send** $a$ **to** right;
 **if** left end **then read** $a$
 **else receive** $a$ **from** left;
 { similar steps are needed to shift *leading_edge* and *traveling_b* }
 **if** not right end **then send** *leading_edge* **to** right;
 **if** left end **then read** *leading_edge*
 **else receive** *leading_edge* **from** left;
 **if** not right end **then send** *traveling_b* **to** right;
 **if** left end **then read** *traveling_b*
 **else receive** *traveling_b* **from** left;
 { now we pass $c$ left, initializing new $c$'s to 0 at the right }
 **if** left end **then write** $c$
 **else send** $c$ **to** left;
 **if** not right end **then receive** $c$ **from** right
 **else** $c := 0$;
 { now we freeze a $b$ if we are at the leading edge of the $a$'s }
 **if** *leading_edge* **then**
  $b := traveling\_b$;
 { regardless, we add the $a * b$ product to $c$; if we are outside the
  range where $ab$ is meaningful, the product will be 0 }
 $c := c + a * b$;
**beat end;**
**beat begin**
 **if** not right end **then send** *traveling_b* **to** right;
 **if** left end **then read** *traveling_b*
 **else receive** *traveling_b* **from** left;
**beat end**

**Fig. 5.2.** Systolic convolution algorithm.

timing of data flow in systolic networks and, more importantly, convert some nonsystolic algorithms into systolic ones. We begin by developing a notation for the timing of data flow and then show how these diagrams can be transformed.

### Timing Diagrams

Let us suppose we have a collection of processors and some named streams of data that flow through them, as did the streams $a$, $c$, *traveling_b* (which we shall here refer to as $b$), and *leading_edge* (here called $\ell$). In each case, the streams remained in a fixed order, although they traveled at different rates. Note that wires carrying the utilities, $VDD$, ground, and clock signals, are not

|  |  |  |  |  |  |  |  |  |
|---|---|---|---|---|---|---|---|---|
| $a_2$ | 0 | $a_1$ | 0 | $a_0$ | 0 | 0 | 0 | 0 |
| $b_4$ | $b_3$ | $b_2$ | $b_1$ | $b_0$ | 0 | 0 | 0 | 0 |
|  |  |  |  | $b_0$ | 0 | 0 | 0 | 0 |
|  |  |  |  | $c_0$ | 0 | $c_1$ | 0 | $c_2$ |

|  |  |  |  |  |  |  |  |  |
|---|---|---|---|---|---|---|---|---|
| $a_2$ | 0 | $a_1$ | 0 | $a_0$ | 0 | 0 | 0 | 0 |
| $b_5$ | $b_4$ | $b_3$ | $b_2$ | $b_1$ | $b_0$ | 0 | 0 | 0 |
|  |  |  |  | $b_0$ | 0 | 0 | 0 | 0 |
|  |  |  |  | $c_0$ | 0 | $c_1$ | 0 | $c_2$ |

|  |  |  |  |  |  |  |  |  |
|---|---|---|---|---|---|---|---|---|
| 0 | $a_2$ | 0 | $a_1$ | 0 | $a_0$ | 0 | 0 | 0 |
| $b_6$ | $b_5$ | $b_4$ | $b_3$ | $b_2$ | $b_1$ | $b_0$ | 0 | 0 |
|  |  |  |  | $b_0$ | $b_1$ | 0 | 0 | 0 |
|  |  |  |  | 0 | $c_1$ | 0 | $c_2$ | 0 |

|  |  |  |  |  |  |  |  |  |
|---|---|---|---|---|---|---|---|---|
| 0 | $a_2$ | 0 | $a_1$ | 0 | $a_0$ | 0 | 0 | 0 |
| $b_7$ | $b_6$ | $b_5$ | $b_4$ | $b_3$ | $b_2$ | $b_1$ | $b_0$ | 0 |
|  |  |  |  | $b_0$ | $b_1$ | 0 | 0 | 0 |
|  |  |  |  | 0 | $c_1$ | 0 | $c_2$ | 0 |

|  |  |  |  |  |  |  |  |  |
|---|---|---|---|---|---|---|---|---|
| $a_3$ | 0 | $a_2$ | 0 | $a_1$ | 0 | $a_0$ | 0 | 0 |
| $b_8$ | $b_7$ | $b_6$ | $b_5$ | $b_4$ | $b_3$ | $b_2$ | $b_1$ | $b_0$ |
|  |  |  |  | $b_0$ | $b_1$ | $b_2$ | 0 | 0 |
|  |  |  |  | $c_1$ | 0 | $c_2$ | 0 | $c_3$ |

**Fig. 5.3.** Details of systolic convolution.

considered data streams, since these signals are either constant, in the case of $VDD$ and ground, or repetitive and not influenced by the time unit at which they arrive at a processor, as in the case of the clock signals.

We can represent the flow if we draw the processors as nodes of a directed graph, with an arc from node $n$ to node $m$ labeled $X(d)$ to signify that the stream of data named $X$ flows from processor $n$ to processor $m$ with nonnegative integer delay $d$. By that we mean the element in the $X$ stream received by $n$ at one beat will be received by $m$ exactly $d$ beats later.

If the value of $X$ is computed or modified at a node, then we require that all the data needed for that computation be available on the beat at which the $X$-value itself arrives. The only case where it makes sense for $X$ to be passed from a node with delay $d = 0$ is if the value of $X$ is passed as received and not modified in any way.

As mentioned, we assume $d \geq 0$. We shall take as a formal definition of "systolic" that a network is *systolic* if and only if $d \geq 1$ for all delays $d$ in the

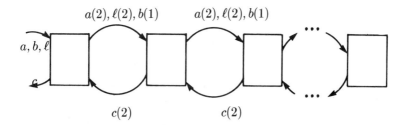

**Fig. 5.4.** Timing diagram for convolution.

network.

There may be gaps in the streams, as represented by the 0's in Fig. 5.3; that is, there may be beats at which the value of the element in that stream is ignored by the processors. Further, we allow streams to branch and join, in the sense that more than one copy of the stream may leave a node, and there may be more than one input to a node that is labeled by the same stream name. In effect, we assume that the processor at the node can distinguish among different inputs with the same name. The stream names are just for convenience, to help the reader follow the data through the network.

Let us also assume that any network we design "works," i.e., it performs whatever function it is intended to perform. We shall deal only with transformations of the network that change the delay on various arcs, and we require that the transformation preserve whatever function the network is required to perform, perhaps after redesigning the processors at the nodes.

**Example 5.2:** In Fig. 5.4 we see the algorithm of Section 5.1 represented by a timing diagram. That is, the streams $a$ and $\ell$ flow right, with delay 2, i.e., they are passed on alternate beats. Stream $b$ flows right on every beat, so it has delay 1. Stream $c$ flows left on alternate beats. We also indicate that the source of streams $a$, $b$, and $\ell$ is on the left, as is the sink (place where the output occurs) for stream $c$. $\square$

**Example 5.3:** Now, let us see the timing diagram of a circuit that is not systolic, yet can be viewed as consisting of several interacting data streams flowing through a network of processors. The example we shall consider is the regular expression recognizing network of Fig. 3.15.

There are two signal streams in this network. One, the *control stream* $c$, is the bit that tells each node whether or not it is "on." Fork and or-nodes pass this signal to their successors with a delay of 0, so we shall see $c(0)$ on every arc out of a fork or or-node. Recognizer nodes also pass the $c$ signal, possibly modified. That is, if the recognizer fails to see its desired input it does not turn its successors on, i.e., it passes $c = 0$, while if it sees its input it passes on whatever value of $c$ it received. There is a unit of delay associated with

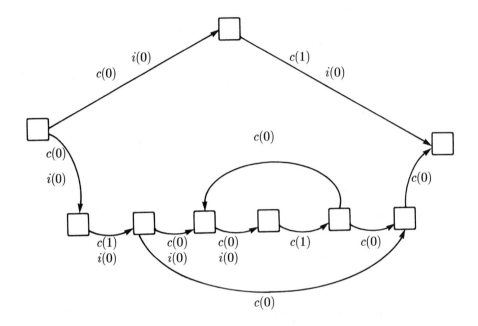

**Fig. 5.5.** Nonsystolic regular expression recognizer.

the recognition process, so each recognizer node will have $c(1)$ attached to its outgoing arc.

The other signal is the input. The input must travel with zero delay from its source, which we assume is the initial node. It must reach every recognizer node, and it should also reach the final node in case the network in question is to be attached to another network, which will also need the input signal. We have choices about the route(s) that the input signal takes, and we shall choose to have it travel from the initial node through the recognizer for $a$, to the final node, and also along the lower path in Fig. 3.19, but only as far as the last recognizer node. Figure 5.5 shows this network.

The reader should note that while we do not consider the distribution of signals like $VDD$, we must consider the distribution of the input explicitly. The reason is that the input changes at each beat, and unlike the utilities, we cannot allow the input intended for one beat to arrive at a given processor during some other beat. □

**Transformation Rules**

We shall consider two rules for eliminating zero-delay arcs, and thus making networks systolic. The first multiplies the delays in any network by a constant, and the second has the effect of causing a given processor to operate with a *lag* $\ell$, that is, in such a way that its activity at time $t$ is what it would be in the untransformed network at some time $t + \ell$; $\ell$ can be positive or negative. The purpose of the first rule is to give us more delay to work with, and the second rule lets us distribute the delays on arcs so none is zero.

When we transform networks, we should consider the effect we have on three quantities.

1.   $D$, the *delay* experienced by a given stream at a given node.
2.   $S$, the *spacing* of real elements of a given stream, as they arrive at a given node.
3.   $P$, the *period* or number of beats between the arrival of real elements from a given stream at a given node.

For example, the $a$-stream in Fig. 5.3 has a delay of 2, a spacing of 2, and a period of 4. Note that the law $DS = P$ must hold in general.

A fundamental observation is that while we may change the delay, spacing, and/or period of a stream at a node, if we are to be sure that the network continues to perform correctly, it suffices to know that the periods of all streams entering a node are either unchanged or are all multiplied by the same constant. In this way we can be sure that the same collection of elements of the various streams arrives at a node simultaneously. While there are transformations that violate this constraint yet manage to preserve the function of a particular network, we shall not attempt to say anything in general about such transformations. We now define two valid transformations and prove that they preserve the function of the network in the following two lemmas.

**Lemma 5.1:** If we multiply all delays on the arcs of a timing diagram by some constant $k > 1$, then the processors may be redesigned so the network will continue to perform its function, although at $1/k^{th}$ the speed (i.e., the period is multiplied by $k$).

**Proof:** One way to make sure that the network continues to perform the same function is to multiply the period of all streams at all nodes by a constant. This may be done by the simple expedient of redesigning the processors to take $k$ beats in place of one. That is, between each two beats of the original processor network, the new processors will count $k-1$ beats where they do nothing. The effect is that in the equation $DS = P$, we multiply $D$ and $P$ by $k$ and leave $S$ fixed. Note that we could contemplate other approaches, such as dividing $S$ by $k$, multiplying $D$ by $k$, and leaving $P$ fixed, provided that all spacings were at least $k$. $\square$

It is interesting to note that the modification similar to Lemma 5.1 where instead of multiplying delays by a constant, we add a constant, does not work. For suppose there are two streams $a$ and $b$ that travel from node $u$ to node $v$ along paths with lengths $\ell$ and $m$. Then two elements of the $a$ and $b$ streams starting out at $u$ at the same time will arrive at $v$ at times that differ by $k(\ell - m)$ more than before. Thus, $v$ will not necessarily have the same collections of elements on which to perform its operations as it did before the additional delay was introduced.

The next transformation involves projecting processors backward or forward in time. That is, suppose we decided to make processor $X$ perform the same operations, but do them one beat later than before. To allow this to happen, we would have to add 1 to all the delays on arcs entering $X$ and subtract 1 from all the arcs leaving. If some arc leaving $X$ already has a delay of 0 or 1, we may not be able to realize such a transformation physically, but otherwise we can.

There are significant modifications to the processors that we must do. For example, suppose some arc leaving $X$ had a delay of $D$, and $X$ needed $D$ beats to compute its output element for that arc. Then we would have to modify our notion of a beat so it consists of enough clock cycles that the output can be computed within $D - 1$ beats. Since we may assume that in the original network, the needed inputs all arrive $D$ beats before the output is produced, and $D > 1$, this transformation is possible.

There is another significant modification we must make to processors, in order that the equation $DS = P$ continue to hold. Note that it is not generally feasible to change $P$ for one stream and one processor, if the network is to continue to work. Moreover, we cannot conveniently change $P$ globally, as we did in the proof of Lemma 5.1. Thus, if $D$ changes and $P$ stays fixed, we are forced to change the spacing; for example the new spacing is $DS/(D-1)$ if the old spacing is $S$, and the delay changes from $D$ to $D - 1$. This spacing need not be an integer, but that problem can be solved by a periodic variation of the spacing; we give an example of a spacing $4/3$ in Example 5.4, to follow. In principle, even a spacing less than one can be accommodated, if we allow several stream elements to be bundled into one.

Similarly, we can advance a processor $X$ in time by subtracting 1 from the delay on each incoming arc, provided that delay is at least 2, and also adding 1 to each outgoing arc's delay. We have thus proven the following lemma.

**Lemma 5.2:** Let $k > 0$. If $v$ is a node whose every outgoing arc has a delay greater than $k$, then we may add $k$ to each incoming arc and subtract $k$ from each outgoing arc. If processors are modified appropriately, the circuit will continue to perform the same function. Similarly, if every incoming arc for $v$ has delay greater than $k$, then we can subtract $k$ from the delays on the incoming arcs and add $k$ to the delays on the outgoing arcs, and preserve the function of the network. $\square$

These two lemmas can be used to give a construction of systolic networks from almost arbitrary ones. The only condition we require is that the original network not have any cycles all of whose arcs have zero delay.

**Theorem 5.1:** Any finite network without cycles consisting solely of zero-delay arcs can be transformed into a network performing the same function, all of whose delays are at least 1.

**Proof:** Let $N$ be the original network, and assume $N$ has no zero-delay cycles. Then for every node $v$, let $lag(v)$ be the length of the longest path consisting of only zero-delay arcs ending at $v$. Since there are no zero-delay cycles, $lag(v)$ is finite for every $v$. Moreover, there must be at least one node of lag 0, or else we could follow every path backwards indefinitely, and since $N$ is finite, we would eventually find a cycle of zero-delay arcs. Intuitively, $lag(v)$ is the amount by which $v$ must lag behind the nodes of lag 0 in the redesigned network.

We intend that an arc from $u$ to $v$ with delay 0 in $N$ will be given a delay $lag(v) - lag(u)$, using Lemma 5.2. Now, the longest zero-delay path entering $u$ must be at least 1 shorter than the longest such path entering $v$; thus the new delay on this arc will be positive. The problem is that $v$ may have an arc with some delay $d > 0$ to some node $w$ with a lag much less than that of $v$, and the condition for Lemma 5.2 to apply would be violated, since we have to make a subtraction of $lag(v) - lag(w)$ from the arc with delay $d$. The solution is first to use Lemma 5.1 to multiply $d$ by some suitably large constant $k$ such that

$$dk > lag(v) - lag(w) \tag{5.1}$$

where $v$ and $w$ range over all pairs of nodes such that there is an arc of some positive delay $d$ from $v$ to $w$. The construction of the revised network is now clear.

1. Select $k$ to satisfy (5.1).
2. Multiply all delays by $k$, using Lemma 5.1 to justify the change.
3. Delay each node of lag $\ell$ by $\ell$ beats. This has the effect of replacing every arc from $u$ to $v$ with delay 0 by an arc of delay $lag(v) - lag(u)$, which is positive by what we have argued above. If the arc from $u$ to $v$ has positive delay $d$ initially, then the new delay is $dk + lag(v) - lag(u)$, which by (5.1) is positive. $\square$

**Example 5.4:** Let us consider the network of Fig. 5.5 again. In Fig. 5.6 it is redrawn with lags indicated in the nodes. The value of $k$ is dictated by the arc from the node of lag 4 to its successor of lag 0. That arc has a delay 1, so $k > (4 - 0)/1$, i.e., $k = 5$.

In Fig. 5.7 we see the result of applying the construction of Theorem 5.1. Notice that an arc from $u$ to $v$ with original delay $d$ is given delay

$$dk + lag(v) - lag(u)$$

in general, or $5d + lag(v) - lag(u)$ in this case. All these delays are positive, so we have constructed a systolic recognizer for this regular expression. While many regular expressions can have their recognition graphs made systolic by Theorem 5.1, it is not clear that all can. The reason is that the feedback arcs

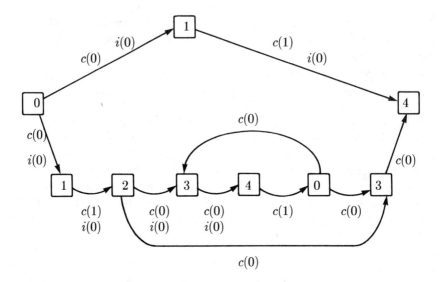

**Fig. 5.6.** Lags of nodes in Fig. 5.5.

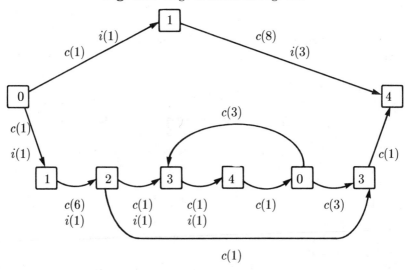

**Fig. 5.7.** Systolic regular expression recognizer.

of zero delay introduced by the construction for * interact with feed-forward
arcs of zero delay, both those carrying control and those carrying the input. □

**Example 5.5:** While in Example 5.4 the constant $k$ by which we had to
multiply the given delays was quite large, leading to a circuit that appears to

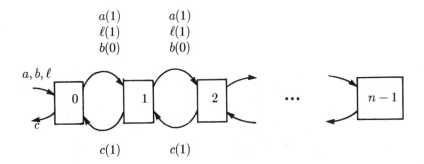

**Fig. 5.8.** Initial stage in the design of systolic convolution algorithm.

operate much more slowly than we might imagine is necessary, such need not always be the case. As an example, let us reconsider the design of a systolic convolution circuit from Section 5.1, picking up the design process at the point where we have deduced the following.

1. The $a$'s must travel leftward, and the $c$'s rightward.
2. There must be placeholders between each two consecutive $a$'s and between each two consecutive $c$'s, so $a$'s and $c$'s do not skip over each other.
3. A signal $\ell$ (*leading_edge*) must travel with the $a$'s to mark $a_0$.

However, we have not yet deduced the way around the fact that at the $j^{th}$ beat, $b_{j-1}$ must be where $a_0$ is, so it can be frozen at that processor.

Our initial solution to the problem might be to broadcast $b_j$ from the leftmost processor to all the others, with zero delay, so the processor that has $\ell = 1$ could freeze $b_j$. This thought gives rise to the network shown in Fig. 5.8.

We have indicated for each of the $n$ processors the lag associated with each one because of the zero-delay path from the leftmost node. Among the positive delay arcs, only the $c$'s go from higher lag nodes to lower lag nodes, but the difference in lag is always 1. That is, the factor $k = 2$ suffices to satisfy (5.1), because the arcs in question, from some node $v$ to the node $w$ to its left, have $d = 1$ and $lag(v) - lag(w) = 1$.

Let us examine how the delay, spacing, and period must change at each node and for each stream. The characteristics for the $\ell$ stream are the same as for the $a$ stream, so we shall ignore the former. For the $a$ stream, we started with $D = 1$, $S = 2$, and $P = 2$. When we use Lemma 5.1, with $k = 2$, we get $D = 2$, $S = 2$, and $P = 4$. Then, when we use Lemma 5.2, $D$ rises to 3 at each node. Since we cannot change $P$, we must lower $S$ to $4/3$ so the equation $DS = P$ will continue to hold. That is to say, in every four positions of the $a$ stream there will be three real elements and one placeholding 0.

For the $c$ stream, we start with $D = 1$, $S = 2$, and $P = 2$, then multiply $D$ and $P$ by $k = 2$ to get $D = 2$, $S = 2$, and $P = 4$, as for the $a$ stream.

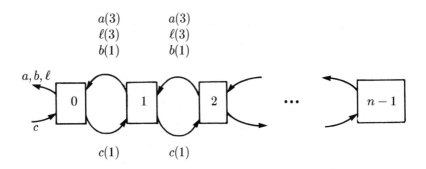

**Fig. 5.9.** Revised network for convolution.

However, the effect of Lemma 5.2 is to set the delay back to $D = 1$ at each node, so we must raise $S$ to 4 in compensation. That is, we insert three dummies between each pair of real values in the $c$ stream.

Finally, for the $b$ stream, the equation $DS = P$ is anomalous, since while $P = 1$ is apparent, i.e., a new $b$-value arrives at each node at each beat, we have $D = 0$, so $S$ must be infinite. The effect of Lemma 5.1 is to set $P = 2$, with $D$ still 0. Then, when we consider the effect of Lemma 5.2 in delaying nodes, the delay for the $b$ stream at each node becomes 1, so $S$ must become 2, i.e., the $b$ stream will alternate real values and placeholders.

Figure 5.9 shows the revision of the network in Fig. 5.8, and Fig. 5.10 shows the first, fourth, and seventh beats of this new algorithm. The $\ell$'s are not shown; they travel with the $a$'s. Neither are the frozen $b$'s shown; they are frozen when they meet $a_0$, which travels with the $\ell = 1$ flag.

The $a$ stream, with spacing 4/3 and delay 3, moves by allowing a real element to move forward when and only when the processor ahead holds a placeholder. Thus, at beat 2, $a_0$ and $a_3$ will advance from the positions they held at beat 1, leaving placeholding 0's in the first and fifth positions. This frees places for $a_4$ and $a_1$ to advance at the third beat, and by the fourth beat, $a_2$ advances to position four, as shown in Fig. 5.10.

To reinforce the notion that the timing is correct, notice that at beat 4, $c_1$ meets $a_0$ and the frozen $b_1$, which is at the same cell as the traveling $b_1$, since the latter value has just met $\ell_0$ along with $a_0$. At the fifth beat, not shown, $c_1$ will move left and meet $a_1$ along with the frozen $b_0$, which is at the center position, where the traveling $b_0$ met $\ell_0$ at beat 1.

Note that although the new algorithm is not identical to the one developed in Section 5.1, whose network is shown in Fig. 5.4, both behave correctly and produce their output at the same rate. $\square$

| $a_3$ | $0$ | $a_2$ | $a_1$ | $a_0$ | | | | |
|---|---|---|---|---|---|---|---|---|
| $b_2$ | $0$ | $b_1$ | $0$ | $b_0$ | | | | |
| | | | | $c_0$ | $0$ | $0$ | $0$ | $c_1$ |

| $a_4$ | $a_3$ | $0$ | $a_2$ | $a_1$ | $a_0$ | | | |
|---|---|---|---|---|---|---|---|---|
| $0$ | $b_3$ | $0$ | $b_2$ | $0$ | $b_1$ | $0$ | $b_0$ | |
| | $c_0$ | $0$ | $0$ | $0$ | $c_1$ | $0$ | $0$ | $0$ |

| $a_5$ | $a_4$ | $a_3$ | $0$ | $a_2$ | $a_1$ | $a_0$ | | |
|---|---|---|---|---|---|---|---|---|
| $b_5$ | $0$ | $b_4$ | $0$ | $b_3$ | $0$ | $b_2$ | $0$ | $b_1$ |
| $0$ | $0$ | $c_1$ | $0$ | $0$ | $0$ | $c_2$ | $0$ | $0$ |

**Fig. 5.10.** Beats 1, 4, and 7 of revised convolution algorithm.

## 5.3 MATRIX MULTIPLICATION AND TRANSITIVE CLOSURE

In this section we shall consider how to perform matrix multiplication and transitive closure in a systolic way. In the next section we give some applications and extensions of these algorithms.

### Matrix Multiplication

Let us begin by designing a systolic algorithm to compute the product of two $n \times n$ matrices $C = AB$. That is, if $A$ has element $a_{ij}$ in row $i$ and column $j$, and $B$ has $b_{ij}$ in that position, then the element $c_{ij}$ in row $i$ and column $j$ of the product matrix $C$ is given by

$$c_{ij} = \sum_{k=1}^{n} a_{ik}b_{kj}$$

Let us follow the same procedure we followed in Section 5.1 to derive one of several possible algorithms. While a linear array of processors was used for convolution in Section 5.1, to process matrices we prefer a two-dimensional array of processors, arranged in a grid. We shall read the elements of $A$ in from the left moving rightward, row above row, and we shall have the elements of $C$ travel leftward to meet them, also in a row-above-row manner. To avoid having $a$'s and $c$'s miss each other, we put placeholders between real values of both streams, as we did for convolution.

When $a_{ik}$ meets $c_{ij}$, the $b$-value they need is $b_{kj}$, because $a_{ik}b_{kj}$ is a term that contributes to $c_{ij}$. Note that there is no need for an $a$-value and a $c$-value from different rows ever to meet each other. Figure 5.11 shows the first three beats of this process and the $b$'s needed at each position.

We notice two facts from Fig. 5.11.

1.   The same $b$-value is needed everywhere in a column at the same beat.

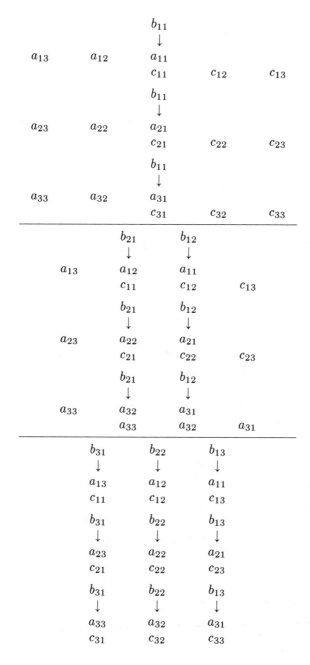

**Fig. 5.11.** Initial plan for matrix multiplication algorithm.

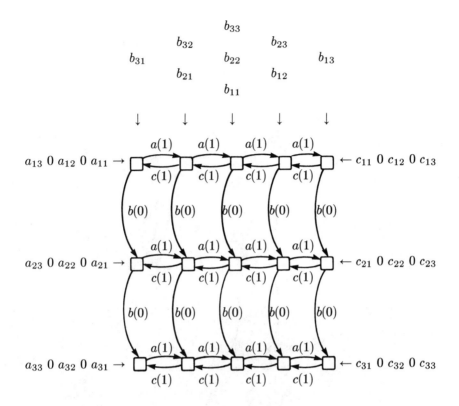

**Fig. 5.12.** Flow of matrix elements through network.

2.  The $b$-values needed on the $i^{th}$ beat have subscripts that sum to $i+1$. That is, $B$ is needed diagonal-by-diagonal.

The network of processors, with delays indicated, is shown in Fig. 5.12. We also show how the matrices $A$, $B$, and $C$ are expected to flow through the network. This algorithm is not yet systolic because of the zero delays for the $b$'s.

We can make the algorithm systolic by the simple expedient of giving the $i^{th}$ row a lag of $i - 1$, for $1 \leq i \leq n$. Then, all the zero-delay arcs in Fig. 5.12 are given delay 1, while the arcs with a delay of 1 in Fig. 5.12 need no be changed. The reason is that all arcs with delay 1 travel between nodes of the same lag, so $k = 1$ suffices in the inequality (5.1). The network of Fig. 5.12 functions just as before, but the $b$-values are passed with delay 1 down the columns. Also, to compensate for the lag of each row, the $a$'s and $c$'s fed to row $i$ must be delayed $i - 1$ beats from the timing shown in Fig. 5.12. The revised picture is shown in Fig. 5.13, with placeholding 0's in the $a$ and $c$ matrices omitted.

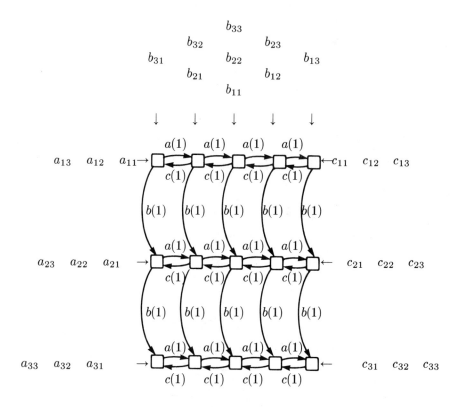

**Fig. 5.13.** Systolic matrix multiplication algorithm.

## Systolic Transitive Closure

A *Boolean matrix* is a matrix each of whose elements is either 0 or 1. A Boolean matrix can represent a directed graph, if we let the nodes of the graph be $1, 2, \ldots, n$ and let the element $a_{ij}$ of the matrix be 1 if there is an arc from $i$ to $j$, and 0 otherwise. The *transitive closure* of a Boolean matrix $A$ representing a graph $G$ is that matrix $A^+$ whose element in row $i$ and column $j$ is 1 if there is a path of length zero or more from node $i$ to node $j$, and 0 otherwise.†

One way to compute the transitive closure involves *Boolean matrix multiplication*, which is defined as is ordinary matrix multiplication, but with logical "and" replacing multiplication of elements and logical "or" replacing sum of elements. That is, the element $c_{ij}$ in the Boolean product of Boolean matrices $A$ and $B$ is defined as

$$c_{ij} = (a_{i1} \wedge b_{1j}) \vee (a_{i2} \wedge b_{2j}) \vee \cdots \vee (a_{in} \wedge b_{nj})$$

---

† Strictly speaking, what we called "transitive closure" is the "reflexive and transitive closure."

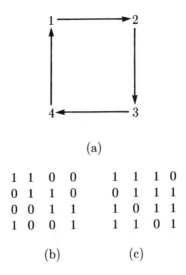

(a)

$$
\begin{array}{cccc}
1 & 1 & 0 & 0 \\
0 & 1 & 1 & 0 \\
0 & 0 & 1 & 1 \\
1 & 0 & 0 & 1
\end{array}
\qquad
\begin{array}{cccc}
1 & 1 & 1 & 0 \\
0 & 1 & 1 & 1 \\
1 & 0 & 1 & 1 \\
1 & 1 & 0 & 1
\end{array}
$$

(b)                    (c)

**Fig. 5.14.** Computation of transitive closure.

We also need the operation of *Boolean matrix addition*, in which the element $c_{ij}$ in the sum $C$ of Boolean matrices $A$ and $B$ is $a_{ij} \vee b_{ij}$.

If $A$ represents graph $G$, then $A + I$, where $I$ is the *identity* matrix, with 1 along the diagonal and 0 elsewhere, represents the paths of length 0 or 1 in $G$. That is, $A + I$ has 1 in row $i$ and column $j$ if and only if there is a path of length zero (i.e., $i = j$), or one (i.e., an arc) in $G$. It is easy to check that $(A + I)^2$, the Boolean product of $A + I$ with itself, represents paths of length two or less, that $((A + I)^2)^2$ represents paths of length four or less, and so on. Thus, with $\log n$ Boolean matrix multiplications, we can produce $A^+$, since if there is a path from $i$ to $j$ then there is a path that has length at most $n$.

**Example 5.6:** Figure 5.14(a) shows a graph; Fig. 5.14(b) shows $A + I$, where $A$ is the matrix that represents Fig. 5.14(a). Then, Fig. 5.14(c) shows $(A + I)^2$. If we square Fig. 5.14(c), we get a matrix of all 1's, which is $A^+$, because the graph is *strongly connected*, i.e., there is a path from every node to every other. $\square$

This algorithm takes time $O(n \log n)$ and area $O(n^2)$, if we use the systolic matrix multiplication algorithm just described, since all registers in processors are one bit long. However, there is a systolic transitive closure algorithm using the same area that runs in time $O(n)$, and we endeavor to describe this now.

For this algorithm, we use a mesh of processors, as depicted in Fig. 5.15. We use $A_{ij}$ both for the processor in row $i$ and column $j$, and for a one-bit value stored there. Eventually, we shall find that this value $A_{ij}$ equals 1 if and only if the transitive closure of a given matrix $A$ has 1 in that position. Initially, $A_{ij}$

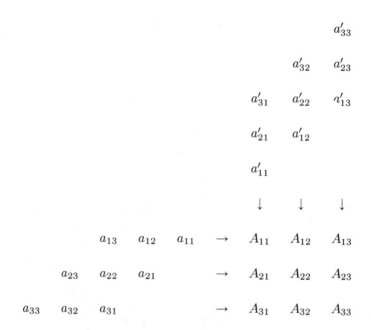

**Fig. 5.15.** Systolic transitive closure.

holds 0, however.

We also pass the given matrix $A$ through the processor mesh, both from the left, and from the top. We refer to the elements of the matrix passed in from the left as $a_{ij}$'s and those from the top as $a'_{ij}$'s. Although initially $a_{ij}$ and $a'_{ij}$ are both the element of $A$ in row $i$ and column $j$, these values change, as do the $A_{ij}$'s, as the algorithm progresses.

**Algorithm 5.1:** *Guibas-Kung-Thompson Transitive Closure Algorithm.*

INPUT: Two copies of an $n \times n$ matrix, with 1's on the diagonal, read into an $n \times n$ array of processors, as suggested in Fig. 5.15.

OUTPUT: The transitive closure of the input matrix. The output is found in the processor array and is also read out of the right and bottom edges in the order indicated in Fig. 5.15.

METHOD: The algorithm consists of three identical passes, in which the $a_{ij}$'s and the $a'_{ij}$'s are passed through the processor mesh. Initially, the matrices read in from the left and top are $A + I$, where $A$ is the matrix whose transitive closure we are to take.

As an element completes one pass, by reaching the right or bottom edge, it is immediately fed back to the left or top edge, respectively, to begin the next pass. Connections between top and bottom and between left and right in the mesh are used to make this feedback possible.

The computations performed at the processors are the same in each pass and at each beat. Suppose that at some beat $a_{ik}$ and $a'_{kj}$ are at processor $A_{ij}$. Note that at every beat the pair of elements reaching the processor $A_{ij}$ must take that form for some $k$, unless no elements at all reach $A_{ij}$ on that beat. In proof, the $a$-element and the processor must have the same first subscript, as we can observe from Fig. 5.15, and similarly, the $a'$-element and the processor must share the second subscript. But, it is not hard to deduce from the pattern of flow in Fig. 5.15 that if $a_{ik}$ meets $a'_{\ell j}$, then $k = \ell$. This fact will be proven in Lemma 5.3.

The steps we perform at the processor $A_{ij}$ are as follows.

1.  $A_{ij} := A_{ij} \vee (a_{ik} \wedge a'_{kj})$. That is, if both $a_{ik}$ and $a'_{kj}$ are 1, set $A_{ij}$ to 1, and otherwise, leave $A_{ij}$ as it is. Intuitively we want, after several passes, for $a_{ik}$ and $a'_{kj}$ to be 1 if there are paths from $i$ to $k$ and from $k$ to $j$, respectively; if that is the case, then there is a path from $i$ to $j$, and we set $A_{ij}$ to 1.

2.  When $a_{ij}$ arrives at $A_{ij}$, we assign $a_{ij} := A_{ij}$. Other $a_{ik}$'s, for $k \neq j$, reaching $A_{ij}$ are passed through unchanged. Similarly, when $a'_{ij}$ arrives at $A_{ij}$ we assign $a'_{ij} := A_{ij}$, and other $a'_{kj}$'s are left alone. These actions enable changes in $A_{ij}$, accumulated in (1), to be transmitted to the corresponding elements $a_{ij}$ and $a'_{ij}$, to further change the $A$'s on the next pass.

We must explain how processor $A_{ij}$ is to know that it is being visited by $a_{ij}$ or $a'_{ij}$, so that it may transmit its value by rule (2). One way would be for each of the $a$'s and $a'$'s to carry the value of their subscripts for comparison with corresponding values kept in the $A_{ij}$'s. However, that would add a factor of $n^\epsilon$ to the time and space taken by the algorithm, and we reject this approach in favor of one that takes only a constant factor extra time and space.

We shall show in Lemma 5.3 that $a_{ij}$ arrives at $A_{ij}$ at beat $i + 2j - 2$, and $a'_{ij}$ arrives there at beat $2i + j - 2$, where "beat 1" is the time that $a_{11}$ and $a'_{11}$ arrive at $A_{11}$. Thus, to signal the arrival of $a_{ij}$, a signal that begins at the upper left-hand corner ($A_{11}$) propagates downward and to the right. It travels downward one row at every beat, and it also propagates to the right, but only every other beat. Thus, many copies of the signal exist after a while, but the first arrival of the signal at a processor $A_{ij}$ is at the same time that its corresponding $a$-value $a_{ij}$ arrives. A similar signal that moves right with every beat and down with every other heralds the arrival of the corresponding $a'$-value, $a'_{ij}$. $\square$

It is hardly obvious that Algorithm 5.1 works, in the sense that after the three passes, $A_{ij}$ will be 1 if and only if $A^+$, the transitive closure of the original matrix, has 1 there. We prove this result in a series of lemmas. In what follows, it is useful to define a *k-path* from node $i$ to node $j$ to be a path from $i$ to $j$ that goes through no node numbered higher than $k$. Note that by "goes through" we explicitly exclude the endpoints, so $i$ and/or $j$ may exceed $k$. In particular, if $i = j$, or if there is an arc from $i$ to $j$ in the original graph, then there is a $k$-path from $i$ to $j$ in that graph for any $k$.

**Lemma 5.3:** Suppose "beat 1" of some pass of Algorithm 5.1 is defined to be when $a_{11}$ and $a'_{11}$ are at $A_{11}$. Then $a_{ij}$ arrives at $A_{ik}$ at beat $i + j + k - 2$, and $a'_{ij}$ arrives at $A_{kj}$ at beat $i + j + k - 2$.

**Proof:** Note that the copy of the matrix read in from the left has columns with constant sum of subscripts. In particular, the column containing $a_{ij}$ will arrive at the leftmost column of the mesh at beat $i + j - 1$. Then, $k - 1$ more beats are necessary to get that element to the $k^{th}$ column, where $A_{ik}$ resides, for a total of $i + j + k - 2$ beats. The argument about the copy of the matrix read from the top is similar. $\square$

Note that as a special case of Lemma 5.3, $a_{ij}$ meets $A_{ij}$ at beat $i + 2j - 2$, and will pick up whatever value of $A_{ij}$ is there at that time. Similarly, $a'_{ij}$ meets $A_{ij}$ at beat $2i + j - 2$.

**Lemma 5.4:** Suppose there is a $min(i, j)$-path from $i$ to $j$, that is, a path that goes through no node as high as its endpoints. Then on pass 1 of Algorithm 5.1, $A_{ij}$, $a_{ij}$, and $a'_{ij}$ are all set to 1.

**Proof:** We must set up an inductive hypothesis stronger than the theorem, so we can deal with the question of when certain assignments occur. In particular, our inductive hypothesis claims that if $i = j$, or there is an edge from $i$ to $j$, then $a_{ij}$ and $a'_{ij}$ are 1 initially, and $A_{ij}$ is set to 1 at beat $i + j + min(i, j) - 2$, when either $a_{ij}$ and $a'_{jj}$ or $a_{ii}$ and $a'_{ij}$, whichever comes first, meet at $A_{ij}$. Note that $a_{ii}$ and $a'_{jj}$ are always 1. If there is a path of length two or more from $i$ to $j$, then our inductive hypothesis is that $A_{ij}$ is set to 1 before beat $i + j + min(i, j) - 2$, which is before either $a_{ij}$ or $a'_{ij}$ reach there. Thus $a_{ij}$ and $a'_{ij}$ are both assigned 1 when they reach $A_{ij}$.

We prove the result by induction on the length of the shortest path from $i$ to $j$. For the basis, paths of length zero or one, there is nothing to prove since $a_{ij}$ and $a'_{ij}$ are 1 initially, and $A_{ij}$ is clearly assigned 1 at the beat

$$i + j + min(i, j) - 2$$

as indicated in the statement of the inductive hypothesis.

For the induction, suppose there is a $min(i, j)$-path of length two or more from $i$ to $j$. Then there exists some other node $\ell$ on the path; let us take $\ell$ to be the highest-numbered node on the path, except for the endpoints. Note that $\ell < i$ and $\ell < j$, because the path is a $min(i, j)$-path. Also note that since $\ell$ exceeds any other node on the path besides $i$ and $j$, there is a $min(i, \ell)$-path from $i$ to $\ell$ and a $min(\ell, j)$-path from $\ell$ to $j$, and both of these paths are shorter than the path from $i$ to $j$.

By the inductive hypothesis, $a_{i\ell}$ is set to 1 at or before the time it meets $A_{i\ell}$ at beat $i + 2\ell - 2$ and $a'_{\ell j}$ is set to 1 at or before the time it meets $A_{\ell j}$ at beat $j + 2\ell - 2$. Then at time $i + j + \ell - 2$, which is later because $\ell < min(i, j)$, Lemma 5.3 tells us that $a_{i\ell}$ and $a'_{\ell j}$ are both at processor $A_{ij}$, where by rule (1) of Algorithm 5.1, they set $A_{ij}$ to 1. Again because $\ell < min(i, j)$, $a_{ij}$ and $a'_{ij}$ have yet to visit $A_{ij}$ on this pass, which they do at time $i + j + min(i, j) - 2$. At that time they pick up the 1 value by rule (2) of Algorithm 5.1. $\square$

**Lemma 5.5:** After pass 2 of Algorithm 5.1:

a)   If there is a $j$-path from $i$ to $j$, then $A_{ij}$ and $a_{ij}$ are set to 1 by beat $i + 2j - 2$.

b)   If there is an $i$-path from $i$ to $j$, then $A_{ij}$ and $a'_{ij}$ are set to 1 by beat $2i + j - 2$.

c)   If there is a $max(i, j)$-path from $i$ to $j$, then $A_{ij}$ is set to 1 at some time.

**Proof:** We proceed by induction on the length of the path in question. If the length is 1, the variables mentioned in the lemma were 1 initially or set to 1 on pass 1. Suppose there is a $j$-path of length at least two from $i$ to $j$. Let $\ell$ be the highest-numbered node on the path, exclusive of the endpoints. Then $\ell < j$, and there is a shorter $\ell$-path from $i$ to $\ell$. By the inductive hypothesis, since an $\ell$-path is surely a $max(i, \ell)$-path, $a_{i\ell}$ will be set to 1 at beat $i + 2\ell - 2$, when it meets $A_{i\ell}$.

By Lemma 5.4, $a'_{\ell j}$ is already 1 at pass 1, since there is a $min(\ell, j)$-path from $\ell$ to $j$ because $\ell$ was chosen largest on the path from $i$ to $j$. Thus, at beat $i + j + \ell - 2$, which is later than $i + 2\ell - 2$, $a_{i\ell}$ and $a'_{\ell j}$ meet at $A_{ij}$, and set the latter to 1. Then, when $a_{ij}$ arrives at $A_{ij}$ at time $i + 2j - 2$, which is later, it too is set to 1.

We have thus proved part (a) of the induction; a similar argument proves (b). These two observations about $A_{ij}$ together imply that it is set to 1 if there is a $max(i, j)$-path, so they prove (c). $\square$

**Theorem 5.2:** After the third pass of Algorithm 5.2, $A_{ij}$ is set to 1 if there is any path from $i$ to $j$, i.e., if the element of the transitive closure $A^+$ in row $i$ and column $j$ is 1.

**Proof:** By Lemma 5.5, if there is a $max(i, j)$-path from $i$ to $j$, then $a_{ij}$ is already 1 after pass 2; thus we are done. The only other possibility is that the highest numbered node $\ell$ on some path from $i$ to $j$ is larger than either $i$ or $j$. If so, then there is an $\ell$-path from $i$ to $\ell$ and an $\ell$-path from $\ell$ to $j$, by the maximality of $\ell$. Then by Lemma 5.5, $a_{i\ell}$ and $a'_{\ell j}$ are set to 1 by pass two, so when they meet at $A_{ij}$ on pass three, they set $A_{ij}$ to 1. $\square$

It should be clear that the time taken by Algorithm 5.1 is $O(n)$, and the area it uses is $O(n^2)$; thus it offers an $O(n^4)$ $AT^2$ product.

## 5.4 OTHER MATRIX AND GRAPH ALGORITHMS

In this section we shall briefly sketch some interesting algorithms that are related to or make use of the systolic matrix multiplication and transitive closure algorithms of the previous section.

### Shortest Paths

A generalization of the problem of finding the transitive closure is that of finding shortest paths in a graph. We assume a directed graph whose arcs are weighted by nonnegative integers. The *length* of a path is the sum of the weights of the edges along that path. The *shortest path problem* is to find, for each pair of nodes $i$ and $j$, the minimum length of any path from $i$ to $j$.

We may modify Algorithm 5.1 to compute shortest paths. First, the input matrix $A$ has $a_{ij}$, the element in row $i$ and column $j$, equal to the weight of the arc from $i$ to $j$. If $i = j$, then the element is 0, because there is always a path of length zero from a node to itself. If there is no arc from $i$ to $j$, then $a_{ij}$ will be a special "infinite" value, which is larger than the weight of any arc or any sum of arcs. We shall probably want to reserve one bit of the registers representing weights to indicate an infinite value. Thus, $a_{ij}$, $a'_{ij}$, and $A_{ij}$ are each values requiring $k$ bits for some fixed $k$, rather than single bits as for transitive closure.

Initially, set all $A_{ij}$'s to infinity. The diagonal elements $A_{ii}$ will be set to 0 on pass 1, when $a_{ii}$ and $a'_{ii}$, which are both 0 initially, meet.

Rule (2) of Algorithm 5.1, where $a_{ij}$ and $a'_{ij}$ pick up the value of $A_{ij}$, does not change. Rule (1) is replaced by

$$A_{ij} = min(A_{ij}, a_{ik} + a'_{kj})$$

That is, if the path from $i$ to $j$ through $k$ is shorter than the shortest path from $i$ to $j$ found so far, then set $A_{ij}$ to that shorter value.

The proof that this modification computes shortest paths is completely analogous to Theorem 5.2. For example, in place of Lemma 5.4 we must prove that if there is a $min(i, j)$-path of length $\ell$ from $i$ to $j$, then after pass 1, $A_{ij}$, $a_{ij}$, and $a'_{ij}$ are no larger than $\ell$. We leave the details as an exercise.

If weights on arcs are $O(n)$, they can be represented by $O(\log n)$ bits, and it is easy to show that the shortest path algorithm takes time $O(n^{1+\epsilon})$ and area $O(n^{2+\epsilon})$.

### Connected Components

We can augment Algorithm 5.1 to compute connected components in time $O(n^{1+\epsilon})$ and area $O(n^{2+\epsilon})$. Since this performance is inferior to Algorithm 4.4, and the latter algorithm shares with the present one the advantage that the adjacency matrix is read in a row at a time, it is not clear what role the present algorithm plays, but it is simple and possibly instructive.

**Algorithm 5.2:** *Systolic Connected Components Algorithm.*

INPUT: Two copies of the adjacency matrix of an undirected graph, with diagonal elements set to 1, read into an $n \times n$ array in the manner of Algorithm 5.1.

OUTPUT: The number of the connected component of node $i$ appears at the right end of row $i$ of the processor array.

METHOD:

1.  Using Algorithm 5.1, compute the transitive closure of the adjacency matrix of the undirected graph in question. Now, $A_{ij}$ is 1 if and only if there is a path between $i$ and $j$. Thus, the group to which node $i$ belongs is that $j$ such that $A_{ij}$ is the leftmost 1 in row $i$, because then $j$ is the lowest-numbered node to which $i$ is connected.

2.  Set integer-valued variables propagating from the left end of each row. Initially, the value is 1, but the value increases by 1 each time the variable is shifted right, until it reaches a processor $A_{ij}$ holding value $A_{ij} = 1$, at which time the count propagates right without further change. Then the value shifted right from the rightmost processor in row $i$ is the group of $i$.

☐

## Transposition

Another matrix operation that will prove important in algorithms to follow is computing the transpose of a matrix $A$. The *transpose* $A^t$ is $A$, with the elements $a_{ij}$ and $a_{ji}$ interchanged. That is, $A_{ij}^t = A_{ji}$ for all $i$ and $j$.

There are a variety of ways to take the transpose of a matrix that is stored in a mesh. We prefer a method that does not depend explicitly on the size $n$ of the matrix, for reasons discussed in Chapter 4; we want the algorithm to work when many chips holding the basic processor pattern are wired together to make a larger pattern, in this case a mesh. The following is one way to solve the problem.

**Algorithm 5.3:** *Transposition in Place.*

INPUT: An $n \times n$ matrix stored in a mesh of processors of the same size.

OUTPUT: The transpose of that matrix, in the same mesh of processors.

METHOD: To begin, we assume that the diagonal, the processors $A_{ii}$ for $1 \leq i \leq n$, are marked by a bit. If that has not occurred prior to the execution of this algorithm, then we may send a signal from the upper left corner, $A_{11}$, propagating southeast along the diagonal at the same time the algorithm begins. That is, the signal propagates downward and to the right on alternate beats, but it only "marks" the processors reached after a move right. If we do so, the diagonal will be completely marked by the time it is needed.

We shall describe how we get the elements in the upper triangle, that is, those $A_{ij}$ for which $i < j$, to migrate to their proper positions in the lower

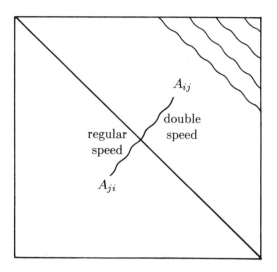

**Fig. 5.16.** Transposition algorithm.

triangle, namely, $A_{ji}$. At the same time that these steps are performed, we also execute analogous steps get the elements of the lower triangle to their transposed positions in the upper triangle.

A signal begins at $A_{1n}$, the processor in the upper right corner. In a four-beat cycle, it propagates both left and down one row or column, so the signal travels in diagonal waves as suggested in Fig. 5.16. As the wave meets an element $a_{ij}$ in the upper triangle, it "bumps" the element and starts it traveling downward and to the left, but at double speed, that is, taking only two beats to move from $A_{ij}$ to $A_{i+1,j-1}$, two more to move to $A_{i+2,j-2}$, and so on. Note that because the elements move faster than the wave, the elements will not bunch up, but travel in a southwest direction in the order in which they were bumped.

When the bumped element reaches the diagonal, it slows to regular speed, that is, four beats to travel one row down and one column left. On the other hand, when the wave reaches the diagonal, two things happen. First, it loses its power to bump elements. Second, it switches to double speed. As a result, the wave catches up to $A_{ij}$ after the latter has traveled as far after the diagonal as it did before it reached the diagonal, that is, when the element initially in processor $A_{ij}$ reaches processor $A_{ji}$, which is its proper position in the transpose. We thus freeze each element at the position it has when it is "bumped" by the wave for the second time. $\square$

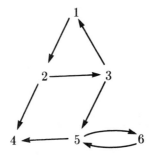

**Fig. 5.17.** Example digraph.

**Strong Components**

A *strong component* of a directed graph is a maximal set of nodes such that there is a path from every node in the strong component to every other node in that strong component.

**Example 5.7:** The digraph of Fig. 5.14(a) consists of a single strong component. The strong components of Fig. 5.17 are $\{1, 2, 3\}$, $\{4\}$, and $\{5, 6\}$. $\square$

As we computed connected components by finding, for each node $i$, the lowest-numbered node in its component, we can find strong components if we compute for each $i$ the lowest numbered node in the same strong component as $i$. The following algorithm is a simple way to use what we have developed for the computation of strong components.

**Algorithm 5.4:** *Strong Components Algorithm.*

INPUT: Same as Algorithm 5.1, the transitive closure algorithm.

OUTPUT: At the right end of the $i^{th}$ row of the mesh of processors will appear the number of the strong component of node $i$, that is, the smallest number $j$ such that there are paths in both directions between nodes $i$ and $j$.

METHOD:

1. Let $A$ be the adjacency matrix of the given graph. Compute $A^+$ using Algorithm 5.1.

2. While remembering $A^+$ in the mesh of processors, also compute its transpose, $(A^+)^t$, in the same mesh.

3. Compute $A^+ \wedge (A^+)^t$ by setting $A_{ij}$ to 1 if and only if both $A^+$ and $(A^+)^t$ have 1 in that position. The result is that processor $A_{ij}$ holds 1 if and only if there is a path from $i$ to $j$ and a path from $j$ to $i$, i.e., nodes $i$ and $j$ are in the same strong component.

4. In the same manner as step (2) of Algorithm 5.2, the systolic connected components algorithm, start a variable at the left end of each row $i$ to find the smallest value of $j$ for which $A_{ij} = 1$. Thus, the value leaving row $i$ at

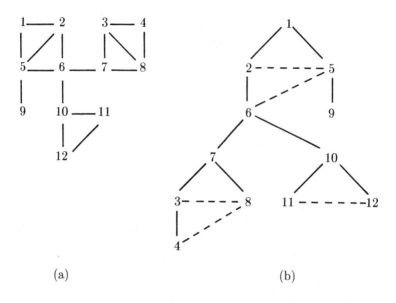

**Fig. 5.18.** Breadth-first spanning tree.

the right end will be the lowest-numbered node in the strong component containing $i$, which we take to be the number of $i$'s strong component. □

## Spanning Trees

Our next algorithm is to find for a connected, undirected graph $G$ a *spanning tree* for $G$, that is, a graph with the same set of nodes as $G$, and with a subset of the edges of $G$ that is sufficient to connect all the nodes of $G$, but that has no cycles. In fact, the spanning tree we construct will be a *breadth-first* spanning tree, meaning that it has a designated root node, and that the path to the root from any node $i$ is as short as any path between the root and $i$.

**Example 5.8:** Figure 5.18(a) shows a graph and Fig. 5.18(b) is a breadth-first spanning tree, with root 1, for that graph. The tree edges are solid and the other edges of the graph are shown by dashed lines in Fig. 5.18(b). Notice how in a breadth-first spanning tree, every non-tree edge goes between nodes that are at the same depth or at depths that differ by one. □

Our next algorithm uses the shortest path algorithm to find a breadth-first tree. It does so by finding the shortest distance from node 1, the root, to every other node, and then letting each node pick a parent whose distance, i.e., its depth in the tree, is one less than its own depth.

**Algorithm 5.5:** *Breadth-First Spanning Tree.*

INPUT: Same as Algorithm 5.1, but with the matrix representing a connected, undirected graph.

OUTPUT: In the processor array will appear the matrix representing an $n$-node graph that is a spanning tree for the graph represented by the input. The spanning tree will be represented as a directed graph, with edges directed from child to parent.

METHOD:

1. Using the generalization of Algorithm 5.1, compute the shortest paths in the given graph $G$ by treating the adjacency matrix of $G$ as if it represented edge weights; a present edge has a weight of one, whereas the weight of an absent edge is infinite. As a result, $A_{ij}$, the value stored at the processor $A_{ij}$ in a mesh of processors, will equal the number of edges on the shortest path between nodes $i$ and $j$.

2. In particular, $A_{1j}$ is the shortest path length from node 1 to node $j$, which, if we take 1 to be the root of the breadth-first tree, must be the depth of $j$, which we denote by $d_j$.

3. For each column $j$, broadcast $d_j$ to all processors in that column.

4. For each row $i$, processor $A_{i1}$ also has $d_i$, since the given graph is undirected. Send $d_i$ right, along the $i^{th}$ row, until it meets a processor $A_{ij}$ holding a value $d_j$ which is one less than $d_i$. Set the variable $A_{ij}$ to 1 and all other $A_{ik}$'s, $k \neq j$, to 0. Note that exactly one processor per row will hold 1, except in the first row, where all are 0. The mesh now holds the adjacency matrix of a directed graph that is a breadth-first tree for $G$, with all arcs directed from child to parent. $\square$

**Example 5.9:** If we compute depths for the graph of Fig. 5.18(a) in step (1) of Algorithm 5.5, we find that node 1 has depth zero, nodes 2 and 5 are at depth 1, 6 and 9 at depth two, 7 and 10 at depth three, 3, 8, 11, and 12 at depth four, and 4 at depth five. In step (4), there is often no choice of parent for a node. In one of the cases where there is a choice, row 6, the value $d_6 = 2$ first meets a column where the depth is 1 at column 2; thus the parent of node 6 in the tree is 2, rather than 5. We have illustrated the choices of parent made by Algorithm 5.5 in Fig. 5.18(b). $\square$

**Finding Bridges**

A *bridge* in an undirected graph is an edge whose removal divides one connected component into two. For example, the bridges in Fig. 5.18(a) are $(5,9)$, $(6,10)$, and $(6,7)$.

There is a clever algorithm for finding bridges, of which the spanning tree construction just presented is one part. Briefly, we construct the transitive closure of the tree alone, with the arcs directed from child to parent. In the transitive closure of the tree, a node is reached by all and only its descendants. We also construct the transitive closure of the graph consisting of all the tree arcs, plus pairs of arcs in both directions in place of every nontree edge of the original graph. The bridges are exactly those edges $(c, p)$ such that $c$ is a child of $p$ in the tree, and the nodes that can reach $c$ in both graphs are the same.

**Example 5.10:** In Fig. 5.18(b), one graph consists of only the solid edges, directed upward, with the dashed edges not present at all. The second graph consists of the solid edges directed upward plus the dashed edges as pairs of arcs; e.g., the dashed edge $(2, 5)$ is replaced by the arcs $2{\rightarrow}5$ and $5{\rightarrow}2$.

For example, $(7, 6)$ is a bridge, and in the graph that is a directed tree, node 7 is reached only from its descendants, as is true for every node in the tree. In the graph augmented by arc pairs for each nontree edge, it is still not possible to reach 7 from outside its descendants. However, for an edge like $(2, 6)$, which is not a bridge, in the augmented graph node 6 can be reached from 2, 5, and 9, which are not its descendants in the tree. That is, there are paths $2{\rightarrow}5{\rightarrow}6$ and $9{\rightarrow}5{\rightarrow}6$ in the augmented graph. $\square$

The construction is formalized in the next algorithm.

**Algorithm 5.6:** *Atallah–Kosaraju Bridge-Finding Algorithm.*

INPUT: Same as Algorithm 5.5.

OUTPUT: At the top edge of the processor array there appears in column $j$ a bit indicating whether node $j$ is one end of a bridge. At the bottom of column $j$ the number of the node at the other end of the bridge appears.

METHOD:
1.  We begin by reading one copy of the input matrix $A$ into the $n \times n$ array of processors without change.
2.  Use Algorithm 5.5 to compute the adjacency matrix $T$ of a breadth-first spanning tree for $G$, represented in the same mesh. Note that $T$ is a directed graph, and its arcs are the edges of the spanning tree, directed upward.
3.  While preserving copies of $A$ and $T$ in the processor mesh, compute the transpose $T$ by Algorithm 5.3. The resulting matrix $T^t$ is the tree edges as arcs directed downward.
4.  Compute $B = A \wedge \neg T^t$. The complementation ($\neg$) is implemented by complementing the bit representing $T^t$ at each processor, and the logical "and" of the two adjacency matrices is likewise performed bit-by-bit at the various processors. The resulting matrix is the adjacency matrix of what we referred to above as the "augmented" graph. That is, it consists of the tree arcs directed upward, and all the other edges of $G$ directed in both directions. Put another way that makes the formula for $B$ more transparent, the matrix $A$, the adjacency matrix of an undirected graph, can be interpreted as a directed graph with arcs in both directions in place of each edge. The term $\wedge \neg T^t$ deletes from this directed graph the arcs formed from the tree edges, but directed downward; these are the only arcs

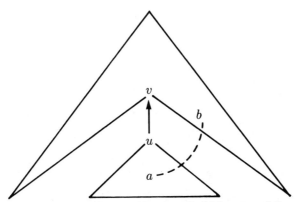

**Fig. 5.19.** Diagram for proof of Theorem 5.3.

of the directed version of $G$ not in $B$.

5.  Compute $T^+$ and $B^+$ by Algorithm 5.1, simulating the arrival of copies of the matrix at the left and top by initially cycling copies of the rows and columns of $T$ and $B$ by appropriate amounts. We leave the details of this operation as an exercise.

6.  For each $j$, compare column $j$ of $B^+$ with that of $T^+$. If they are the same, then the tree edge with child $j$ is a bridge; if not, this edge is not a bridge. We may easily generate a bit at the top of each column to indicate which nodes are, with their parents in $T$, the ends of a bridge. Then, by passing this bit vector through the matrix $T$, we can also deduce the parent ends of the bridges if desired. $\square$

**Theorem 5.3:** Algorithm 5.6 correctly finds the bridges of a graph.

**Proof:** First, it is easy to see that if $(u, v)$ is an edge of graph $G$ but not selected for the spanning tree $T$, then $(u, v)$ cannot be a bridge. The reason is that $u$ and $v$ are surely connected via the tree to its root, and the edge $(u, v)$ cannot be part of these paths, because it is not part of the tree. Then deletion of $(u, v)$ does not disconnect $u$ from $v$, because they are still connected via those edges of $G$ that are tree edges. (Recall that although we represent $T$ by arcs, the arcs all correspond to undirected edges that are present in $G$.)

Now consider the case where $(u, v)$ is a tree edge with $u$, say, the child. If $(u, v)$ is a bridge, it must separate the descendants of $u$ from the other nodes of $G$; thus there cannot be an edge like $(a, b)$ of Fig. 5.19. As a result, there is no way for a node like $b$ to reach the descendants of $u$, and in the graph $B$ of Algorithm 5.6, it will be only the descendants of $u$ that can reach $u$. Thus, the columns for $u$ in $B^+$ and $T^+$ will agree. Conversely, if $(u, v)$ is not a bridge, then there must be at least one nontree edge of $G$ that, like $(a, b)$ in Fig. 5.19, connects a descendant $a$ of $u$ (possibly $u$ itself) to a node $b$ that is not a descendant of $u$. But then, the node $b$ can reach $u$ in $B$ but not in the

```
(1)  for i := 1 to n do begin
(2)      for j := i + 1 to n do aᵢⱼ := aᵢⱼ/aᵢᵢ;
(3)      aᵢᵢ := 1;
(4)      for k := i + 1 to n do
(5)          for j := i to n do
(6)              aₖⱼ := aₖⱼ - aₖᵢ * aᵢⱼ
     end
```

(a) Triangulation algorithm.

```
for i := 1 to n do begin
    for j := i + 1 to n do aⱼᵢ := aⱼᵢ/aᵢᵢ;
    for j := i + 1 to n do
        for k := i + 1 to n do
            aⱼₖ := aⱼₖ - aⱼᵢ * aᵢₖ
end
```

(b) Determinant calculation.

**Fig. 5.20.** Two algorithms to systolicize.

directed version of $T$. $\square$

### EXERCISES

5.1: In Fig. 5.20(a) there is a program to triangulate an $n \times n$ matrix. The exercise is to convert it into a systolic algorithm.

    a)    Begin by writing a nonsystolic timing diagram, with nodes in an $n \times n$ array corresponding to the matrix in an obvious way. Use zero-delay arcs to propagate $a_{ii}$ instantaneously along row $i$ at step (2) and to propagate $a_{ki}$ along row $k$ for step (6).

    b)    Convert your answer to (a) into a systolic algorithm using the techniques of Section 5.2.

5.2: Repeat Exercise 5.1 for the "determinant calculator" in Fig. 5.20(b). That algorithm leaves the diagonal so that the determinant of the original matrix equals the product of the diagonal elements.

\* 5.3: Design a linear systolic array to perform integer bitwise multiplication in essentially the way the algorithm of Section 5.1 performs convolution. That is, the bit strings of the two numbers to be multiplied are read in from the left, low-order bits first, and the output bits come out of the left end, with the first bit appearing a few beats after the first input bits.

\* 5.4: There are two ways to compute $c$, the convolution of streams $a$ and $b$, besides the arrangement discussed in Sections 5.1 and 5.2.

a)   The $a$ and $b$ streams enter from one side, say the left, and the $c$ stream exits from the opposite side.

b)   The $a$ and $b$ streams enter from opposite sides, with the $c$ stream exiting from one of those sides.

Investigate the possibility of systolic convolution in each of the above two ways. How does the speed compare with that of the approach in Section 5.1, where $a$ and $b$ are read from one side and $c$ comes out the same side?

** 5.5: Show that if a constant number of bits are read at any time, then convolution of streams of length $n$ requires area $\Omega(n)$.

** 5.6: Characterize the class of regular expression graphs that can be made systolic.

** 5.7: Give the best possible lower bound on $AT^2$ for transitive closure.

5.8: Give a systolic algorithm to compute shortest paths in a graph by generalizing the transitive closure method in Algorithm 5.1. Prove that your algorithm is correct.

* 5.9: Give an example of a graph where all three passes of Algorithm 5.1 are needed to compute the transitive closure.

* 5.10: Modify Algorithm 5.1 to compute the strict transitive closure, where the diagonal element $A_{ii}$ is not set to 1 unless there is a path of length one or more from node $i$ to itself.

5.11: Give a formal specification for step (2) of Algorithm 5.2, using the language of Section 4.2.

* 5.12: The *biconnected components* of an undirected graph are those maximal sets of nodes that cannot be disconnected by the deletion of any one of them. For example, it is not possible that two ends of a bridge are in the same biconnected component, unless the biconnected component is the bridge alone. Give a systolic algorithm to find the biconnected components of a graph.

5.13: Modify Algorithm 5.5 so the parent numbers of the nodes in the spanning tree appear at the right edge.

5.14: Show how it is possible to take one copy of a matrix, stored in an $n \times n$ array of processors, with cyclic connections from bottom to top and from right to left, and cycle copies of the rows and columns so that we may then perform Algorithm 5.1 as if two copies of the data were being read in from the outside.

5.15: Show that we may take any systolic network, add a new node that broadcasts data to all the nodes with a delay of 0, and retime the network to be systolic and running at half the speed of the old network.

* 5.16: Design circuits that implement the following data structures.

a)   A counter. That is, the circuit must be able to count the number of 1 inputs that it receives and indicate when it has reached a preset limit with only a fixed number of bits delay, independent of how large the count is.

b)  A *priority queue*. That is, the circuit accepts (key, value) pairs, and at any time can produce the value associated with the smallest key stored, and then delete that (key, value) pair from its memory, using only a fixed number of beats to do so.

c)  A stack. The circuit must accept requests to push or pop values and use only a fixed number of beats to do so.

## BIBLIOGRAPHIC NOTES

The notion of a systolic array was popularized by H.-T. Kung. Early papers with such designs include Kung and Leiserson [1978], Kung [1979], Foster and Kung [1980], and the chapter by Kung and Leiserson in Mead and Conway [1980]. Leiserson [1983] also contains many of the fundamental results on the subject. However, note that Exercise 5.3 on integer multiplication is from Atrubin [1965].

Algorithm 5.1, for transitive closure, is from Guibas, Kung, and Thompson [1979]. The algorithms of Section 5.4 are from Atallah and Kosaraju [1982]. The development of systolic algorithms discussed in Section 5.2 is from Leiserson and Saxe [1981]. See also Leiserson, Rose, and Saxe [1983].

The paper by Hong and Kung [1981] discusses the input/output complexity of algorithms and the way that systolic algorithms are often well tuned to the demands of speed and pin limitation.

There have been a great number of recent papers about systolic algorithms for performing various calculations. For example, see Fisher [1981] on statistical calculations or Savage [1981] on graph connectivity. Leiserson [1979] and Guibas and Liang [1980] discuss simple systolic data structures, e.g., those of Exercise 5.16. Basic database operations are covered in Kung and Lehman [1980] and Lehman [1981]. Two-dimensional convolution is discussed by Kung and Song [1981] and Kung, Ruane, and Yen [1981].

Chang and P. Fisher, and A. Fisher et al. [1983], describe the design of general-purpose chips for use in systolic arrays.

# 6

# ORGANIZATIONS WITH
# HIGH WIRE AREA

In this chapter we shall consider algorithms for certain processor organizations in which the wire area dominates the processor area, unlike the organizations such as the mesh that have become familiar over the previous two chapters. Two interconnection patterns that interest us are the "shuffle exchange" and the "butterfly." We shall also discuss the $k$-dimensional cube organization, which is actually closely related to the butterfly. These organizations are especially suited for sorting and computing the Fast Fourier Transform, among other problems.

Two valuable properties of the organizations we discuss are the limited degree of nodes, and the fact that there are enough wires to move data rapidly across any given partition that divides the nodes into two roughly equal sets. The former property simplifies the design of processors, while the latter is essential for critical operations such as sorting.

We believe that the organizations discussed in this chapter are among those best suited for the design of "supercomputers," or networks of many processors connected to perform tasks in a highly parallel way. We shall find that these organizations require almost $\Omega(n^2)$ area when $n$ nodes are laid out in the plane, e.g., if many processors are laid out on one chip. Furthermore, even if processors are on individual chips or individual circuit boards connected by "real" wires, we find that $\Omega(n^{3/2})$ volume is needed in three-space. On the other hand, their ability to transmit large amounts of data across arbitary boundaries quickly gives these organizations an essential advantage for many important applications, over organizations like the tree and rectangular grid.

Another important aspect of these high-interconnect networks concerns the simulation of algorithms written to run on a hypothetical parallel computer of a type that it seems cannot be built directly. Processors organized in one of these organizations can simulate a machine in which many processors and many memories are all allowed to talk to each other at once, in arbitrary combinations.

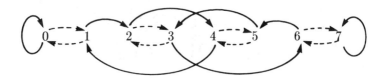

**Fig. 6.1.** Shuffle-exchange interconnections for $n = 8$.

## 6.1 THE SHUFFLE-EXCHANGE ORGANIZATION

Suppose we have $n$ processors numbered $0, 1, \ldots, n-1$, where $n$ is a power of 2. The *exchange* interconnection consists of wires in both directions between each even numbered processor and the next higher odd numbered processor. Thus 0 is connected to 1, 1 to 0, 2 to 3, 3 to 2, and so on. We use $EX(i)$ to stand for the processor to which processor $i$ is connected. Then $EX(i) = i + 1$ if $i$ is even, and $EX(i) = i - 1$ if $i$ is odd.

The *perfect shuffle* interconnection connects each processor $i$ to processor $2i \pmod{n-1}$. As a special case, if $i = n - 1$, even though

$$2(n-1) = 0 \pmod{n-1}$$

we connect processor $n-1$ to itself, rather than to processor 0. We use $PS_n(i)$, or $PS(i)$ when $n$ is understood, to be the processor to which processor $i$ is connected by the perfect shuffle on $n$ elements. Thus, $PS_n(i)$ is $2i \pmod{n-1}$ if $0 \leq i < n - 1$, $PS_n(n-1) = n - 1$, and $PS_n(i)$ is undefined if $i \geq n$.

**Example 6.1:** The perfect shuffle (solid) and exchange (dashed) interconnections for $n = 8$ are shown in Fig. 6.1. $\square$

The reason the perfect shuffle interconnection received its name will be clear from the following experiment. Suppose we have a deck of cards, numbered $0, 1, \ldots, n-1$, in order. Suppose also that we cut the deck exactly in the middle, so $0, 1, \ldots, n/2 - 1$ are in one half and $n/2, n/2 + 1, \ldots, n - 1$ are in the second half. Finally, let us shuffle the deck perfectly, that is, taking a card from the first pile then the second, alternately, until the packs are merged. The resulting order of the cards is $0, n/2, 1, n/2 + 1, \ldots, n/2 - 1, n - 1$. That is, card $i$ has moved to position $PS(i)$.

### Memory Maps

Suppose we have a network of processors with the perfect shuffle and exchange interconnections; we call this a *shuffle-exchange network*. That is, each processor $i$, $0 \leq i < n$, is able to pass data to processors $EX(i)$ and $PS(i)$. Note that although the exchange interconnection is symmetric, the perfect shuffle is not;

although processor 2 can pass data to processor 4 (since $PS(2) = 4$), it does not follow that 4 can pass data directly to 2. We may also consider networks where the perfect shuffle interconnection is bidirectional; i.e., processor $i$ can pass data to $PS^{-1}(i)$ as well as to $PS(i)$ and $EX(i)$, where $PS^{-1}(i)$ is defined to be that $j$ such that $PS(j) = i$.

Let each processor $i$ initially hold a datum $d_i$. Suppose that at each step we either

1.   Pass the datum in each processor $j$ to $EX(j)$,
2.   Pass the datum in each processor $j$ to $PS(j)$, or
3.   Pass the datum in each processor $j$ to $PS^{-1}(j)$.

Then we can keep track of the current position of each $d_i$ by listing in a column the binary number of the current processor holding $d_i$.

The table listing the current position of the $d_i$'s we call the *memory map*. Surprisingly, the set of possible memory maps after arbitrary sequences of shuffles and exchanges is not quite as extensive as we might expect. The reason is not hard to see. First, observe that if we apply the exchange operation, the current location of $d_i$ changes only in that the least significant bit is complemented.

Suppose that we apply the perfect shuffle operation. Let $i$ be $a_1 \cdots a_k$ in binary, where $n = 2^k$. If $a_1 = 0$, which implies that $i < n/2$, then

$$2i \ (\mathrm{mod} \ n - 1)$$

is $2i$, which has the binary representation $a_2 \cdots a_k 0$. If $a_1 = 1$, meaning that $i \geq n/2$, then $2i \ (\mathrm{mod} \ n - 1)$ is $2i - n + 1$. Since $n = 2^k$, $2i - n$ has binary representation $a_2 \cdots a_k 0$, and $2i - n + 1$ has representation $a_2 \cdots a_k 1$. Thus, whether $a_1$ is 0 or 1, $PS(i)$ has binary representation $a_2 \cdots a_k a_1$. That is, when we apply the perfect shuffle operation, the memory map has the bits in each row cyclically shifted left one position. Similarly, one can show that applying the inverse perfect shuffle, if the interconnection network allows it, has the effect of cyclically shifting the bits of the memory map one position right.

**Example 6.2:** Let $n = 8$. In Fig. 6.2(a) we see the initial memory map, which is the three-bit binary numbers written in numerical order. Figure 6.2(b) shows what happens after applying the perfect shuffle operation; the leftmost column in Fig. 6.2(a) has become the rightmost in Fig. 6.2(b), and the other two columns have shifted left. Next, Fig. 6.2(c) shows the result of applying the exchange operation; the rightmost column of Fig. 6.2(b) has been complemented in 6.2(c). Finally, Fig. 6.2(d) illustrates the result of applying the inverse perfect shuffle to Fig. 6.2(c). There, the data in processor $i$ has been transferred to that $j$ such that $i = PS(j)$, and in terms of the memory map, the columns of Fig. 6.2(c) have been cyclically shifted right.

Note that the net effect of these three operations is to exchange the first and last halves of the data. That is, $d_i$ has been moved to processor $i + n/2$

| 000 | 000 | 001 | 100 |
|-----|-----|-----|-----|
| 001 | 010 | 011 | 101 |
| 010 | 100 | 101 | 110 |
| 011 | 110 | 111 | 111 |
| 100 | 001 | 000 | 000 |
| 101 | 011 | 010 | 001 |
| 110 | 101 | 100 | 010 |
| 111 | 111 | 110 | 011 |
| (a) | (b) | (c) | (d) |

**Fig. 6.2.** Effect of shuffle-exchange network on memory map.

if $i < n/2$ and to $i - n/2$ if $i \le n/2$. Algebraically, we can write the identity $PS^{-1}(EX(PS(i))) = (i + n/2) \,(\text{mod } n)$. $\square$

### Necklaces

Another interesting fact about the shuffle-exchange graph is that repeated shuffling rapidly takes us back to where we started. For a quick proof, note that if $n = 2^k$, then $k$ shuffles will cycle the memory map completely, restoring each $d_i$ to processor $i$. Algebraically, $PS^{\log_2 n}(i) = i$.

The processors that $d_i$ travels through in response to repeated shuffles we call the *necklace* of $i$. What we have just said should convince us that each necklace is no longer than $\log_2 n$, but it could be shorter. If so, we call it a *short necklace*.

**Example 6.3:** Let $n = 16$. Then 0 and 15 are in necklaces by themselves. The necklace containing 1 is $\{\,1, 2, 4, 8\,\}$, since these are the numbers whose four-bit binary representations are cyclic shifts of that for 1, i.e., 0001, 0010, 0100, and 1000. Similarly, the numbers whose binary representations are cyclic shifts of 0011, namely, $\{\,3, 6, 9, 12\,\}$ form a necklace, as do the cyclic shifts of 0111, i.e., $\{\,7, 11, 13, 14\,\}$. The remaining numbers, $\{\,5, 10\,\}$ form a short necklace; these are the cyclic shifts of 0101. Notice that the reason 0101 gives a short necklace is that it is a pattern formed by the repetition of a smaller pattern, 01. $\square$

It turns out to be important in what follows, when we try to lay out shuffle-exchange graphs, to bound the number of necklaces. A simple way to do so is the following.

**Lemma 6.1:** The number of necklaces of the $n$-processor shuffle-exchange graph, for $n = 2^k$, is $O(2^k/k)$.

**Proof:** In order for $i$ to be on a short necklace, there must be some integer $1 \le m < k$, such that when we rotate the binary representation $a_1 \cdots a_k$ $m$ positions, we get what we started with, that is, $a_1 \cdots a_k = a_{m+1} \cdots a_k a_1 \cdots a_m$.

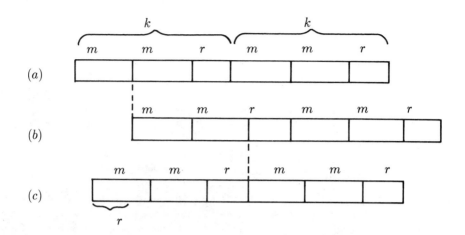

**Fig. 6.3.** Cyclic shifts of a binary word.

Assume without loss of generality that $m$ is as small as possible.

We must prove that $m$ divides $k$ evenly. Suppose that the remainder when $k$ is divided by $m$ is $r$, $1 \le r < m$. It helps to imagine that $a_1 \cdots a_k$ is an infinitely repeating sequence, with $a_i = a_{i(\bmod k)}$ for all $i \ge 1$. Then this infinite word, drawn as (a) in Fig. 6.3, matches itself shifted $m$ places right, as shown in Fig. 6.3(b). If we compare bits $a_{k+1} \cdots a_{k+m-r}$ of Fig. 6.3(a) with the corresponding bits of 6.3(b), we discover that $a_1 \cdots a_{m-r} = a_{r+1} \cdots a_m$. If we compare the next $r$ bits of Fig. 6.3(a) and (b), we find that $a_1 \cdots a_r = a_{m-r+1} \cdots a_m$.

Let $a_1 \cdots a_{m-r} = \alpha$ and $a_{m-r+1} \cdots a_m = \beta$. From the two equalities mentioned above, we can deduce that $\alpha\beta = \beta\alpha$, and in particular, each group of length $m$ shown in Fig. 6.3(a, b) is both $\alpha\beta$ and $\beta\alpha$. Also, each group of length $r$ shown explicitly in those figures is $\beta$.

We may thus deduce that the word in Fig. 6.3(c), which is our original word shifted $m - r$ places, agrees in corresponding positions with the other two words. In proof, suppose that $m$ divides $k$ twice with remainder $r$, as suggested in Fig. 6.3. Then we must show that $\beta\alpha\beta\beta\alpha$, which is positions $m - r + 1$ through $k + m - r + 1$ of Fig. 6.3(a), equals $\alpha\beta\alpha\beta\beta$, which is the first $k$ positions of Fig. 6.3(c). But the equality $\alpha\beta = \beta\alpha$ makes this step easy. Similarly, if $k = pm + r$ then we must show $\beta(\alpha\beta)^{p-1}\beta\alpha = (\alpha\beta)^p\beta$, which again requires only the identity $\alpha\beta = \beta\alpha$.

Unless $r = 0$, we have contradicted the minimality of $m$, and conclude that $m$ divides $k$. Our conclusion is that for $i$ to be on a short necklace, the binary representation of $i$ must be the repetition $k/m$ times of some pattern of length $m$, where $m > 1$, and $m$ is a divisor of $k$.

The number of possible values of $i$ is thus no greater than

$$2^{k/2} + 2^{k/3} + 2^{k/4} + \cdots$$

In general it is less, since not all integers will divide $k$. However, this sum is surely upper bounded by $2^{k/2+1}$. Thus, all but this number of the $2^k$ integers are on necklaces of size exactly $k$. There can be no more than $2^k/k$ full size necklaces, and the number of short necklaces is surely no greater than the number of integers on short necklaces; in fact it will be less. We conclude that the number of necklaces is upper bounded by $2^k/k + 2^{k/2+1}$, which is $O(2^k/k)$, or $O(n/\log n)$. $\square$

## Area of Shuffle-Exchange Layouts

We shall now consider the area needed for the layout of a shuffle-exchange graph, using the model of Chapter 3, where processors, or nodes, are assumed to occupy one grid point, and wires have the width of one grid line. Our analysis will indicate the actual area required for a network of real processors if we multiply height and width by the larger of the number of wires connecting processors and the height or width, respectively, of a processor. In particular, we shall discover that the area used by wires will generally dominate the area used for the processors.

The following lemma turns out to be useful in the analysis.

**Lemma 6.2:** Consider the problem with input bits $x_0, \ldots, x_{n-1}, c_1, \ldots, c_{\log n}$, and output bits $y_0, \ldots, y_{n-1}$, in which each $y_i$ must be set equal to a certain $x_j$ determined from the $c$'s as follows. Let $i = a_1 \cdots a_k$ in binary, where $k = \log n$. Then the binary representation of $j$ is $b_1 \cdots b_k$, where for $1 \le \ell \le k$,

$$b_\ell = a_\ell \oplus c_\ell$$

Here, $\oplus$ is the modulo 2 sum. Then this problem requires $AT^2 = \Omega(n^2)$.

**Proof:** The problem is essentially that of Exercise 2.1(c), and its proof is very much like the lower bound proof for the barrel shift problem. $\square$

It turns out that the problem of Lemma 6.2 is very easy to solve on a network of processors organized as a shuffle exchange graph. This fact, together with the lower bound of that lemma on the $AT^2$ product, implies a lower bound on the area of the layout.

**Theorem 6.1:** Any shuffle-exchange graph layout requires area $\Omega(n^2/\log^2 n)$.

**Proof:** We shall use an $n$-node shuffle-exchange graph to solve the problem of Lemma 6.2 in the following way. Variable $x_i$ is read at processor $i$ at the first time unit, and at the end, $y_i$ is written by that processor. We shall assume that $c_1, \ldots, c_k$ are read and distributed to all the processors, in that order. We can use another layer for the wires that distribute these control bits, then invoke Theorem 1.1 to say that this extra layer does not change the growth rate of the required area, just the constant factor.

At alternate beats, we shall apply the shuffle permutation to the data in the processors. After each shuffle, if the current control bit $c_i$ is 1, then we apply the exchange permutation. After $k$ shuffles, the memory map will have had its bits cycled completely around, but some of the positions will have been complemented by the exchange permutations. To be precise, the $i^{th}$ bit position has been complemented if and only if $c_i = 1$.

The effect on the memory map therefore has been to send the datum read by the processor whose number in binary is $a_1 \cdots a_k$ to the processor whose number is $(a_1 \oplus c_1) \cdots (a_k \oplus c_k)$, which is exactly the index of the $y$ output that must receive that input in the problem of Lemma 6.2.

Our conclusion is that $O(\log n)$ time suffices to solve the problem of Lemma 6.2 with a circuit whose area is, to within a constant factor, the area of the layout of the shuffle-exchange graph. We thus have, by Lemma 6.2,

$$A \log^2 n = \Omega(n^2)$$

from which the theorem follows. $\square$

It is a result of Kleitman et al. [1981] that Theorem 6.1 is the best possible, i.e., there is a way to lay out the shuffle-exchange graph of $n$ nodes in $O(n^2/\log^2 n)$ area. We shall not prove this result here, but we shall show some of the important ideas in an example.

**Example 6.4:** We shall lay out the shuffle-exchange graph of 16 nodes in a grid. All the nodes of a given necklace are found in a single column, and their order must allow them to be linked in a circle, that is, by a wire that runs down the column and then reverses and runs up the column and back to its starting point. For example, the necklace $\{1, 2, 4, 8\}$ could appear in the order 1248, or the order 1842; in the latter case we would link 1 and 2 going down, then reverse and link 4, 8, and back to 1, on the way up. We could not, however, allow the order 4182, because three wires would be necessary in the region between 1 and 8. Furthermore, if two necklaces share a column, then one must be completely above the other. As a result, there will never be more than two wires running along any column.

The exchange permutation is associated with the rows, in the sense that each even node and the next odd numbered node must appear on the same row. Moreover, if two pairs share the same row, then one must be to the left of the other. As a consequence, there will be at most two wires at any point along a row.

In Fig. 6.4 we see a dense packing of the necklaces for the 16-node shuffle-exchange graph. $\square$

Note that Example 6.4 generalizes in the sense that by Lemma 6.1 we never need more than $O(n/\log n)$ columns. However, it is subtle that one can always do with $n/2$ rows, since it is not clear that the constraints on the order of the elements of the necklaces can be honored simultaneously, and still allow the

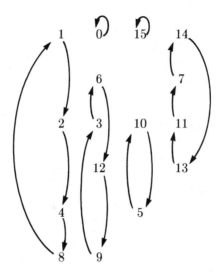

**Fig. 6.4.** Layout for shuffle-exchange graph.

exchange pairs to share rows.

## 6.2 THE BUTTERFLY ORGANIZATION

The *butterfly network* is a connection system most frequently associated with the Fast Fourier Transform; an example of the network is shown in Fig. 6.5. In general, it consists of $(k + 1)2^k$ nodes, organized as $k + 1$ *ranks* of $n = 2^k$ nodes each. Optionally, we shall identify the top and bottom ranks, so there is no rank 0, and the processors on ranks 1 and $k$ are connected directly.

Let us denote the $i^{th}$ node on the $r^{th}$ rank by $p_{ri}$, $0 \le i < n$, $0 \le r \le k$. Then node $p_{ri}$ on rank $r > 0$ is connected to two nodes on rank $r - 1$, the two nodes $p_{r-1,j}$ such that either $j = i$ or the binary representation of $j$ differs from that of $i$ in only the $r^{th}$ place from the left. Thus, if the binary representation for $i$ is $a_1 \cdots a_{r-1} 0 a_{r+1} \cdots a_k$, and that for $j$ is $a_1 \cdots a_{r-1} 1 a_{r+1} \cdots a_k$, then the nodes $p_{ri}$ and $p_{rj}$ are each connected to $p_{r-1,i}$ and $p_{r-1,j}$. These four connections form a "butterfly" pattern, from which the name of the network is derived. Note also that as the ranks go down, which in our notation means that the bits in which $i$ and $j$ differ increase in significance, the difference between $i$ and $j$, that is, the width of the butterfly's wings, increases exponentially, and in general is equal to $2^{k-r}$.

The butterfly turns out to be another network whose wire area must dominate its processor area. The exact analysis of the area requirement for this network turns out to be considerably easier than the analysis for the shuffle-exchange network.

**Theorem 6.2:** The best layout for the butterfly network with $n \log n$ nodes

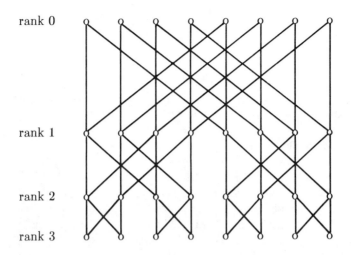

**Fig. 6.5.** Butterfly connection for $k = 3$ ($n = 8$).

has area that is $O(n^2)$ and $\Omega(n^2)$.

**Proof:** The straightforward layout suggested by Fig. 6.5 has area $O(n^2)$. Note that if we let $k = \log n$, as previously, and assume that $n$ is a power of 2, then this network has $k + 1$ ranks of width $n$, but the separation between the ranks goes down exponentially. That is, the distance between ranks 0 and 1 is $n$, but between ranks 1 and 2 is only $n/2$ space, between ranks 2 and 3 is $n/4$ space, and so on, which sums to a height of $O(n)$. The network is similar to the network of Example 2.7 (Fig. 2.3), whose area was discussed in detail.

For the lower bound, let us first observe that we may assume without loss of generality that there are connections between corresponding nodes of the first and last ranks, that is, edges between $p_{0i}$ and $p_{ki}$ for $0 \le i < n$. The reason is that, given a layout for the butterfly as formally defined, we could add wires for these edges by following the path made by a column, from $p_{0i}$ to $p_{1i}$, to $p_{2i}$, and so on, to $p_{ki}$. By Theorem 1.1, we can restore the number of layers with only a constant factor increase in area.

To prove the lower bound, we consider $k = \log n$ copies of the problem of Lemma 6.2. The inputs for the $r^{th}$ copy of the problem are read at the beginning at the nodes of the $r^{th}$ rank, $1 \le r \le k$; $x_i$ is read at $p_{ri}$. The control bits will be read at appropriate times that the reader may deduce, and when a control bit for a copy of the problem is read, it is distributed to all the nodes of the rank at which the data for that copy currently resides.

The copies of the problems are cycled around the ranks, so the copy starting at rank $r$ moves at the next stage to rank $r - 1$, then to rank $r - 2$, and so on, then to rank 0 and back to ranks $k$, $k - 1$, and finally back to $r$, where the output for that copy is made. When moving from some rank $r$ to $r - 1$, we have the opportunity to exchange variables at nodes $p_{ri}$ and $p_{rj}$ if $i$ and $j$ differ only in the $r^{th}$ position of their binary representations. That is the same opportunity we have in the shuffle-exchange network when the memory map has been cycled so that the $r^{th}$ position was rightmost. In the shuffle-exchange network we complement the $r^{th}$ bit by applying the exchange operation, and we omit that operation if we do not wish to complement. In the case of the butterfly, we complement by using the diagonal connections, and if we do not wish to complement we use the vertical connections.

Let us, therefore, at this time read the control bit for this problem corresponding to position $r$. If it is 1, then we perform the exchange, sending the data in $p_{ri}$ to $p_{r-1,j}$ and sending $p_{rj}$ to $p_{r-1,i}$. If the control bit is 0, then we do not exchange, sending $p_{rm}$ to $p_{r-1,m}$ for all $m$. As a result, when the inputs for the $r^{th}$ copy return to the $r^{th}$ rank, the bits have been permuted according to the dictates of the control bits for that copy.

It is left as an exercise that the problem consisting of $\log n$ copies of the problem of Lemma 6.2, each with $n$ input bits plus control bits, has information content $\Omega(n \log n)$. That is not paradoxical, since $n \log n$ is the number of inputs that this problem has. Given this fact, we can conclude that $AT^2 = \Omega(n^2 \log^2 n)$ for this problem. We have just given a way to solve the problem in $O(\log n)$ time, so we conclude that $A = \Omega(n^2)$. $\square$

## The Cube-Connected Cycles Interconnection

It is often convenient to view the butterfly in quite a different way, one where the processors are seen as the vertices of a $k$-dimensional cube for some $k$. Actually, the vertices of the cube correspond to entire columns of the butterfly of Fig. 6.5. The processors at each vertex of the cube (column of the butterfly) are connected in a cycle, in the order in which they appear in the column, with the processor in rank 0 identified with the processor in rank $k$, so rank 1 is connected in the cycles with rank $k$. Hence the organization has been known as the *cube-connected cycles* or *CCC* organization.

In addition to connections around the cycles, for each cycle and each dimension there is one processor that communicates with the neighboring cycle in that dimension. Specifically, processor $p_{ri}$ communicates with processor $p_{rj}$ if and only if $i$ and $j$ differ in the $r^{th}$ bit from the left, in their binary representations. Thus, the "dimensions" of the cube correspond to adjacent ranks in the butterfly. The connections are shown in Fig. 6.6 for the case $k = 3$.

The only detail necessary to relate the butterfly to the CCC is that in the CCC, $p_{ri}$ is connected to $p_{r-1,i}$ (around the cycle) and to $p_{rj}$ (across the cube), while in the butterfly it was connected to $p_{r-1,j}$ instead. However, by following a pair of edges in the CCC network, we can get to $p_{r-1,j}$ from $p_{ri}$; we go across

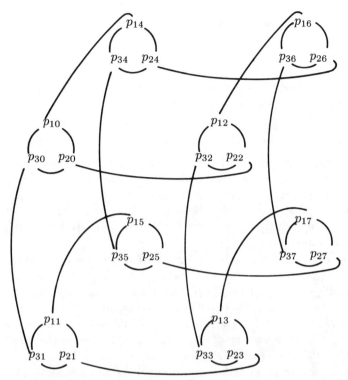

**Fig. 6.6.** The cube-connected cycles organization.

the cube to $p_{rj}$ and then around the cycle to $p_{r-1,j}$. As an alternative, we could run pairs of wires along each edge of the cube, connecting $p_{ri}$ to $p_{r-1,j}$ and $p_{rj}$ to $p_{r-1,i}$ directly.

## 6.3 ALGORITHMS ON BUTTERFLY NETWORKS

In this section, we shall show how to use the butterfly for sorting and Fast Fourier Transform algorithms. It should be apparent that any such algorithm can also be implemented on the CCC, but we shall prove more—that they can be implemented on the shuffle-exchange network and another network, called the hypercube, mentioned in Exercise 3.7. For the proof, we shall consider a restricted class of algorithms on butterfly networks.

### Normal Butterfly Algorithms

Let us call an algorithm running on a butterfly network of processors *normal* if at each beat, all the relevant data resides in the processors of a single rank $r$. Moreover, at each beat, either

1.   No data is passed between processors,

2.   The processors at rank $r$ pass their data to the two processors at rank $r-1$ to which they are connected, or

3.   They do the same with their neighbors at rank $r+1$.

Those familiar with the FFT, for example, will note how the data of that algorithm flows through the ranks from top to bottom and is, therefore, normal; we shall consider this algorithm in detail shortly.

**Theorem 6.3:** Any normal algorithm that runs in $T(n)$ time on a butterfly network can be made to run in $O(T(n))$ time on a shuffle-exchange network.

**Proof:** We shall assume that the shuffle-exchange network has the $PS$, $PS^{-1}$, and $EX$ interconnections, and we shall similarly assume that in the butterfly, data can flow upwards or downwards between adjacent ranks. We also assume that ranks 0 and $k$, the top and bottom ranks, are identified and therefore connected. Analogous results for networks that are unidirectional can be proved in a similar fashion.

Let us assume that the data is initially read into rank $k$ of the butterfly. If so, we can read the same data that was read into processor $p_{ki}$ into processor $i$ of the shuffle-exchange network. If the data is initially read into some other rank $r$, then we read the data into the shuffle-exchange network in a permuted order, so that, in effect it appears that the memory map of the shuffle exchange network has been cycled so the $r^{th}$ bit is rightmost. We leave the details of how to select the initial order of data as an exercise.

We now perform an induction on the number of moves made by the butterfly network. We claim as an inductive hypothesis that if the data is at rank $r$ after some step, then in the shuffle-exchange network, the same data appears, but the positions are those determined by the memory map in which the initial memory map has been cycled so that the $r^{th}$ bit is rightmost. That, is, the data in processor $p_{ri}$ of the butterfly is, if $i = a_1 \cdots a_k$ in binary, found in the processor of the shuffle-exchange network with binary representation $a_{r+1} \cdots a_k a_1 \cdots a_r$.

If at the next beat, the data stays at the $r^{th}$ rank, then the shuffle-exchange network can simulate the butterfly in a straightforward way. Suppose instead, that the data is passed to rank $r-1$. Then each pair of processors $p_{ri}$ and $p_{rj}$, where $i$ and $j$ differ only in the $r^{th}$ bit, pass some or all of their data to $p_{r-1,i}$ and $p_{r-1,j}$.

In the shuffle-exchange network, each processor $\ell$ keeps its data and also passes it to processor $EX(\ell)$. If $\ell$ is the processor holding the data of $p_{ri}$ above, then $EX(\ell)$ represents processor $p_{rj}$, since those are two processors whose numbers differ only in the $r^{th}$ bit. Thus, each processor of the shuffle-exchange with binary representation $a_{r+1} \cdots a_k a_1 \cdots a_r$ holds the data that was found in the processors in rank $r$ of the butterfly with numbers, in binary, $a_1 \cdots a_{r-1} 0 a_{r+1} \cdots a_k$ and $a_1 \cdots a_{r-1} 1 a_{r+1} \cdots a_k$. At the next beat of the butterfly, the data at the processors with the same numbers, but in rank $r-1$, will be computed from that data.

Thus, let each processor $\ell$ of the shuffle-exchange pass this data to processor $PS^{-1}(\ell)$. Now, processor $a_r \cdots a_k a_1 \cdots a_{r-1}$ can compute the data that belongs in processor $p_{r-1,i}$, where $i$ in binary is $a_1 \cdots a_k$. This observation completes the inductive hypothesis in the case that data is passed from rank $r$ to $r-1$.

In the case that data goes from rank $r$ to $r+1$, we instead first pass the data in each processor $\ell$ of the shuffle-exchange to processor $PS(\ell)$, and then we retain the data there while passing a copy to processor $EX(PS(\ell))$. Then, each processor $a_{r+2} \cdots a_k a_1 \cdots a_{r+1}$ can compute the data for processor $p_{r+1,i}$ of the butterfly, where $i$ is $a_1 \cdots a_k$ in binary. Thus, the inductive hypothesis is proved for the next beat. Since in each case, at most two steps of passing data are used by the shuffle-exchange to simulate one step of the butterfly, the former operates at at least half the speed of the latter. □

### The Hypercube

Another useful interconnection network is the *hypercube*, where for some $d$, the nodes are the vertices of a hypercube of dimension $d$. That is, there is a node numbered $i$, for $0 \le i < 2^d$, and two nodes are adjacent if and only if the binary representations of their numbers differ in exactly one bit position. The cube connected cycles interconnection is a hypercube where each node has been replaced by a cycle of length $d$, incidentally.

While the CCC is isomorphic to the butterfly, the hypercube is really the butterfly with the columns collapsed. That is, the edge in the hypercube between nodes with binary representations $i$ and $j$ that differ only in bit $r$ can be identified with the butterfly attached to the nodes in columns $i$ and $j$ between ranks $r$ and $r-1$. Thus, the following is an easy exercise.

**Theorem 6.4:** Any normal algorithm that runs in time $T(n)$ on a butterfly network can be made to run in time $T(n)$ on a hypercube. □

Note that, unlike Theorem 6.3, here we do not even need a constant factor slowdown.

### The Fast Fourier Transform

The *Discrete Fourier Transform* of a sequence $a_0, \ldots, a_{n-1}$ is the sequence $b_0, \ldots, b_{n-1}$, where for $0 \le j < n$,

$$b_j = \sum_{i=0}^{n-1} a_i \omega^{ij}$$

Here, $\omega$ is a *primitive $n^{th}$ root of unity*. That is, $\omega^n = 1$, but $\omega^i \neq 1$ for $1 \le i < n$. Often, the $a$'s are complex numbers, and $\omega$ is taken to be the complex number of magnitude 1 and angle $2\pi/n$, which is what we normally

| rank 1 | $\omega^0$ | $\omega^0$ | $\omega^0$ | $\omega^0$ | $\omega^4$ | $\omega^4$ | $\omega^4$ | $\omega^4$ |
|--------|------------|------------|------------|------------|------------|------------|------------|------------|
| rank 2 | $\omega^0$ | $\omega^0$ | $\omega^4$ | $\omega^4$ | $\omega^2$ | $\omega^2$ | $\omega^6$ | $\omega^6$ |
| rank 3 | $\omega^0$ | $\omega^4$ | $\omega^2$ | $\omega^6$ | $\omega^1$ | $\omega^5$ | $\omega^3$ | $\omega^7$ |

**Fig. 6.7.** Pattern of $exp(r, i)$ values.

regard as the "$n^{th}$ root of 1." However, we sometimes find the Discrete Fourier Transform used where the $a$'s are integers modulo some integer $m$, and $\omega$ is an integer with the properties of a primitive $n^{th}$ root of unity when computation is done modulo $m$. The reader is referred to Aho et al. [1974] for details regarding this and other aspects of the subject of Fourier transforms not covered here.

The Discrete Fourier Transform has numerous applications in signal processing; a key property is that the transform of the convolution of two sequences is the componentwise product of the transforms of the sequences, which gives us a fast way to compute convolutions.

The *Fast Fourier Transform*, or FFT, is a particular algorithm for computing the Discrete Fourier Transform. The FFT takes time $O(n \log n)$ on a serial computer, compared with $O(n^2)$ if we applied the definition directly. Moreover, we shall see that $O(\log n)$ time suffices on a butterfly or shuffle-exchange network, where we are, in effect, computing the FFT with great parallelism.

Before giving the FFT algorithm, we must learn a notation that is essential. Let $n = 2^k$ be understood. Then $exp(r, i)$ is that $k$-bit integer $j$ such that if $i$ in binary is $a_1 \cdots a_k$, then $j$ in binary is $a_r a_{r-1} \cdots a_1 00 \cdots 0$. That is, we take the first $r$ bits of $i$, reverse them, and replace the remaining bits by 0's to get $j$. We use "$exp$" for this operation because it turns out to be the appropriate exponent of $\omega$ to use in certain steps of the FFT algorithm.

**Example 6.5:** Let $k = 4$, $r = 3$, and $i = 13$, or 1101 in binary. Then to find $exp(3, 13)$ we take the first three bits of $i$, 110, reverse them, and pad with zeroes (one zero in this case), to get 0110, i.e., $exp(3, 13) = 6$.

We shall, in a moment, find it useful to think of $\omega^{exp(r,i)}$ as residing at processor $p_{ri}$ of the butterfly. Figure 6.7 indicates the pattern of resident values for $k = 3$. □

We can now state the FFT algorithm simply, on the assumption that the constant $\omega^{exp(r,i)}$ is available at processor $p_{ri}$. The algorithm appears in Fig. 6.8. It involves computing a value $d_{ri}$ at each processor $p_{ri}$ of the butterfly. Initially, the $a$'s are read in parallel at rank 0, with $d_{0i} = a_i$. At successive beats, the $d$'s are computed in parallel for ranks 1, 2, and so on. Finally, it turns out that $d_{ki}$ is $b_j$, one of the desired outputs of the FFT. In particular, $j = exp(k, i)$, i.e., the binary representations of $i$ and $j$ are the reverses of each other.

(1)  **for** $i := 0$ **to** $n-1$ { in parallel } **do**
(2)       **read** $d_{0i}$;
(3)  **for** $r := 1$ **to** $k$ **do**
(4)       **for** each pair $i$, $j$ differing only in bit $r$,
(5)            with $i$ having 0 in that bit **do** { in parallel } **begin**
(6)                 $d_{ri} := d_{r-1,i} + \omega^{exp(r,i)}d_{r-1,j}$;
(7)                 $d_{rj} := d_{r-1,i} + \omega^{exp(r,j)}d_{r-1,j}$
          **end**

**Fig. 6.8.** FFT Algorithm.

We shall omit a proof that the algorithm in Fig. 6.8 does compute the Discrete Fourier Transform; again the reader is referred to Aho et al. [1974]. However, let us observe that the algorithm can be executed in $O(\log n)$ time on the butterfly network, if we take the computation of lines (6) and (7) to be unit time operations. In practice, these computations may take time that depends on $n$, so we must multiply the time by whatever the time for these operations turns out to be. Steps (1–2) can be executed in parallel in rank 0, taking one beat. The inner loop of lines (4–7) can also be done in $O(1)$ time, since the butterfly connections between ranks $r-1$ and $r$ are exactly what we need to pass the values $d_{r-1,i}$ and $d_{r-1,j}$ to $p_{ri}$ and $p_{rj}$, where $d_{ri}$ and $d_{rj}$ are computed according to lines (6) and (7). Finally, the outer loop beginning at line (3) is executed $\log n$ times, so the total time spent is $O(\log n)$.

However, it is not necessarily reasonable to require that at each processor $p_{ri}$ the constant $\omega^{exp(r,i)}$ be read in. One alternative is to read in only the values $\omega^{exp(k,i)}$ at the processors $p_{ki}$ of the last rank. Then, we can, in $\log n$ beats, compute all the desired values by passing data to progressively lower ranks. The rule to follow is this. Suppose we have computed the values for rank $r+1$, and wish to compute them for rank $r$. Let $i$ and $j$ differ only in bit $r$, with $i$ having the 0 there. Then $exp(r,i) = exp(r,j)$, and both are $exp(r+1,i)$ shifted one place left, as the reader may show easily. Thus, both $\omega^{exp(r,i)}$ and $\omega^{exp(r,j)}$ are the square of $\omega^{exp(r+1,i)}$.

Even if we do not wish to read the various powers of $\omega$ in at the $k^{th}$ rank, we can compute them all in $O(\log n)$ steps beginning only with 1 and $\omega$ at the lower left hand processor, $p_{k0}$. We leave this result as an exercise. A consequence of the analysis we have made is the following theorem.

**Theorem 6.5:** If arithmetic operations on data take time $t(n)$ each, then we can implement the FFT on a butterfly, shuffle-exchange, or hypercube in $O(t(n)\log n)$ time.

**Proof:** The result for the butterfly is a consequence of the analysis of the program in Fig. 6.8. The result for the shuffle-exchange then follows from Theorem 6.3, and for the hypercube we use Theorem 6.4. $\square$

Note that the shuffle-exchange, using its best possible layout area of

$$O(n^2/\log^2 n)$$

gives us an $AT^2$ product for the FFT of $O(s(n)^2 t(n)^2 n^2)$, where $s(n)$ is the side of a square area holding a processor for the arithmetic operations on data. If $O(\log n)$-bit words suffice to represent data, then $AT^2 = O(n^{2+\epsilon})$. However, there are applications where the number of bits, and hence $s(n)$ and $t(n)$, are $\Omega(n)$; see Aho et al. [1974].

**Sorting**

We shall implement a variant of the odd-even merge sort discussed in Section 4.4, by a normal butterfly algorithm; this algorithm will take $O(\log^2 n)$ steps.

Before describing the algorithm, let us consider some interesting and important structure possessed by the butterfly network. First, if we remove rank 0 and the connections between ranks 0 and 1 from a butterfly of width $n = 2^k$, then we separate the left and right halves of the butterfly, and each half is a butterfly of width $n/2$, in which ranks 1 through $k$ of the original butterfly play the roles of ranks 0 through $k - 1$. Similarly, if we remove rank $k$ and the connections between ranks $k$ and $k - 1$, we divide the butterfly into two butterflies of width $n/2$. In this case, the two halves are interlaced; one consists of the even columns, the other of the odd columns.

Some notation regarding ways to transfer data in the butterfly will prove useful. First, we shall assume as before that the relevant data at processor $p_{ri}$ is $d_{ri}$. We *copy rank $r$ to rank $r - 1$* by doing $d_{r-1,i} := d_{ri}$ for $0 \leq i < n$. Similarly, we can copy rank $r$ to rank $r+1$. If $k$ is the last rank, so the butterfly connections are between adjacent columns, then we *copy-exchange from rank $k$ to rank $k - 1$* by doing in parallel:

> **if** odd($i$) **then** $d_{k-1,i} := d_{k,i-1}$
> **else** { $i$ is even } $d_{k-1,i} := d_{k,i+1}$

That is, even and next odd pairs are exchanged through the butterfly going from rank $k$ to rank $k - 1$.

Another useful preliminary concerns a slight variation on odd-even merging, in which instead of first merging odds with odds and evens with evens, we merge the odds of one list with the evens of the other. The purpose is so that the final comparisons can be between even and next odd pairs (starting the count at 0), rather than odd and next even pairs as in Lemma 4.1.†

**Lemma 6.3:** Let $a_0, \ldots, a_{n-1}$ and $b_0, \ldots, b_{n-1}$ be two sorted lists. Suppose we merge $a_0, a_2, \ldots, a_{n-2}$ with $b_1, b_3, \ldots, b_{n-1}$ to get list $c_0, \ldots, c_{n-1}$ and we

---

† Note that Lemma 4.1 referred to "even and next odd" pairs, because the count started at 1, not 0 as we do here.

**procedure** $merge(k)$;
(1)  **if** $k = 1$ **then** { lists are one element each; compare them in parallel }
(2)      **begin** $d_{00} := min(d_{10}, d_{11})$; $d_{01} := max(d_{10}, d_{11})$;
(3)      $d_{10} := d_{00}$; $d_{11} := d_{01}$ **end**
     **else**{ $k > 1$ } **begin**
(4)      copy-exchange $d_{k0}, \ldots, d_{k,n/2-1}$ into $d_{k-1,0}, \ldots, d_{k-1,n/2-1}$;
(5)      copy $d_{k,n/2}, \ldots, d_{k,n-1}$ into $d_{k-1,n/2}, \ldots, d_{k-1,n-1}$;
(6)      $merge(k-1)$; { in parallel for both halves, using ranks $0, \ldots, k-1$ }
(7)      **for** $i := 0$ **to** $n - 1$ **do** { in parallel } **begin**
(8)          **if** odd($i$) **then** $d_{ki} := max(d_{k-1,i-1}, d_{k-1,i})$;
(9)          **else** $d_{ki} := min(d_{k-1,i}, d_{k-1,i+1})$
         **end**
     **end**

**Fig. 6.9.** The modified odd-even merge.

merge $a_1, a_3, \ldots, a_{n-1}$ with $b_0, b_2, \ldots, b_{n-2}$ to get list $d_0, \ldots, d_{n-1}$. Then the complete sorted list is obtained by meshing the $c$'s and $d$'s to form the list

$$c_0, d_0, c_1, d_1, \ldots, c_{n-1}, d_{n-1}$$

and then, if necessary, exchanging pairs $c_i, d_i$.

**Proof:** The proof is similar to Lemma 4.1, and we leave it as an exercise. $\square$

**Algorithm 6.1:** *Normal Butterfly Algorithm for Sorting.*

INPUT: A list of elements to be sorted, one in each of the processors in the top rank (rank 0) of a butterfly network.

OUTPUT: The same list, in sorted order, at the bottom rank of the butterfly network.

METHOD: An important subroutine is the procedure *merge* in Fig. 6.9, which takes two sorted lists, in rank $k = \log_2 n$ of a butterfly of width $n$, with the first list being the data $d_{k0}, \ldots, d_{k,n/2-1}$ in the left half of the rank, and the second list being the data $d_{k,n/2}, \ldots, d_{k,n-1}$ in the right half. The result is one sorted list consisting of the merger of these two half-size sorted lists; the resulting list, $d_{k0}, \ldots, d_{k,n-1}$, appears in rank $k$. An explanation of how *merge* works is given in the proof of Theorem 6.6.

We use this odd-even merge in a recursive, parallel sorting routine shown in Fig. 6.10. The procedure $sort(k)$ takes an unsorted list in rank 0 and produces a sorted list in rank $k$. The algorithm then consists of loading the data into rank 0 of the full butterfly, calling $sort(k)$, and then reading the sorted list from rank $k$. $\square$

**Theorem 6.6:** Algorithm 6.1 is a normal butterfly algorithm. It sorts correctly, and if the data to which it is applied requires $t(n)$ to compare elements, and

```
procedure sort(k);
begin
    if k = 0 then return { only one element to sort }
    else begin
        copy rank 0 to rank 1;
        sort(k − 1); { in parallel, using ranks 1, . . . , k as two
            butterflies. The two sorted lists wind up in the left and
            right halves of rank k }
        merge(k)
    end
end
```

**Fig. 6.10.** Normal butterfly sorting algorithm.

requires processors of side $s(n)$, then it takes time $O(t(n)\log^2 n)$ and has an $AT^2$ product of $O(s^2(n)t^2(n)n^2\log^2 n)$ on a shuffle-exchange network.

**Proof:** It is easy to check that the algorithm is normal, because each step transfers all data from one rank to an adjacent rank. Note, however, that we must regard the two statements on line (2) as taking place simultaneously, in one step, and line (3) must be regarded similarly. Further, lines (4) and (5) also take place in one step. We can make these stipulations because none of the actions we wish to do in parallel try to use the same processors.

Note that as *merge* and *sort* make recursive calls, the number of smaller butterflies that are operating independently grows exponentially, yet they all operate in parallel and in lock-step, each doing the same thing at the same time, and having their data at the same rank of the full butterfly.

It is also easy to check that *merge* implements the variant odd-even merge of Lemma 6.3, with lines (4–5) of Fig. 6.9 setting up two subproblems where the odds of the left half are merged with the evens of the right, and vice versa. The two subproblems are solved recursively at line (6) on the two butterflies that result form the removal of rank $k$. Clearly, the time spent on $merge(k)$ is $O(1)$ plus the time for the recursive call to $merge(k − 1)$. Thus, the total time spent is $O(k)$, which is $O(\log n)$.

If *merge* works correctly, so does *sort*. For the time taken by $sort(k)$, we observe that all steps are $O(1)$, except for the call to $merge(k)$, which takes $O(k)$, and except for the recursive call to $sort(k − 1)$. Thus, the time taken by $sort(k)$ is $O(k^2)$, or $O(\log^2 n)$. All this analysis assumes that comparisons take one step. If comparison takes time $t(n)$, then the obvious multiplication of the time by that factor must occur.

We may then invoke Theorem 6.3 to say that this algorithm can be implemented in the same time on a shuffle-exchange network, and invoke the result of Kleitman et al. [1981] to say that area $O(n^2/\log^2 n)$ times the square of the side of a processor (which is also a bound on the number of wires represented by an edge of the network) suffices. This proves that $AT^2 = O(s^2(n)t^2(n)n^2 \log^2 n)$. □

## 6.4 IDEAL PARALLEL COMPUTERS AND THEIR SIMULATION

A large literature concerning "parallel algorithms" exists; some of it is mentioned in the bibliographic notes. A significant aspect of these algorithms is that in many cases, the model of a computer on which they run is not physically realizable directly in present day hardware. Typically, an "ideal" parallel computing engine is postulated, where there are some large number of processors and some large number of memories. In one step, each processor can access (read from or write into) one memory.

Processors are assumed to act under control of a program local to that processor, although typically the programs executed by the processors act in concert toward some goal. Models differ in regard to simultaneous access of a memory by more than one processor. For example, we may allow

1. No simultaneous access. The processors have to resolve contention for memories among themselves, perhaps by executing programs that are carefully designed that no contention occurs.

2. Simultaneous reads are allowed but no simultaneous writes are allowed. This model makes sense if each memory stores a single datum, which could be placed on a bus for all readers to see at once.

3. Simultaneous reads and writes are allowed. When several processors try to write simultaneously, there are various assumptions that could be made about what happens. For example, we could assume that garbage is written, that one write request is honored and the others ignored, or that the requests are queued and handled one at a time.

**Example 6.6:** The study of parallel algorithms is generally directed toward the solution of enormous problems, e.g., "weather prediction." However, we can illustrate some of the issues with a simpler problem; we shall consider convolution, the problem introduced in Section 5.1. There, we are asked to compute the sequence $c_0, \ldots, c_{2n-1}$ from the sequences $a_0, \ldots, a_{n-1}$ and $b_0, \ldots, b_{n-1}$, where

$$c_j = \sum_{i=0}^{n-1} a_i b_{j-i}$$

We might, for example, store $a_i$ and $b_i$ at memory $i$, and we might compute $c_j$ at processor $j$. Each $c_j$ is initially 0. At the first step, processor $j$ requests $a_0$ and $b_j$ (assuming $j < n$; if $n \leq j < 2n$, nothing is requested), multiplies them, and adds them to the sum $c_j$. At the second step, the $j^{th}$ processor requests $a_1$ and $b_{j-1}$, and so on. After $2n - 1$ steps, all the $c_j$'s are computed.

This algorithm is actually not very good. With roughly the same number of processors, we can compute convolutions by taking the FFT of the two sequences, multiplying corresponding terms, and inverting by a process very much like that discussed in the previous section, taking time $O(\log n)$, rather than $O(n)$. Further, it possibly suffers from the defect that all processors want read access to the same memory at each step, since they all want $a_t$ at step $t + 1$. The clever reader can see how to fix the algorithm so no two processors access any memory simultaneously. However, we shall see that there is a still more fundamental problem with this algorithm. $\square$

## Connection Networks

As was mentioned, none of these models can be implemented directly in existing or contemplated hardware. The problem is that we need a *switching network* to connect processors to memories; such a network must allow, by setting certain control parameters ("switches") properly, that any such interconnection may be realized. It is not feasible, for example, to have all the processors and memories talk simultaneously on a bus. Neither will putting the processors at the top of a butterfly and the memories at the bottom work; a similar comment applies to networks such as the shuffle-exchange or the hypercube. Example 6.8 discusses why the networks studied so far are not adequate to implement arbitrary processor-memory interconnections. Furthermore, if we allow simultaneous read or write of a memory by more than any fixed number of processors in one "step," it is not clear that any interconnection network could ever be built.

However, even if we focus our attention on a computer model where only one read or write of any memory at one step is permitted, the design of an interconnection network is not trivial. An interconnection network in which each switch has a fixed number of other switches, memories, or processors to which it is connected requires at least $O(\log n)$ levels in a computer with $n$ processors and $n$ memories. If not, then some processor cannot even connect to some memory. As a consequence, $\Omega(\log n)$ time must elapse between the request for data by a processor and the receipt of that data.

This lower bound can be achieved if we allow *global* control of the network, that is, all the switches are set by a controller that knows exactly which memory each processor wishes to access. This situation may be appropriate if the processors access memories in a regular pattern, such as for the convolution example, but it is more likely that the setting of switches in the network will have to be under *local* control; that is, switches are set by signals from the individual processors, indicating which memory the processor wants to connect to. It is not known whether $O(\log n)$ delay is attainable with only local control. However, we shall mention a scheme that offers $O(\log n)$ delay per step of the ideal computer with probability approaching 1. The conclusion is that simulation of an ideal computer, at least one without simultaneous reads or writes in any memory, can be achieved by real hardware with a factor of $\log n$ slowdown, where $n$ is the number of processors and memories.

### Packet Switching Networks

A useful model of the switching network that connects the processors and memories is the *packet switching network*. In such a network, processors send *packets*, which are messages containing a request to read or write a particular memory, together with some *routing information*, telling, or suggesting, what route through the network this message should take to arrive at the desired memory. Similarly, the memories send packets containing the data satisfying a read request and with routing information back to the processors.

**Example 6.7:** Suppose we have a 32-node shuffle-exchange network, with a processor and memory at each node. Suppose also that the processor at node 13 wishes to send a packet to the memory at node 25. This packet will at least have the source (13) and destination (25) along with its information, but we may wish to have processor 13 compute the complete route for this packet, and include the route with the packet's information.

For the shuffle-exchange network, the algorithm to compute routes is simple. Look at the binary representations of the source and destination. We may cycle the bits of the source left by following the shuffle interconnection, that is, we move from the node numbered $a_1 \cdots a_k$ to the node numbered $a_2 \cdots a_k a_1$. If desired, we complement the least significant bit by following the exchange interconnection. In this manner, by $k$ shuffles and at most $k$ exchanges, i.e., at most $2 \log_2 n$ steps in an $n$-processor, $n$-memory network, we can send a packet from any processor to any memory.

In some cases, we can be more clever in choosing routes. For example, we can arrive at processor $25 = 11001$ from processor $13 = 01101$ by exchanging, shuffling, and exchanging again, thereby traveling through the sequence of nodes

$$01101 \rightarrow 01100 \rightarrow 11000 \rightarrow 11001$$

☐

The reader should beware, however, that the machine we contemplate, does not route one packet at a time, but may be routing simultaneously, one packet from each processor to some memory or vice versa. Thus, the ability of a node to receive a packet at any time may depend on whether it has room to store the packet, i.e., whether it can get rid of packets it had already stored and pass them further along their routes. The reason a "logjam" can occur is that, in the shuffle-exchange for example, a node can receive packets from both its neighbors, while it may be able to get rid of only one packet, because all packets currently stored at the node want to go to the same neighbor.

### Overview of the Section

We are going to study methods for efficiently simulating progressively more general "parallel computers." To perform the simulation, we give algorithms for increasingly general kinds of routing. These forms of routing are:

1. *Permutation Routing.* Here, every processor sends a request to exactly one memory, and there is a request for each memory. This process is relatively easy to perform, but does not simulate machines with simultaneous reading permitted. In fact, it does not even represent machines with no simultaneous access very well, since it requires that at each time, each processor generate a unique request, whether or not it wants to access data.

2. *Partial Routing.* A subset of the processors each send a request to one memory, but two processors are not permitted to request access to the same memory. This form of routing is a good representation for a machine with no simultaneous access permitted, but in which processors can request no data at a given time unit.

3. *Many-One Routing.* Here, each processor requests access to some memory, and more than one processor may request the same memory. As we can make any request be "null," many-one routing is easily seen to be a generalization of partial routing. This form of routing models the sort of parallel computer in which simultaneous reads are allowed. It could even be said to model machines with simultaneous writes, provided we specified an arbitration rule when two processors wished to write different values for the same object.

### Permutation Routing

We shall begin our study of the simulation of "ideal" computers by considering how to simulate a step of the ideal machine in which each processor sends one packet to some memory, and each memory is the target of one of the processors. We assume that there are an equal number of processors and memories, and that the locations of the memories and processors are nodes of a network, numbered $0, 1, \ldots, n - 1$. That is, processor $i$ and memory $i$ are each at node $i$, and for some permutation $\pi$, processor $i$ sends a packet to memory $\pi(i)$. We call the operation of sending packets from processors to memories (or vice versa) under the above constraints a *permutation routing*. Shortly, we shall relax the assumption that the processor-to-memory assignment is a permutation.

A fundamental relationship, that any sorting algorithm serves to perform a permutation routing, is expressed in the next theorem.

**Theorem 6.7:** There is an interconnection network that allows us to perform any permutation routing in $O(\log^2 n)$ steps, using only local control.

**Proof:** Let each packet carry its destination as routing information. Sort the packets by their destinations, so the lowest numbered destination goes to node

0, the next lowest to node 1, and so on. Because we have a permutation routing, we know that the $i^{th}$ destination (counting from 0) is in fact node $i$. Thus, the sort routes each packet exactly where it wishes to go. The sort of Algorithm 6.1 requires only local control, since the only decisions regarding the route of a packet are made when data, i.e., destination tags, are compared at individual processors, in lines (2), (8), and (9) of Fig. 6.9.

We can implement the sort on a hypercube or shuffle-exchange network in $O(\log^2 n)$ steps, so in the same time we can effect any permutation routing between processors and memories residing at the nodes of this network. We can also use the butterfly interconnection, if we keep all the processors on one rank, say $k$, and all the memories on one rank, perhaps rank $k$ or rank 0, and use the other nodes of the butterfly simply to pass packets. The advantage of the butterfly over the shuffle-exchange is not clear, but since the cost of the network will most likely be dominated by the cost of the processors and memories, not by the switching nodes, the disadvantage is not clear either. □

### Partial Routing and a Fast Expected Time Router

We shall now consider a method for routing that takes $O(\log n)$ time with probability approaching 1, where $n$ is the number of nodes in the network. In fact, this method does not even require that the routing be a permutation routing; it is adequate that it be a *partial routing*, where some subset of the processors send packets to some memories (or vice versa), but no memory receives more than one packet. We shall describe the algorithm in terms of the hypercube, although the same idea appears to work on many different networks.

In essence, the idea is for each processor to launch its packet, and have each packet travel towards its destination. If two or more packets currently at a node compete for the same edge out, some choice of which to send first must be made. If more packets arrive at a node than can be sent out immediately, then the packets are stored at the node until they can move.† It might appear that this simple idea will always be successful in routing all the packets with little, if any, delay because of competition for the right to traverse an edge. However, the next example shows that this does not have to be the case.

**Example 6.8:** Suppose we have a 10-dimensional hypercube, with 1024 nodes corresponding to the 10-bit binary integers. Suppose the desired routing sends a packet from the processor at each node with binary representation $a_1\cdots a_{10}$ to the memory at the node with the reverse representation, i.e., $a_{10}\cdots a_1$. Let us also choose a particular routing strategy, where a packet at a node travels next to the neighbor whose number differs from that of the current node in the leftmost bit where the current and destination nodes differ (and no other, of

---

† Thus, there must be a significant amount of storage available at each node, or the scheme will not work.

course). Thus the route from 1000101010 to its reverse, 0101010001, would be first to 0000101010, then to 0100101010, then 0101101010, and so on; the first three moves change the bits in the first, second and fourth positions, a change in the third not being necessary.

Consider what happens to the thirty-two packets that start from nodes of the form $a_1 \cdots a_5 00000$. Since the reversals of these integers all begin with five 0's, all thirty-two of these packets will, in their initial moves, change their locations to be at nodes whose numbers begin with five 0's. Furthermore, until the first five bits become 0, the last five remain 0. We are led to the conclusion that all thirty-two packets try to go through node 0000000000 within the first five moves. Also, half of these, or sixteen, will try to take the edge to 0000010000; thus this edge forms a bottleneck. It will take at least sixteen steps for the packets to reach their destinations, even though none is more than ten hops away from that destination.

Worse, this example easily generalizes to show that for particular permutations and particular local routing strategies, it can take time $2^{d/2-1}$ to complete the routing in a hypercube of dimension $d$. That is, the time is almost the square root of the number of nodes $n = 2^d$. □

Considering Example 6.8, it is perhaps surprising that a strategy can be given that will perform any routing, including the reversal routing mentioned in that example, and with probability arbitrarily close to 1 will finish within $O(\log n)$ steps. Further, the route selection algorithm is almost the one given in Example 6.8, with one additional detail. Instead of sending the packet directly to its destination, we send it to a random node, and then send the packet from the random node to its true destination.

In each step, we use the route selection strategy expounded in Example 6.8; if the packet is at a node with binary representation $a_1 \cdots a_d$, and it wishes to go to node $b_1 \cdots b_d$, then find the lowest $i$ for which $a_i \neq b_i$, and send the packet along the edge in dimension $i$, that is, to node $a_1 \cdots a_{i-1} b_i a_{i+1} \cdots a_d$. Call this strategy a *left-to-right* routing. Then the algorithm can be summarized succinctly as follows.

**Algorithm 6.2:** *Valiant-Brebner Routing Algorithm.*

INPUT: A destination for one packet at each node of a hypercube.

OUTPUT: The algorithm makes no explicit output, but results in the routing of the packets to their destinations.

METHOD:

1.    For each node $i$ that wishes to send a packet, select a random target node $t(i)$ by picking each of the $d$ bits in the binary representation of $t(i)$ to be 0 or 1, independently, with probability $1/2$ of each outcome. Make the route of the packet from $i$ to $t(i)$ be the left-to-right routing; that is, the packet must traverse at each step the edge whose dimension corresponds to the

leftmost bit position in which $t(i)$ and the current position of the packet differ. The routing information now placed in each packet is not only this route, but also its true destination.

2.  Let the packets follow their routes as fast as possible. If there is competition for an edge leaving a node, any one packet that needs to traverse that edge may be allowed to do so at a time unit. Packets unable to continue their route at a step are queued for transmission at later steps.

3.  When the packet from $i$ reaches $t(i)$, it is given a left-to-right route to its true destination, and sent on its way. Note that several packets may have the same random target, so if there is a choice, only one packet can be sent at a step, and the others must be queued.† While on its way to the true destination, a packet must yield to packets still trying to reach their random target, if they compete for an edge. Conflicts among packets heading to their true destination can be resolved arbitrarily, however. □

Valiant and Brebner [1981] prove the following theorem about Algorithm 6.2. We shall omit the rather complex proof.

**Theorem 6.8:** No matter what partial routing is specified, the probability is less than $(.74)^d$ that Algorithm 6.2, run on a hypercube of $d$ dimensions, takes more than $8d$ steps to complete. □

Thus, for large dimensions, the probability of more delay than eight times the theoretical minimum goes exponentially rapidly to zero. Experiments described in the paper just cited indicate that on the average, the time to completion is much less than Theorem 6.8 would suggest. In practice, a little more than $2d$ steps suffice on the average.

Algorithms with good expected performance, such as Algorithm 6.2 and a method of sorting on the butterfly sketched in Appendix 4, that finishes in $O(\log n)$ time with probability close to 1, represent a promising new direction for the design of algorithms on networks. However, the reader should be aware of several issues.

1.  Because of overhead, it is not clear that Algorithm 6.2 will be superior in practice to the algorithm for permutation routing described in Theorem 6.7. On the other hand, if we want a partial routing, sorting does not appear to be of significant help, and Algorithm 6.2, which does not require that each node be the target of a packet, must be looked on more favorably.

2.  Algorithm 6.2 requires storage for an arbitrary number of packets at each node, while the sorting-based algorithm does not. We must deal with the possibility that, however infinitesimal the probability is, it is possible for so many packets to accumulate at a node, that there will not be room to hold them. There is even the possibility of deadlock, where packets cannot leave

---

† A key observation in the proof that the algorithm has a fast expected time is that the probability of any node being the target for more than $c \log_2 n$ nodes decreases exponentially with $c$. Thus, bottlenecks will tend not to develop, or to be of short duration.

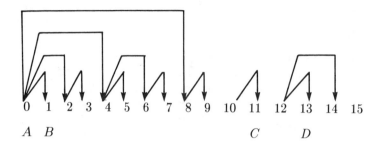

**Fig. 6.11.** Data distribution.

any of a subset of the nodes, because all these nodes are full, and all the packets at these nodes have next destinations that are in the same subset. In this case, we must abort the routing; presumably we shall restart the process, and with high probability, nothing will go wrong the second time.

## Data Distribution

We intend to show how to simulate a parallel computer where many processors can read the same memory at once. A crucial subroutine in this simulation is what we call the *data distribution* problem. Suppose we are given nodes $0, 1, \ldots, n-1$ arranged in some network. Certain of the nodes are *leaders*; they possess data that they must share with all the higher numbered nodes, up to but not including the next leader; nodes receiving the data of a given leader are called its *followers*.

**Example 6.9:** In Fig. 6.11 we see sixteen nodes, of which numbers 0, 1, 11, and 13 are leaders, holding data $A$, $B$, $C$, and $D$, respectively. Nodes 2–10 must receive $B$, 12 receives $C$, and 14 and 15 receive $D$. In this figure we see a particular distribution pattern, which we shall describe later. $\square$

We shall now sketch an algorithm for distributing the data of a leader to all of its followers—an algorithm that can be implemented efficiently on the networks that we have been discussing.

**Algorithm 6.3:** *Efficient Data Distribution Algorithm.*

INPUT: Information at a subset (the leaders) of the nodes $0, 1, \ldots, n-1$, of some network. The network must have the property that there is an edge from any node $j$ to node $j + 2^i$, whenever $i$ is an integer and $j + 2^i$ is less than $n$.

OUTPUT: There is no specific output. Rather, the information at each leader is distributed to its followers.

METHOD: Let us say that a node is *live* if it has received the data that its leader is trying to broadcast. Initially, only the leaders are live. If $n = 2^d$, then the algorithm takes $d$ phases, where in the $i^{th}$ phase, each live node tries to pass its

**for** $i := 1$ **to** $d$ **do**
   **for** $j := 0$ **to** $n - 1$ { in parallel } **do**
      **if** node $j$ is live **then begin**
         pass the data and leader number from node $j$
           to node $j + 2^{d-i}$;
         **if** the leader number at $j + 2^{d-i}$ agrees with that
         passed **then begin**
           make $j + 2^{d-i}$ live;
           record at node $j + 2^{d-i}$ the data just passed it
      **end**
   **end**

**Fig. 6.12.** Data distribution algorithm.

data $2^{d-i}$ positions right, i.e., node $j$ passes to node $j + 2^{d-i}$. The algorithm is shown in Fig. 6.12.

This version assumes that each node knows the number of its leader, a condition that will hold in the application we have in mind. However, the reader can easily modify the algorithm for the case where nodes cannot identify their leaders. Nodes broadcast to the right all data they receive, but remember only the data that came from the highest-numbered leader (which will be lower than the number of the node itself), so at the end, each node knows the data broadcast from the closest leader to its left. $\square$

**Example 6.10:** Fig. 6.11 illustrates Algorithm 6.3. There, $d = 4$. When $i = 1$, node 1 passes data to node $1 + 2^{d-i} = 9$. Node 9 accepts this data because it knows that 1 is its leader. In contrast, although 0 passes to 8, the information is not accepted by 8, because 8 is not a follower of 0.

When $i = 2$, 1 passes to 5, and when $i = 3$, 1 passes to 3, 5 passes to 7, and 13 passes to 15. The other nodes receive data from their immediate neighbor on the left when $i = 4$. $\square$

**Lemma 6.4:** Algorithm 6.3 distributes the data from each leader to all its followers.

**Proof:** We shall prove the following by induction on the number of bits in which the binary representation of $j - \ell$ is 1. The inductive hypothesis is that if $\ell$ and $j$ are any two nodes with the same leader (neither is necessarily that leader), $\ell < j$, and $\ell$ becomes live before phase $i$, where $i$ is the leftmost position in which the binary representation of $j - \ell$ is 1, then $\ell$ passes its data to $j$. The basis, where $j - \ell$ has one 1, in position $i$, is trivial, since we are given that $\ell$ will be live at the $i^{th}$ iteration; therefore it will pass its data to $j$.

For the induction, suppose that $j - \ell$ has two or more 1's, the leftmost of which is in position $i$. Then at the $i^{th}$ iteration, $\ell$ will pass its data to node

$m = \ell + 2^{d-i}$, where $d = \log n$, and $n$ is the number of nodes in the network. Now $m$ lies between $\ell$ and $j$, so it has the same leader as these. Since

$$j - m = j - \ell - 2^{d-i}$$

and $j - \ell$ has 1 in the position with value $2^{d-i}$, it follows that the binary representation of $j - m$ has one fewer 1 than that of $j - \ell$. Moreover, all those 1's are to the right of the $i^{th}$ position. The inductive hypothesis thus applies to $m$ and $j$, and we conclude that $m$ passes its data to $j$; therefore, $\ell$'s data, which is the same, is also passed to $j$. $\square$

It is not immediately obvious that Algorithm 6.3 can be implemented on the networks such as the butterfly to run in $O(\log n)$ time. The reason is that each iteration of the outer loop in Fig. 6.10 must wait until the previous iteration has completed passing data, and it is possible that about $d = \log n$ steps must take place before data can travel from some node $j$ to $j + 2^{d-i}$. Thus, our first guess might be that $O(\log^2 n)$ steps are needed in the worst case.

For that matter, it is far from obvious that starting from a node $j$, we can send a message to node $j + 2^i$ in time $O(\log n)$. We shall show a way of doing so on the butterfly; even though the algorithm we present is not a normal butterfly algorithm in the sense of Section 6.3, it is possible to implement this algorithm on a hypercube or shuffle-exchange network without more than one packet traversing an edge at any time unit.

Suppose we are at a node with number $a_1 \cdots a_d$ in binary, in rank $d-i+1$, and we wish to send a packet to the node whose number is higher by $2^i$. If $a_i = 0$, we use the butterfly between ranks $d - i$ and $d - i + 1$ that connects column $a_1 \cdots a_{i-1} 0 a_{i+1} \cdots a_d$ to $a_1 \cdots a_{i-1} 1 a_{i+1} \cdots a_d$. On the other hand, $a_i$ may be 1. In that case, let $a_1 \cdots a_i$ be of the form $w01^k$,† where $w$ is a string of 0's and 1's of the appropriate length, $i - k - 1$. Then we pass the data through the sequence of nodes $w01^k x, w01^{k-1}0x, w01^{k-2}00x, \ldots, w0^{k+1}x, w10^k x$. Note that we pass the data to the left by an amount that doubles each time we move, until when we hit a 0 in the binary representation, we can send the packet right by an amount that exceeds the sum of the leftward moves by exactly $2^{d-i}$.

Figure 6.13 illustrates the path of the packet, which we shall refer to as its *trajectory*, through the hypercube. The packet originates at $A$ and goes through $B$, $C$, and $D$, finally arriving at its destination $E$.

If we start the data at rank $i$ for the $i^{th}$ iteration of the outer loop in Fig. 6.12, then we can send data along the trajectory for each node and then move the data down to rank $i + 1$ in $O(\log n)$ time.‡ That still does not guarantee us even an $O(\log^2 n)$ data distribution algorithm, because we have to verify that

---

† Note that in the special case where $a_1 \cdots a_i$ is all 1's, there is no node with number $2^i$ higher, so it is not necessary to pass data in this case.

‡ If the network is the hypercube or shuffle-exchange, then there is no need to move data down the columns, because each column is a single node.

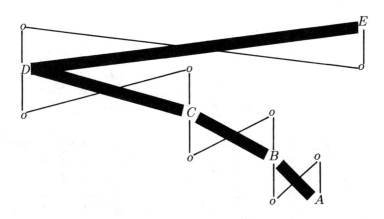

**Fig. 6.13.** Trajectory of a packet.

there is a limit to the number of packets passing through any node at any time.

However, we shall prove considerably more. With a carefully choreographed strategy, we can have the packets complete all iterations in $O(\log n)$ time. Moreover, we can even do so on the hypercube, so not only is there a limit to the number of packets at any node of the butterfly at any time, there is a limit to the number of packets at any column of the butterfly at any time. The key to the proof is defining a *launch time* for each packet that is sent in Algorithm 6.3. If $0 \le j < n = 2^d$, $1 \le i \le d$, and the binary representation of $j$ is $a_1 \cdots a_d$, then $time(j, i)$ is $3r + 4s$, where $r$ is the number of 1's among $a_1, \ldots, a_{i-1}$ and $s$ is the number of 0's among $a_1, \ldots, a_{i-1}$.

**Example 6.11:** $time(j, 1) = 0$ for any $j$, which correctly implies we can launch packets for the first iteration of Algorithm 6.3 at time 0. Also, $time(100110, 5) = 14$, since there are two 1's and two 0's among the first $5 - 1 = 4$ bits of the binary string 100110. □

We need to prove two things about the schedule implied by the function *time*.

1.   If at the $i^{th}$ iteration of Algorithm 6.3, node $j$ is live, then it receives the data it must transmit before $time(j, i)$.

2.   If we represent the butterfly by a hypercube, i.e., columns are collapsed into single nodes, then there are never more than two packets at any node (and in fact, these packets are headed for different edges out of the node).†

We prove these results in the next two lemmas. These lemmas are oriented to the situation where the nodes are organized as a hypercube, so the time to traverse a trajectory is equal to the number of sideways hops, e.g., the trajectory in Fig. 6.13 takes time 4. We do not count moves within a column,

---

† In fact, by a slightly more generous formula for *time*, we can guarantee that there is never more than one packet at a node of the hypercube. We leave this observation as an exercise.

since the columns of the butterfly are single nodes in the hypercube. A simple modification of the *time* function suffices to prove that Algorithm 6.3 can also be implemented in $O(\log n)$ time on a butterfly.

**Lemma 6.5:** Suppose that the time to traverse a trajectory is equal to the number of sideways hops in that trajectory. Then for every phase $i$ and node $j$, the packet that node $j$ receives in phase $i - 1$ arrives before $time(j, i)$.

**Proof:**

*Case 1:* $j$ in binary is of the form $0^{i-1}x$, for some binary string $x$ of length $d - i + 1$. Then no packet is received at $j$ in phase $i - 1$, because there is no node whose number is $j - 2^{d-i+1}$. Thus, there is nothing to prove in this case.

*Case 2:* $j$ in binary is of the form $w1x$, where $w$ is of length $i - 2$ and $x$ is of length $d - i + 1$. Then on phase $i - 1$, the packet for $j$ comes from the node with number $w0x$ in binary. The time of the launching of this packet, which is $time(w0x, i - 1)$, depends only on the first $i - 2$ bits, i.e., on $w$. Moreover, the time to launch the packet from $j$ for phase $i$ is

$$time(w1x, i) = time(w0x, i - 1) + 3$$

Since the packet takes only one time unit to reach $j$ from $w0x$, we have proved the lemma in this case.

*Case 3:* $j$ in binary is $w10^k x$, where $k \geq 1$, $w$ is of length $i - k - 2$, and $x$ of length $d - i + 1$. Then on phase $i - 1$, the packet for $j$ comes from node number $w01^k x$. Let $q$ be the sum of three times the number of 0's in $w$ plus four times the number of 1's in $w$. Then the launch time for the packet from $j$ on phase $i$ is $time(j, i) = q + 4k + 3$. The launch time on phase $i - 1$ for the packet from $w01^k x$ is $q + 4 + 3(k - 1)$; note that the last explicit 0 does not contribute in phase $i - 1$, since it is in position $i - 1$. This packet takes $k + 1$ hops to reach its destination; thus it arrives at $j$ at time $q + 4 + 3(k - 1) + k + 1 = q + 4k + 2$, just in time for the launch from node $j$ for the $i^{th}$ phase. $\square$

**Lemma 6.6:** If packets are launched at the launch times defined by the *time* function, and they travel along one edge at each time unit until they reach their destination in the hypercube, then there are never more than two packets at a node at any time unit, and these are intending to traverse different edges at the next time unit.

**Proof:** First, let us consider Fig. 6.13, and the trajectories that can go through a node like $D$, which we shall suppose is on rank $i$. Also, let the number of the column containing $D$ be $j$. Then for any $i$, there is a packet launched from some node in column $j$ of the butterfly (i.e., from node $j$ of the hypercube) at time $time(j, i)$. There are also packets launched from other nodes, like $C$, $B$, $A$, or perhaps nodes at higher ranks, located to the right of $A$.

However, in order for a packet launched in phase $i + k$ to reach node $D$, $j$ in binary must be of the form $w0^k x$, where $w$ is of length $i$, and $k \geq 1$. The source of the packet must be $w1^k x$. The launch time of this packet is $time(w1^k x, i+k)$ and it travels for $k$ time units, so it arrives at column $j$ at $time(w1^k x, i+k)+k$, which is $time(j, i+k)$, as the reader may easily verify by noting that $j$ is $w0^k x$ in binary. Thus, all the packets that travel through column $j$ on their trajectory do so at a different time, those of the form $time(j, p)$ for various values of $p$.

We conclude that there is at most one packet traveling through node $j$ of the hypercube, and one being launched at any time. Moreover, the packet launched at $time(j, i+k)$ follows the edge in dimension $i+k$, while the packet traveling through node $j$ will follow the edge in dimension $i$, so there is never any competition for edges. $\square$

We have thus proved the following important fact.

**Theorem 6.9:** We can perform data distribution on the hypercube in $O(\log n)$ time, where $n$ is the number of nodes.

**Proof:** Lemma 6.5 says that if there is no delay of packets, then the schedule of packet launches given by the *time* function will implement Algorithm 6.3 correctly. Lemma 6.6 tells us that these packets are never delayed. The last time at which a packet is launched is $time(00\cdots 0, d) = 4(d-1)$, where $d = \log n$, and each packet arrives after $d$ steps at most. Thus, after $O(\log n)$ steps, all packets have arrived at their destinations, bearing the required data. $\square$

### Many-One Routing

Now we shall consider how to simulate an ideal parallel computer where at each step, any number of processors can read any memory, although only one memory can be read in a step by any processor. The data read from a memory by all processors reading it must be the same (or we must regard the requests from a memory as if all were asking for the entire set of data requested by the individual processors). Thus, a "memory" might best be thought of as a register or single value, as in Example 6.6. Our rule for writing remains the same; only one processor can write a memory at one step, and a processor can write only one memory at a step.

The direction of packets from processors to memories with no constraint on how many packets can be destined for one memory is called *many-one routing*, because many processors can send to one memory. The next algorithm uses Algorithms 6.2 and 6.3 to perform a many-one routing and to get the replies to the processors in time $O(\log n)$ with probability approaching 1. The key idea is that requesters for a memory elect a leader, who deals with the memory and then distributes the result rapidly, using the data distribution algorithm just covered. It also depends for its $O(\log n)$ speed on the "Flashsort" sorting algorithm that we sketch in Appendix 4.

**Algorithm 6.4:** *Many-One Routing and Reply.*

INPUT: A request by each of $n$ processors to read a datum from some memory, with multiple requests for the same memory allowed.

OUTPUT: There is no formal output. Rather the requests are satisfied by transmitting the contents of the desired memory to each processor.

METHOD:

1.  Let each processor $i$ have a request to read memory $m_i$. Processor $i$ generates a packet containing $i$, $m_i$, and the name of the data it wishes to read from $m_i$.

2.  Sort the packets by the value of $m_i$, i.e., by destination. Now, all the packets destined for a particular memory are at consecutive nodes, although a node holding $m_i$ may bear no relation to the node at which the memory $m_i$ is located.

3.  Determine which nodes are *leaders*. A node $j$ is a leader if $j = 0$ or the destination $m_i$ in the packet it received at step (2) differs from the destination received by node $j - 1$. This step is performed by passing all destination tags one position right, by an algorithm that is a special case of Algorithm 6.3. The leaders will act as representatives for all requests to read a particular memory; after reading the data from that memory, they will broadcast it to all their *followers*, the nodes numbered consecutively, up to but not including the next leader.

4.  Each leader sends a read request for data from the memory $m_i$ that is the destination of the packet it holds. The memories then reply to the nodes sending the request. These two operations are partial routings.

5.  The leaders broadcast the data they received to their followers, using Algorithm 6.3. So that a node can recognize data coming from its leader, the memory from which the data came is indicated in the message. Thus, a node can compare the memory number received with the second component in its own packet; if they match then the data comes from its leader.

6.  All nodes return their data to the nodes placing the requests. That is, a packet of the form $(i, m_i)$ is returned to node $i$, along with the data requested from memory $m_i$. □

**Example 6.12:** Suppose we have eight nodes, and all but 2 and 4 wish to read memory 3; processors 2 and 4 wish to read memory 0. Figure 6.14(a) shows the request packets generated by the processors at step (1) of Algorithm 6.4. Figure 6.14(b) shows the result of sorting by second components in step (2), and the leaders, nodes 0 and 2, found in step (3), are starred. Node 0 is always a leader, and node 2 is a leader because node 1 holds a different second component.

Figure 6.14(c) shows the read requests generated in step (4); (d) shows them reaching their destination, and (e) shows the data returned in reply. Note that the packets of line (b) remain at the nodes during this step. Finally, Fig. 6.14(f) shows the result of distributing the data at step (5), and Fig. 6.14(g) shows the result of the permutation routing of step (6). □

|  | 0 | 1 | 2 | 3 | 4 | 5 | 6 | 7 |
|---|---|---|---|---|---|---|---|---|
| (a) | (0,3) | (1,3) | (2,0) | (3,3) | (4,0) | (5,3) | (6,3) | (7,3) |
| (b) | (2,0)* | (4,0) | (0,3)* | (1,3) | (3,3) | (5,3) | (6,3) | (7,3) |
| (c) | read 0 |  | read 3 |  |  |  |  |  |
| (d) | read 0 |  |  | read 3 |  |  |  |  |
| (e) | $A$ |  | $B$ |  |  |  |  |  |
| (f) | (2,0,$A$) | (4,0,$A$) | (0,3,$B$) | (1,3,$B$) | (3,3,$B$) | (5,3,$B$) | (6,3,$B$) | (7,3,$B$) |
| (g) | (0,3,$B$) | (1,3,$B$) | (2,0,$A$) | (3,3,$B$) | (4,0,$A$) | (5,3,$B$) | (6,3,$B$) | (7,3,$B$) |

**Fig. 6.14.** Example of Algorithm 6.4.

**Theorem 6.10:** Algorithm 6.4 runs in time $O(\log n)$ on a hypercube, where $n$ is the number of nodes in the network, with probability that approaches 1 in the sense of Theorem 6.8.

**Proof:** Step (2) is a general sorting step. We have given only an $O(\log^2 n)$ upper bound for sorting, in Theorem 6.7. To claim that in the probabilistic sense of Theorem 6.8, $O(\log n)$ suffices, we must invoke the "Flashsort" algorithm of Reif and Valiant [1983], from Appendix 4. The partial routing of step (4) and the permutation of step (6) are accomplished by Algorithm 6.2. The data distribution of step (5) takes $O(\log n)$ by Theorem 6.9, and the leadership detection of step (3) can be done as a special case of data distribution. Thus, all steps require either $O(\log n)$ time in the worst case or $O(\log n)$ time in the probabilistic sense of Theorem 6.8. □

## EXERCISES

6.1: Illustrate the shuffle-exchange interconnection for 16 and 32 nodes.

6.2: Show the steps made by Algorithm 6.1 when sorting the sequence

$$6, 2, 5, 4, 1, 8, 3, 7$$

on a butterfly of width eight.

6.3: Give a normal butterfly algorithm to reverse a sequence of numbers found in the top rank of a butterfly. If we implement this algorithm on a shuffle-exchange network, what does the sequence of memory maps look like?

6.4: What are the necklaces for the perfect shuffle on 32 and 64 nodes?

6.5: Give the memory map states starting from Fig. 6.2(a) for the sequence of operations $EX$, $PS$, $EX$, $PS$, $PS$.

** 6.6: Show that $O(n^2/\log^2 n)$ area suffices to lay out a shuffle-exchange graph on $n$ nodes. If you cannot achieve this best upper bound, any growth rate less than $n^2$ is a hard exercise.

* 6.7: Suppose we have a butterfly of width $n$, in which the top and bottom ranks have been identified. Prove that if we rotate the ranks so rank $r$ is on the bottom, i.e., the order from the top is $r+1, r+2, \ldots, k, 1, 2, \ldots, r$, then it is possible to permute the columns so the resulting network is congruent to the original butterfly.

6.8: Show that a shuffle-exchange network in which data is transmitted only along the $PS$ and $EX$ interconnections (never $PS^{-1}$) can simulate a normal butterfly algorithm where data is passed in one direction only.

** 6.9: Prove that the problem consisting of $\log n$ independent copies of the problem of Lemma 6.2 requires $AT^2 = \Omega(n^2 \log^2 n)$.

6.10: Show how to compute the $n$ powers of $\omega$ on a butterfly in $O(\log n)$ time, given only 1 and $\omega$. Specifically, $\omega^{exp(k,i)}$ must appear at processor $p_{ki}$.

6.11: Prove Theorem 6.4, that a normal butterfly algorithm can be implemented without slowdown on a hypercube.

6.12: Show how to simulate a normal butterfly algorithm on the shuffle-exchange if the relevant data is initially at rank $r$ of the butterfly.

6.13: Prove Lemma 6.3, the correctness of the variant of odd-even merge.

6.14: Redo Example 6.6, on convolution, so no simultaneous access to any one memory is needed.

6.15: Show that Algorithm 6.3, the data distribution algorithm, can be made to work correctly even if nodes do not know their leaders, but remember the information received from the highest numbered leader that sends information (possibly indirectly).

** 6.16: Implement Algorithm 6.3 on a shuffle-exchange network.

* 6.17: Redefine launch times for Algorithm 6.3 so there is never more than one packet at any node of the butterfly.

6.18: Suppose that each node on the bottom rank of a butterfly of width $n$ holds a number. Show that in time $O(\log n)$ we can arrange that every node knows the sum of the numbers to its left. Note that this is the essence of step (3) in the Flashsort algorithm discussed in Appendix 4.

## BIBLIOGRAPHIC NOTES

The use of the shuffle-exchange graph as a parallel computation medium was first discussed by Stone [1971]. The cube connected cycles was first used explicitly by Preparata and Vuillemin [1979]. Wise [1981] discusses the layout of butterfly-like networks.

The result that $\Omega(n^2/\log^2 n)$ is required for the layout of the shuffle-exchange is from Thompson [1980], while the fact that this growth rate is an upper bound was proved by Kleitman, Leighton, Lepley, and Miller [1981]. Earlier, weaker upper bounds were shown by Hoey and Leiserson [1980] and by Steinberg and Rodeh [1980].

The design of supercomputers, in particular perfect-shuffle oriented machines, was discussed by Schwartz [1980]. Bianchini and Bianchini [1982] is a discussion of how one might actually wire a collection of processors and memories on circuit boards using this organization. As predicted by the theory, this wiring task becomes progressively harder as the size of the network increases.

Shaw [1982] and Stolfo and Shaw [1982] describe the design of a tree-oriented supercomputer, and we should probably regard the systolic array ideas discussed in the previous chapter as another form of supercomputer organization. The variety of computation networks and the extent to which they could simulate one another was investigated by Siegel [1977, 1979].

The material in Section 6.4 on simulating ideal parallel computers is based on material in Borodin and Hopcroft [1982], with two principal exceptions. Algorithm 6.2 is from Valiant and Brebner [1981], with a similar result being found in Galil and Paul [1981]. Also, the Flashsort algorithm outlined in Appendix 4 but used in Section 6.4, was first expressed by Reif and Valiant [1983]. The initial attempts to classify parallel computer models by the way they accessed data, e.g., whether simultaneous read and/or write was allowed, were by Fortune and Wyllie [1978] and Goldschlager [1978].

A great number of papers have been written on algorithms for such ideal parallel machines. Sorting, merging, and routing algorithms, on which the theory of Borodin and Hopcroft [1982] is based, can be found in Lev, Pippenger, and Valiant [1981], Preparata [1978], Shiloach and Vishkin [1981], and Valiant [1982a, b]. Other problems were considered in Borodin, von zur Gathen, and Hopcroft [1982], Chin, Lam, and Chen [1982], Chin and Wang [1982], Hirschberg and Wolper [1982], and Dekel, Nassimi, and Sahni [1979]. See also Chandra, Stockmeyer, and Vishkin [1982] for a characterization of problems solvable on such models.

There has also been recent effort on development of algorithms that run quickly with high probability on parallel computers, such as the butterfly, that have high information transfer rates. In addition to the algorithms of Valiant and Brebner [1981] and Reif and Valiant [1983] just mentioned, there has been recent work by Aleliunas [1982], Reif and Spirakis [1982], and Upfal [1982]. There is, in fact, an algorithm due to Ajtai, Kolmos, and Szemeredi [1983] for sorting on a parallel machine that uses comparison elements and takes $O(\log n)$ time in the worst case, not just with high probability. Because of the large constant factors involved, it is unclear whether this algorithm is superior to Flashsort in practice.

# 7

# OVERVIEW OF VLSI
# DESIGN SYSTEMS

We now begin the last of the three major units in this book: the algorithmic aspects of VLSI design systems. In this chapter we shall discuss in broad outline the nature of a system that supports the design of integrated circuits. The similarities to and differences from the design of large software systems are discussed, and we illustrate the levels of abstraction found in languages used to specify designs. In subsequent chapters we shall discuss some of the algorithms that figure prominently in a design system. These algorithms are divided into two groups: compiling algorithms that are used to lower the level of abstraction of a design automatically, and algorithms used in utilities connected with the design system. We make no attempt to be exhaustive or encyclopedic; rather we have selected from among existing systems for some that illustrate interesting algorithms and/or interesting possibilities.

The reader should beware that most of the design languages and facilities discussed in this and the following chapters are in a state of ongoing development. Their details at any given time may differ from those described here, and even for those systems that are "frozen," we have selected portions for discussion, and there may be much more to the language or system than appears here.

## 7.1 DESIGN LANGUAGES

While the design of single chips with the complexity of a microprocessor is a relatively recent phenomenon, people have been designing software systems of equal or greater complexity for many years. One might naturally wonder whether the experience and tools used in software design could be carried over directly to the world of VLSI. Certainly there are several factors in common. In both the circuit world and program world, the problem can be characterized as one of implementing algorithms. For example, the fetch-decode-execute cycle of a microprocessor is principally an algorithm, one of deciding what instruction to execute and performing the required operations.

Another similarity is that in each case, a large design must be carried out by several workers, and coordinating pieces of a project is a major problem. A third similarity is that designers are much more productive when they are given high-level languages in which to express their designs. For example, programmers almost never work in machine code, and they generally prefer high-level languages of the nature of Pascal, say. Many even prefer much higher level languages like APL. There are, however, several reasons why these similarities have not been exploited as far as would seem possible.

1. The process of running a program, seeing if it works, discovering that it does not, finding the bug, fixing it, recompiling, and running again takes a matter of minutes in the software world. In comparison, the delay between the time a design for a chip is sent to a fabrication facility and the time the actual chip is returned is measured in days to months. Further, while the novice programmer is told that "when a program does not work it is not the machine's fault," the analogous admonition is not valid where fabrication of chips is concerned. Often, the returned chip fails to implement the design properly. This phenomenon, as well as the need for specialized testing hardware, makes the process of testing and debugging considerably more difficult for chips than for programs.

2. As discussed in Section 1.4, the cost of a chip is an exponential function of the area it requires. In the program domain, we are often willing to trade a factor of ten or more in the running time of our programs for the privilege of writing in a high-level language, one that makes the job of the programmer easy. Frequently, we cannot afford such factors in the area or time taken by a circuit design. Unfortunately, implementation of high-level design languages is still in its infancy, and often languages that allow the design of VLSI circuits at a high level do not produce circuits whose area and speed compare favorably with that of a design for the same problem in a low-level language.

3. Syntax checking for "machine code" is much harder in the circuit world than in programming languages. If we produce a program in machine or assembly language, either by writing it directly or as the output of a compiler, the problem of syntax checking is trivial; we have only to look at each instruction to see that we have not accidentally entered a nonexistent operation code or provided the instruction with the wrong number of operands.† The analogous syntax checking for the "machine language" of circuit design is not nearly so simple. This language is one whose primitives are colored rectangles, such as CIF, to be discussed in the next section. The analog of syntax rules for such a language involves the design rules discussed in Chapter 1, and these cannot be checked locally, i.e., on an instruction-by-instruction basis. Rather, in order to check whether two

---

† Of course that does not guarantee the program works properly, only that it obeys the syntax rules of its language.

rectangles are too close, we need to know whether they are electrically connected, and we cannot know the answer to that question unless we examine the entire design. We shall discuss design rule checking in Section 9.3.

The consequences of these differences are several. First, circuit design systems rely heavily on low-level input. Since at the lowest level, designs are geometric patterns, the use of graphics and graphics editors to enter and modify designs is very important. Since it is not possible to design, test, and modify in a rapid cycle, there is great reliance on ways to simulate the chip that would result from a design without actually manufacturing the chip. Thus, various kinds of simulators also play an important role in the design process. We shall discuss algorithms used by some design tools of this class in Section 9.4.

## Levels of Abstraction

In order to understand the overall design system, it helps to be aware of the many levels of abstraction that various design languages use. Of course, not all levels may be available or necessary in any one design system. The principal sorts of languages that appear are the following.

### Geometry Languages

These, like CIF mentioned above, use colored rectangles, or more generally, colored shapes, as their primitives. It is possible to fabricate a chip directly from a design expressed in this language.

### Sticks Languages

In these languages, transistors and vias (places where wires change layer) are represented by points of a grid, and wires are represented by lines. Information regarding the thickness of wires and the exact positions of the points is supplied by the compiler that translates the sticks language into geometry. The relative position of the various circuit elements is supplied by the designer.

A significant advantage of this sort of language is that it is almost impossible to make a design rule error, since it is the responsibility of the compiler to position elements and wires so the rules are respected. The disadvantage is that some increase in area over a design at the geometric level is to be expected. We shall discuss such a language, LAVA, in Section 7.6, and in Section 8.2 we shall discuss an algorithm for doing the *compaction* of sticks diagrams, that is, converting the abstract coordinates of the points into concrete geometry.

### Switch Languages

Here, the primitives are transistors and nodes, which are points connected to one of the three terminals of one or more transistors. These three terminals are generally called the *source* and *drain*, which are the two ends, and the *gate*, which separates the source and drain. In an *enhancement mode* transistor, which we have called a pulldown or pass transistor in Chapter 1, the source and drain are connected if the gate is "on," and the source and drain are disconnected if the gate is "off." In a *depletion mode* transistor, of which the primary example is the pullup discussed in Chapter 1, the same three terminals are attached, but the transistor is always "on." Often, as in the pullup, the source and gate are attached, and the depletion mode transistor acts like a resistor, which is an essential part of most gates used in a circuit.

A switch level language is not necessarily suitable for design. Unlike sticks, it does not specify the relative positions of the transistors. A compiler to convert a switch language specification into a geometry specification will produce area-efficient layouts only with great difficulty. However, a switch language is very appropriate as the input language for a simulator; a common pattern is to *extract* a switch level representation from a geometry specification and simulate the latter by simulating the former. We shall discuss circuit extraction and switch simulation in Sections 9.2 and 9.4, while esim, a switch language, is mentioned in Section 7.4.

## PLA Personalities

The *programmable logic array*, or PLA, is a specialized layout for implementing switching logic and sequential machines, such as the control portion of a microprocessor or many similar sorts of chips. The *PLA personality* language is a succinct way of specifying the exact function and layout of the particular PLA desired. This sort of language leads to area-efficient layouts and allows design to be carried out at a relatively high level. However, it is specialized, since it is useful only for control-like aspects of a design, where the PLA is often the desired form of layout. We shall discuss PLA's and their implementation later in this chapter.

## Logic

Ordinary Boolean or switching logic, perhaps augmented with a notion of sequentiality so we can talk about the "current" or "previous" value of a variable, forms a high-level way to specify designs. It is more versatile than PLA personalities since it cannot only specify the logic of a control unit, but can specify aspects of the *data path*, that is, the circuitry used to store and manipulate data, such as registers and adders. However, with a language this high-level, the designer has lost all control over the layout and relative positioning of circuit elements, and must rely on a compiler to select a reasonable implementation of the logic, such as a collection of PLA's. We shall discuss the logic language lgen in Section 7.5 and in Section 8.3 consider one way to compile it into geometry.

A *schematic* language, often referred to as "logic," is a related form of input, actually midway between a switch language and what we call a logic language. Schematic diagrams have nodes, just as in switch languages, but in place of transistor switches are gates chosen from some large repertoire, e.g., "and" and "or" gates, and perhaps storage elements. Thus, schematics are more specific than logic in that they fix a particular set of gates which will implement each logic statement. In comparison, a logic language gives us many options, including the discovery of common subexpressions among logical expressions, for example.

*Finite State Machine Languages*

These languages are designed to specify the control of a microprocessor or similar chip. They can be compiled into logic or PLA personalities, for example. While a PLA is often a good, area-efficient method of implementing a sequential machine, the compiler for the language must decide on the coding of states of the machine, and this step is hard to do optimally, i.e., in a way that minimizes the area of the PLA. Thus in general, heuristics must be employed in the compilation process. We shall discuss a sequential machine language called SLIM in Section 7.8, and in Sections 7.9 and 8.4 discuss some compilation issues for a language based on regular expressions as a way to specify sequential machines.

*Procedural Languages*

The extreme of high-level design languages is an ordinary programming language, in which the algorithm to be performed by the circuit is written. Compilers for such languages frequently wind up generating the layout of a microprocessor intended to implement this program and only this program. The sequencing rules for statements of the program are embedded into the control of the "microprocessor," while the variables are represented by registers. Often, one or several busses connect the registers. Operations on variables are performed by arithmetic units attached to the busses; the registers, arithmetic units, and busses connecting them form the *data path.*

Procedural languages that are somewhat more specialized to circuit design than are ordinary programming languages also exist. These languages, often called *register transfer* languages, deal with registers (variables of a programming language, in effect), but specify the sequencing of actions involving these registers in terms of events at the input and events at other registers, rather than in terms of "if," "while," and other sequencing primitives of ordinary programming languages.

## Design System Organization

There are several ways a design system can be organized. In the remainder of this section, we shall sketch four different approaches and show how the various parts of the system fit together. These four are not intended to be exclusive or exhaustive; sometimes systems borrow from several of these approaches. The

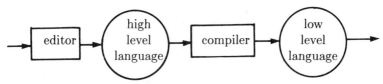

**Fig. 7.1.** A silicon compiler.

approaches are

1. A "silicon compiler."

2. A conventional system based on support for design at a low level.

3. A system based on automated comparison of designs carried out at several different levels.

4. A "designer's assistant," that supports design through several levels of abstraction.

*The Silicon Compiler*

This system is an ideal one, whose production use has not arrived in the early 1980's, but that is very possibly what the future will bring. Figure 7.1 shows a simple silicon compiler, in which all design is done at a high level and compiled into a low-level language. There could be several stages of compilation; for example, the "low-level" language could be logic, which is compiled further into PLA personalities, which in turn is compiled into geometry.

*A Low-Level Oriented System*

In Fig. 7.2 we see a fairly conventional system, in which most design is done at a low level, although there may be some specialized languages, such as PLA personalities, or sticks that can be converted to the geometry language by a compiler. Entry of data is through either a graphics editor that deals with a geometry language of colored shapes, or a conventional text editor.

The various pieces of geometry language are put together much as procedures are combined to form a system in the software world; we shall discuss this issue when we discuss CIF in the next section. The pieces, combined, form the design, and this design must be tested as best we can before fabrication of the chip. We, therefore, see in Fig. 7.2 that the geometry language is subjected to a design rule checker, and violations of the design rules require the designer to modify the design, using the editor that created the design in the first place.

Similarly, the geometry language is passed to an extractor, which turns it into a language, such as a switch language, suitable for simulation of the design. The designer then determines the response of the design to sequences of inputs selected by the designer, and if the design does not meet his expectations, a modification of the design, followed by retesting, occurs. Of course, no amount of simulation can assure us that the design works under all possible conditions.

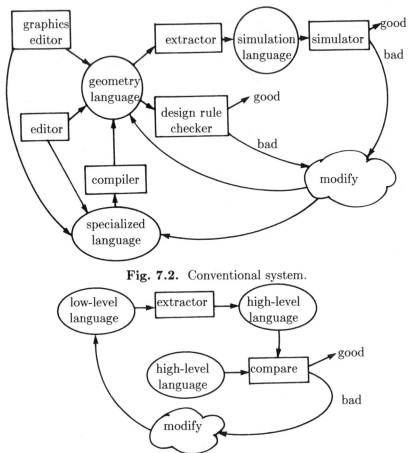

**Fig. 7.2.** Conventional system.

**Fig. 7.3.** A comparison-based approach.

*A Comparison-Based System*

Figure 7.3 shows a system that assists the designer in verifying that his design meets some high-level specifications. This arrangement is primarily experimental, but could be used as part of the conventional system of Fig. 7.2.

The designer specifies his design in both a high-level and a low-level language. Presumably, he can express the design at a high level with fewer opportunities to go wrong, so that specification is more likely to be the correct one. To make the low-level design correct, an extractor converts the low-level language to the high level, and then a comparison algorithm checks that the two high-level specifications are functionally equivalent.

Note that it is not sufficient just to determine whether the two high-level designs are identical; we must check that they do the same thing. Thus,

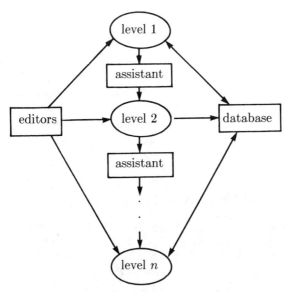

**Fig. 7.4.** The designer's assistant.

the comparison algorithm may be very hard to write, or it may involve a computationally intractable problem. Similarly, there is no guarantee that extraction from a low-level language to a language at a much higher level is easy or even possible, if the jump in level is great.

*A Designer's Assistant*

Figure 7.4 shows a system where several levels are linked together by a database and by "assistants," which are routines that in some way help the designer to fill in the details of his high-level design to make a design at a lower level. The "assistant" may be an "Artificial Intelligence" oriented program that incorporates rules that a design must follow (analogous to design rules for geometry languages) and heuristics that help the designer select a good instantiation for his high-level design.

Alternatively, the assistant may be a routine that performs certain tasks for the designer. For example, in Section 9.5 we shall discuss automatic placement and routing. The input to such a tool is a collection of designs, say, at the geometry level (although it could be higher), together with information about the wires connecting the *ports*, which are named points on the edges of the various circuits. The "assistant" could perform the task of selecting the exact placement of circuits and selecting the routes that the wires connecting ports will follow.

The database plays an important role, not only in this sort of system but any system where designs for the same circuit expressed in different languages coexist, or where different designers work on pieces of the same design together.

Thus, we would expect to find at least a rudimentary database facility, even if it is only a hierarchical directory and file organization, connected with any design system.

In the style of Fig. 7.4, the database is especially important, because the design can be regarded as a single object, whose various pieces are instantiated at different levels at the same time. A change at a high level necessitates a change at all lower levels, and the database manager should be responsible for flagging all pieces of the design that are no longer valid. Under the most ideal circumstances, the "assistants" can redesign at the lower levels automatically, if the changes at higher levels are not too great.

## 7.2 CIF: A GEOMETRY LANGUAGE

CIF, the *Caltech Intermediate Form* is a language whose primitives are colored shapes. It is called "intermediate" because it is both a language in which designs of circuits can be couched and a language from which the masks (like negatives) from which a chip is fabricated can be made automatically. It is a fairly powerful language, and we shall not attempt to define the full language here, but rather show its essential nature. The best documentation on the language is probably Chapter 4.5 of Mead and Conway [1980].

### The Box Statement

The most basic statement is one defining a rectangle, called a *box*. We place a box at a particular place on a grid by a statement of the form

B <xdist> <ydist> <xcent> <ycent>

where <xdist> and <ydist> are the lengths of the sides along the $x$ and $y$ axes, and (<xcent>, <ycent>) are the $x$ and $y$ coordinates of the center of the rectangle. Thus

B 6 2 1 3

represents the box shown in Fig. 7.5.

We can also express the box more verbosely, as

Box Length 6 Width 2 Center 1, 3

As a rule, in CIF only upper case letters have significance, and the lower case letters and comma in the above line are "comments." Thus, yet another way to specify the box of Fig. 7.5 is

B L 6 W 2 C 1 3

We can also specify a rotation of the rectangle with two additional components. The clause

Direction *a b*

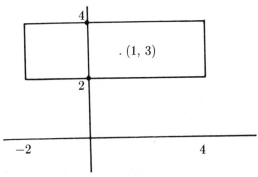

**Fig. 7.5.** A box.

calls for the $x$ axis to be rotated until it has slope $a/b$. Thus

Direction 1 1

calls for a rotation of forty-five degrees counterclockwise. As with the other parameters, the word "Direction" can be abbreviated as "D" or omitted altogether, if the integers defining the slope are understood to come fifth and sixth in the list. Thus, we can rotate the box of Fig. 7.5 forty-five degrees clockwise with any of the statements

Box Length 6 Width 2 Center 1 3 Direction −1 1
B L 6 W 2 C 1 3 D −1 1
B 6 2 1 3 −1 1

Note that whatever rotation we choose, the term "Length" refers to distance along the (rotated) $x$ axis, and "Width" to the distance along the (rotated) $y$ axis.

## The Layer Statement

In addition to size, position, and orientation, a box must be assigned a "color" or layer. The statement

L <layer designation>

specifies a layer. As with the box statement, "L" may be written out as "Layer".† The <layer specification> is a code for the name of one of the layers used in the design. The most commonly used ones in the NMOS technology are the following.

---

† Or for that matter, as "Licketty split," since lower case letters are ignored.

| ND | Diffusion (green) |
| NP | Polysilicon (red) |
| NM | Metal (blue) |
| NC | Contact (black) |
| NB | Buried Contact (brown) |
| NI | Implant (yellow) |

A layer statement causes all subsequent boxes (and other shapes defined by other types of statements, which we shall not discuss) to be of that layer, until the next layer statement is met.

**Example 7.1:** We shall design a simple inverter in CIF, using the box and layer statements. The design is shown in Fig. 7.6. The inverter consists of a pullup above a pulldown, with the input at the gate of the pulldown and the output taken from the bottom of the pullup. The pullup consists of a thin ($2\lambda$ wide) diffusion channel, widening at the base to width $4\lambda$ to form a "butting contact." The channel is covered by a polysilicon rectangle $6\lambda$ wide, and the region for $2\lambda$ on all sides of the place where the channel and the polysilicon cover overlap is implanted to make the transistor be a depletion mode transistor, i.e., one that is always on and functions as a resistor.

The butting contact connects the polysilicon cover to the diffusion channel, and also involves a $4\lambda \times 6\lambda$ piece of metal with a $2\lambda \times 4\lambda$ contact cut in its center. Since the butting contact involves all three layers, the voltage at the bottom of the pullup is available on all three layers. We choose, in this example, to take the output of the inverter, which is this voltage, on the diffusion layer.

Below the butting contact is the pulldown. We have chosen to make the diffusion channel wider, $4\lambda$, in this area, so the pulldown is twice as wide as it is long. Such a pulldown has only half the resistance of a pulldown that is as wide as long. Recall that the ratio of the resistance of the pulldown to the resistance of the pullup must be a small fraction, usually $1/4$ or $1/8$. The resistance of a transistor, when on, is proportional to the ratio of the length of its channel to the width of its channel. In Fig. 7.6, the pullup has a ratio of about 8 units to 2 units, or 4, while as we mentioned, the pulldown has a ratio of 0.5, so the whole inverter has a resistance ratio of $4/0.5$, or 8.

In Fig. 7.7 we see the CIF specification of the inverter of Fig. 7.6. Note that all CIF statements are separated by semicolons. Also, there is an intentional error in Fig. 7.7, since decimal points are not permitted in constants, as in 13.5 and 4.5 on lines (2) and (3). It is required in CIF that all constants be integers. Moreover, the units of dimensions are not $\lambda$, as we have implied. We shall clear up this matter of scaling of constants after we discuss "symbols," next. □

**Defined Cells**

CIF has a mechanism much like a procedure call, that enables us to define a

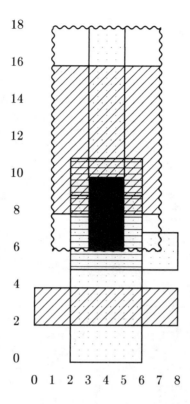

**Fig. 7.6.** An inverter.

| (1)  | L ND;            |
|------|------------------|
| (2)  | B 2 9 4 13.5;    |
| (3)  | B 4 9 4 4.5;     |
| (4)  | B 2 2 7 6;       |
| (5)  | L NP;            |
| (6)  | B 6 8 4 12;      |
| (7)  | B 8 2 4 3;       |
| (8)  | L NC;            |
| (9)  | B 2 4 4 8;       |
| (10) | L NM;            |
| (11) | B 4 6 4 8;       |
| (12) | L NI;            |
| (13) | B 6 12 4 12;     |

**Fig. 7.7.** CIF statements.

cell, or *symbol*, to be some collection of shapes, and then to "call" that cell
as many times as we wish. Each call to a cell places a copy of that cell with
the origin at a designated position in the grid of the cell doing the call. When
placing a cell, we may rotate its $x$ axis to a designated slope, as for boxes, and
we may *mirror* the cell about the $x$ and/or $y$ axis, which negates all $x$ and/or
$y$ coordinates, respectively. These operations are in addition to a *translation*,
which moves the origin of the grid for the cell to a designated point in the grid
of the calling cell. The order of rotation, translation, and mirroring can make a
difference in the ultimate position of the cell, and the user indicates the proper
order by the order in which the transformation designators appear in the call
statement.

The definition of a cell is introduced by a statement of the form

DS <symbol number> <scale>

where the <symbol number> is an integer that serves as the "name" of the
cell, and <scale> is a pair of integers $a$ and $b$, such that all dimensions and
coordinates of boxes are multiplied by $a/b$.† The DS stands for "Definition
Start," and we can use that more verbose form if we choose. The end of a
definition is marked by the statement DF, or "Definition Finish."

The fundamental unit of dimensions is not $\lambda$, as we tacitly implied in
Example 7.1, but rather .01 micron. Thus, if we wish to write dimensions in
terms of $\lambda$, we must place the statements in a symbol definition and use a scale
$a$:1, where $a$ is $\lambda$ in hundredths of a micron.

**Example 7.2:** Suppose we wish to make a cell of the inverter of Fig. 7.6.
Remember that the half-integer numbers 13.5 and 4.5 in lines (2) and (3) of
Fig. 7.7 are not legal CIF, so let us double all the numbers in Fig. 7.7. Then if,
say $\lambda = 1.5$ microns, we can scale the numbers by another factor of 75 to give
the dimensions their correct values in terms of hundredths of microns, which is
the only valid way to give dimensions in CIF. Thus, we can declare symbol 123
to be an inverter cell with $\lambda = 1.5$ microns by

DS 123 75 1;
    statements of Fig. 7.7, with all numbers multiplied by 2
DF;

☐

The call of a symbol is effected by the statement

C <symbol number> <list of transformations>

where <list of transformations> is a list of elements, each with one of the

---

† Dimensions in other shape defining statements that we have not covered are similarly scaled,
but if the definition of the symbol includes a call to some other symbol, the dimensions of
boxes in the definition of that symbol are not multiplied by $a/b$. Rather, they are scaled by
the ratio in the definition of the symbol called.

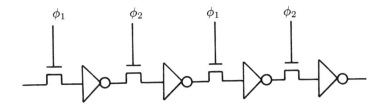

**Fig. 7.8.** Shift register.

forms

1. T <xorigin> <yorigin>. This transformation places the origin for the grid of the called cell at the point with $x$ coordinate <xorigin> and $y$ coordinate <yorigin> in the grid of the calling cell.

2. R $a$ $b$. The $x$ axis in the called cell is rotated to have slope $a/b$, just as the Direction parameter in the box command rotates the $x$ axis of the rectangle.

3. MX and/or MY. MX mirrors in $x$, by multiplying each $x$ coordinate in the definition of the called cell by $-1$, and MY does the same with the $y$ coordinates.

As usual, letters like C, R, T, and M can be written more verbosely, as Call, Rotate, Translate, and Mirror, for example.

**Example 7.3:** Suppose we want to design a cell that is a simple four stage shift register, with alternate stages controlled by alternate clock phases $\phi_1$ and $\phi_2$, as suggested in Fig. 7.8, which uses the notation described in Section 1.2. Thus, at phase $\phi_1$, the input is let into the first stage inverter and the inverted output appears. However, that output cannot reach the input to the second stage until $\phi_2$. Similarly, the output of stage two does not reach the input to stage three until the second $\phi_1$ cycle, and so on.

We have defined a cell suitable for the four inverters; it is the symbol 123 defined in Example 7.2. Let us suppose that another cell is defined to be a pass transistor controlled by a wire that we might attach to one of the clock signals. This cell must connect with the green output of the symbol 123, whose layout we saw in Fig. 7.6; that green wire must be extended to pass under a red wire holding the clock signal to make the pass transistor, and we must then convert the green wire to red, through a butting contact for example, and adjust its vertical position so it lines up with the red input on the left of Fig. 7.6. Let us call this cell by symbol number 456, and suppose it is defined by something of the form

```
DS 789 150 1;
C 456;
C 123 T 11 0;
C 456 T 19 0;
C 123 T 30 0;
C 456 T 38 0;
C 123 T 49 0;
C 456 T 57 0;
C 123 T 68 0;
    box statements to bring power to the tops of the
        inverter pullups, ground to the bottoms, and
        clock signals to the pass transistors
DF;
```

**Fig. 7.9.**  Sketch of a shift register definition.

```
DS 456 75 1;
    suitable list of boxes; dimensions in units of λ/2
DF;
```

Let us further suppose that symbol 456 is $11\lambda$ wide, just as symbol 123, the inverter, is $8\lambda$ wide. Then we can specify a cell that acts as a four stage shift register by defining a new symbol, say 789, by the sequence of calls and other box statements sketched in Fig. 7.9.

Notice how the first copy of the clocking transistor, symbol 456, has its origin at the same place as the origin of the symbol 789 being defined. The inverter to its right, symbol 123, is translated $11\lambda$ to the right to account for the width of the first cell, the next copy of symbol 456 is translated $19\lambda$ right to account for the widths of the first two cells, and so on. Also note that the boxes alluded to in Fig. 7.9 are specified in units of $\lambda$, since we declared a scale of 150:1 for symbol 789, and $\lambda$ was taken to be 1.5 microns in our example. Finally, observe that the dimensions of boxes in cells 123 and 456 are only multiplied by 75, not by 75 $\times$ 150, since scaling of the symbol 789 does not affect the scale of dimensions in symbols called. $\square$

## 7.3 CHISEL: A PREPROCESSOR FOR GENERATING CIF

CHISEL, styled as a "tool for making chips" by its author (Karplus [1982]), is a preprocessor for the C programming language. The primitives of CHISEL are not much higher than those of CIF, although there is some "syntactic sugaring" in CHISEL. The important things that CHISEL provides include:

1.   A powerful library facility, and the ability to link libraries of cells together in the way a loader links procedures of an ordinary programming language

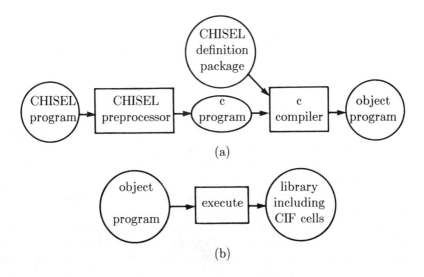

**Fig. 7.10.** Use of CHISEL.

together.

2. A method for managing *ports*, which are named points in cells, for allowing ports in called cells to become ports of the calling cell, and for handling the names of ports when a single cell is replicated many times in one larger cell.

3. A facility for doing (somewhat) automatic river routing, i.e., routing of wires in a single layer, as well as several more complicated kinds of routing. We shall not discuss this facility here, although river routing in general is discussed in Section 9.6.

The user of CHISEL writes a program in the language c, using certain dictions that are not part of c, but that are handled by the CHISEL preprocessor, which converts them into standard C code. In addition, the CHISEL user has available a number of C type declarations and C procedures, defined in a file to be included with the user's program. The steps needed to produce CIF from a CHISEL program are illustrated in Fig. 7.10. In Fig. 7.10(a), we see the compilation of a CHISEL program. In Fig. 7.10(b), the object program that is produced by the compilation phase is executed. The result of running this program is the generation of a CHISEL library; this library includes CIF cell definitions as well as information about them, such as the names and locations of ports. It is possible to take the CIF from the library and fabricate a chip therefrom, or we may use the library as a resource for other CHISEL programs, by having those programs open the library and refer to named cells found there.

One might naturally wonder why it is necessary or desirable to embed geometric design primitives in a general-purpose programming language, or why it is desirable to write a program that generates CIF rather than simply writing the CIF in the first place. The answer is that there are many situations where it is important that the user do a substantial amount of computation before deciding what CIF to generate.

**Example 7.4:** As one example, chips designed to do encoding or decoding may be required to multiply or divide streams of bits by large, fixed constants. Each output bit is some Boolean function of the bits in the input stream, and we could determine exactly what those functions are by a computation involving the multiplier or divisor. However, the calculation of the Boolean functions is arduous, and even if we know that the constants involved are never going to change, we might be much better off writing a computer program to generate the functions than doing the calculation by hand. The computational power of the programming language will also most likely be of use when deciding how to draw the gates that will implement the computed Boolean expressions. The advantage of the CHISEL approach will become even clearer when the plans change, and the chip must be designed using different constants as multipliers and divisors.

Another place in which the power of a programming language is important is in the writing of other tools for circuit design. For example, a good PLA generator takes a PLA personality, which is a matrix of bits, as input, and generates a CIF layout. There is considerable calculation involved in generating the right PLA and designing it to have a small area.

A third place where calculation is needed is when we wish to place cells from a library somewhere on a grid and then wire together corresponding pairs of ports. The distance between cells needed to run the necessary wires is a complicated function of the relative positions of the ports; we shall discuss how to do such calculations in Section 9.6. $\square$

## Points

In CHISEL, *point* is a data type, a record consisting of two fields, $x$ and $y$. Thus, if $p$ is a variable of type *point*, then $p.x$ and $p.y$ are the coordinates of the point. There is a built-in variable *At* of type *point*. The value of *At* changes automatically as we lay down wires and transistors, always being the last point reached.

The notation $[[x, y]]$ is used to designate a point. For example, $[[3, 4]]$ designates the point with $x$ coordinate 3 and $y$ coordinate 4. $[[At.x, At.y + 1]]$ designates the point one unit above the point *At*. Depending on the context, the value of *At* may or may not be changed. The grid to which point coordinates refer is usually in units of $\lambda$, but can be defined to be any fraction of $\lambda$.

The use of the commercial at-sign forces a reassignment of the value of *At*. Thus @$[[3, 4]]$ is an expression whose value is the point $[[3, 4]]$, but with the side effect, when evaluated, of setting $At.x$ to 3 and $At.y$ to 4.

## Boxes

In CHISEL, all boxes and wires are assumed to be rectilinear, that is, their edges run only horizontally or vertically. This restriction is not generally severe, since many fabrication processes can make chips only from rectilinear CIF anyway, and there are but a few cases where nonrectilinear boxes are important for saving a significant amount of area; for example, memory chips often rely on nonrectilinear geometry. Corresponding to the Box statement in CIF is the statement

RectI($<$x1$>$, $<$y1$>$, $<$x2$>$, $<$y2$>$)

where ($<$x1$>$, $<$y1$>$) gives the coordinates of one corner of the rectangle, and ($<$x2$>$, $<$y2$>$) gives the coordinates of the opposite corner. Another form of the Rect statement, RectP($<$p1$>$, $<$p2$>$) exists; this form takes pointers $<$p1$>$ and $<$p2$>$ to points that are the opposite corners.

## Layers

The statement

Layer($<$layer name$>$)

sets the current layer to that given by the layer name. It also sets a variable WireWidth to the minimum width for wires in that layer, e.g., 2 for poly or diffusion, and 3 for metal. The $<$layer name$>$ can be any of DIFF, POLY, METAL, CUT, IMPLANT, or several others. These correspond to ND, NP, NM, NC, and NI in CIF, respectively. Actually, DIFF, and the other layer values are defined constants, and each is a power of 2. Thus, we can add layers and the CHISEL system will treat the result as the set of layers in the sum. When we discuss transistors we shall see how the ability to write expressions like DIFF+POLY for layers is useful.

## Cells and Cell Libraries

The typical CHISEL program generates a library, and it may also open one or more libraries to use some cells of those libraries to help define its own cells. A library consists of two files, one containing CIF cells and the other containing header information. We shall not go into the variety of information found in the header file, except to say that it includes names for cells and ports and information about those ports such as the layer on which we can connect to them. CHISEL programs refer to cells by a name, which is a quoted string of characters, and the header file associates with that name the number of the CIF symbol defining that cell. A utility program provided with CHISEL renumbers CIF symbols of several libraries so the same number is not used for two different

```
Cell "butcon";
    Layer(POLY); RectI(−2, 0, 2, 3);
    Layer(DIFF); RectI(−2, −3, 2, 1);
    Layer(METAL); RectI(−2, −3, 2, 3);
    Layer(CUT); RectI(−1, −2, 1, 2);
EndCell;
```

**Fig. 7.11.** CHISEL definition of a butting contact cell.

symbols.

To define a cell we begin with

Cell <cell name>;

and end with

EndCell;

We shall not discuss the syntax for opening and closing libraries.

The statement that corresponds to Call in CIF is

put <cell name> @<point> <options>

The cell called <cell name> is placed in the grid of the calling cell with its origin at <point>. The <options>, whose syntax we shall not discuss, include the ability to rotate the cell (but only by a multiple of ninety degrees), mirror the cell, and replicate the cell any designated number of times. Replication allows us to make an array of copies of a cell, optionally with alternate copies mirrored, which is a common way to save space because adjacent copies can then share power or ground wires.

For example, the code in Fig. 7.11 defines a cell that is a butting contact, with the polysilicon piece at the top and the origin for the cell in the center of the butting contact. When used, this cell can be rotated, using a syntax we shall not discuss, to have the poly on the bottom or either side.

## Wires

CHISEL provides several wire commands, such as

CWire(@<initial point>, <x1>, <y1>, <x2>, <y2>, ...)

This command defines a line that is the centerline of a wire in the current layer. The width of the wire is equal to the value of variable WireWidth. Recall that WireWidth is set to the minimum allowable widths of wires in the current layer whenever a Layer command is executed, but it can be assigned by the user if desired. The line begins at the point <initial point> and proceeds in the $x$

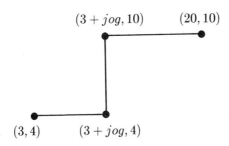

**Fig. 7.12.** The centerline of a wire.

direction to the $x$ coordinate $<x1>$.† Then we move in the $y$ direction until the $y$ coordinate becomes $<y1>$, then in the $x$ direction until the $x$ coordinate becomes $<x2>$, and so on.

The value of $At$ follows each point along the path. Thus, $At$ is initially set to $<$initial point$>$, then $At.x$ is set to $<x1>$, while $At.y$ is unchanged; then $At.y$ is set to $<y1>$ while $At.x$ is unchanged, and so on. This fact allows us to have the wire make steps for a fixed distance in one of the four directions, without necessarily knowing where it is when it begins the move, and it also allows wires to be run to a known point from an unknown one.

**Example 7.5:** Let *jog* be a variable name. We can run a wire from the point $[[3, 4]]$ to $[[10, 20]]$ moving up after moving right for distance *jog*, if we issue the command

> CWire(@$[[3, 4]]$, $At.x + jog$, 10, 20)

The centerline of the wire is shown in Fig. 7.12. Note that we could have used $3 + jog$ in place of $At.x + jog$, since we know that after evaluating the expression @$[[3, 4]]$, $At.x$ will have the value 3. The value of $At$ after the first leg is $[[3 + jog, 4]]$; after the second it is $[[3 + jog, 10]]$; and after the completion of the wire it is $[[20, 10]]$. □

A number of alternative ways to specify wires are available in CHISEL. We can use LWire in place of CWire to have the wire extend entirely to the left of the line. That is, if we traveled along the wire in the direction in which the command makes it grow, the line would be on our right. Similarly, the command RWire places the wire to the right of the line.

In addition, it is permissible to omit the term @$<$initial point$>$ from the list of arguments of any of these wire drawing commands, in which case the initial point is taken to be $At$. Note that we can always tell whether the first argument is an initial point or $<x1>$, because $<$initial point$>$ must be preceded by @, while $<x1>$ is of type integer.

---

† If we wish to begin by moving in the $y$ direction, we must let $<x1>$ be the $x$ coordinate of the $<$initial point$>$, which, since $At$ is then the initial point, can be expressed as $At.x$.

```
        Cell "inverter";
(1)         Layer(DIFF+POLY+IMPLANT);
(2)         CWire(@[[4, 16]], At.x, At.y − 8);
(3)         put "butcon" @[[At.x, At.y]];
(4)         Layer(DIFF);
(5)         CWire(@[[At.x + 2, At.y − 2]], At.x + 2);
(6)         Layer(POLY+DIFF);
(7)         CWire(@[[2, 3]], At.x + 4);
        EndCell;
```

**Fig. 7.13.** CHISEL definition of inverter.

**Transistors**

CHISEL takes the point of view that the user should explicitly say that he wishes a transistor drawn, rather than simply placing red and green wires or boxes in the same place. While the crossing of red and green wires is not explicitly forbidden, it is considered by the system to be worthy of a warning message, since it may be an accidental crossing that will ruin the design.

To make an enhancement mode transistor, one gives the command

   Layer(DIFF+POLY)

and draws a wire across the channel of the intended transistor. The polysilicon gate of the transistor will be drawn along this wire and extended $2\lambda$ in both directions along the wire, so it overlaps the place of the transistor by that amount, as required by the Mead-Conway design rules. The diffusion channel is extended in the direction orthogonal to the direction of the wire, both to the left and to the right, by $2\lambda$.†

Similarly, depletion mode transistors can be specified by setting the layer to DIFF+POLY+IMPLANT and drawing a wire along the channel of the transistor, i.e., the place where the diffusion wire is covered by the polysilicon. The diffusion channel is extended $2\lambda$ in both directions along the channel, and the polysilicon is extended $2\lambda$ to the left and right of the wire drawn. The implant is extended for $2\lambda$ in all directions around the wire.

**Example 7.6:** Let us design the inverter cell of Fig. 7.6, using CHISEL. The code for this cell is given in Fig. 7.13. We have assumed that the cell "butcon", defined in Fig. 7.11, is available in a library that has been opened by the current program. We have omitted declaration of ports for the cell, a subject we shall cover shortly.

   Lines (1)–(2) define the pullup. The centerline of its channel is declared to

---

† The amounts by which the gate and channel are extended can be adjusted by user-set parameters.

run from $(4, 16)$ down to $(4, 8)$. Note that the term $At.x$ in line (2) does nothing; it is an artifact of the requirement that the first move be in the $x$ direction. The effect of line(2) is to draw a box $2\lambda$ wide in diffusion, running from $y = 18$ down to $y = 6$, i.e., $2\lambda$ beyond the specified wire, at both ends. It also draws a $6\lambda \times 8\lambda$ polysilicon box and a $6\lambda \times 12\lambda$ implant box; these are the boxes in lines (6) and (13) of Fig. 7.7, the CIF description of the same cell.

Line (3) copies the butting contact of Fig. 7.11, with its origin (the center) at the point currently denoted by $At$. That point is the end of the wire drawn at line (2), the point $[[4, 8]]$. Lines (4) and (5) draw the stub of diffusion for the output; they correspond to line (4) of Fig. 7.7.

Lines (6)–(7) draw the pulldown. The actual line drawn is from $[[2, 3]]$ to $[[6, 3]]$, but this line results in a $2\lambda$ wide polysilicon wire centered on that line but extended $2\lambda$ to the left and right, i.e., the box defined in line (7) of Fig. 7.7. The diffusion channel runs along that line, but is extended an extra $2\lambda$ on either side to make a box $6\lambda$ high and $4\lambda$ wide with corners at $[[2, 0]]$ and $[[6, 6]]$. Note that line (7) uses an absolute point value, rather than a relative value defined in terms of $At$. There is some disadvantage in doing so, because bugs in the design due to modifications may be less likely if positions of important features are defined relative to other features. However, in this simple case, there is little harm in laying out the inverter from two independent points. $\square$

**Ports**

*Ports* are points of reference in cells. They have a number of attributes, including a name, a layer or layers in which connections can be made, and a direction in which we can expect to run a wire connecting to the port, without accidentally meeting anything internal to the cell. Ports may be copied from a called cell to a calling cell, by a syntax omitted from this discussion.

We define a port by the statement

DefinePort $<$port name$>$ @$<$point$>$ dir $<$direction$>$

The $<$port name$>$ is a quoted character string that identifies the port among all ports in the current cell, although ports in different cells may have the same name. The $<$point$>$ is the reference point for the port, and the $<$direction$>$ is the direction in which we expect a wire to leave the port safely, e.g., +x means to the right, that is, in the positive $x$ direction. The port acquires a layer from the layer currently in effect.

**Example 7.7:** The inverter cell would logically be given four ports, the input, the output, and connections for $VDD$ and ground at the top and bottom, respectively. We could append the code of Fig. 7.14 after line (7) of Fig. 7.13 to define these four ports. Alternatively, we could introduce the ports at times when the value of $At$ made a definition of the point for the port in relative terms convenient. For example, we could append after line (5) of Fig. 7.13

Layer(DIFF);
DefinePort "VDD connect" @[[4, 18]] dir +y;
DefinePort "ground connect" @[[4, 0]] dir −y;
DefinePort "output" @[[8, 6]] dir +x;
Layer(POLY);
DefinePort "input" @[[0, 3]] dir −x;

**Fig. 7.14.** Port definitions for inverter cell.

DefinePort "output" @[[$At.x, At.y$]] dir +x

or simply

DefinePort "output" dir +x

since the point $At$ may be taken as a default value of <point>. The point in the code after line (5) is the desired one because the position of $At$ after executing that line is exactly where we want the port to be. Note that the proper direction for the port is +x, since it is only safe to bring a wire to touch that port from the right. □

To access the points that ports represent, or to access other information about ports, such as the layer or layers to which they may be connected, or their direction, we use two built-in functions, Ports and PortValue. The call

Ports(<port name>, <cell name>, 0)

returns an object of type *ports* corresponding to the port whose name is <port name> in the cell <cell name>. The third parameter is not discussed here. The type *ports* is a pointer to a record whose fields give relevant information about the port.

We shall not discuss accessing any information but the point for a port. To find this point, we call the function

PortValue(<port>, <index>)

Here, <port> is an object of type *ports*, returned by function Ports. To understand the role of <index>, we must realize that a port may be an array of points. For example, if we replicate a cell, a port of that cell may become an array of points, one for every copy of the cell. These points share the same name; they are distinguished by the second component in calls to PortValue. The value returned by PortValue is a pointer to a point.

**Example 7.8:** Suppose that cell1 and cell2 in Fig. 7.15 are parts of a data path connected by a bus. We have placed the cells at points origin1 and origin2, and wish to wire together the points on the right edge of cell1 and the left edge of cell2 that are the connection points for the bus wires. However, we do not

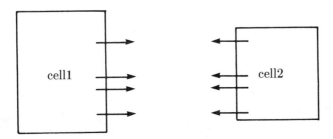

**Fig. 7.15.** Two cells to be connected.

know exactly where the connection points are. We know only that there is a port "busright" of cell1 that is an array of BUSSIZE points, indexed 0 through BUSSIZE−1, as all arrays are in the language c. Similarly, a port "busleft" in cell2 is the corresponding array of points on the left edge of cell2.

The code to perform the wiring is shown in Fig. 7.16. To make sense of some of the C syntax, it helps to know that

**for** $(i = 0; i < \text{BUSSIZE}; i++)$

is c's way of saying

**for** $i := 0$ **to** BUSSIZE−1 **do**

and the * operator performs dereferencing, so

*PortValue(bus1, $i$)

is equivalent to the Pascal

PortValue(bus1, $i$)↑

and in particular, the two uses of the * set leftend and rightend to be the relevant points themselves, rather than pointers to points, as returned by PortValue. Also, curly braces are used for **begin** and **end**, and != is the symbol for "not equal to."

Notice the test to see whether corresponding points can be wired together with a horizontal wire. If the cells are not aligned, it might be feasible to jog the wires as suggested in Fig. 7.12; but as we do not know exactly where the points are, we run the risk of having wires cross accidently if we do so. The proper way to handle such a situation is with CHISEL "fringes," which we shall not cover here. In practice, many more tests would be made by accessing data about the ports "busright" and "busleft", such as checking that they could be accessed from the directions +x and −x, respectively, and that they could be attached in the POLY layer. □

```
ports bus1, bus2; /* variables to hold objects returned by Ports */
point origin1, origin2, leftend, rightend;
bus1 = Ports("busright", "cell1", 0);
bus2 = Ports("busleft", "cell2", 0);
Layer(POLY);
for(i = 0; i<BUSSIZE; i++) {
    leftend = *PortValue(bus1, i);
    rightend = *PortValue(bus2, i);
    if(origin1.y + leftend.y != origin2.y + leftend.y)
        error("misaligned cells");
    else CWire(@[[origin1.x + leftend.x, origin1.y + leftend.y]],
    origin2.x + rightend.x);
```

**Fig. 7.16.** CHISEL program using port information.

## 7.4 ESIM: A SWITCH LEVEL LANGUAGE

We shall now briefly mention an example of a switch language, called esim. This language is part of a system intended to simulate designs written in CIF. That is, esim is an example of the simulation language mentioned in Fig. 7.2. It is intended to be generated by an extractor, from a CIF file. The esim file, in turn, is input to a simulator, which takes as input both the switch description generated by the extractor and a sequence of inputs to be applied to the circuit. Input-generating commands are also part of the esim language, but we shall not discuss their details here. Neither shall we discuss the algorithms underlying the operations of extraction or simulation, which are left for Sections 9.2 and 9.4.

While we might consider using a switch language as a design language, that does not seem to be a very effective approach to design. The switches make no reference to placement of the individual circuit elements, so automatically generating a good layout from switches is not easy. Further, the logic language to be described in the next section appears to have significantly easier-to-use input style, while presenting a compilation problem that is little more difficult.

The two elements of circuit description in esim are nodes and transistors. Nodes represent collections of CIF boxes that are electrically connected. Two boxes are connected if they are in the same layer and overlap or if they are juxtaposed, i.e., they share a border. Boxes are also electrically connected if they overlap, one is a contact, and the other is any of the three layers: metal, poly, or diffusion. Finally, this notion of "connected" must be extended transitively, so two boxes $B_1$ and $B_n$ are connected if there is a sequence of boxes $B_2, \ldots, B_{n-1}$ such that $B_i$ is connected to $B_{i+1}$ for all $i$, $1 \leq i < n$.

Transistors are three terminal devices, the terminals being called the gate, source, and drain. Conventionally, the source and drain, which are opposite ends of the diffusion channel through which current flows when the gate is open (high), are defined so current normally flows from drain to source.† In many transistors, especially pass transistors, or pulldowns, current may flow through the channel in either direction, and the assignment of source and drain is arbitrary.

In esim, we represent nodes by integers. Transistors are represented by the statements

> e <gate> <source> <drain>

and

> d <gate> <source> <drain>

The d and e stand for depletion mode (e.g., a pullup) and enhancement mode (a pass transistor or pulldown), respectively. The <gate>, <source>, and <drain> are integers, indicating the nodes to which each of the three transistor terminals are connected.

**Example 7.9:** In Fig. 7.17 we see the simple shift register of Fig. 7.8 drawn in terms of switches, i.e., transistors and numbered nodes. Nodes 1, 2, 3, and 4 are $VDD$, ground, $\phi_1$, and $\phi_2$, respectively; node 5 is the input and 13 the output.

In Fig. 7.18 we see the definition of the twelve transistors of Fig. 7.17. First come the four pullups, from the right; note that their gates and sources are the same node in each case. Then come the eight pass transistors and pulldowns, in order from the left.

For example, the first depletion mode transistor has gate and source at node 7, which is the output of the first inverter, and drain connected to node 1, which is $VDD$. The first enhancement mode transistor (e 3 5 6) has gate connected to node 3, which is $\phi_1$, source connected to the input, and drain connected to the gate of the pulldown for the first inverter. □

## 7.5 LGEN: A LOGIC LANGUAGE

The language lgen is a logic language developed at Bell Laboratories by S. C. Johnson. It provides a significant amount of optimization of the layout that is generated from logic equations, although the layouts from which it chooses the best are restricted to arrangements with several long rows of pullups. There is evidence, however, that the area produced by the lgen compiler is often fairly close to what would be produced by a hand design of a circuit to implement the

---

† That sounds backwards, but recall that electrons, being negative charges, move in the direction opposite to that of the current. The source is the "source" of electrons, which are then "drained" away by the drain.

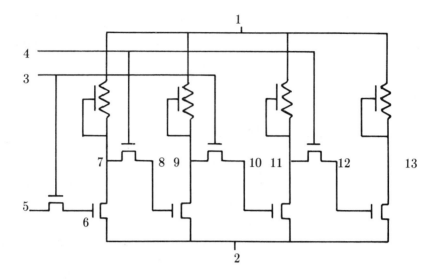

**Fig. 7.17.** A switch representation of a shift register.

| | | | |
|---|---|---|---|
| d | 7 | 7 | 1 |
| d | 9 | 9 | 1 |
| d | 11 | 11 | 1 |
| d | 13 | 13 | 1 |
| e | 3 | 5 | 6 |
| e | 6 | 2 | 7 |
| e | 4 | 7 | 8 |
| e | 8 | 2 | 9 |
| e | 3 | 9 | 10 |
| e | 10 | 2 | 11 |
| e | 4 | 11 | 12 |
| e | 12 | 2 | 13 |

**Fig. 7.18.** esim representation of Fig. 7.17.

same function. We shall discuss some of the algorithms involved in optimizing logic in Section 8.3.

The form of a logic equation is

$$<variable> = <expression>$$

where the $<variable>$ is a name, like the variables of a programming language, and the $<expression>$ is a logical expression involving the logical connectives "and," "or," and "not." The operators used are those of the language c; & is used for "and," + for "or," and ! for "not." As usual, the precedence of

LEFT phi1, phi2, input;
RIGHT output;
CLOCK phi1, phi2;
middle = LAST input;
output = LAST middle;

**Fig. 7.19.** lgen specification of two-bit shift register.

operators is ! (highest), then &, then + (lowest). Thus

$$w = x\&y + x\&z + y\&z$$

says $w$ is true if two or more of $x$, $y$, and $z$ are true.

Lgen allows us to declare certain variables to be inputs or outputs, by indicating an edge of the circuit at which the value of the variable is to be made available (if an output) or taken (if an input). The declaration form is

&lt;edge&gt; &lt;list of variables&gt;

Here, &lt;edge&gt; can be any of TOP, BOTTOM, LEFT, or RIGHT. For example,

LEFT input1, phi1, phi2

declares three variables, input1, phi1, and phi2, to have their connections on the left edge of the circuit.

There is a notion of two-phase clocking built in, and one can declare certain variables to be clock phases by

CLOCK phi1, phi2

Of course we need not call the clock signals phi1 and phi2.

To allow a circuit to be sequential, that is, to remember the value of a variable from the previous clock cycle, we may declare, for example,

$x$ = LAST $y$

The value of $x$ at one occurrence of clock phase 1 will be the same as the value that $y$ had at the previous phase 1.

**Example 7.10:** The shift register of Fig. 7.17 really holds two bits, not four, since in two-phase operation we would make inputs only in phase 1. Put another way, the output at one occurrence of phase 1 is the same as the value in the second inverter, the node numbered 9 in Fig. 7.17, of the previous phase 1. The value at node 9, in turn, is the input present at the phase 1 previous to that.

Thus, in lgen, we may describe the operation of Fig. 7.17 with only three variables, which we might call input, middle, and output, corresponding to nodes 5, 9, and 13 of Fig. 7.17, respectively. If we follow the selections of edges for the clock, input and output signals that we made in Fig. 7.17, we could specify the function performed by that figure by the lgen program of Fig. 7.19.

```
LEFT current, phi1, phi2;
RIGHT output;
CLOCK phi1, phi2;
oldout = LAST output;
previous = LAST current;
nexttoprev = LAST previous;
output = oldout & (current + previous + nexttoprev) +
    current & previous & nexttoprev;
```

**Fig. 7.20.** lgen specification of the bounce filter.

There is no guarantee, of course, that the output of the lgen compiler would look like Fig. 7.17, just that it would perform the same function. However, in this case there would be a strong resemblance. □

**Example 7.11:** The shift register actually uses no logic; it just remembers bits. We shall close with an example that uses the logic equations of lgen; this example will appear several times subsequently. The problem is to design a "bounce filter," a device that samples its input, which is 0 or 1, at each phase 1 of the clock and generally transmits its input to the output. However, the filter must ignore "bounces," which we shall define here to mean one or two input bits that disagree with their surrounding bits in the input sequence. That is, the device always has a notion of what the input should be, either 0 or 1. If it sees one or two bits that disagree with its notion of what the input should be, it ignores them, continuing to produce what it thinks is the proper output. If it sees three consecutive inputs that disagree with what it thinks the input should be, then it revises its notion of the "right" input and proceeds to output the new value, until such time as its mind is again changed by three consecutive bits.

Thus, for example, if the filter initially thinks 1 is the proper output, and it receives 10010100010, it would continue emitting 1's until it received the ninth bit, and would then start emitting 0's.

Our program uses three variables, current, previous, and nexttoprev, to remember the three most recently received bits. The variable output remembers the present notion of the proper output to make. The value of output is, therefore, 1 if it was 1 at the previous clock cycle, and the three variables current, previous, and nexttoprev are not all 0. The output is also 1 whenever these three variables are all 1, regardless of the previous output. The lgen program to describe this bounce filter, neglecting some details of initialization for variables, is shown in Fig. 7.20. □

## 7.6 LAVA: A STICKS LANGUAGE

We shall now consider an example of a language that deals with circuits on the switch level, like esim. However, this language, LAVA, is a "sticks" type language, meaning that it also deals with the relative positions of the circuit elements and the wires that connect them. It is, therefore, much more suitable for a design language than is esim, which was never intended to be used for that purpose.

The term "sticks" refers to the fact that the wires are drawn as lines, with the appropriate thickness usually chosen by the system. However, the term "sticks" has come to mean much more. Transistors are represented by nodes, with very little detail about their shape, usually just the all-important width/length ratio of its channel. A sticks compiler must provide some mechanism for selecting the exact geometry for the transistors, as well as their relative placement and the paths taken by the connecting wires. A sticks compiler is not required to select layers for wires nor, as we shall see when we discuss the compilation of sticks languages in Section 8.2, is it required to make very imaginative transformations of the positions of circuit elements specified in its input.

### Points

The design of a cell in LAVA presumes the existence of an integer grid, as in CIF or CHISEL. However, unlike those languages, the grid coordinates are abstract; they will be translated into real coordinates by the LAVA compiler so that all design rules will be followed, yet the resulting circuit is as compact as the compiler can manage. We shall discuss the details of cell compaction in Chapter 8, but for the moment, the reader should simply bear in mind that the compilation process will not bend abstract grid lines; objects declared to be at the same horizontal or vertical position will be implemented with the objects aligned in the horizontal or vertical dimension, respectively. Thus, if we wish there to be freedom in, say, the relative vertical position of two points, they should be declared to live on different abstract horizontal grid lines.

The following is the syntax for declaring a point.

<point name> = <object type> [<xpos>, <ypos>]

(<xpos>, <ypos>) will be the coordinates of the point called <point name> in the abstract grid. There are several different options for <object type>, and in some cases, information about that object may be added to the above statement form. The types of objects that may be found at a point are the following.

1.  The point may be nothing but a reference point, perhaps a bend in a wire. In this case the <object type> is

point <layer>

where <layer> is blue, red, or green.

2.  The point may be a via, in which case <object type> is

    via <layers>

    The <layers> are the abbreviations for the layers connected by the via. For example, a contact between metal and poly (blue and red) would be specified by "via br", and the effect of a butting contact, whose purpose is to connect green and red could be specified by "via gr". Such a via might or might not be implemented as a butting contact by the LAVA compiler.

3.  An enhancement mode transistor (pulldown or pass transistor) is denoted by putting "tran enh" as the <object type>. In this case, and for the other transistor-declaring statements, we must follow the coordinates of the point by certain information about the transistor that we shall discuss in a moment.

4.  A depletion mode transistor is represented by <object type> "tran dep".

5.  However, if the depletion mode transistor is a pullup, i.e., its source and gate are electrically connected, then the <object type> is "load dep".

## Transistors

In the case of items (3)–(5) above, we must specify a width/length ratio for the transistor channel, and we may, if necessary, specify a rotation from the conventional orientation for transistors that LAVA assumes. Thus, we use point defining statements of the form

<point name> = <transistor type> [<xpos>, <ypos>]
(<ratio>) rotate <rotation>

The <ratio> is of the form $a{:}b$, where $a/b$ is the ratio of the width of the channel to its length. The clause "rotate <rotation>" is optional. If used, <rotation> is an integer, which indicates the angle of clockwise rotation in units of thirty degrees.

The normal orientation for an enhancement mode transistor is with the gate input to the left and the channel running vertically. Thus, if we wanted to connect the gate to a wire running vertically from above, we would rotate the transistor ninety degrees clockwise by saying "rotate 3". The normal orientation for a pullup is with the channel running vertically, with the drain at the top and the source and gate connection at the bottom. Again, rotation in units of thirty degrees may change this orientation.

When connecting to a transistor at a point, we must specify which of the several terminals we wish to connect to. In the case of an enhancement mode transistor, there are three terminals, the gate, source, and drain. It is

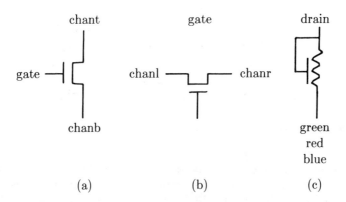

**Fig. 7.21.** terminal names.

not normally useful to distinguish the source from the drain, although it is important to tell on which side of the gate we wish to make a connection. Thus, LAVA uses the term "gate" for the gate terminal, but uses "chan," for channel, followed by one of the letters "l", "r", "t", or "b" for left, right, top, and bottom, whichever is most appropriate. Thus, an enhancement mode transistor'in its normal orientation would use the terminal names suggested in Fig. 7.21(a), while one rotated ninety degrees would use the terminal names in Fig. 7.21(b).

In the case of a pullup, we use the terminal names indicated in Fig. 7.21(c), regardless of orientation. The point where the gate and source are tied together can be connected to in any of the three layers, so the same node is referred to by any of the colors red, blue, and green. The opposite end of the channel, which is the drain, is referred to by that name. It, as well as the channel ends in an enhancement mode transistor, are diffusion, and therefore can be connected to directly only by a green wire. Other wires must go through an explicit via before connecting to the drain of a pullup or the channel ends of an enhancement mode transistor.

We specify a terminal of a transistor by appending a dot and the terminal name to the point name. Thus, if p1 is a point at which there is an enhancement mode transistor in the normal orientation of Fig. 7.21(a), we can connect to the gate from the left by referring to p1.gate, and we can connect to the channel from the top by specifying p1.chant or from the bottom by specifying p1.chanb.

**Wires**

Wires run horizontally or vertically between two points. Thus, a horizontal wire must have endpoints whose coordinates have the same $y$ value, and a vertical wire's endpoints require the same $x$ value. There is an exception if one of the

endpoints is on the boundary of the cell. We shall cover the declaration of
boundary points later, but for the moment, let us just assume that certain
points are declared to be left, right, up, or down. Any wire connecting to a
point that is left or right must be horizontal, and any wire connecting to a point
that is up or down must be vertical. Thus, no coordinates need be specified
for boundary points; their position can be inferred from the points they are
connected to.

We declare a wire with the statement

    <point1> # <point2>

<point1> and <point2> are names of points. If the point is a transistor,
however, then instead of a point name, we use a terminal name, like p1.gate.
The meaning of the # operator is that a wire is to be drawn between <point1>
and <point2>.

Certain specifications, surrounded by parentheses, may follow the #. For
example, if we wish the wire to be of more than the minimum width, we
may follow it by (width $n$), where $n$ is the desired width. Another possible
specification is the layer of the wire, which is indicated by the layer's color: red,
blue, or green. Often, we shall be able to deduce the layer of a wire without such
a specification, e.g., because one end of the wire is a green point. However, we
shall generally include a layer specification with all declared wires, for clarity.

Thus, if v1 is the name of a point that is a red-blue via, and t1 is the name
of a point with an enhancement mode transistor, we might use a statement like

    v1 #(red) t1.gate

to run a polysilicon wire between the via and the transistor gate. We must
deduce from the coordinates of v1 and t1 whether the wire is vertical or
horizontal. We could also omit "(red)" after #, and deduce that the wire in
question is red because one end is a gate.

Although we shall not do so here, it is possible to save a considerable
amount of effort, at the cost of making the LAVA code somewhat less transparent, if we follow the C convention† that an assignment (a point declaration
in the case of LAVA) is an expression whose value is the same as the right side
of the assignment symbol. Thus, in a single statement, a point may be declared
and immediately used as an argument to a # operator. For example, the points
v1 and t1 referred to above could have been declared at the same time a red
wire between then was drawn if we had said

    (v1 = via br [10, 15]) #(red) (t1 = tran enh [20, 15] (1:1)).gate

As another shorthand, we can connect more than two points in one statement
by using more than one # operator, as

---

† LAVA, like CHISEL, is oriented toward the language C, from which much of its style is
taken and in which it is implemented.

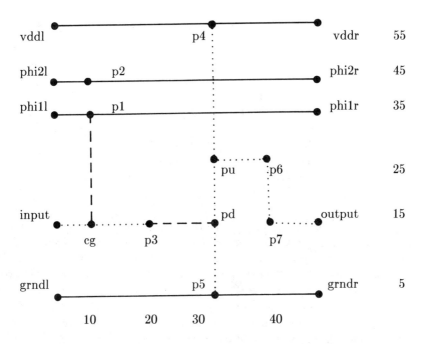

**Fig. 7.22.** Stick diagram of inverter cell.

p0 # p1 # ⋯ # p*n*

## Conditionals

We should think of the design of a cell as if we were writing a procedure. Some of the parameters to the cell/procedure are the names of the boundary points, while other parameters can be integer variables that influence the design of the cell. Thus, it is possible to generate slightly different versions of cells at different places in the overall layout, the actual version being determined by the values of these control parameters when the cell is placed, i.e., the procedure for that cell is called by another cell. We shall cover matters concerning cells shortly.

The control parameters may be used in conditional statements with the style of the language C. The major difference between the C syntax for conditionals and that of most algol-like languages is that === is the operator for "equals," and != the operator for "not equal to." Also remember that the word "then" is not used, and curly braces stand for "begin" and "end."

**Example 7.12:** We shall design a LAVA cell that is an inverter; four of them could serve to build the shift register of Fig. 7.17, and in fact, we shall show in a subsequent example how to build that circuit. Note that there are really two different flavors of inverter used in Fig. 7.17, depending on whether the clocking gate controlling input to the pulldown is attached to $\phi_1$ or $\phi_2$. Thus, we shall use a control parameter, *phase*, that tells us which clock wire to connect to. There are also ten variables that represent points at the left and right boundary, as shown in Fig. 7.22. We shall assume that each of these external points is blue except the input and output, which are green.

Figure 7.23 gives LAVA code that defines the stick diagram of Fig. 7.22. To help follow the code, we have used multiples of 10 for abstract horizontal coordinates and numbers ending in 5 abstract vertical coordinates. We have also generated the various points and wires column by column, first declaring the points in each column, then drawing the vertical wires in that column, and finally the horizontal wires that come from the left and connect to a point in that column.

For example, line (1) of Fig. 7.23 declares cg to be a pass transistor at the point $(10, 15)$ in the abstract coordinate space. The rotate-clause declares the gate to be connectable from the top, while the channel runs left-to-right. Finally, the ratio $(1:1)$ says that the place where the gate and channel cross is square; presumably it will be $2\lambda$ on a side because wires are given their minimum possible width as a default.

The gate of cg will be connected either to point p1 at line (5) or to p2 at line (8), depending on the value of the parameter *phase*. At line (11), the left end of the channel of cg is connected to the port called input, while at line (13), the right end of the channel is connected to the point p3, which was declared to be a red-green via at line (12). The point p3 is at abstract coordinates $(20, 15)$, which agrees with cg in the $y$ direction; thus the wire drawn at line (13) is horizontal. $\square$

## Cells

The declaration of a cell in LAVA has the form

    cell $<$name$>$($<$port list$>$; $<$control parameter list$>$)

where $<$name$>$ is the name by which we refer to the cell, the $<$port list$>$ is a sequence of variables that represent the points of attachment on the boundary of the cell, and the $<$control parameter list$>$ is a list of other variables whose values, when established by calls to the cell, can influence the design of the cell as the parameter *phase* did in Example 7.12.

Following this cell header is a list of declarations for the points on the $<$port list$>$. Each such declaration is of the form

    $<$name$>$: $<$layer$>$: $<$border$>$

$<$name$>$ is the name of the port, while the $<$layer$>$ is one of the colors: blue, red, or green. The $<$border$>$ is one of up, down, left, or right and indicates an edge of the rectangle bounding the cell, on which the port lies. Following

```
(1)      cg = tran enh [10, 15] (1:1) rotate 3;
(2)      if (phase == 1) {
(3)          p1 = via br [10, 35];
(4)          p2 = point blue [10, 45];
(5)          p1 #(red) cg.gate;
         }
         else { /* phase is 2 */
(6)          p1 = point blue [10, 35];
(7)          p2 = via br [10, 45];
(8)          p2 #(red) cg.gate;
         }
(9)      phi1l #(blue) p1;
(10)     phi2l #(blue) p2;
(11)     input #(green) cg.chanl;

(12)     p3 = via rg [20, 15];
(13)     cg.chanr #(green) p3;

(14)     p4 = via bg [30, 55];
(15)     pu = load dep [30, 25] (1:4);
(16)     pd = tran enh [30, 15] (2:1);
(17)     p5 = via bg [30, 5];
(18)     p4 #(green) pu.drain;
(19)     pu.green #(green) pd.chant;
(20)     pd.chanb #(green) p5;
(21)     p3 #(red) pd.gate;
(22)     vddl #(blue) p4;
(23)     grndl #(blue) p5;

(24)     p6 = point green [40, 25];
(25)     p7 = point green [40, 15];
(26)     p6 #(green) p7;
(27)     pu.green #(green) p6;

(28)     p4 #(blue) vddr;
(29)     p2 #(blue) phi2r;
(30)     p1 #(blue) phi1r;
(31)     p7 #(green) output;
(32)     p5 #(blue) grndr;
```

**Fig. 7.23.** Design of inverter cell in LAVA.

```
cell shiftcell (vddl, vddr, phi1l, phi1r, phi2l, phi2r,
    input, output, grndl, grndr; phase)
        vddl: blue: left;
        vddr: blue: right;
        phi1l: blue: left;
        phi1r: blue: right;
        phi2l: blue: left;
        phi2r: blue: right;
        input: green: left;
        output: green: right;
        grndl: blue: left;
        grndr: blue: right;
        {
            statements of Fig. 7.23
        }
```

**Fig. 7.24.** Cell declaration.

these declarations comes the body of the cell, surrounded by curly braces.

**Example 7.13:** The cell that we started to design in Example 7.12 can be specified completely by the additional statements shown in Fig. 7.24. Those statements declare the ten ports on the left and right of the cell, of which all but *input* and *output* are blue. □

## Calling a Cell

Cells, once declared, can be used by calling them in the definition of other cells. The syntax of a call is

<cellname> (<actual parameter list>) [<origin>] (<size>)

Here, <cellname> is the name of the cell, of course. The <actual parameter list> is a list of values to be given to the ports and control parameters of the cell. The actual parameter corresponding to a port is a variable name, which becomes a point name of the calling cell. The actual parameter corresponding to a control parameter is an expression, such as a constant or a formula involving control parameters of the calling cell.

The <origin> is a pair of integers giving the $x$ and $y$ coordinates of the origin of the called cell in the abstract grid for the calling cell. The <size> is another pair of integers giving the length in the $x$ and $y$ directions, respectively, of the rectangle occupied by the called cell. These dimensions are in the units of the abstract grid for the calling cell.

Specifying the dimensions of the cell in the abstract grid of the calling cell gives us some useful information about the coordinates of ports. For example, the $x$ coordinate of the origin plus the width of the cell gives the number of the vertical grid line on which the ports on the right border lie. The horizontal grid lines for these ports will be specified when we abut them to ports of another cell or wire them to another point.

**Example 7.14:** We can place a copy of the cell of Fig. 7.24, in its version where $\phi_1$ controls the input, with its origin at the point (100, 200), and with the assumed height and width of the placed cell equal to 10 and 20, respectively, by

> shiftcell (vleft, vright, p1left, p1right, p2left, p2right,
>     in, out, gleft, gright; 1) [100, 200] (20, 10);

Note that the control parameter *phase* has been given the value 1, and new names have been used for all of the port parameters of the called cell. Also note that the grid scale for the called cell does not influence the size of the cell when placed in the abstract grid of the calling cell. □

### Abutment of Cells

Frequently, we want the ports of two cells to touch. For example, if we make a shift register out of the cell of Example 7.12, we shall want to create an array of such cells and have the ports on the right border of one line up with the corresponding ports on the left border of its neighbor to the right. For this purpose we introduce another modifier for the # operator, in which we specify the horizontal or vertical alignment of two ports at a designated grid line. The syntax for this operator is

> <point1> #(<axis>[<coordinate>]) <point2>

The <axis> is either "|", representing the abutment of points on the top and bottom borders of two cells at a particular vertical grid line, or "—", representing the abutment of two points on the left and right borders at a designated horizontal grid line. In either case, the <coordinate> gives the designated grid line. Thus,

> q1 #(—[10]) q2

causes points q1 and q2 to become the same point and to live on horizontal grid line 10.

### Iteration

A powerful operator provided by the LAVA system is an "iterate" command that causes a cell to be repeated some number of times. The syntax is

> iterate <count> <direction> <cell call>
>     <body>

cell twophaseshift (vddl, vddr, phi1l, phi1r, phi2l, phi2r,
   input, output, grndl, grndr)
     vddl: blue: left;
     vddr: blue: right;
     phi1l: blue: left;
     phi1r: blue: right;
     phi2l: blue: left;
     phi2r: blue: right;
     input: green: left;
     output: green: right;
     grndl: blue: left;
     grndr: blue: right;
     {
        shiftcell(avl, avr, ap1l, ap1r, ap2l, ap2r, ain, aout, agl, agr; 1)
         [0, 0] (10, 10);
        shiftcell(bvl, bvr, bp1l, bp1r, bp2l, bp2r, bin, bout, bgl, bgr; 2)
         [10, 0] (10, 10);
        vddl $\#(-[9])$ avl; avr $\#(-[9])$ bvl; bvr $\#(-[9])$ vddr;
        phi2l $\#(-[7])$ ap2l; ap2r $\#(-[7])$ bp2l; bp2r $\#(-[7])$ phi2r;
        phi1l $\#(-[5])$ ap1l; ap1r $\#(-[5])$ bp1l; bp1r $\#(-[5])$ phi1r;
        input $\#(-[3])$ ain; aout $\#(-[3])$ bin; bout $\#(-[3])$ output;
        grndl $\#(-[1])$ agl; agr $\#(-[1])$ bgl; bgr $\#(-[1])$ grndr;
     }

**Fig. 7.25.** Two phase shift cell in LAVA.

The $<$count$>$ is the number of copies of the cell we desire. The $<$direction$>$ is up, down, left, or right, and tells us the direction in which the array of copies of the cell extends. The $<$cell call$>$ is a call with the information required in calls to cells; it includes the name of the cell, parameters for the call, a position for the origin, and dimensions for the cell in the grid of the calling cell. The origin is the origin for the first copy only. The position of successive copies is determined by the $<$direction$>$ and the size of the cell. For example, if the direction is *down*, the value of the origin in the $<$cell call$>$ is [0, 100], and the cell size is (20, 10), then the origin of the copies all have $x$ coordinate 0, and they have $y$ coordinates 100, 90, 80, ... .

The $<$body$>$ of the iteration command is a sequence of steps to be performed after laying down the second and subsequent copies of the cell. For example, commands in the body of a downward iteration would connect the ports on the bottom of one copy to the ports on the top of the next copy, using the $\#(-[i])$ operator.

**Example 7.15:** Let us use cell calls and iteration to design a cell that is a complete shift register for $n$ bits, where $n$ is a control parameter of the cell. We first need a cell that is two copies of the cell *shiftcell* of Fig. 7.24; this cell shifts a bit through two inverters at the two clock phases. The specification of this cell, called *twophaseshift*, is given in Fig. 7.25. Note that we have elected to use the same parameter names for this new cell as we did for *shiftcell*, and in calls to *shiftcell* we invent new actual parameter names; these names begin with $a$ for the first copy and $b$ for the second copy. Also, observe that the value of *phase* in the first call to *shiftcell* is 1, and in the second call it is 2. Note that *twophaseshift* has no control parameter analogous to *phase*.

Following the call to two copies of *shiftcell*, there are fifteen statements, spread over the last five lines of Fig. 7.25; the effect of these is to abut the ports that should be at the same points. Since we have declared the two copies of *shiftcell* to be 10 units high, we may use 1, 3, 5, 7, and 9 as the horizontal grid lines for the five wires that run through the cells. For example, the next-to-last line in Fig. 7.25 says that the port *input* of *twophaseshift* is to be the same point as *ain*, and the point is on horizontal grid line 3. We also know it is on vertical grid line 0, since the first call to *shiftcell* places that cell with its left border at $x = 0$.

On that same line we are also told that the output of the first copy of *shiftcell*, which we called *aout*, abuts the input *bin* of the second copy, and these points are also on horizontal grid line 3. Their vertical grid line is 10, since that is the width of the cell in the first call. The last statement on the line says that point *bout* of the second call is the same as port *output* of *twophaseshift*; the coordinates of that point are (20, 3).

Two copies of *twophaseshift* form the two-bit shift register of Fig. 7.22. We shall design a cell *shiftreg* that takes parameter $n$ and generates an $n$-bit shift register for any $n \geq 1$. As with the previous cell, we shall use the parameter names from Fig. 7.24 for the parameters of *shiftreg*, and we invent new names, beginning with $c$, for the actual parameters of *twophaseshift*. We do not run $VDD$, ground, or the clock signals through to the right border of *shiftreg*, although the output is taken at the right edge.† There is one control parameter, $n$ for *shiftreg*. The specification of this cell, using the iterate command, is in Fig. 7.26. □

## 7.7 PLA'S AND THEIR PERSONALITIES

A PLA (programmable logic array) is a compact structure for implementing *two level logic*, that is Boolean expressions that are sums (logical "or") of products (logical "and") of literals; *literals* are variables or their complements. An example of how a PLA works is shown in Fig. 7.27. There are two groups of

---

† This decision is taken only for variety. It costs us nothing to invent port names for these wires on the right, and it might be useful to have them, so they could feed some other cell, or even so these signals could be taken for *shiftreg* from the right.

```
cell shiftreg (vddl, phi1l, phi2l, input, output, grndl; n)
    vddl: blue: left;
    phi1l: blue: left;
    phi2l: blue: left;
    input: green: left;
    output: green: right;
    grndl: blue: left;
    {
        iterate n right twophaseshift (cvl, cvr, cp1l, cp1r, cp2l, cp2r,
            cin, cout, cgl, cgr) [0, 0] (10, 10)
        {
            /* connect successive copies of twophaseshift */
            cvr  #(−[9]) cvl;
            cp2r #(−[7]) cp2l;
            cp1r #(−[5]) cp1l;
            cout #(−[3]) cin;
            cgr  #(−[1]) cgl;
        }
        /* now connect first and last copies to outside.
            note parameters in copy i can be referred to by [i] */
        vddl  #(−[9]) cvl[1];
        phi2l #(−[7]) cp2l[1];
        phi1l #(−[5]) cp1l[1];
        input #(−[3]) cin[1];
        output #(−[3]) cout[n];
        grndl #(−[1]) cgl[1];
    }
```

**Fig. 7.26.** $n$-bit shift register cell in LAVA.

wires running vertically; the *and-plane* is the group on the left and the *or-plane* appears on the right. The terms "and-plane" and "or-plane" come from the fact that the former implements the product terms and the latter the sums.

Running horizontally across the and-plane are *term wires*, two of which we have shown in Fig. 7.27 and named $A$ and $B$. In the normal NMOS implementation of a PLA, the term wires start out as metal, with a pullup at the left end. They change to polysilicon when they cross into the or-plane. The vertical wires in the and-plane are polysilicon, and between each pair of vertical wires runs one diffusion ground wire†.

---

† The normal prohibition against running power or ground in anything but metal can be violated here, because each diffusion wire carries only a small amount of current for a short distance.

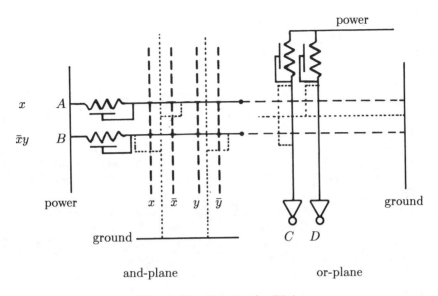

**Fig. 7.27.** Sketch of a PLA.

At certain places, a diffusion *tap* is connected to a term wire; it runs under a crossing polysilicon wire, and connects to the adjacent ground wire. The effect is that a transistor is formed, and if the vertical polysilicon wire is on (logical 1), any current that passes through the pullup from $VDD$ will run to ground through the tap. The only way the term wire can be high is if all poly wires crossing where there are taps are low. Thus, in the and-plane, each term wire implements the nor function; it is on if and only if none of a selected set of vertical wires is on.

For example, in Fig. 7.27, wire $A$ is high if and only if the vertical wire labeled $\bar{x}$ is low, since there is a tap under that wire only. Thus $A$ implements the term $x$. Similarly, the $B$ term wire is high if and only if the $x$ wire and the $\bar{y}$ wires are both low, i.e., $B$ implements the term $\bar{x}y$.

In the or-plane, the role of rows and columns is reversed, with the term wires becoming poly and the vertical wires running in metal. Each vertical wire in the or-plane begins at the top with a pullup, and at the bottom it is the input to an inverter. Diffusion taps at various places along each vertical wire will short that wire to ground if the term wire crossing at that point is high. Thus, the input to the inverter is high if and only if none of the term wires crossing taps are high, and, therefore, the output of the inverter will be high if one or more of these term wires are high. That is, each vertical wire in the or-plane implements the sum (logical or) of some subset of the terms represented by the term wires—those that cross the vertical wire where there are taps.

For example, the vertical wire $C$ implements $x + \bar{x}y$, because both $A$ and $B$ cross at taps. On the other hand, $D$ implements the function $x$, since only $A$ crosses it where there is a tap.

We should now see the general rule for implementing a collection of sums of terms as a PLA.

1.  For each sum, make one vertical wire in the or-plane.
2.  For each term appearing in one or more sums, make one term wire running horizontally across the planes.
3.  For each literal used in one or more terms, we require a vertical wire in the and-plane. However, the wire needed for an uncomplemented literal $x$ actually carries signal $\bar{x}$, and the wire for a complemented literal $\bar{x}$ carries signal $x$.
4.  Each term is the product of certain literals. For each of those literals, tap the term's wire at the points in the and-plane where the wires for each of those literals (i.e., wires carrying signals that are the complements of those literals) cross the term's wire.
5.  For each sum of terms, tap the vertical wire for that sum at the points in the or-plane where the wires for the terms in the sum cross.

**Example 7.16:** Let us design a PLA for the bounce filter introduced in Example 7.11. First, let us note that the circuit described in Fig. 7.20 in the lgen language is sequential; that is, certain values computed as outputs are fed back to become inputs at the next clock cycle. As a consequence, we shall need to introduce clocking into the PLA. The normal way to do so is to have one phase, say $\phi_2$, control inputs to the inverters that are attached to the bottoms of the vertical wires in the or-plane, and have the other phase control inputs to the vertical wires in the and-plane.

One input to the and-plane is the variable *current*; that signal represents the current input to the bounce filter. The other inputs are the variables *oldout*, *previous*, and *nexttoprev*, whose values are, respectively the values of *output*, *current*, and *previous* from the last clock cycle, as indicated by the LAST operators in Fig. 7.20. For succinctness, and to help us distinguish the names of vertical wires in the and-plane from vertical wires in the or-plane, we shall abbreviate *current*, *oldout*, *previous*, and *nexttoprev* by $c_{in}$, $o_{in}$, $p_{in}$, and $n_{in}$, respectively.

The output variables in Fig. 7.20 are those that are either computed and present on a border (*output* is the only such example) or are fed back by being the argument of a LAST operator. In the latter category are *output*, *current*, and *previous*, whose values must, therefore, be computed in the or-plane. We shall call the wires computing these $o_{out}$, $c_{out}$, and $p_{out}$, respectively; these are the vertical wires in the or-plane.

We must write formulas for each of the three output variables in sum of products form. The formulas for two of these are quite easy:

$$c_{out} = c_{in}$$
$$p_{out} = p_{in}$$

The third requires a small amount of work, since we must expand some terms in the logic equation on the last line of Fig. 7.20 to get the desired sum-of-products form

$$o_{out} = o_{in}c_{in} + o_{in}p_{in} + o_{in}n_{in} + c_{in}p_{in}n_{in}$$

Finally, the rules for feeding back wires from the or-plane to the and-plane are dictated by the LAST statements in Fig. 7.20. The feedback wiring is

$$o_{out} \rightarrow o_{in}$$
$$c_{out} \rightarrow p_{in}$$
$$p_{out} \rightarrow n_{in}$$

In Fig. 7.28 we see the PLA that implements these equations. Taps are represented by circles at the point of intersection of two wires. Note that each of the four input wires appears in the and-plane after going through an inverter, because they appear only uncomplemented in terms. If some or all of the input variables had appeared complemented in terms, we would have to feed back the uncomplemented version of those variables as well, so there would be two wires for some inputs, as for $x$ and $y$ in Fig. 7.27. □

Incidentally, there is an easy rule for reading diagrams of the taps in a PLA such as Fig. 7.28. Assume each horizontal wire starts out "on" at the left end. The "on" signal tries to propagate to the right, but each time it meets a tap in the and-plane, it can pass only if the vertical wire at that tap is on. If the signal reaches the or-plane, then it turns on all the vertical wires that it meets at a tap.

Let us also remark on the scale represented by the grid of Fig. 7.28. If we pack wires and taps very tightly, we need only a square $7\lambda$ on a side to represent one grid square in Fig. 7.28. We must arrange that the blue-green vias attaching taps are shared by two adjacent squares, and as we mentioned, two red wires share one green ground wire. As a result, if all taps are present, each square is the mirror image of its horizontal and vertical neighbors. In Fig. 7.29 we see six such squares, representing the intersection of three adjacent vertical lines and two adjacent horizontal lines in the and-plane. They cover an area $14\lambda$ high and $21\lambda$ wide. Of course, in practice, some or all of the taps may be missing.

It is not necessarily desirable that the squares of the PLA grid be implemented with the most compact possible design, because to do so will force the elements on the periphery of the PLA, the pullups, inverters, and perhaps clock gates, to be designed in a long, skinny fashion, possibly increasing the total area of the PLA. Alternative cell designs for PLA's are found in Mead and Conway [1980] and Newkirk and Mathews [1983].

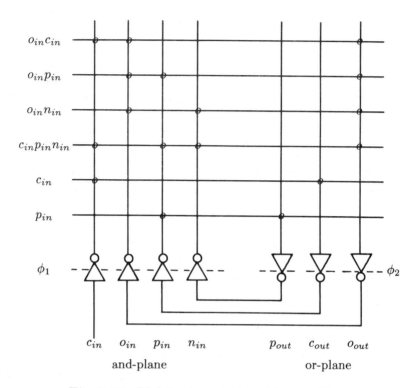

**Fig. 7.28.** PLA implementation of bounce filter.

**The PLA Personality**

If we examine Figs. 7.27 and 7.28, it should be clear that we can summarize all
we need to know about a given PLA by listing

1. Names for each of the output wires, i.e., the vertical lines in the or-plane,
   in the order of their occurrence.

2. Names for the input variables, i.e., the vertical lines in the and-plane, in
   the order of their occurrence. If the circuit is sequential, we might also
   want to know the correspondence of variables fed back, but this information is generally considered to be outside the domain of PLA description
   languages, so all PLA's appear to be implementing purely combinational
   logic.

3. The number of term wires, and the positions of the taps on each.

   Various syntaxes have been used to specify wire names, and we shall not
fix on one. It is useful for the PLA description to mesh with some other design
language, and whatever notation for ports is used in that design language should
be used to specify these wire names, and perhaps their positions on the lower
border of a cell that is the PLA. For example, we could use CHISEL or LAVA
notation to specify the input and output wire names and their relative or exact

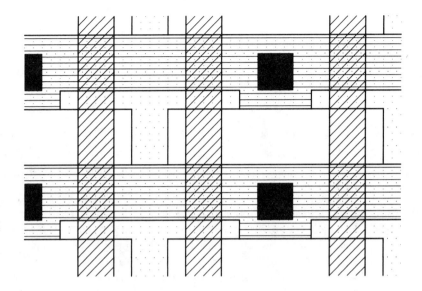

**Fig. 7.29.** Six squares of a compact PLA design.

positions.

The essential information about the PLA is the tap positions. A matrix notation for this information has become known as a *PLA personality*. In the or-plane, we use 1 for a tap and 0 for no tap. In the and-plane, a character in the column for an input variable $x$ has the following meanings.

1.  A 0 means there is a tap under the vertical wire for $\bar{x}$. Note that this wire is the one that feeds signal $x$ uncomplemented to the and-plane.

2.  A 1 means there is a tap under the vertical wire for $x$, that wire being the one where $x$ is complemented before being run through the and-plane.

3.  A 2 means that neither wire for $x$ is tapped for the term line at hand.

Note that we do not allow both the $x$ and $\bar{x}$ wires to have taps in the same row, since this would surely cause the value for that term to be low.

Obviously, other codes for the same conditions may also be used. Note, however, that this scheme is preferable to one in which we distinguish columns for an input wire $x$ and one for $\bar{x}$, since one of the functions of a PLA generator might be to decide from the given personality which input wires need to be fed in both true and complemented versions, and to save space for those inputs that are needed in only one polarity.

We shall exhibit the personality as two matrices, side by side. The one on the left is for the and-plane, and the other for the or-plane. In each matrix the $i^{th}$ row corresponds to the $i^{th}$ term.

**Example 7.17:** The PLA of Fig. 7.27, if that were the entire PLA, could be

$$
\begin{array}{ll}
1122 & 001 \\
2112 & 001 \\
2121 & 001 \\
1211 & 001 \\
1222 & 010 \\
2212 & 100
\end{array}
$$

**Fig. 7.30.** PLA personality for bounce filter.

represented by the matrices

$$
\begin{array}{ll}
12 & 11 \\
01 & 10
\end{array}
$$

Note, for example, that the wire labeled $A$ in Fig. 7.27 is tapped by the vertical wire labeled $\bar{x}$, which is the wire that represents $x$ uncomplemented when it appears as a factor in terms, so the character in the upper left is 1, not 0.

The personality for the PLA in Fig. 7.28 is shown in Fig. 7.30. Note that there are no 0's in the and-plane matrix, because no inputs are used complemented in the terms. □

## 7.8 FINITE AUTOMATON LANGUAGES: SLIM

We shall discuss several languages for specifying finite state, or sequential, machines.† The first is SLIM, a language developed by J. Hennessy for specifying deterministic finite automata and translating them into a variety of types of circuits, principally circuits the heart of which is a PLA. In the next section we shall mention a system developed by H. Trickey and the author for specifying sequential machines by regular expressions. This system also uses a language based on nondeterministic finite automata, and we shall discuss that language briefly, as well.

### PLA Implementation of Finite Automata

Before illustrating the languages for defining automata, let us mention two different approaches to implementing automata as PLA's. The first is a straight-forward approach, where we select a binary code for states, with the intent of minimizing the number of term wires. That is, we wish to minimize the number of different terms found among all the sum-of-products Boolean expressions used to describe the outputs and the bits of the next state in terms of the inputs and the bits of the current state. This minimization problem is $\mathcal{NP}$-complete, but in Chapter 8 we shall discuss some heuristics that have been used to come up

---

† We shall use the term "finite automaton" in this section for what is variously called a "sequential machine," "state machine," "finite state machine," or "deterministic finite state machine." A formal definition of the notion can be found in Hopcroft and Ullman [1979].

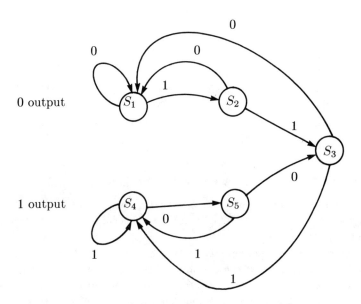

0 output

1 output

**Fig. 7.31.** Finite automaton for the bounce filter.

with small sets of terms. While this approach is a very conventional one, we shall work out one example, the "bounce filter" from Example 7.11.

**Example 7.18:** A finite state machine for the bounce filter is shown in Fig. 7.31. Transitions from states $S_1$ and $S_2$ give output 0, while transitions from $S_4$ and $S_5$ give output 1. Transitions from $S_3$ give the same output as the input. State $S_1$ is the initial state, in which the automaton starts before any input is made. There is a natural interpretation for each state. We are in state $S_1$ whenever 0 is the correct output, and we can tolerate two consecutive 1's without changing our mind about the output. In $S_2$, we can tolerate only one 1 before changing our mind. States $S_4$ and $S_5$ are analogous, but the output is 1 and we can tolerate some 0's before changing our mind. In $S_3$ we are forced to make a decision based on the next input, because we either think the output should be 1, but we have seen two 0's, or we think the output should be 0 and we have seen two 1's.

Let us represent the state by three bits, $x_1x_2x_3$, with the obvious correspondence that state $S_i$ is represented by $i$ in binary; so, for example, we are in $S_4$ if $x_1$ is 1, while $x_2$ and $x_3$ are 0. In general, the PLA that implements the output and next state functions will require one term per transition of the automaton. However, depending on the particular code chosen for the states, several transitions may be expressed by one term. The code we have chosen is actually quite good in this regard. The three transitions into $S_1$ on input 0 can be expressed by one term. That is, if the input is 0, and the bit $x_1 = 0$, then the

```
0  0  2  2          0  0  1  0
1  0  0  1          0  1  0  0
1  0  1  0          0  1  1  0
1  0  1  1          1  0  0  1
1  1  2  2          1  0  0  1
0  1  0  0          1  0  1  1
0  1  0  1          0  1  1  1

in x₁ x₂ x₃        x₁ x₂ x₃ out
```

*in* $x_1$ $x_2$ $x_3$              $x_1$ $x_2$ $x_3$ *out*

**Fig. 7.32.** PLA personality for finite automaton.

output is 0 and the next state is $S_1$. In that case we want to set $x_1 = x_2 = 0$ and $x_3 = 1$. There is a similar coincidence concerning the transitions into $S_4$ on input 1 from $S_4$ and $S_5$.

We show the PLA personality for the bounce filter derived in this way in Fig. 7.32. We have labeled the columns, with $x_1$, $x_2$, and $x_3$ in the and-plane representing the current state, and the same names in the or-plane represent the next state. Note that we have kept the same order of the state bits in both planes. This choice was made so state codes could be read easily. In practice, we would reverse the order of the state bits in the or-plane, so they could be fed back to the and-plane, as in Fig. 7.28.

For example, the first row says that if the input is 0 and the first state bit is 0 (i.e., we are in $S_1$, $S_2$, or $S_3$), then emit 0 output, and set the state to $S_1$, or 001 in binary. The second line represents the transition from $S_1$ to $S_2$ on input 1. it says that with input 1, in state 001, make output 0 and set the state to $S_2$, or 010. □

## Microcode Implementation of PLA's

A somewhat different way to implement a finite automaton with a PLA is to fix an order for the states, and let the default transition be from a state to the next state in order. This is the *microcode* viewpoint, where each state may be thought of as a line of microcode; each line may test for certain conditions and jump if they are met. If no jump is called for, then the next state in the selected ordering is executed.

The code for the current state is maintained by a counter that is not part of the PLA, and in place of the state-out bits in the implementation just described, the PLA produces

1.  A bit telling whether or not there is a jump. We shall refer to the bit that tells whether or not we must change to a nondefault state as *jump*.

2.  If *jump* is 1, then a sequence of bits gives the code for the new next state. The counter adds 1 to the state number if *jump* = 0, and copies the new next

| 0 | 0 | 2 | 2 |     | 1 | 0 | 0 | 1 | 0 |
|---|---|---|---|-----|---|---|---|---|---|
| 1 | 1 | 2 | 2 |     | 1 | 1 | 0 | 0 | 1 |
| 0 | 1 | 0 | 1 |     | 1 | 0 | 1 | 1 | 1 |
| 1 | 0 | 1 | 1 |     | 0 | 0 | 0 | 0 | 1 |
| 2 | 1 | 2 | 2 |     | 0 | 0 | 0 | 0 | 1 |

$in \ x_1 \ x_2 \ x_3$        $jump \ y_1 \ y_2 \ y_3 \ out$

**Fig. 7.33.** PLA for microcoded version of Fig. 7.31.

state if $jump = 1$. The PLA inputs include bits that indicate the current value of the counter.

**Example 7.19:** For the bounce filter automaton of Fig. 7.31, the natural order of states to take is $S_1, \ldots, S_5$. Then, of the ten transitions, four are to the next state in order, and the remaining six require a jump. However, of these six, three are jumps to $S_1$ on 0 and another two are jumps to $S_4$ on 1, and we can express these succinctly as we did in Example 7.18. Figure 7.33 gives one possible PLA to implement the microcoded version of the automaton. There, $x_1$, $x_2$, and $x_3$ are the inputs from the program counter, and $y_1$, $y_2$, and $y_3$ are the outputs to the counter. The variable $jump$ indicates whether a jump is to be made by the counter, and $in$ and $out$ are the input and output of the bounce filter.

In Fig. 7.33, the first row says that when the input is 0 and the state code begins with 0, i.e., the state is $S_1$, $S_2$, or $S_3$, we make a jump, and the jump is to state 001, i.e., $S_1$. The second row similarly says that if the input is 1 and the state is $S_4$ or $S_5$, we make a jump to state 100, i.e., $S_4$; in this case we also make the output 1. Row three says that in state $S_5$ with input 0, we jump to state $S_3$ and make the output 1. These three rows together express all the six transitions that are jumps to a state out of sequence.

We must also express conditions, other than those of rows two and three, under which the output is 1. Row four covers the transition from $S_3$ to $S_4$ on input 1 that causes a 1 output, while row five says that the output is 1 whenever the state is $S_4$ or $S_5$, regardless of input. This row captures some of the cases of a 1 output that are also covered by rows two and three. □

Note that the number of entries in Fig. 7.33 is somewhat less than in Fig. 7.32, which indicates that the former PLA is smaller than the latter. The difference is mitigated somewhat by the fact that the PLA of Fig. 7.33 needs the counter, with circuitry to add 1 and to load new values in response to jumps. However, in practical situations, it appears that a microcoded approach requires less total area than a PLA implementing all state transitions directly.

## SLIM

SLIM (Stanford Language for Implementing Microcode) is a language based on Pascal for specifying finite automata, and translating them to logic. Its output can be of several forms, such as those suggested in Examples 7.18 and 7.19, or a microcoded ROM (read-only memory). In its most rudimentary form, a SLIM program consists of:

1. A list of input and output variables.
2. A set of functions that return the value of an input variable and thus represent the elementary input conditions that assert a particular wire is on or assert that it is off.
3. A set of procedures that each generate one or more output signals. These are called by the various states when the input conditions warrant them. The syntax for the input and output lists is

> **inputs** <variable list> : <attribute list>
> **outputs** <variable list> : <attribute list>

4. The keyword **fsm** followed by a list of states and their actions.

The variables on the <variable list> for input and output are each either single variable names, or vectors. A vector variable is indicated by

> <name>[<range>]

For example, $zap[1..8]$ denotes eight input bits named $zap[1], \ldots, zap[8]$. If $zap$ is an output variable, it can be assigned bit by bit, or by assigning an integer to $zap$ itself, so $zap := 3$ sets $zap[1], \ldots, zap[6]$ to 0 and $zap[7] = zap[8] = 1$.

There are various attributes that may appear in the <attribute list>, some of which involve features of SLIM we shall not cover. For the moment, the only attributes we shall mention are "bottom" and "top", which refer to the side of the PLA on which the input or output appears. That is, in the previous section we tacitly assumed that all inputs and outputs appeared at the bottom. In reality, the wires could run from either top or bottom, and different wires could run from different directions. We shall even discuss in Chapter 8 ways to make wires from both directions appear to occupy the same column.

The declaration of input functions has a very powerful syntax. Here we shall consider only the most rudimentary of such functions. The simplest form is

> **function** <name>: Boolean;
> **definition** <literal>;

The <literal> is an input variable or its negation, i.e., an input variable preceded by "not". Thus, if $i$ is an input variable, we might declare

> **function** $in$: Boolean;
> **definition** $i$;

However, much more complicated situations are possible, including functions that have parameters and change their meaning according to the value of those parameters. SLIM even allows the names of the variables in the **definition** part to be computed as a function of the parameters, using a string concatenation operator.

Further, the function can have a body following the definition part, including statements that simulate the action of the circuit being specified, so we can test the circuit before implementing it. Of course, that test is correct only if the code to do the simulation matches the action taken by the finite automaton being defined. However, since each function and procedure defined represents a very small step, the reliability with which we can write such simulation code is high.

In their most rudimentary form, the output procedures are similar. They look like

> **procedure** <name>;
> **definition** <list of outputs>;

The <list of outputs> is a sequence of terms separated by the word "and". Each term can be an output variable, or it could be an assignment of an integer to an array-valued output variable, so

> **definition** $x$ and $zap = 3$;

causes the output signal $x$ to become 1 and also causes the last two wires representing the vector $zap$ to become 1.

## Specification of States

As we mentioned, the keyword **fsm** introduces a list of the states of the finite state machine being defined. The declaration of a state is of the form

> <name>: [ <list of clauses> ]

where <name> is the name of the state. The clauses in the list are separated by semicolons and are of the following form:

> **if** <condition> => <action>

followed by an optional

> **else** <action>

The actions are any of
1. **next** <statename>, meaning that if the condition is met, go next to state <statename>.
2. A call to an output procedure.
3. [ <action1>; <action2>; $\cdots$ ; <actionn> ], where each <actioni> is an action of one of the forms above; presumably only one will be a

**program** bounce;

**inputs** *in*: bottom;
**outputs** *out*: bottom;

**function** intest: Boolean;
    **definition** *in*;
**procedure** output;
    **definition** *out*;

**fsm**
    $S1$: [ **if not** intest $=>$ next $S1$ ];
    $S2$: [ **if not** intest $=>$ next $S1$ ];
    $S3$: [
            **if not** intest $=>$ next $S1$;
            **else** output
      ];
    $S4$: [
            output;
            **if** intest $=>$ next $S4$
      ];
    $S5$: [
            output;
            **if** intest $=>$ next $S4$;
            **else** next $S3$
    ]

**Fig. 7.34.** SLIM program for bounce filter.

next state action and the others will be output actions. The outputs are executed simultaneously, and their order does not matter.

The $<$condition$>$ is a logical expression whose operands are function calls. Each function call is true or false, depending on the input variable that it refers to, so we can evaluate the condition. If the condition is true, then we execute the associated output procedures and next state transitions. If no next state transition is triggered by the truth of some condition for the state, then the next state is the one following the current state in the order that states are listed.

**Example 7.20:** Let us sketch the bounce filter example in SLIM. We should emphasize that there are normally many other components to a SLIM program, for example, those concerned with simulation of the circuit being specified. However, a sketch of the relevant portion of the code is shown in Fig. 7.34.

The keyword **program** introduces the code, which is ended by a final dot.

We see the declaration of one input variable *in* and one output variable *out*. Their values are embodied in the function intest, which returns the value of *in* and the procedure output, which sets the value of *out* to 1.

The action for state $S1$ is to jump to state $S1$ if the condition **not** intest is true, i.e., *in* is 0. If $in = 1$, the default next state $S2$ is taken. The action for state $S2$ is analogous; neither $S1$ nor $S2$ makes the output 1 under any circumstance. In $S3$, we jump to $S1$ if $in = 0$, and set $out = 1$ if $in = 1$. In the latter case, we take the default next state, $S4$, because no next state action is executed. In states $S4$ and $S5$ we make $out = 1$ regardless of the input, and we jump back to $S4$ whenever $in = 1$. If we are in $S5$ we must jump to $S3$ if $in = 0$, and to $S4$ if $in = 1$. □

### Additional Features of SLIM

There are a number of interesting features of SLIM that we shall only allude to; the bibliographic notes mention the places where more details can be found. We shall simply enumerate some of these ideas, and then discuss "pipelining," a useful and original idea found in SLIM.

1.  Simulation code. What we showed in Fig. 7.34 was only those parts of the language that serve as data; this data is fed to a routine that produces a PLA, a microcoded PLA, or other implementation of a finite automaton, as desired by the user. In reality, the functions and procedures that define input and output also include bodies that are executable Pascal code. For example, input functions include code to return true if and only if the input condition it is intended to detect is met. Thus, the function intest from Fig. 7.34 might be completed with

    **begin** intest := $(in = 1)$ **end**

    The SLIM program itself can include other data type definitions, procedures to be executed at each time unit (e.g., to read input data), and auxiliary variables, so a complete simulation of the automaton specified can be carried out by compiling and running the Pascal portions of the SLIM program. The treatment of SLIM programs as both data and (Pascal) program is illustrated in Fig. 7.35.

2.  To help check that the automaton under design is correct, we may embed within the executable portions of SLIM programs assertions about relationships among variables that we expect to hold. These are checked when the simulation program is run.

3.  In addition to state transition actions indicated by the keyword "next", control can flow by call and return actions. A state $s$ can call a state $t$, which passes control to $t$ as if "next" had been used. However, at some later time, the "return" action will be encountered, whereupon control will return to the state following $s$. Only one level of call is allowed, unless an

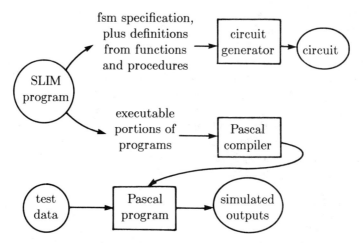

**Fig. 7.35.** SLIM program as data and executable code.

external stack, not provided by SLIM, is used.

## Pipelining

There are situations, such as in the design of the controller for a microprocessor or similar device, when it is convenient for an input to trigger a sequence of events. For example, if the controller discovers that an add operation must be performed, we wish to generate a signal to clear a bus. Then, at the next clock cycle we wish to load the bus from a designated register, perform the addition at the cycle after that, and so on. We may even wish that some of these events, especially clearing (*enabling*) the bus, had already occurred.

The obvious way to handle a sequence of actions is for the output signals that cause them to be associated with the states that follow (in the actual execution) that state at which the need to perform an addition was first discovered. SLIM provides a different and much more convenient way, which also encourages the use of parallelism in designs. All these output signals are defined to be part of the output procedure called when the need to add is discovered. Then, in the **outputs** list, we associate with certain of these output variables the attribute earlier($i$) or later($i$).

If output variable $x$ is declared later($i$), then the output is made $i$ clock cycles later than it would be if it did not have that declaration.

**Example 7.21:** Suppose that *out* in Fig. 7.34 were declared

> **outputs** *out*: bottom, later(2);

Also, suppose we are in state $S_3$, and intest is true, i.e., $in = 1$. Then *out* will not become 1 immediately, but rather after:

1.  We first jump to $S_4$, as required because intest is true, then
2.  We either stay in $S_4$ or go to $S_5$, depending on the next value of $in$, and finally,
3.  In one of those states, we read the next value of $in$.

At that time, the value of $out$ becomes 1, as if it were required by this most recent input and the current state. $\square$

Note that the attribute earlier($i$) associated with output variable $x$ causes that output to be made in some state $S$ if the future inputs might possibly lead from $S$, in $i$ clock cycles, to a state in which $x$ is mentioned in an output action. This feature of SLIM is quite handy for controller design, as the next example will show. It does, however, lead to many opportunities for errors, since the output action many not be appropriate in all predecessors of $S$ from $i$ cycles ago. SLIM does possess additional debugging aids, which we shall not discuss, to help avoid the problems introduced by careless use of the pipelining feature.

**Example 7.22:** We shall exhibit the use of pipelining in the design of a microprocessor. Let us suppose that this microprocessor allows certain of its instructions to apply binary operations, in particular addition, to the contents of two registers, and to store the result in a third register. We also suppose that there are three busses, each connected to all the registers and to an adder unit. Some of the instructions are of the form

$$<\text{binary op}>\ R_1\ R_2\ R_3$$

The bits of the $<$binary op$>$ are fed to the controller, which is a finite state machine specified in the SLIM language, while the groups of bits $R_1$, $R_2$, and $R_3$ are stored outside the controller when the instruction is decoded, and used to determine which of the various registers hold the operands and the result. For example, we shall want the controller to issue a command to load the contents of register $R_1$ onto bus 1, letting the actual bits $R_1$ determine which register it is that gets copied onto the bus.

In order to load a value onto a bus, at the previous cycle the bus must be enabled, which in effect means that its old value is erased and it is made ready to accept a new value. Thus, the sequence of events that the controller must cause, if the contents of registers $R_1$ and $R_2$ are to be added and their sum placed in register $R_3$, is:

1.  Busses 1 and 2 are enabled.
2.  $R_1$ is loaded onto bus 1, $R_2$ is loaded onto bus 2, and at the same time, bus 3 is enabled.
3.  The adder performs the addition of the values on busses 1 and 2, with the result going onto bus 3.
4.  Bus 3 is copied into $R_3$.

Observe it is not until step (2) that the controller needs to know that registers $R_1$ and $R_2$ are the ones involved, and it is not until step (3) that it needs to know that the operation is addition. Thus, we are safe in assuming it is between steps (1) and (2) that the instruction is actually made available to the controller. The actions of enabling busses 1 and 2 can be performed in advance of the controller's knowing that it must load $R_1$ and $R_2$ onto these busses.

The only catch is that whatever instruction is actually executed must not depend on these busses not being enabled; e.g., no instruction executed instead of the addition could use the values found on those busses. Further, we must be sure that when we enable the busses in anticipation of a possible add instruction, we are not destroying data that was needed for a previous instruction. In other words, if we are going to pipeline actions into the past, we had better be very careful we are not either interfering with something that took place in the past, or fouling up an alternative present whose existence can be known only after our actions in the past have been completed.

We shall use the following output signals to represent these actions.

1.  $en_i$ causes bus $i$ to be enabled.
2.  $load_i$ causes the register indicated by the group of bits $R_i$ to be loaded onto bus $i$, for $i = 1, 2$.
3.  $add$ causes the contents of busses 1 and 2 to be added and the result to appear on bus 3.
4.  $store_3$ stores the contents of bus 3 into the register indicated by the group of bits $R_3$.

The relevant portion of the SLIM code to cause these actions is shown in Fig. 7.36. Note that the pipelining of the output signals is part of their declaration, so they are delayed or advanced by the same amount no matter what output procedure uses them. A renaming feature of SLIM allows the same signal to be declared several times, with different pipelining or other attributes. Also, observe how the procedure do_the_add appears to call for all the signals to be raised at once, while in reality, because of their different pipelining attributes, they are actually raised according to the proper schedule, as described at the beginning of this example. □

## 7.9 A REGULAR EXPRESSION LANGUAGE

While describing sequential processes, such as controllers, by a finite automaton is quite effective, we should also be aware of an alternative approach, one where we specify patterns representing the events that cause the outputs. The language used is that of regular expressions;† it is useful for specifying certain types of patterns, although in principle it can be used for any sequential process, with varying degrees of succinctness.

An implementation of such a language has been done by A. Karlin, H.

---

† The reader not familiar with this notation should read Appendix 3.

**inputs** opcode[8]: bottom;
    { assuming eight bits for the operation }
**outputs**
    en1, en2: bottom, earlier(1);
    en3, load1, load2: bottom;
    add: bottom, later(1);
    store3: bottom, later(2);
**function** addop;
    **definition** opcode=43;
        { assumes 43 is the operation code for an addition }
**procedure** do_the_add;
    **definition** en1, en2, en3, load1, load2, add, store3;
**fsm**
    . . .
    { state 123 is assumed to be the state in which an operation
       is chosen }
$S_{123}$: **if** addop => do_the_add;
    { actions for other operations }
    . . .

**Fig. 7.36.** Use of pipelining in SLIM.

Trickey, and the author, in which we found that the size of PLA's we generated could be made comparable to, although usually somewhat greater than, the size of PLA's generated by hand or by a language like SLIM. The work on optimizing the output of the language is considerable, because the regular expression patterns do not directly indicate the states of a finite automaton recognizing the patterns; we shall discuss in Section 8.4 some of the compilation and optimization techniques used.

Whether the expressive power of the regular expression language is worth the added area, or whether the area penalty exists after improved optimization efforts, remains to be seen. We did find that to make the language usable in a reasonable number of situations, we had to add the capability of declaring states, which behave like states of a finite automaton.

### Declarations

A program in the regular expression language consists of a declaration portion followed by a semicolon and a single regular expression, which may involve the generation of many output signals and even involve certain state transition actions. The classes of symbols that may be declared are the following.

1.   line <list of names>. Each name in the list is declared to be the name

of an input wire, from which we shall build the definitions of abstract input symbols that are the operands of the regular expression. Names may represent single wires or collections of wires. A name $zap[i]$ declares $zap$ to be a collection of wires, named $zap[1], \ldots, zap[i]$.

2.  symbol <list of symbol declarations>. Each symbol declaration is of the form

    <symbol name> (<list of wires>)

    A wire in the list can be either a wire name $x$, or its negation, $-x$. In the former case, the wire $x$ must be on for the symbol to be recognized; in the latter case it must be off. Thus, we might represent an ascii character by seven wires and declare the abstract symbol $a$ to be true if and only if the character 'a' (octal 141) is seen, by:

    line c[7]
    symbol a(c[1], c[2], −c[3], −c[4], −c[5], −c[6], c[7])

    Of course, not every wire must be specified as being on or off for each symbol, so the symbols can represent single wires, their negation, binary codes for wires, or anything in between. As a consequence, more than one symbol may be seen on the input at a given time, in the sense that the wires defining both symbols are simultaneously on. Note that this situation is at variance with the common view that the input symbols for a regular expression or finite automaton are mutually exclusive.

3.  output <list of outputs>. Each name on the list of outputs is declared to be an output signal.

4.  state <list of states>. Each name on the list of states is declared to be a state.

5.  subexp <name> = <regular expression>. The <name> is declared to stand for the <regular expression>.

## Regular Expressions

The operands of regular expressions are the names declared to be symbols, outputs, states, or previously declared subexpressions. Subexpressions are expanded in the obvious way. There are two special operands, the dot, which is matched by any input, and #, which is never matched. The latter symbol is useful when we introduce explicit state transitions into the regular expression.

The operators are + (union), * (closure), and concatenation (represented by juxtaposition, just as in the usual regular expressions). The language has two additional postfix operators, ++, which stands for "one or more occurrence of," and ?, which stands for "zero or one occurrence of." That is, if $R$ is a regular expression, then $R++ = RR^*$, and $R? = R + \epsilon$.

We may regard control as residing in one or more operands of the expression at all times. If at one clock cycle, control resides at an input symbol $a$, and the input wires are set so that $a$ is recognized, then for the next clock cycle, all the "successors" of that operand $a$ receive control. The notion of successor in a regular expression is a technical one, and we defer the details of the subject to our discussion of compilers for regular expressions in Section 8.4. However, we shall give the intuitive idea momentarily.

If one of the successors is an output symbol, then that output signal is made at the same clock cycle at which the $a$ was recognized. Outputs are transparent as far as the successor relationship is concerned, so if a string of output symbols is preceded by input symbol $a$ and followed by $b$, then that $b$ is a successor of that $a$, as are all the output symbols in that string. If a successor is a state name, say $S$, then we look for the definition point for $S$, which is denoted by $S$ followed by a colon. For the next clock cycle, all the successors of the definition point of $S$ receive control.†

## Positions and Successors

Now let us describe the successor relationship in greater generality. To begin, we must identify each position in a regular expression, which we shall do by subscripting each operand by a unique subscript. Thus, the regular expression $((a + b)Ua) + +$, where $a$ and $b$ are input symbols and $U$ an output symbol, would be viewed thusly.

$$((a_1 + b_2)U_3a_4) + + \tag{7.1}$$

Suppose some string $c_1c_2\cdots c_n$, consisting of input symbols only, is in the set of strings defined by this expression. Then each $c_i$ must be an occurrence of the symbol at some position $j_i$ of the regular expression. Of course, $c_i$ must be the symbol at position $j_i$, but there is the additional requirement that the matching of symbols follow the meaning of the expression. Thus, if $c_1 = a$, the value of $j_1$ could only be 1, not 4, because the operand $a$ in position 4 could not be the first symbol in a string generated by that regular expression. For the details concerning this idea, the reader is referred to Hopcroft and Ullman [1979] or to the more formal treatment of the subject in Section 8.4.

We say position $j$ is a *successor* of position $\ell$ if there is some input string $c_1\cdots c_k$ such that for some $i$, $c_i$ is an instance of position $\ell$ and $c_{i+1}$ is an instance of position $j$. Position $j$ is *initial* if it is an instance of the first input symbol in some string generated by the expression. We generalize the above definitions to include the case where the symbol in position $j$ (but not $\ell$) is an output symbol or a state symbol. Position $j$ is then said to be initial, or a successor of $\ell$, if that would be the case had the operand at position $j$ been an input symbol.

---

† We can now see the use of the # symbol. If we precede $S$ : by #, then since # is never matched, we shall never "accidentally" pass control to $S$ : .

```
line input
symbol zero(−input), one(input)
output OUT
;
.* one one (one zero? zero?)++ OUT
```

**Fig. 7.37.** Regular expression definition of the bounce filter.

**Example 7.23:** For the expression (7.1), positions 1 and 2 are initial, because they match input strings like *aaba* and *baaa* in the first position. That is, the four symbols in *aaba* match only positions 1, 4, 2, and 4, respectively, while *baaa* matches 2, 4, 1, 4. We also see from these strings that position 4 is a successor of positions 1 and 2, while positions 1 and 2 are each successors of 4. That makes intuitive sense, since strings generated by the expression (7.1) consist of one or more pairs of symbols, the first being an *a* from position 1 or a *b* from position 2, and the second being an *a* from position 4. Thus, we oscillate between position 4 and positions 1 or 2.

The output symbol *U* at position 3 is a successor of positions 1 and 2, since if *U* were an input symbol, we would generate strings like *aUabUa* and *bUaaUa*. Note however, that we do not view position 4 as a successor of position 3, because output and state positions do not have successors. □

## Translation of Regular Expressions to PLA's

We shall now take up the bounce filter as an example, considering not only how it can be specified in the regular expression language, but how it could be translated into a PLA. We leave the subject of optimal PLA generation to Section 8.1.

**Example 7.24:** In Fig. 7.37 we see a declaration of the bounce filter as a regular expression. Line (1) declares the name *input* to be the name of the single wire that is input to the circuit, and line (2) declares *zero* and *one* to be abstract symbols that are seen when this wire is off and on, respectively. The name *OUT* is declared to be the lone output signal, at the third line.

To see why the expression works, note that in order to generate the signal OUT, control must reach the position of the operand OUT. For that to happen, we must see an input sequence that can begin any way we like, since .* matches any number of arbitrary inputs, then two 1's, and one or more groups consisting of a 1 and up to two 0's. That is, we are sure to see three 1's in a row, and thereafter, all groups of 0's are of length at most two.

To see how this expression could be transformed into a PLA, we shall first write down a nondeterministic finite automaton (NFA) for the expression.† The

---

† The reader unfamiliar with nondeterministic automata may wish to read Appendix 3.

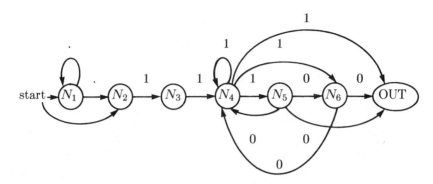

**Fig. 7.38.** NFA for bounce filter.

states of this NFA correspond to the positions of the expression, and there is a transition from position $i$ to position $j$, labeled $a$, if $j$ is a successor of $i$, and the operand at position $i$ is the input symbol $a$. The initial states of the NFA are those that correspond to initial positions.

The nondeterminism in the automaton comes from the fact that a state may have several successors on the same input symbol, and in general, the NFA is in many states after processing a given input sequence, just as we viewed ourselves as being "at" several of the positions of the regular expression at the same time.

If we annotate the regular expression from Fig. 7.37 with subscripts to name positions, we have

$$._1{}^*\ one_2\ one_3\ (one_4\ zero_5?\ zero_6?)++ OUT_7$$

The NFA derived from the successor relationship among positions is shown in Fig. 7.38.

Since a nondeterministic automaton can be in more than one state at a time, we cannot implement the NFA of Fig. 7.38 as we did the deterministic finite automaton of Fig. 7.31. One simple approach to implementing NFA's is to use a *one-hot code*, where each state corresponding to an input symbol has a private feedback wire. For each state, there is a row of the PLA that says: if the wire for the state is on, and the input is the one corresponding to that state,† then turn on the wires in the or-plane corresponding to all the successors of that state, including output wires. The output wires correspond to states of the NFA associated with output symbols, like $OUT$ in Fig. 7.38. Such a PLA for Fig. 7.38 is shown in Fig. 7.39. Note that the order of the state wires in

---

† Note that in the expression-to-NFA translation scheme suggested above, all transitions out of a state are associated with the same input symbol, and that symbol is the one at the position of the regular expression corresponding to the state. We shall discuss in Section 8.4 other ways to build NFA's from expressions not having this property, and there are both advantages and disadvantages to the scheme we use here.

| 2 | 2 | 1 | 2 | 2 | 2 | 2 | 2 |   | 1 | 1 | 0 | 0 | 0 | 0 | 0 |
|---|---|---|---|---|---|---|---|---|---|---|---|---|---|---|---|
| 2 | 1 | 2 | 1 | 2 | 2 | 2 | 2 |   | 0 | 0 | 1 | 0 | 0 | 0 | 0 |
| 2 | 1 | 2 | 2 | 1 | 2 | 2 | 2 |   | 0 | 0 | 0 | 1 | 0 | 0 | 0 |
| 2 | 1 | 2 | 2 | 2 | 1 | 2 | 2 |   | 0 | 0 | 0 | 1 | 1 | 1 | 1 |
| 2 | 0 | 2 | 2 | 2 | 2 | 1 | 2 |   | 0 | 0 | 0 | 1 | 0 | 1 | 1 |
| 2 | 0 | 2 | 2 | 2 | 2 | 2 | 1 |   | 0 | 0 | 0 | 1 | 0 | 0 | 1 |
| 1 | 2 | 2 | 2 | 2 | 2 | 2 | 2 |   | 1 | 1 | 0 | 0 | 0 | 0 | 0 |
| 1 | 1 | 2 | 2 | 2 | 2 | 2 | 2 |   | 0 | 0 | 1 | 0 | 0 | 0 | 0 |

*start input* $N_1$ $N_2$ $N_3$ $N_4$ $N_5$ $N_6$            $N_1$ $N_2$ $N_3$ $N_4$ $N_5$ $N_6$ *OUT*

**Fig. 7.39.** PLA from NFA of Fig. 7.38.

the or-plane is reversed (for readability) from what it should be if we wish to
to draw the feedback wires without crossovers.

For example, the first row says that if the input is arbitrary (because dot
matches any input), and the wire for state $N_1$ is on, then for the next cycle,
turn on the wires for states $N_1$ and $N_2$, because these are the successors of $N_1$.
The fourth line says that if the input is 1, and the wire for state $N_4$ is on, then
for the next cycle turn on the wires for states $N_4$, $N_5$, $N_6$, and the wire for the
output, because these are all the successors of $N_4$.

The last two rows represent the initialization of the NFA; without them,
we could never get the PLA to turn on any feedback wires. The penultimate
row is justified by the fact that $N_1$ is an initial state, and says that if the start
signal is on, and the input matches dot, i.e., any input, then we are in states $N_1$
and $N_2$ at the next cycle, while the last line reflects the fact that $N_2$ is initial
by saying that if the input is 1 and the start signal is on, then we are in $N_3$,
the successor of $N_2$, for the next cycle.

The reader should compare Fig. 7.39 with the PLA for a deterministic finite
automaton shown in Fig. 7.32. Clearly the NFA-based PLA is larger, yet there
are some mitigating factors. One is that the feedback wires in the and-plane
of Fig. 7.39 have no 0's, and therefore, no columns for their complement are
needed when we implement the PLA personality as a real circuit. Secondly, we
could arrange to initialize the PLA in a more direct way in Fig. 7.39, thus saving
the last two rows and the leftmost column. In comparison, Fig. 7.32 contains
no provision for initialization, it being assumed that on startup, three 0's will
get the circuit into state $S_1$ of the deterministic finite automaton. Lastly, Fig.
7.39 is not the best possible PLA; we have used essentially no optimization or
care in state representation. $\square$

**Example 7.25:** To illustrate the state feature of the regular expression language, we shall translate the SLIM finite automaton of Fig. 7.34 into that language; the result is shown in Fig. 7.40. Note that since there is but one regular

line input
symbol zero(−input), one(input)
output OUT
state S1, S2, S3, S4, S5
;
S1: (zero S1 + one S2) +
# S2: (zero S1 + one S3) +
# S3: (zero S1 + one OUT S4) +
# S4: (zero S5 + one S4) OUT +
# S5: (zero S3 + one S4) OUT

**Fig. 7.40.** Deterministic finite automaton in regular expression language.

expression, the various states and their actions are written as the sum of five terms. Since technically, each state name followed by a colon would be in an initial position, we precede all but the initial state $S1$ by the symbol #, which is never matched. Thus, the only way for a symbol like $S2$: to be activated is for the symbol $S2$ to be the successor of an active position. Note that in states $S3$, $S4$, and $S5$, both the output $OUT$ and the state symbol $S4$ are activated when the input symbol *one* is seen. □

**Example 7.26:** As our last example, let us consider Example 7.22, which specified part of a microprocessor control using SLIM. The pipelining of SLIM can be expressed by the order in which symbols appear in the expression, with the dot, the symbol that matches any input, used to separate outputs that come at successive clock cycles. Thus, we might declare the input and output signals from Example 7.22 by

symbol *addop*( suitable setting of input wires )
output $en_1$, $en_2$, $en_3$, $load_1$, $load_2$, $add$, $store_3$

Then, at a position in the regular expression that will become active one cycle before we are ready to decode the next instruction, we say

$en_1$ $en_2$ (*addop* $load_1$ $load_2$ $en_3$ . $add$ . $store_3$ +
other actions for other operations )

Note that input *addop* and the two dots are three input symbols that separate the groups of output signals that are generated; recall that each of a sequence of output signals is a successor of the preceding input symbol. □

## EXERCISES

7.1: The following are simple circuits that can be designed using many of the languages described in this chapter.

a)   A modulo 2 counter.

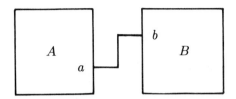

**Fig. 7.41.** Diagram for Exercise 7.10.

b)  A modulo 5 counter that produces three output bits, representing in binary the number of times that the input has been 1, modulo 5.

c)  A one-hot modulo 5 counter, with five output wires $y_0, \ldots, y_4$; $y_i$ is on whenever the number of input 1's seen so far is congruent to $i$ modulo 5.

d)  A multiplexor, taking inputs $x$, $y$, and $z$, and producing output equal to $x$ if $z = 0$ and equal to $y$ if $z = 1$.

e)  An adder for two-bit numbers, producing a sum and carry.

f)  The bounce filter introduced in Example 7.11.

Design CIF layouts for each of these problems.

7.2: Design the circuits of Exercise 7.1 in CHISEL.

7.3: Design the circuits of Exercise 7.1 in LAVA.

7.4: Design the circuits of Exercise 7.1 in esim.

7.5: Design the circuits of Exercise 7.1(a–e) in lgen.

7.6: Give PLA personalities for the functions described in Exercise 7.1(a–e). For 7.1(b, c), draw the finite automaton and give both ordinary and microcoded (using an external counter) PLA's.

7.7: Write SLIM programs for the circuits of Exercise 7.1(a–c).

7.8: Write regular expression programs in the notation of Section 7.9 for Exercises 7.1(a–c).

7.9: Design the cell 456 discussed in Example 7.3. It must fit between two copies of cell 123 without causing any design rule errors.

* 7.10: Suppose we have two CHISEL cells $A$ and $B$, as shown in Fig. 7.41. $A$ has its origin at the lower right, and $B$ has its origin at its lower left. There is a port $a$ of $A$ that is to be wired to a port $b$ of $B$; they are on the sides shown, but we do not know if $b$ is above, below, or level with $a$. Write a CHISEL program to place $A$ and $B$ as close as possible, e.g., abutting if $a$ and $b$ are at the same height, and to run a wire between $a$ and $b$ if necessary, as suggested in Fig. 7.41.

** 7.11: Repeat Exercise 7.10, but assuming that there are two ports $a_1$ and $a_2$ on the right side of $A$ and ports $b_1$ and $b_2$ on the left side of $B$. You may assume that $a_1$ is above $a_2$ and $b_1$ is above $b_2$, but no other relationships involving relative height are known in advance. You must write a CHISEL program to place $A$ and $B$ as close as possible and wire $a_1$ to $b_1$ and $a_2$ to

$b_2$ without any design rule violations.

7.12: Which of the layer specifications in the LAVA program of Fig. 7.23 are really needed, assuming the layers of the ports are given?

* 7.13: Informally describe an algorithm to translate LAVA into esim.

* 7.14: For Example 7.22 (concerning SLIM pipelining), suppose that $R_1$, $R_2$, and $R_3$ need not all be distinct. Under what circumstances with the resulting circuit fail to work correctly?

## BIBLIOGRAPHIC NOTES

Overviews of VLSI design systems can be found in Batali et al. [1981], Parker et al. [1979], and the survey by Sequin [1982]. Systems of the "designer's assistant" type are discussed by Brown and Stefik [1982] and Stefik et al. [1982]. While the Palladio project described there served as an instructive vehicle, its orientation toward "AI" paradigms probably prevents it from running sufficiently fast. The bibliographic notes in Chapter 9 gives references to the PI system, which can be viewed as another sort of "designer's assistant." References on silicon compilers are given in Chapter 8.

The most accessible description of CIF is in Mead and Conway [1980]. CHISEL is described in Karplus [1983], while some other design languages that embed in a general-purpose programming language are described in Johnson and Browning [1980], Batali et al. [1981], and Goates and Patil [1981].

The idea of a sticks-level language is attributed to Williams [1978]; LAVA is described by Matthews, Newkirk, and Eichenberger [1982]. The language esim was implemented by C. Terman. Baker and Terman [1980] describe the system, but the language itself is described in the manual pages accompanying the system. A brief introduction to the lgen language appears in Johnson [1983]. SLIM is described in Hennessy [1981], and the regular expression language of Section 7.9 in Trickey and Ullman [1981].

# 8

# COMPILATION AND OPTIMIZATION ALGORITHMS

We now shall consider methods for compiling high-level VLSI design languages into lower level languages. Often it is not difficult to compile a high-level design into *some* low-level design; the problem is producing a low-level design that is roughly as good as we would get if we started by designing at the low level. Here, a "good" design is one that takes little more area than the low-level design and whose speed is comparable to that of the low-level design. There is some advantage to high-level design even if we cannot always produce "good" circuits, because high-level designs tend to be portable. They can be implemented quickly on a new process that becomes available after the design is finished, perhaps yielding circuits that are smaller and faster than the ones we would have obtained had we designed at a low level, but then been unable to take advantage of the new technology.

In this chapter we shall consider the algorithms behind a number of compilation and optimization processes. We begin with a discussion of the optimization of PLA personalities. We may view this subject as one of producing good circuits from logic, that is, compiling a language like lgen discussed in Section 7.5, or we may view it as compiling finite automaton languages like those discussed in Sections 7.8 and 7.9. Then we consider compilation of sticks languages, of which LAVA from Section 7.6 is our principal paradigm. Again, we are not concerned only with translating from a high-level language (sticks) to a low-level one (CIF), but with producing good quality circuits from the high-level description.

Next, we consider the lgen compiler, which does not translate into PLA personalities as we might have expected, but into arrays of pullups and pulldowns, often called *Weinberger arrays*. A number of interesting combinatorial algorithms must be solved to compile lgen into good layouts. Then we discuss some of the techniques used to compile and optimize the regular expression language of Section 7.9, and we close with a discussion of how ordinary programming languages might be compiled into good circuits.

## 8.1 OPTIMIZATION OF PLA PERSONALITIES

In this section we shall consider how to take a PLA personality, which is really just a notation for specifying several Boolean expressions as sums of products of literals, and producing a PLA of relatively small area. There are two directions we could take here, and we shall touch upon them both. First, we can improve the expressions themselves, i.e., transform the personality into another personality that is smaller, yet defines the same input/output relationships among the variables. The second is to build the PLA in a different way from that suggested in Section 7.7. For example, it is possible to run certain inputs to the and-plane from the top, while others run from the bottom. If a pair of inputs meets certain criteria, it may be that the pair, one running from the top and the other from the bottom, can share a column, thus compacting the PLA. Another idea we shall discuss briefly is that it is sometimes possible to decompose one PLA into two that take less total area than the original.

### Personality Manipulation

The title sounds ominous, but all we wish to do is develop some rules for changing a PLA personality in ways that preserve the functions computed, yet reduce the number of rows. In general, PLA personalities are too large, and the combinatorial problems involved too time consuming,† for us to expect to find optimal designs very often. Thus, we are usually content with algorithms that produce a good, even if not optimal, solution; for example, the problem of "minimizing" the rows of a PLA is really that of finding an equivalent PLA with relatively few rows, and doing so in an affordable amount of time.

The first thing we might check for when trying to minimize a PLA personality is if any row is redundant. Suppose that the inputs to the and-plane are $x_1, \ldots, x_n$ and the outputs from the or-plane are $y_1, \ldots, y_m$. Then each term represents a set of points in the $n$-dimensional Boolean cube, those points that are input assignments for which the term is 1 (true). For example, if $n = 4$, a row with 1021 in the and-plane represents the term $x_1 \bar{x}_2 x_4$, which in turn represents the vertices of the 4-cube with coordinates $(1, 0, 0, 1)$ and $(1, 0, 1, 1)$.

To find out whether a row $r$ is redundant, we must consider every output $y_i$ in which $r$ has a 1 in the or-plane, looking at the set $S_i$ consisting of other rows that have 1 for output $y_i$. The set of points covered by one or more of the rows in $S_i$ must be a superset of the points covered by row $r$. Row $r$ is redundant, and may be eliminated, if this containment holds for every output in which $r$ has a 1 in the or-plane.

We can express the containment condition in Boolean algebra conveniently. Let $f$ be the term for the row $r$ whose redundancy we are considering. Focus on

---

† Almost all the optimization problems we encounter in VLSI design, as elsewhere, are $\mathcal{NP}$-complete, or worse.

a particular output for which this row has a 1, and let $g_1, \ldots, g_k$ be the terms of all the other rows with 1 in that output. We are asking if

$$f \subseteq g_1 + \cdots + g_k$$

That inclusion must be taken to refer to the sets of points that satisfy the expressions on the left and right. Put an algebraically equivalent way, we ask whether

$$\bar{f} + g_1 + \cdots + g_k$$

is true for every assignment of values to the input variables, i.e., whether this expression is a tautology.

To answer this question, we need algorithms to do two things. First, we must take the complement of a product of literals, namely $f$ in the above formula. Second, we must test whether a sum of products of literals is a tautology.

Taking the complement of a product of literals is easy. Assume that $f$ is expressed as a sequence of 0's, 1's, and 2's, as in PLA personalities. Then $\bar{f}$ is the sum of a collection of terms, one term for each 0 and each 1 of $f$. If there is a 1 in position $i$, then we take the term that has 0 in position $i$ and 2's elsewhere, while if position $i$ has 0, the term we need has 1 there and 2's in positions other than $i$. No term is generated if there is a 2 in position $i$. For example, if $f$ is represented by the row 012, then we generate 122 from position one, 202 from position two, and nothing from position three. In the usual notation for Boolean expressions, row 012 is $\bar{x}_1 x_2$, and the formula we obtain for its complement is $x_1 + \bar{x}_2$.

A point covered by $f$ cannot be covered by any of those terms, because it must disagree with the term for position $i$, if there is one, in its own $i^{th}$ position. Conversely, a point not covered by $f$ must disagree with the sequence for $f$ in some position $i$ where the sequence has 0 or 1. Then the term for position $i$ will cover that point.

Testing whether a Boolean expression is a tautology is most surely not doable in polynomial time. However, the following heuristic seems to be fairly effective. It is based on the idea that any expression $f(x_1, \ldots, x_n)$ can be written $x_1 f_1(x_2, \ldots, x_n) + \bar{x}_1 f_0(x_2, \ldots, x_n)$. If $f$ is a sum of terms, represented as in PLA personalities, then $f_1$ is formed from $f$ by taking the rows with 1 or 2 in the first column, then removing the first column, and $f_0$ is constructed similarly from the terms with 0 or 2 in the first column. We thus convert an $n$-column matrix into two of $n - 1$ columns each. The original matrix is a tautology if and only if both of the new matrices are. The sum of the numbers of rows in the two could be almost twice the number of rows of the original, because rows beginning with 2 appear in both new matrices. Thus, both the number of matrices, and their total size, can grow exponentially, as we proceed to eliminate

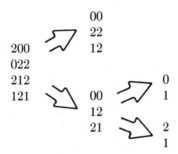

**Fig. 8.1.** Recursive tautology test.

columns.

However, we often do not have to extend the method recursively until we reach matrices of one column. There are many heuristics that can be used to tell in certain cases what the answer to a subproblem is. Two useful ones are:

1.  If a matrix has a row with all 2's, then it surely is a tautology, because this one term covers all the points in the Boolean cube.

2.  Suppose a matrix has $n$ columns, so its rows are terms representing points in the Boolean $n$-cube. A term $t$ with $i_t$ 2's represents $2^{i_t}$ points, so we can easily obtain an upper bound on the number of points all the terms cover, by summing $2^{i_t}$ over all terms $t$. If this sum is less than $2^n$, the matrix cannot represent a tautology. In this case, we not only answer "no" for the particular matrix at hand; we also know that the original matrix is not a tautology.

**Example 8.1:** In Fig. 8.1 we show a sum of four terms, $\bar{x}_2\bar{x}_3 + \bar{x}_1 + x_2 + x_1x_3$, represented in PLA personality form. It is split into two matrices, the upper being the second and third columns of those rows with 0 or 2 in the first column and the lower being the same columns for rows with 1 or 2 in column one. The upper matrix contains a row 22, so it is clearly a tautology. The lower one needs to be split again. Of the two matrices that result, the lower has a row of all 2's, i.e., a single 2. The upper has rows 0 and 1, which represents the expression $x_3 + \bar{x}_3$; it is clearly a tautology also. We conclude that the original matrix is a tautology. $\square$

We shall now summarize the above ideas by giving a test for redundant terms in a PLA.

**Algorithm 8.1:** *Testing for Redundant PLA Terms.*

INPUT: A PLA personality.

OUTPUT: An equivalent PLA personality with no redundant rows.

METHOD: The discussion above gives us the way to test whether a row representing term $f$ in the and-plane has all its points covered by the set of rows

(1)  **for** each row $r$ **do begin**
(2)        **for** each column $c$ of the or-plane in which row $r$ has 1 **do begin**
(3)                let $R$ be the set of rows other than $r$ with 1 in column $c$;
(4)                **if** the set of points covered by $r$ is not a subset of the points
                        covered by the union of the members of $R$ **then**
(5)                        **goto** 999 { row $r$ is not redundant; consider next row }
        **end**;
        { if we reach here, $r$ is redundant }
(6)        delete $r$ from PLA personality;
(7)        999:
    **end**

Fig. **8.2.**  Elimination of redundant rows.

$\{ g_1, \ldots, g_k \}$. That is

1.    Compute $\{ h_1, \ldots, h_r \}$, the set of terms in the complement of $f$.

2.    Test whether $h_1 + \cdots + h_r + g_1 + \cdots + g_k$ is a tautology.

The algorithm for testing each row of the PLA personality and eliminating it
if it is redundant, is shown in Fig. 8.2. The test at line (4) is decided using the
steps just outlined above. Note that if we eliminate a row, subsequent tests in
Fig. 8.2 refer to the revised matrix. $\square$

**Example 8.2:** In the following PLA personality

$$
\begin{array}{cc}
200 & 11 \\
102 & 01 \\
121 & 01 \\
122 & 10
\end{array}
$$

we claim that the second row is redundant. Note that does not follow immediately by comparing the second and fourth rows, because even though the
fourth row covers every point of the 3-cube that the second row covers, the
latter has a 1 in the column for the second output, while the fourth row does
not.

    If we apply the test of Algorithm 8.1, we shall eventually consider $r$ to
be row 2. To test row 2 we have only to consider column $c$ to be the second
output column, because row 2 has no other 1's in the or-plane. Thus, the set
$R$ is rows 1 and 3, because these are the other rows with 1 in the last column.
We, therefore, test whether the complement of row 2, plus rows 1 and 3, is a
tautology. The complement of the second row, which is 102, consists of the
two rows 022 and 212. These, plus the first and third rows, which are 200 and
121, are the four rows that we tested for tautologyhood in Example 8.1. We,
therefore, conclude that row 2 is redundant, and eliminate it to get the PLA
personality

$$\begin{array}{ll} 200 & 11 \\ 121 & 01 \\ 122 & 10 \end{array}$$

☐

## Raising of Terms

Elimination of redundant terms gives us a PLA personality that is minimal in the obvious sense that we cannot eliminate any more terms from it. It is not, however, guaranteed to have the minimum number of rows of all PLA personalities that are equivalent to it in the sense of computing the same output functions. To find this minimum-row equivalent is not computationally feasible when there are hundreds of rows involved. However, there are some manipulations of the PLA personality that sometimes lead to the elimination of more rows by the redundancy test.

The idea, called *raising*, is to take a term, say $x_1 f$ (or equivalently $\bar{x}_1 f$), where $f$ is the product of some of the other variables besides $x_1$, and see if it can be replaced by $f$. The motivation for doing so is that if a term covers more points, then we have better chances that some other term will become redundant. Evidently, $f$ covers twice as many points as $x_1 f$ does.

To test whether $x_1 f$ can be replaced by $f$, we have to know that the points covered by $\bar{x}_1 f$ are also covered by other terms in the set of available terms, for then and only then will the replacement of $x_1 f$ by $f = x_1 f + \bar{x}_1 f$ not change the function computed. If there are several output functions, we must consider each output column $c$ in which the row for term $x_1 f$ has a 1, and then compare $\bar{x}_1 f$ with $g_1 + \cdots + g_k$, the sum of the other terms with 1 in column $c$. Of course, there is nothing special about $x_1$; in practice we can attempt to raise by eliminating any factor from any term. In the PLA personality, raising is represented by replacing a 0 or 1 by a 2. The algorithm for raising is summarized as follows.

**Algorithm 8.2:** *Raising of Terms.*

INPUT: A PLA personality.

OUTPUT: An equivalent PLA personality, where no terms can be raised and no terms are redundant.

METHOD: We consider each row $r$ and each column $i$ of the and-plane in which $r$ has 0 or 1, and we do the following for each column $c$ of the or-plane where $r$ has 1. If for all $c$ the test below is positive, then we may raise position $i$ of row $r$ by placing 2 there.

1.  Find the set $R$ of rows other than $r$ with 1 in column $c$. Let $\{g_1, \ldots, g_n\}$ be the terms corresponding to the rows of $R$.

2.  Complement row $r$ in position $i$ by replacing 0 by 1 there or vice versa.

| 2101 | 1 | 2122 | 2221 | 1 |
|------|---|------|------|---|
| 0002 | 1 | 2212 | 0002 | 1 |
| 0211 | 1 | 2220 | 2221 | 1 |
| 1112 | 1 | 0002 | 1112 | 1 |
| 1021 | 1 | 0211 | 2221 | 1 |
|      |   | 1112 |      |   |
|      |   | 1021 |      |   |

(a)                         (b)                         (c)

**Fig. 8.3.** Raising terms.

Then let $h_1 + \cdots + h_m$ be the complement of the term represented by this modified row $r$.

3.  Test whether $h_1 + \cdots + h_m + g_1 + \cdots + g_n$ is a tautology.

After raising all the terms we can, we may then apply Algorithm 8.1 to the PLA personality with all terms raised as far as possible. If a term was redundant to begin with, it will still be redundant after raising, and with luck, there will now be more redundant terms than there were before we applied the raising algorithm. □

**Example 8.3:** Consider the PLA personality in Fig. 8.3(a). We can try to raise the second position of the first row, for example. If we complement that position, the and-plane portion of the first row, which is 2101, becomes 2001. We then wish to take the complement of this term, which is $2122 + 2212 + 2220$. Then we form the expression consisting of these three terms plus the terms of the second through fifth rows of Fig. 8.3(a); this expression is shown in Fig. 8.3(b). We find it is a tautology, using the test outlined in Algorithm 8.1, so position two of row one may be raised, making that row 2201.

We can also raise the third position of row 1, making that row 2221. The third and fifth rows of Fig. 8.3(a) also raise to 2221, while the second and fourth rows do not raise at all. After raising, we obtain the personality of Fig. 8.3(c), and it is easy to discover that any two of the first, third, and fifth rows are redundant and can be eliminated by Algorithm 8.1. The final, minimal PLA personality is thus

| 2221 | 1 |
|------|---|
| 0002 | 1 |
| 1112 | 1 |

□

The reader should remember that Algorithms 8.1 and 8.2 do not necessarily produce a minimum-row PLA; they only produce one that is minimal in the sense that further raising and elimination of redundant terms is not possible. It is also worth noting that the order in which raising is applied to the positions and terms can make a difference in the outcome, since a term such as 112 might be raisable to 212 or to 122, but not to 222. Similarly, the order in which we eliminate terms can make a difference in the final outcome.

## PLA Folding

Having improved the set of Boolean formulas we wish to implement as a PLA, as far as is feasible considering the complexity of the problem, we still have some opportunities to save space, if we modify the details of the PLA structure. There are a great many ways we could design circuits that were essentially PLA's, yet that differed in significant ways from the organization assumed in Section 7.7. Here, we shall discuss only one such reorganization, called "folding."

When we fold a PLA, we allow and-plane inputs to arrive at either the top or bottom of the PLA, and we hope that two inputs arriving from different directions can share a column, thus saving the width of a column of the PLA personality, which consists of two vertical wires of the PLA, or at least $14\lambda$. Before proceeding, we shall make the very reasonable assumption that if both an input and its complement are needed (i.e., the PLA personality has both 0's and 1's in the column for that input), then we keep both complemented and uncomplemented wires for that input together, running in adjacent columns of the PLA, as usual.

The motivation for this assumption is that should we separate the wires representing $x$ and $\bar{x}$, we would have to run one from its source to the point where it entered the PLA, which would waste some of the area we are trying to save. Surely there are situations where we may separate the $x$ and $\bar{x}$ wires without incurring a significant area cost, and the reader is invited to modify the method we discuss here to take advantage of such situations.

Similarly, there are times when it is not reasonable to alter the direction from which certain inputs reach the PLA, because when designing a PLA we may not have control over where on the chip the sources of the inputs are placed. On the other hand, there are situations where one can select the edge at which inputs appear, such as when the PLA is implementing a sequential machine, and many of its inputs are fed back from the or-plane. While we have assumed that the vertical wires in the or-plane have their pullups at the top and outputs at the bottom, we could selectively reverse some of them with little or no extra wiring required. In fact, to do so may actually save space, since inverters for the outputs, which have to be wider than the pullups, will not appear in a row at the bottom, but rather in two rows, at top and bottom.

In Fig. 8.4 we see the way that two inputs $x$ and $y$ might be paired. Note that the $x$ and $y$ wires cannot overlap, nor can $\bar{x}$ and $\bar{y}$ overlap, although the breakpoints need not be the same. Also note that we could just as well have paired $x$ with $\bar{y}$ and $\bar{x}$ with $y$, but we have only two ways to pair these two

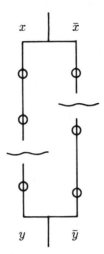

**Fig. 8.4.** Pairing of inputs in a folded PLA.

inputs, since we are not considering, say, pairing $x$ with $y$ but $\bar{x}$ with some $z$ other than $\bar{y}$.

Should we choose to pair $x$ and $y$ in the manner shown in Fig. 8.4, we put certain constraints on the order of the wires for the terms. First, if the wires for $x$ and $y$ both need to cross the same term wire, then there is no way we can pair $x$ and $y$, although we might consider pairing $x$ with $\bar{y}$ and $\bar{x}$ with $y$. Second, assuming we can pair these wires, we see that every term that requires $x$ must be above every term that requires $y$, and every term that requires $\bar{x}$ must be above every term that requires $\bar{y}$.

It is not feasible to check all possible sets of pairings of a large number of columns, including the possibility that any two columns can be paired with either on top, and either with the complemented and uncomplemented wires facing each other as in Fig. 8.4 (*straight*), or with the complemented wires facing uncomplemented ones (*twisted*). Thus, we shall attempt only to get a pairing that is maximal, in the sense that there are no two unpaired wires that could be paired without violating the constraints on the order of the term wires. We give the ideas in the next algorithm.

**Algorithm 8.3:** *PLA Folding.*

INPUT: A PLA personality.

OUTPUT: A legal pairing of some of the columns, indicating which of each pair is above the other, and whether each pairing is straight or twisted. Also provided is an ordering of the rows that will make this set of column pairs legal.

METHOD: We consider all pairs of columns $C$ and $D$, in some order. Assuming that these columns of the PLA personality can be paired when they are considered, i.e., they do not have 1's in the same row and they do not have 0's

|   | A | B | C | D | E |
|---|---|---|---|---|---|
| a | 0 | 0 | 1 | 2 | 2 |
| b | 1 | 0 | 2 | 0 | 2 |
| c | 2 | 1 | 2 | 2 | 0 |
| d | 2 | 2 | 0 | 0 | 1 |
| e | 0 | 2 | 2 | 2 | 2 |
| f | 2 | 2 | 2 | 1 | 1 |

**Fig. 8.5.** PLA personality.

in the same row, and neither has yet been paired with another column, we consider pairing $C$ and $D$, with either above the other† and either straight or twisted.

At all times, we keep a directed graph representing the constraints on the order of the term wires induced by whatever pairings we have made so far. Initially, the graph consists of the nodes for the terms, and no arcs. Suppose we decide to pair columns $C$ and $D$, with $C$ above $D$, straight. We create two new nodes, $p$ and $q$. Thus, $p$ represents the constraint that the terms using the $C$ wire must be above the terms using the $D$ wire, while $q$ represents the fact that terms using the $\bar{C}$ wire must be above those using the $\bar{D}$ wire.

1. If row $r$ has 1 in column $C$, then draw arc $r{\rightarrow}p$.
2. If row $r$ has 1 in column $D$, then draw arc $p{\rightarrow}r$.
3. If row $r$ has 0 in column $C$, then draw arc $r{\rightarrow}q$.
4. If row $r$ has 0 in column $D$, then draw arc $q{\rightarrow}r$.

Of course, if both (1) and (2) or both (3) and (4) hold, we cannot make the pairing.

If we pair $C$ and $D$ twisted, we do essentially the same thing, but the roles of $p$ and $q$ are exchanged in (2) and (4).

Having tentatively made this pairing, we consider whether we introduce any cycles in the graph. For if a cycle is introduced, then there is no way we can order the terms and still run the paired columns from top and bottom without wires occupying the same space. If we find no cycles in the graph, we make the pairing of $C$ and $D$ permanent, while if cycles are introduced, we remove the arcs just added and consider the next possible pairing, whether of $C$ and $D$ in another arrangement, or of another pair of columns.

At the end, we have made a maximal set of pairings that gives us an acyclic graph. The graph represents a partial order on the term wires, and we pick a total order consistent with the partial order, a process known as "topological sort" (see Aho et al. [1983], e.g.). □

---

† But if there are no pairs selected yet, then consider only $C$ above $D$, since any order we pick for the term wires can be either a top-to-bottom order or a bottom-to-top order.

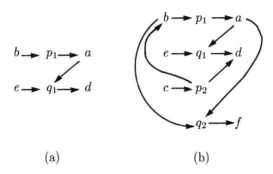

(a)                                (b)

**Fig. 8.6.** Graph showing permissible row orders.

**Example 8.4:** Consider the and-plane portion of a PLA personality shown in Fig. 8.5. We cannot consider pairing columns $A$ and $B$, because row $a$ prevents their being paired straight, and row $b$ prevents their being paired twisted. However, we can pair $A$ and $C$, and we shall do so, with $A$ above $C$, and straight. We introduce nodes $p_1$ and $q_1$; the former is the head of arcs from all the rows with 1 in column $A$ and the tail of arcs to all the rows with 1 in column $C$. There is only one row of each class, $b$ for the first and $a$ for the second. Similarly, $q_1$ is the head of arcs from $e$ and $a$, the rows with 0 in column $A$, and the tail of an arc to $d$, the lone row with 0 in column $C$. The graph so far is shown in Fig. 8.6(a).

Now, suppose we attempt to pair $B$ above $D$, twisted. We introduce nodes $p_2$ and $q_2$. An arc from $c$ to $p_2$ is introduced because column $B$ has 1 in that row, and we have arcs from $p_2$ to $b$ and $d$ because $D$ has 0 in columns $b$ and $d$. Note that because the pairing is twisted, $p_2$ has arcs to the 0's and $q_2$ has arcs to the 1's, instead of vice versa. Thus, there are arcs to $q_2$ from $a$ and $b$, the 0's of column $B$, and an arc from $q_2$ to $f$, the lone 1 in column $D$. We cannot perform any more pairings because there is only one unpaired column, $E$. The complete graph is shown in Fig. 8.6(b); it is acyclic, and one possible order is $c, e, b, a, d, f$. The resulting layout of the PLA is shown in Fig. 8.7. $\square$

### Running Time of Algorithm 8.3

Unlike Algorithms 8.1 and 8.2, which depend on testing tautologies and thus are inherently exponential, Algorithm 8.3 runs in time that is polynomial in the size of its input, the PLA personality. Let $c$ be the number of columns and let $w$ be the "weight" of the personality, i.e., the total number of 0's and 1's. It is not clear how $w$ relates to $c$, or to the number of rows, but it is generally believed that dense PLA's, those with a large fraction of their entries in the and-plane being 0's and 1's, are rare. Thus, we might suppose that $w$ is roughly proportional to the number of rows, rather than to the product of

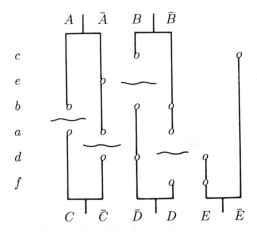

**Fig. 8.7.** Folded PLA.

rows and columns.

Since each 0 and 1 can account for only one arc of the graph under construction, the number of arcs is certainly no more than $w$; we might expect it to be considerably less on average, because dense columns are less likely to be pairable. We can test for acyclicity of the graph in time proportional to the number of arcs, by constructing a depth-first spanning forest† and checking for the existence of backward arcs. This common test for acyclicity is covered in Aho et al. [1983], e.g. There are $4\binom{c}{2}$ possible pairings to consider, i.e., $\binom{c}{2}$ pairs of columns, which may be paired with either above, and may be paired straight or twisted. Thus, Algorithm 8.3 is no worse than $O(wc^2)$.

However, it may in practice turn out much better than that, because having constructed a depth-first spanning forest to verify that a particular pairing leaves an acyclic graph, we may retain this forest, and attach to each node its preorder and its postorder numbers. A node $v$ is a descendant of node $u$ in this forest if and only if

1. $v$ follows $u$ in preorder, and

2. $v$ precedes $u$ in postorder.

Thus, when we consider adding new node $p$ or $q$ in Algorithm 8.3, with the nodes in set $S$ as predecessors and the nodes in set $T$ as successors, we can detect many cycles by using the above test to determine if some node in set $S$ is a descendant of a node in set $T$. If so, we can immediately reject the pairing, although if not, we still must perform the full check for cycles, taking $O(w)$ time.

---

† See Appendix 2 for definitions concerning depth-first search.

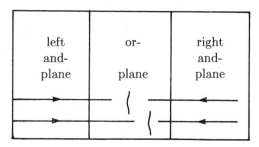

**Fig. 8.8.** PLA with two and-planes.

## Or-Plane Folding

The term wires that form the output of the and-plane are the inputs to the or-plane. We may thus apply the same ideas that we used to fold columns in the and-plane, to the purpose of folding rows of the or-plane, as suggested by Fig. 8.8. There, term wires feed the or-plane from both sides, so we may pair terms using a modified version of Algorithm 8.3. We revise that algorithm to take into account the fact that the or-plane inputs are really single wires, rather than the pairs of wires that are input to the and-plane. However, this change results only in simplification, since one new node, rather than the two nodes $p$ and $q$, must be introduced into the graph.

There is the added problem that depending on how we partition the terms between the two and-planes, some input wires may have to be provided to both sides. Since we would presumably like to minimize the duplication of inputs, we must incorporate this factor into our optimization strategy.

## Partitioning PLA's

A rather different approach to decreasing the area required by a PLA is to partition it into two or more. While one might expect that not to help in general, we should remember that the area of a PLA is essentially proportional to the product of the number of its rows and the number of its columns. Suppose we could somehow divide the input and output wires into two groups, such that each term was computable from the inputs of one group and was needed only for the outputs of the same group. Suppose also that the two groups of inputs, outputs, and terms were each of equal size. Then we could replace the large PLA by two, each of which is half as wide and half as high, and therefore each takes only one quarter the area. In this hypothetical best case we could, merely by splitting one PLA into two, save half the area.

We might consider PLA partitioning to be a matter of examining all possible partitions of the outputs into two or more sets, calculating the number of terms used to compute each set of outputs, and the number of inputs required to compute each set of terms, and selecting the best of these partitions. An alternative, and probably superior method was suggested by Kang and vanCleemput [1981]. It, as should be expected, does not produce the optimum partitioning, but it produces a partitioning that is locally optimal, in the sense that we cannot improve matters by merging the sets of outputs computed by two PLA's into one.

We begin by choosing a formula that represents the estimated area of a PLA with $T$ terms, $I$ inputs, and $U$ outputs. The formula must take into account the fact that the area of a PLA includes not only the matrix where rows and columns intersect, but also "overhead" in each direction, representing the requirements of pullups on wires, drivers, inverters, clock gates, and wires to supply power and ground. We ought to include also a charge to each PLA for the fact that it will probably not fit very snugly into the chip, leaving some unused space on its border, and another charge for the area taken bringing input, output, and utilities (power, etc.) to the PLA. However, we do not concern ourselves with these charges, since they are hard to estimate, and we take as an example estimated cost

$$(10 + T)(10 + 2I + U) \tag{8.1}$$

Now, we begin with every output alone, in its own PLA, with all and only the terms it needs, and with all and only the inputs they need. Then, in some particular order, we consider a pair of PLA's and whether we could save space by merging them; if so we merge them and if not we consider another pair, until all possible mergers result in an area loss.

**Example 8.5:** Suppose that at some point we have two PLA's, each with 5 outputs. The first has 10 inputs and 20 terms; the second has 15 inputs and 30 terms. Using (8.1), the areas of these two PLA's are estimated to be $(10 + 20)(10 + 2 \times 10 + 5) = 1,050$ and $(10 + 30)(10 + 2 \times 15 + 5) = 1,800$.

Suppose also that the two PLA's compute 10 terms in common, so the combined PLA would have 40 terms. Likewise, suppose they share 5 inputs in common, so the combined PLA would have 20 inputs. Then by (8.1), the area of the combined PLA is estimated as $(10 + 40)(10 + 2 \times 20 + 10) = 3,000$. We conclude that the combined PLA would use more area than the two smaller ones, so we do not combine them. $\square$

## 8.2 COMPILATION OF STICKS LANGUAGES

We now turn to another important compiler-like process, that of converting sticks diagrams into layout. In particular, we shall discuss the compilation of LAVA, introduced in Section 7.6, but the ideas apply to sticks languages in general.

The process of sticks compilation is fundamentally that of translating abstract coordinates of points into real coordinates, in such a way that no design rule violations occur, yet the area used is as small as possible. In principle, we could produce any circuit layout whatsoever from a sticks diagram, as long as the layout and the diagram performed the same function, but even to attempt that would involve us in combinatorial problems we could not hope to solve efficiently. What is normally done to make the compilation tractable is to take the sticks diagram more seriously than it needs to be taken; specifically, topological constraints are not violated. Thus, points drawn on one side of a wire will remain on that side when the circuit is drawn.†

Further, the options involved in moving points horizontally and vertically still may lead to too much combinatorics; thus we generally simplify the interactions between selection of the horizontal coordinates and the selection of vertical coordinates. The result is that space-consuming objects often line up along one dimension, making the layout less compact than it could be. LAVA reports this *critical path*, in the hope that the designer can redesign the layout so some of the elements on the critical path can be moved to adjacent lines. Some attempts to do this modification automatically have been made, but the combinatorics involved limit the size of the designs that can be handled.

To see what is involved in sticks compilation, consider the simple stick diagram of Fig. 8.9(a). There, we see two metal wires meeting poly wires at red-blue vias located at points $p_1 = (10, 25)$ and $p_2 = (20, 15)$, the coordinates of the points being the abstract coordinates used in LAVA programs.

Figure 8.9(b) shows the sort of layout we expect to come from the sticks. The abstract coordinates, such as 15, have been replaced by concrete, but as yet unknown coordinates of the layout, such as $y_{15}$. In transforming Fig. 8.9(a) to (b), we assume for specificity that the concrete coordinates are the coordinates of the baseline, which is the bottom or left edge of every feature appearing on the corresponding abstract coordinate. Thus, for example, vias are drawn with their lower left corners at the point of intersection of the concrete coordinates. Note that we do not consider moving point $p_1$ below $p_2$, which may or may not be feasible depending on what else appears along the metal wires in Fig. 8.9(a).

We must develop a set of inequalities among the concrete coordinates, so that all solutions to the inequalities represent possible layouts in which there are no design rule violations. There may be other layouts that are free of design rule errors, yet do not satisfy the constraints, because of our insistence that the topology of the sticks diagram be respected.

For example, in Fig. 8.9(b) the metal-metal design rule requires that $y_{25}$ must be at least $y_{15} + 6$. Note we do not even consider layouts where the metal wire at coordinate $y_{25}$ is placed below the other, even though it could be

---

† However, metal wires, which do not interact with wires in the other layers, can sometimes be moved across poly or diffusion wires.

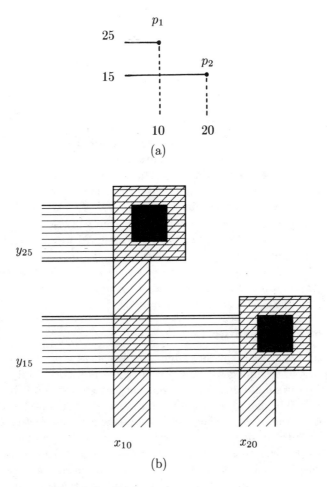

$p_1$

25

$p_2$

15

10        20

(a)

$y_{25}$

$y_{15}$

$x_{10}$                    $x_{20}$

(b)

**Fig. 8.9.** Sticks-to-layout conversion.

placed sufficiently far below that the metal-metal design rule would be satisfied. However, if $y_{25}$ is exactly $y_{15}+6$, and the two vias are too close in the horizontal direction, then there will be a metal-metal separation violation between the top of one via and the bottom of the other. Thus, if $y_{25} = y_{15} + 6$, $x_{20}$ must be at least $x_{10} + 7$, so the two vias are at least $3\lambda$ apart in the horizontal direction. On the other hand, if $y_{25} \geq y_{15} + 7$, then $x_{20}$ can be as low as $x_{10} + 4$, since we then have only to avoid a poly-poly separation violation. Thus, the exact constraint implied by the sizes of objects and our intent to respect the topology of Fig. 8.9(a) is

either $y_{25} \geq y_{15} + 7$ and $x_{20} \geq x_{10} + 4$
or $y_{25} \geq y_{15} + 6$ and $x_{20} \geq x_{10} + 7$

Solving conditional inequalities such as the above may lead to a very good layout, but it is not generally considered feasible to do so if the stick diagram has more than a few hundred points. Instead, systems like LAVA work first in one direction, say vertical, finding all the constraints that the concrete $y$-coordinates must satisfy. Having set up all the vertical constraints, we can solve them for their minimum possible values. Then, with vertical positions of all circuit features fixed, we work on the other direction, setting up inequalities concerning the concrete $x$-coordinates. These are also solved to compact the circuit as much as possible in the horizontal direction, subject to the decisions already made about vertical coordinates.

Thus, in Fig. 8.9, if there were no other reason why $y_{25}$ had to exceed $y_{15}$ by more than 6, we would find that the value assigned to $y_{25}$ would in fact be $y_{15} + 6$. Then, when setting up the horizontal constraints, we would find that $x_{20}$ had to exceed $x_{10}$ by at least 7. We never get to consider that there might be a layout of less total area if we separated $y_{25}$ and $y_{15}$ by one more unit and were thereby able to squeeze $x_{20}$ and $x_{10}$ by $3\lambda$.

## Setting Up the Inequalities

Let us now consider how we might systematically generate the inequalities in the vertical direction, assuming we have not yet fixed any horizontal coordinates. We may make the simplifying assumption that elements with different abstract $x$-coordinates will be placed sufficiently far apart in the horizontal direction that they will not interact. Thus, when considering how low a concrete $y$-coordinate can be, we have to consider only the vertical lines represented by the abstract $x$-coordinates. For example, if there is a point on horizontal line $y_{100}$ at vertical line $x_{50}$ and at the same $x$-coordinate there is a point on a lower horizontal line, say $y_{80}$, and further, there is a layer conflict between the two points, then we shall generate a constraint of the form

$y_{100} \geq y_{80} + c$

for some constant $c > 0$. The same remark is true if one or both of the horizontal lines does not have a point at $x_{50}$, but rather a horizontal wire that crosses $x_{50}$.

By saying that the points or wires have a "layer conflict" we mean that objects of the two layers cannot overlap. Thus, two features of the same layer cause a layer conflict, and features in the poly and diffusion layers also conflict. The only case in which we do not have a conflict is between metal and the other two layers.

If we generated inequalities between every pair of $y$-coordinates, we would

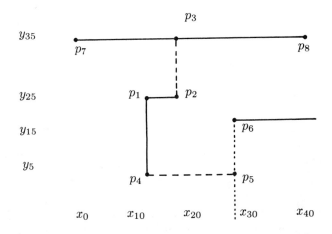

**Fig. 8.10.** A stick diagram.

probably be doing much more work than necessary. Thus, in the example above, we would wish not to generate a constraint between $y_{100}$ and $y_{80}$ based on the features at $x_{50}$ if that constraint, or a stronger one, were implied by constraints among $y_{100}$, $y_{80}$, and some intermediate horizontal line, say $y_{90}$.

A systematic approach to constraint generation is to work upward, considering each $y$-coordinate in turn. We generate a set of constraints saying only that for each layer, and for each abstract $x$-coordinate, this $y$-coordinate must be suitably larger than the highest $y$-coordinate considered so far that has a conflicting layer at that $x$-coordinate. To do so, we must record for each layer $L$ its *fringe*, which is a table giving, for each abstract $x$-coordinate:

1. The name of the highest $y$-coordinate seen so far with a conflicting layer at that $x$-coordinate,† and

2. An amount above that $y$ coordinate that is the closest a feature in layer $L$ can come without causing a design rule violation.

We shall not attempt to write down a complete set of rules, based on particular designs for all the circuit elements, such as vias, gates, and pullups. Rather, we hope an example will make the ideas clear enough.

**Example 8.6:** In Fig. 8.10 we see a stick diagram; what it does is of no significance. The intent is that $p_3$ and $p_6$ are blue-green vias, $p_2$ is a pullup, with the channel running vertically, $p_1$ is a blue reference point, $p_4$ a blue-red via, and $p_5$ a pulldown, with the gate connected from the left. All wires in this

---

† Strictly speaking, LAVA takes into account the electrical connectivity of the wires. If two wires are electrically connected, then there need not be any constraint between them; they can even overlap. Thus, in LAVA, fringes actually record the two highest $y$-coordinates that are not electrically connected. The subject of finding electrically connected objects is discussed in Section 9.2.

|        | $x_{10}$       | $x_{20}$        | $x_{30}$       |
|--------|----------------|-----------------|----------------|
| blue   | $y_{25} + 6$   | $y_{25} + 9$    | $y_{15} + 7$   |
| red    | $y_5 + 6$      | $y_{25} + 15$   | $y_{15} + 5$   |
| green  | $y_5 + 5$      | $y_{25} + 16$   | $y_{15} + 7$   |

**Fig. 8.11.**  Fringes.

diagram are assumed to have minimum width.

Suppose we have processed $y_5$, $y_{15}$, and $y_{25}$, and we wish to generate the constraints that put lower bounds on $y_{35}$. Then Fig. 8.11 shows the fringes for the three layers; the columns correspond to the $x$-coordinates $x_{10}$, $x_{20}$, and $x_{30}$ mentioned in Fig. 8.10. There are not yet any constraints that concern the columns for $x_0$ or $x_{40}$, so the fringes are undefined there. For example, in the $x_{10}$ column, any blue object must be sufficiently above $p_1$ that no metal-metal conflict occurs. Since the horizontal metal wire from $p_1$ to $p_2$ is assumed $3\lambda$ high, and $y_{25}$ represents the low edge of that wire, another blue object cannot lie below $y_{25} + 6$, which explains the first entry in the blue fringe in Fig. 8.11.

The blue point $p_1$ does not affect the red or green fringe at $x_{10}$. Only $p_4$, which is a blue-red via, influences the position of red or green objects. The via at $p_4$, extending $4\lambda$ above the baseline $y_5$, requires a red object to be no lower than $2\lambda$ above the top of the via, i.e., at least $y_5 + 6$. Green objects can be one unit lower than this, because the red-green separation is only one, while the red-red separation is two. We have now explained the first column of Fig. 8.11.

The second column, for $x_{20}$, is explained by the pullup at $p_2$. Since a pullup involves pieces in all three layers, objects of any layer will be at least some distance above the baseline of the pullup, $y_{25}$. However, the exact distance depends on the design used for the pullup and on the length/width ratio specified for the channel by the designer. We have used hypothetical offsets for the three layers, although there is some rationale. For example, the constant 9 for the metal offset is based on the assumption that the pullup has a $6\lambda$ high butting contact, with lower edge on the baseline, and no other metal.

Finally, consider the third column. The lower bound $x_{15} + 7$ for blue and green is based on the fact that objects in these layers must be at least $3\lambda$ above the $4\lambda$-high blue-green via at $p_6$. Similarly, the lower bound on red is based on the fact that a red object must lie at least one unit above this via.

Since the horizontal line at $y_{35}$ has only blue at $x_{10}$ and $x_{30}$, we generate the two constraints that say $y_{35}$ must be at least equal to the blue fringe value at these two $x$-coordinates. That is

$$y_{35} \geq y_{25} + 6$$
$$y_{35} \geq y_{15} + 7$$

Note that these are independent constraints, since we cannot assume any rela-

| | $x_0$ | $x_{10}$ | $x_{20}$ | $x_{30}$ | $x_{40}$ |
|---|---|---|---|---|---|
| blue | $y_{35} + 6$ | $y_{35} + 6$ | $y_{35} + 7$ | $y_{35} + 6$ | $y_{35} + 6$ |
| red | | $y_5 + 6$ | $y_{35} + 5$ | $y_{15} + 6$ | |
| green | | $y_5 + 5$ | $y_{35} + 7$ | $y_{15} + 7$ | |

**Fig. 8.12.** Fringe after consideration of $y_{35}$.

tionship between $y_{25}$ and $y_{15}$.

Next, consider the constraints introduced by the column $x_{20}$. The horizontal line at $y_{35}$ has both blue and green there, so $y_{35}$ is limited by the fringes for both colors. These inequalities are

$$y_{35} \geq y_{25} + 9$$
$$y_{35} \geq y_{25} + 16$$

The last of these is strictly stronger than the first, and also is stronger than the first inequality generated when we considered $x_{10}$. Thus, we may trim the set of constraints generated when we consider $y_{35}$ to

$$y_{35} \geq y_{15} + 7$$
$$y_{35} \geq y_{25} + 16$$

Finally, we must update the fringes to take into account the features at $y_{35}$. The blue wire running all across the figure says that the value of the blue fringe will be $y_{35} + 6$ at all $x$-coordinates (including those of the endpoints $p_7$ and $p_8$), except at $x_{20}$, where it is $y_{35} + 7$, because of the blue-green via there. This via also affects the red and green fringes at $x_{20}$, setting lower limits of $y_{35} + 5$ and $y_{35} + 7$, respectively. The new fringes are shown in Fig. 8.12. $\square$

The process of constraint generation in the horizontal direction is similar to what we have seen. The major difference is that the vertical coordinates are now concrete, and when considering how far left an object can be placed, we must consider not only what is on the same vertical coordinate, but also what is above and below it a sufficiently small distance that design rule violations might occur.

The chief modification we must make is in how we represent fringes. Of course, fringes now represent the leftmost position at which an object of given layer could be placed, but it is no longer sufficient to define the value of this function at the abstract vertical coordinates only. In principle, the value must be defined for every $\lambda$ unit in the vertical direction. In practice, it is generally more succinct to record intervals over which the value is a constant. For example, after considering $x_{10}$ in Fig. 8.10, the blue fringe has value $x_{10} + 6$ for $y$-coordinates $y_5 + 7$ through $y_{25} + 6$, and value $x_{10} + 7$ for $y$-coordinates $y_5 - 3$ through $y_5 + 7$. For example, the latter bound is due to the fact that moving a blue object any further left within the $y$ range $y_5 - 3$ to $y_5 + 7$ will cause a design rule violation with the blue-red via at $p_4$ in Fig. 8.10. The value is "minus infinity" for other $y$-coordinates, i.e., no reason exists why blue objects cannot be moved to the left edge of the cell.

## Solving Sets of Inequalities

If we remember the style of LAVA programs, we note that large cells are typically built from smaller ones, in a hierarchy of several levels. Expanding an entire design to get all the constraints in one direction, and then solving them can be quite time-consuming, since there could be tens of thousands of points, and hundreds of thousands of constraints.

Solving inequalities is not all that difficult, in principle, since we may create a graph, called the *constraint graph*, whose nodes are the variables, with an arc from node $u$ to node $v$ labeled $c$ if there is a constraint of the form

$$v \geq u + c$$

for some constant $c$. From what has gone before, we expect that all such constraints will have $c \geq 0$, although that does not seem to be very significant. Far more important is the fact that the graph will normally be acyclic. That follows because in Example 8.6, we only generated constraints $y_i \geq y_j + c$ if $i > j$. LAVA actually introduces constraints that may form cycles by several mechanisms. In our treatment here, we assume that constraint graphs are acyclic; at the end of the section, we shall mention cycle-forming constraints and what to do about them.

Assuming the constraint graph is acyclic, we may topologically sort† the nodes in time proportional to the number of arcs, i.e., the number of constraints. Then, every node with no predecessors may be taken to have value 0, and we can go down the list of nodes in order. When reaching node $u$ with some predecessors, we have already assigned values for those predecessors. We may assign a value for $u$ by evaluating the constraints connecting $u$ to its predecessors and taking the lowest value consistent with all. That is, if $V(w)$ is the value assigned to node $w$, then $V(u)$ is the maximum over all predecessors $v$ of $u$, of $V(v) + c_v$, where $c_v$ is the label of arc $v \rightarrow u$.

However, there is a better approach for designs specified hierarchically, as they will normally be in LAVA or similar languages. We begin by reducing the constraints for the bottom-level cells, so that we are left only with constraints involving ports on the borders of the cell. The border points are part of some cell $C$ at the next level, and recursively we shall solve for all the points in $C$, as if it were a bottom-level cell. Having completely solved $C$, we have concrete values for points on the border of the original bottom-level cell $D$. By making

---

† A *topological sort* of an acyclic graph is the reverse of a depth-first ordering of the graph. Recall from Appendix 2 that this ordering has the important property that if there is an arc from $u$ to $v$, then $u$ precedes $v$ in the topological ordering.

coordinates for the internal points of $D$ as low as possible, consistent with the constraints and the values of the border points, we may solve for the internal points of $D$.

The key steps in this process are:

1.  The elimination of those constraints within a cell that do not involve the border points, and

2.  The combination of constraints from several cells into one set of constraints.

Item (1), the partial solution of constraints, is actually more complex than simply solving the constraints. However, we cannot solve constraints in several cells independently, as at a higher level, their border points may be connected. If we have found incompatible values for the corresponding border points, the cells cannot be abutted as intended. The steps necessary to solve a set of inequalities about the coordinates of points in one dimension are described in the next algorithm.

**Algorithm 8.4:** *Solution of Constraints in One Dimension (Assumed Vertical).*

INPUT: We are given a hierarchy of cells, as in LAVA, with the following conditions.

1.  Certain points in a cell are designated border points. Presumably, these are the points on the left and right edges of the cell in question, not the top and bottom edges, since we assume that we are solving the vertical constraints.

2.  Except for the top-level cell, which has no border points, each use of a cell $D$ in a cell $C$ at the next level designates a correspondence between the border points of $D$ and points in the grid of $C$.

3.  All constraints within a cell form an acyclic constraint graph.

4.  We require that each cell $D$ used in a higher-level cell $C$ have a designated position for the bottom of $D$ in the grid of $C$.

5.  Finally, we shall assume that the constraint graph is such that there is a path from the bottom coordinate of each cell to every other point.† For convenience, we take the bottom of the cell to be a "border point," which we may do because our assumptions guarantee that the cell bottom will be given an associated coordinate in the surrounding cell.

OUTPUT: A value for the vertical coordinate of each point in each use of each cell, such that all values satisfy the constraints given in input item (3) above and the point correspondences mentioned in input item (2).

---

† The latter property can be assured if, when we compute constraints as in Example 8.6, we start with all fringes having value $y_0$, where $y_0$ is the abstract coordinate of the bottom of the cell in question. In fact, a better strategy may be to design all cells so there can be no design-rule errors between internal features of the cell and any wire running outside the cell, even if the cell and the wire touch at other than a border point. We can achieve that by starting the fringe at $y_0 + 3$ for blue and green, and $y_0 + 2$ for red.

(1) **for** cells at level 0, 1, . . . , level of root **do begin**
(2)      eliminate interior nodes from constraints;
(3)      **if** level $> 0$ **then**
(4)           combine constraints involving border points of subcells
      **end**;
(5) solve constraints for the root cell, by finding a topological
      sort of the nodes in the constraint graph;
(6) **for** all instances of cells at level of root $-1$ down to 0 **do**
(7)      solve constraints for interior points, given values for border points

**Fig. 8.13.** Sketch of hierarchical constraint solution algorithm.

METHOD: The basic algorithm is to eliminate the constraints involving the interior points of a given cell, those that are not designated border points. Levels of the cell hierarchy are defined as follows. At level 0 are cells that have no subcells. Level-1 cells call only level-0 cells; level-2 cells call only cells at levels 0 and 1, and so on. At the top sits one cell that is called by no other cell. This cell represents the design as a whole, and we shall call it the *root* cell. A sketch of the algorithm appears in Fig. 8.13.

There are three steps that require more detail: step (2) where we eliminate interior constraints, step (4) where we combine constraints, and step (7), where we finally solve the constraints involving the interior points of cells.

In step (2), we consider each border point $v$, from the bottom of the cell, in turn. Recall that the cell bottom is regarded as a border point. We perform from each border point a depth-first search of the constraint graph, but when we reach another border point we do not follow arcs out. Thus, we reach all and only the nodes accessible from $v$ without going through any border point. By visiting the nodes in the resulting topological order, we can discover for each node the length of the longest path from $v$ to that node $u$. It is easy to prove by induction on the number of nodes considered that the length $\ell$ of this longest path is the least difference between the values of $u$ and $v$ permitted by the constraints. That is, the constraint $u \geq v + \ell$, but no stronger constraint, is implied by the given constraints. If we focus on the cases where $u$ is a border point only, then we have the constraints between $v$ and all the other border points.

We must repeat this process beginning at all border points. If we meet border point $u$ when searching from border point $v$, then we shall not meet $v$ when searching for $u$, because if we did, then the constraint graph would not be acyclic. The result is that we can list the border points in some order $v_1, \ldots, v_n$ (which may or may not be the bottom-up order of the points in the cell), such that all the constraints involving the border points are $v_i \geq v_j + c_{ij}$, where $i > j$ and $c_{ij}$ is some positive constant.

Note that it is not sufficient to represent the constraints as $v_{i+1} \geq v_i + c_i$ for $1 \leq i < n$, since, for example, we might have $v_2 \geq v_1 + 10$, $v_3 \geq v_2 + 10$, but $v_3 \geq v_1 + 30$, which is not implied by the first two constraints. We shall give a more concrete example where this situation occurs shortly.

Also note that the depth-first search from each border point may involve visiting all, or almost all, the points of the cell; thus in the worst case, the time to eliminate interior points is on the order of the product of the number of border points and the number of constraints. This figure is substantially more than the time to solve the constraints for one cell exactly, which is on the order of the number of constraints, if the constraints are acyclic. However, in practice, eliminating interior points is not as time-consuming as the worst case, since many interior points will be sandwiched between two border points and, therefore, will appear in the depth-first search of only the border point immediately below it. Further, we do this step only once per cell, not per use of that cell, so the overall cost is not great if typical cells are used more than once.

Step (4) of Fig. 8.13 is straightforward. When subcells are called by another cell, their border points will be identified with some other point, either a point of the calling cell or a border point of another subcell. After identification, there could be two or more constraints involving the same pair of points. We may sort the constraints to find such coincidences and eliminate the one that implies the other; that is, eliminate $u \geq v + c$ in favor of $u \geq v + d$ if $c < d$.

Lastly, let us consider step (7) of Fig. 8.13. The idea is similar to the direct solution of constraints. We consider the points in topological order, setting the values of interior points to the smallest value that is implied by the externally defined values of the border points and the constraints. Since the bottom of the cell is a border point, and there is assumed to be a path in the constraint graph from the bottom point to every other point, we know that every point's predecessors in the constraint graph will be considered before that point, when points are visited in topological order. $\square$

**Example 8.7:** Consider the constraint graph shown in Fig. 8.14. This graph represents a hypothetical collection of constraints; e.g., the arc $a \rightarrow f$ represents the constraint $f \geq a + 5$, since the label of the arc is 5. We shall suppose that points $a$, $b$, and $c$ are border points, with $a$ the bottom point, and the others are interior points.

In Fig. 8.15 we see the three depth-first searches, starting from $a$, $b$, and $c$. Note that the search from $a$ does not go past $b$ or $c$, so, for example, the arc $b \rightarrow e$ is not included in Fig. 8.15(a). The order of retreat in Fig. 8.15(a) is the same as a postordering of the nodes of the tree, that is, $b$, $c$, $e$, $f$, $d$, $a$; thus the topological order we choose is the reverse of this order: $a$, $d$, $f$, $e$, $c$, $b$.

We use Fig. 8.15(a) to calculate the constraints between $a$ and the other border points. We start by setting $V(a) = 0$. Note that the value 0 is arbitrary.

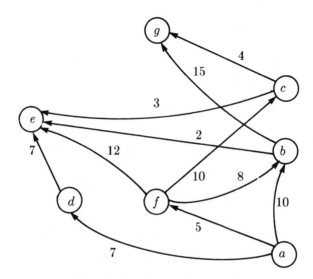

**Fig. 8.14.**  Constraint graph.

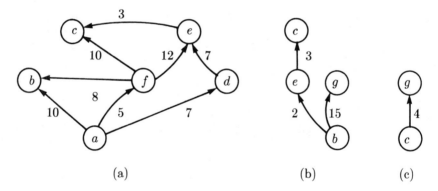

**Fig. 8.15.**  Depth-first searches from border points.

The output of this stage is a set of new constraints involving the border points, and the constants found in these constraints will be the same whatever value of $V(a)$ is chosen.

After $a$ in the topological order comes $d$, so we look at the arcs into $d$ in Fig. 8.15(a). There is only one, from $a$, with label 7, so we set $V(d) = V(a) + 7 = 7$. Similarly, we set $V(f) = 5$. Next in topological order comes $e$, and that node has two predecessors in Fig. 8.15(a). Thus, we set $V(e)$ to the larger of $V(f) + 12$ and $V(d) + 7$; that is the former, or 17. Then, we set

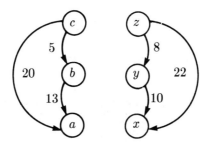

**Fig. 8.16.** Combination of constraints.

$$V(c) = max(V(f) + 10, V(e) + 3) = 20$$
$$V(b) = max(V(f) + 8, V(a) + 10) = 13$$

From the values of $V(b)$ and $V(c)$ we deduce the constraints between $a$ and the other border points; these are:

$$b \geq a + 13$$
$$c \geq a + 20$$

A similar process in Fig. 8.15(b) lets us deduce the constraint between $b$ and $c$, namely,

$$c \geq b + 5$$

since 5 is the length of the longest path from $b$ to $c$ in Fig. 8.15(b). Note that the constraint $c \geq a + 20$ is not implied by the constraints $b \geq a + 13$ and $c \geq b + 5$, so all three constraints must be retained. Also note that Fig. 8.15(c) yields no constraints, because $c$ happens to be the highest of the border points, a fact that we could not tell in advance without computing the transitive closure of the constraint graph of Fig. 8.14.

Now let us give an example of step (4) of Fig. 8.13, where we must combine the constraints just derived about the border points $a$, $b$, and $c$, with constraints at the next higher cell level. Suppose that at that level, the cell represented by the constraint graph of Fig. 8.14 is abutted with another cell, the points $a$, $b$, and $c$ lining up with points $x$, $y$, and $z$ of the latter cell. Let the constraints involving those points be as illustrated in Fig. 8.16.

In this case, the constraint between $x$ and $y$ becomes a constraint between $a$ and $b$. However, the constraint $b \geq a + 13$ is stronger than the constraint $y \geq x + 10$, so the latter may be ignored. However, the constraint $z \geq y + 8$ is stronger than $c \geq b + 5$, so we use the constraint $c \geq b + 8$ henceforth. Similarly, we use the constraint $c \geq a + 22$.

Lastly, let us give an example of step (7) of Fig. 8.13. Let us suppose that after solving the root cell of the design, actual values for unknowns filter down to the lower levels, and we establish that $a = 100$, $b = 113$, and $c = 122$; that is, by coincidence, $b$ and $c$ each take their minimum possible value, given the value of $a$.

We now use these values to solve for the ($y$-coordinate of) every point in Fig. 8.14. First, we select a topological order of the nodes in Fig. 8.14,

$a, d, f, b, e, c, g$

The values $V(.)$ we compute now are the desired concrete values for the $y$-coordinates. In the first pass of Fig. 8.14, we set $V(a) = 100$, as given from the level above. Then we set $V(d)$ to 107, because $a$ is its only predecessor, and 107 is $V(a)$ plus the label of arc $a{\to}d$. Similarly, $V(f) = 105$. We set $V(b)$ to 113, because that value is given from the level above. Next,

$$V(e) = max(V(d) + 7, V(f) + 12, V(b) + 2) = 117$$

$V(c)$ is given as 122. Finally,

$$V(g) = max(V(b) + 15, V(c) + 4) = 128$$

☐

### Cyclic Constraint Graphs

Sometimes LAVA and similar languages allow the introduction of constraints that cause cycles in the constraint graph. If the constant $c$ were always positive in constraints of the form $x \geq y + c$, then a cycle would imply that there is no way to lay out the circuit so the constraints are satisfied. However, when we consider some of the sources of constraints, we find that constraints introducing cycles are likely to have negative constants $c$. That in itself does not guarantee that there will be a solution to the constraints, but it at least makes the existence of a solution possible.

A major source of cyclic constraints is when two wires that theoretically line up have a small amount of freedom in the way they attach. For example, suppose we have a horizontal metal wire, with a via in its middle. We could regard the pieces of the wire to the left and right of the via as separate wires, and we could, therefore, jog the wire up or down by $\lambda$ at the via. A similar situation arises in LAVA when boundary points that end wires of more than minimum thickness are abutted. Thus, in Fig. 8.17 we see two points, $a$ and $b$, that must abut when we place a cell in its surroundings. However, the two wires to which $a$ and $b$ are attached are metal and of width 5. If $u$ and $v$ are the bottoms of the two wires, then $u$ and $v$ may differ, but by no more than 2. Thus, the two constraints introduced form a cycle in the constraint graph; they are

$v \geq u - 2$
$u \geq v - 2$

When there are cycles in the constraint graph, we can no longer use Al-

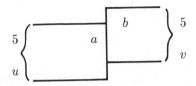

**Fig. 8.17.** Abutment of two wires.

gorithm 8.4 as written. The problem is that we can no longer topologically order the graph so that all predecessors of a node in the constraint graph also precede it in the ordering. One reasonable improvement is to continue to use a node ordering that is the reverse of some depth-first ordering of the graph. As there are cycles, the depth-first spanning forest will now have some back edges, those that go from a node to an ancestor in its spanning tree. However, on the reasonable assumption that there will be few constraints causing cycles, there will not be too many back edges.

We may thus visit the points in the reverse of their depth-first ordering, just as we did in Algorithm 8.4. The only difference is that we may attempt to assign a value to a point $x$ before we have assigned values to all its predecessors. Specifically, if there is a back edge from $y$ to $x$, then the value for $y$ will be unknown when $x$ is considered.

A simple strategy is to assign a value to $x$ as if $y$ did not exist. Then, after assigning values to all the points, visit them all again in the reverse depth-first order. We recompute the value for each point, making use of the current value for each of its predecessors in the constraint graph; presumably, when we consider $x$ again, there will be a value of $y$ available, which may or may not change the value of $x$, depending on the constraints. If any changes are made in one pass, another pass must be made. Note that if $x$'s value changes, then the value of some of $x$'s successors in the constraint graph may change. If these in turn are tails of back edges, then the effect of their changes will not be discovered until the next pass; this pattern may repeat for many passes. As long as there is a solution to the constraints, this process will eventually converge, and if the number of back edges is small, the number of passes may be expected to be small, although there is no guarantee.

There are some more complex algorithms and data structures that will lead to greater efficiency. For example, we can arrange that on the second and subsequent passes, we have to visit only points one or more of whose predecessors have changed on this or the previous pass. The need for such structures when there are few cycles, and the number of points in a cell tends to be small, is not clear.

It is when there are cycles that the hierarchical approach of Algorithm 8.4 really demonstrates its superiority to the straightforward approach of expanding all the cells and then solving one large set of constraints. If there are no cycles, either approach will take time on the order of the number of constraints, and Algorithm 8.4 could even be slower than the straightforward approach, if cells tended to be used few times. However, if there are cycles, and therefore multiple passes over the points are required, the hierarchical approach will tend to use many passes only on the few cells where there are lots of cycles, while the number of passes required by the straightforward approach will be roughly the maximum of the number required by any cell in the hierarchy.

## 8.3 COMPILATION OF LOGIC

When we try to compile logic, such as the language lgen discussed in Section 7.5, we face much more complex problems than when compiling sticks languages or when transforming PLA personalities. The principal cause of the complexity is that we have considerably more freedom in what the final layout should look like, than we do with PLA personalities or sticks. Some of the choices we face are the following.

1.   What should be the overall strategy of the layout? For example, we could convert the logic into sum-of-products form and implement it with a PLA or a collection of PLA's, but there are many other approaches as well, and we shall discuss one shortly.

2.   Given an overall strategy, how should we select an optimal implementation within that strategy? In a sense, we covered this aspect of the problem, for the case where the PLA is the overall strategy, in Section 8.1.

3.   How should we transform the logic to improve the quality of its best possible implementation? Again, Section 8.1 covered some possible transformations within the PLA framework.

### Weinberger Arrays

In this section, we shall consider an alternative to PLA's as a method of implementing logic, called *Weinberger arrays*. We shall also consider some variants of the idea at the end of the section. An advantage of Weinberger arrays is that their area does not generally grow as fast as that required for PLA's, as the size of the problem grows. Moreover, they can accommodate forms of logic other than sums of products, while PLA's are limited to that form.† A disadvantage of Weinberger arrays is that the optimal design problem seems considerably more complex than with PLA's. Also, for large problems, Weinberger arrays tend to take on an awkward shape, being very much wider than they are high. However, modifications of the basic idea can compensate for this problem fairly well.

---

† Variations on the PLA idea can accommodate other forms of logic, but we are always limited to logic of a restricted form.

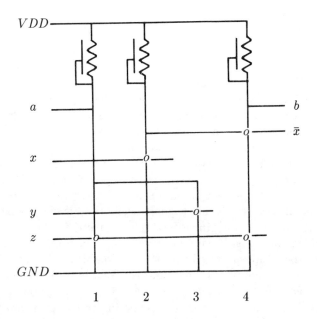

**Fig. 8.18.** A Weinberger array.

An example of a Weinberger array is shown in Fig. 8.18. First, let us observe the basic pattern, that the circuit consists of columns, many or all of which have pullups at the top. Below the pullups, columns are diffusion wires, while in rows we see metal wires. Where a metal wire crosses a diffusion wire, there may be a tap, which we indicate by circles, as we did for PLA's. Just as in PLA's the taps are pieces of polysilicon attached to the metal wire and crossing the diffusion. Thus, if a metal wire is high, current will be allowed to flow down the columns at which it has taps. On the other hand, if the metal wire is low, current will be blocked from traveling down the column, and the output of the column, taken from any point between its pullup and the highest of the taps, will be high.

Thus, a single column implements the NAND function; it is low if and only if all the metal wires crossing it with taps are high. For example, column 2 computes $\bar{x}$, since it has only one tap, at the point where the wire for input $x$ crosses. The output of the second column, which must be taken above any taps on that column, is used as a tap on the fourth column, and is also an output, taken at the right edge. The fourth column computes $\neg(\bar{x}z) = x + \bar{z}$, which is the output $b$.

Several columns, only one of which has a pullup, can work together to implement NOR, or more generally, any expression of the form

$\neg(<\text{sum of terms}>)$

The idea is shown in Fig. 8.18, where columns 1 and 3 implement the function $\neg(y+z)$ of the input wires $y$ and $z$. The first and third columns are tied together by a metal wire, which must run above any taps in the first or third columns. The output of the pair of columns, denoted $a$, happens to be taken from the first column. It must be connected above any taps on the first column, but can be above, below, or level with the connector between columns one and three.

In general, we implement the expression $\neg(t_1 + \cdots + t_n)$, where each $t_i$ is a product of literals, by creating a column for each $t_i$. The "literals" may be inputs, or they may represent other expressions computed within the Weinberger array. We place a pullup on top of one column, and run a single horizontal metal wire connecting a point below the bottom of that pullup to the tops of the wires running in each of the $n-1$ other columns. The column for $t_i$ has a tap where each wire corresponding to a literal in the product $t_i$ crosses it; all these taps must be below the horizontal connecting wire. We may choose the column from which the output is taken, but we shall assume that it is always the column to which we have chosen to attach the pullup.

The reader should observe that by choosing this implementation, we are limiting somewhat the expressions that can be implemented in our Weinberger arrays, using a single pullup. As an example of possible generalizations, we could connect the columns for the terms by several different horizontal wires, and we could have taps above some or all of those horizontal wires. Another limitation we place is that we assume each horizontal wire runs in a single track from the leftmost to rightmost place where it is needed. A more general scheme, actually implemented in the lgen compiler, allows track changes, and tries to minimize those changes. Without such changes, the requirement that the output of a pullup appear above all of its pulldown taps will tend to force the use of too many tracks, as results always require tracks higher than their inputs. The use of a few columns that allow a horizontal wire to switch tracks costs space horizontally, but may save much more space vertically. The motivation for all these limitations, incidentally, is to keep the combinatorics of the logic-to-Weinberger-array problem within manageable bounds.

## Logic Optimization

The first stage of a design of a Weinberger array from logic is to manipulate the logic into a form that requires few columns. Unlike the minimization problem for sum-of-products logic, discussed in Section 8.1, here we have almost unlimited opportunities to change the form of the expressions we are given. The goal is to represent the logic expressions as a directed acyclic graph, where the basic units are trees of the form shown in Fig. 8.19.

Note that special cases of Fig. 8.19 are the NAND function where there is

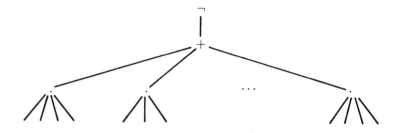

**Fig. 8.19.** General form of operator.

only one term, the NOR function, where all terms consist of one literal, and even the inversion operation, where there is one term with one literal. Also remember that the operands do not actually have to be literals; they can be results of other operators that are in the form of Fig. 8.19. Finally, note that the results of these operators, just like the inputs, can have their values used many times.

The cost of Fig. 8.19 is proportional to the number of terms in the sum, i.e., to the number of columns it requires. However, we should note that there is an undesirable effect of having terms that are the product of many literals. The reason is that if current is to flow through the column, it must flow through each of the taps along that column, with the resistances of each adding to form the resistance of the column. If, say, we wish to maintain a 4:1 pullup/pulldown resistance ratio, and the taps are $2\lambda \times 2\lambda$ transistors, then a column with three taps requires a pullup with a 12:1 channel ratio. This pullup will take much vertical space. An alternative, also space-consuming, is to make the column wider so the pulldown transistors can have smaller channel ratios, thereby consuming horizontal space.

One obvious transformation we can make to our logic is to find common subexpressions, that is, expressions that appear in more than one of the output expressions. The basic idea behind common subexpression detection is to build expression trees from each expression and search for nodes that have the same operator and the same children. The idea, and an efficient algorithm for finding common subexpressions, is discussed in Aho and Ullman [1977].

**Example 8.8:** We must take care first to put the expressions in the form indicated by Fig. 8.19. That is, contrary to expectation, $x+y$ is not a subexpression of $\neg(x+y)$; rather it is the other way around. The reason is that $\neg(x+y)$ is in the desired form, and we can compute from it $x + y = \neg(\neg(x + y))$ by use of a second operator of the form of Fig. 8.19. In this way, we get both expressions using only three columns, two to compute $\neg(x + y)$ and one more to invert it.

On the other hand, if we had decided to compute $\neg(x+y)$ by first computing $x+y$ and inverting it, we would probably come up with an expression like $\neg(\bar{x}\bar{y})$

for $x + y$, which requires three columns, two to compute $\bar{x}$ and $\bar{y}$, and a third to compute the expression itself. To this we must add a fourth column to compute $\neg(x + y)$. $\square$

We shall not attempt to catalog the many transformations on logic expressions that can help transform them into equivalent expressions built from relatively few of the operations illustrated in Fig. 8.19. To begin, any sum $E_1 + E_2$ can be written $\neg(\neg(E_1 + E_2))$, and a similar transformation can be made for any product. Thus, any expression involving Boolean sum, product, and negation can be transformed into some expression involving only operators of the desired form. The use of DeMorgan's laws $[\neg(E_1 + E_2) = (\neg E_1)(\neg E_2)$ and $\neg E_1 E_2 = \neg E_1 + \neg E_2]$ and the law $\neg\neg E = E$ will generally reduce the number of operators and their cost in terms of columns required for the Weinberger array. These rules can be applied locally, wherever we see an opportunity to reduce the cost of the expression. So doing will not necessarily produce a global optimum, since the expression with which we begin may be radically different from its optimal form.

**Example 8.9:** Let us consider the expressions involved in a full adder, where we wish to compute a sum $s$ and carry-out $d$, given two operand bits, $x$ and $y$, and a carry-in $c$. The usual way to write the expressions is in sum-of-products form:

$$d = xy + xc + yc$$
$$s = xyc + \bar{x}\bar{y}c + \bar{x}y\bar{c} + x\bar{y}\bar{c}$$

However, a heuristic that can be built into the logic-to-Weinberger-array compiler is that when we are given two-level logic, and many variables appear both complemented and uncomplemented, we prefer to start with the product-of-sums form. For the full adder, these expressions are:

$$d = (x + y)(x + c)(y + c)$$
$$s = (x + y + c)(\bar{x} + \bar{y} + c)(\bar{x} + y + \bar{c})(x + \bar{y} + \bar{c})$$

Just as the sum-of-products form is built by writing down the situations under which the output 1 is desired, the product-of-sums form is easily discovered by taking the product of the situations under which the output 0 is desired, but then complementing the literals and taking the sum instead of product. Thus, the first factor $(x + y + c)$ in the formula for $s$ comes from the fact that if $\bar{x}$, $\bar{y}$, and $\bar{c}$ are all true, then $s = 0$.

We begin to transform the expression for $d$ by introducing a double negation, i.e., $d = \neg\neg((x + y)(x + c)(y + c))$. Then we use DeMorgan's law to push one negation inside, as $d = \neg(\neg(x + y) + \neg(x + c) + \neg(y + c))$. Finally, we push the inner $\neg$'s down one more level, to get the formula $d = \neg(\bar{x}\bar{y} + \bar{x}\bar{c} + \bar{y}\bar{c})$; the tree for this expression is shown in Fig. 8.20(a). Observe that this expression can be broken into four trees of the form of Fig. 8.19. Three of them are trivial

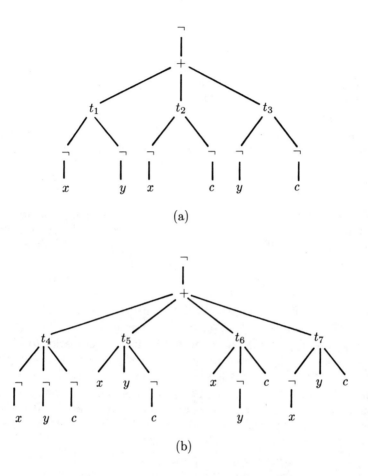

**Fig. 8.20.** Transformed expressions for full adder.

trees that invert $x$, $y$, and $c$. The fourth is the sum of three terms, each of which is the product of two "literals," which are each results of two of the inverters. Also observe that the same general technique will serve to transform any product-of-sums expression into one negated sum of terms, each of which is a product of inputs and/or their negations.

A similar transformation is performed on the expression for $s$, and the result is shown in Fig. 8.20(b). This expression uses the three inversions that Fig. 8.20(a) uses, so we make heavy use of each of the common subexpressions $\bar{x}$, $\bar{y}$, and $\bar{c}$. There is one other operation of the type of Fig. 8.19 present; this one is the negated sum of four terms, so it has a cost of four columns. □

## Ordering the Columns

Having decided on a form of the expressions that, as best we can, reduces the number of columns required, we are ready to select an order in which the columns are to appear. There are several conflicting requirements for such an ordering.

1.  We would like the expressions that are outputs to appear as close as possible to the side on which their port is located. There may also be constraints concerning the order of output ports along the borders, or even constraints on the absolute horizontal or vertical position at which certain outputs must appear, and these may have subtle effects on the desired column order.

2.  Suppose column $C$ computes a value that is used in taps appearing in columns $D_1, \ldots, D_n$. Then we would like the distance between the leftmost and rightmost of $C, D_1, \ldots, D_n$ to be as short as possible. By doing so, we maximize our chances that the horizontal wire connecting the output of $C$ to the taps in columns $D_1, \ldots, D_n$ can share its track with other horizontal wires. Note that there is no particular reason why $C$ has to appear leftmost in the order, since it can just as easily transmit its value to $D_1, \ldots, D_n$ from the right end, or from somewhere in the middle.

3.  When several columns are connected by a horizontal wire to represent a sum of terms, the columns so connected should be as close together as possible. The remarks made under (2) apply in this case as well.

4.  Items (1)–(3) are merely heuristics for obtaining the true goal, that is, minimizing the number of tracks needed for all the horizontal wires. Thus, an ordering of the columns might make the wires short on average, but they might not pack very well into tracks.

Despite what item (4) says, it appears sensible to decouple the complexity of selecting tracks from the complexity of ordering columns, in the hope that the result of solving two hard problems as best we can will be better than trying to solve one nearly impossible problem. Given that we are going to look at the column ordering problem in isolation, we are still faced with a wide variety of choices of strategy.

The lgen compiler uses a recursive partitioning method, where we divide the columns into two groups of about equal size; one group will be the left half of the columns and the other the right. The goal is to minimize the number of wires that must connect columns of the two groups. Then we recursively partition the two halves, until we get down to sets of size one.

There is a nontrivial algorithm that can be used to select each partition of the recursive partitioning process. We shall give the algorithm in its pure form, where it is required to divide, into two parts of exactly equal size, the nodes of a weighted undirected graph. The goal is to arrange that the weights of the edges connecting nodes in the two halves sum to as small a total as possible. The algorithm to be described has been found superior to more straightforward approaches such as dividing the nodes at random and exchanging pairs of nodes until no improvements can be obtained. In contrast, the algorithm to be described allows arbitrary sets of nodes to be swapped, yet takes time that, experimentally, has been found to grow far more slowly than an exponential.

After describing the basic algorithm, we shall mention how it was modified as required by lgen. First, we can allow some flexibility in the sizes of the two sides; they need not be precisely equal. Second, certain nodes must appear on one designated side—those that correspond to inputs and outputs of the circuit.

**Algorithm 8.5:** *Kernighan-Lin Partitioning Algorithm.*

INPUT: A weighted, undirected graph with an even number of nodes, $n$.

OUTPUT: A partition of the nodes into two equal-sized sets, so that the *cut*, that is, the sum of the weights of all edges that connect nodes in both sets, cannot be reduced by the operation described in Fig. 8.21.

METHOD: The algorithm is sketched in Fig. 8.21. We shall not go into detail regarding the computation of $cost_i$ at line (4). Rather, it suffices to observe that we can keep track, for each node $w$, of the sum $s_w$ of the weights of the edges between $w$ and nodes in $S$, and the sum $t_w$ of the weights of the edges between $w$ and nodes in $T$. Thus, if we move $u_i$ of line (3) of Fig. 8.21 from $S$ to $T$, we shall decrease the cut by $t_{u_i} - s_{u_i}$, which could be positive or negative. If we then move $v_i$ from $T$ to $S$, we decrease the cut by $s_{v_i} - t_{v_i}$ minus twice the weight of the edge between $u_i$ and $v_i$, if any. The reason this subtraction is required is because this edge now contributes in the wrong direction, assuming we have not changed $s_{v_i}$ or $t_{v_i}$ after moving $u_i$. After we exchange $u_i$ and $v_i$, we must recalculate $s_w$ and $t_w$ for all the nodes $w$ that have not yet been selected, by a straightforward incremental formula taking into account only the edges from $w$ to $u_i$ and/or $v_i$. $\square$

**Example 8.10:** In Fig. 8.22 we see a graph of sixteen nodes, six of which are designated with letters $a, b, \ldots, f$ for the sake of later discussion. We assume that all edges shown have a weight 1. The initial partition is also shown in Fig. 8.22.

If we consider all possible pairs of a node from the left group and a node from the right group, we quickly discover that the least cost swap is to exchange one of $a$, $b$, or $c$, with one of $d$, $e$, or $f$. However, each such exchange increases the cut from six to eight. Suppose we decide to let $u_1 = c$ and $v_1 = f$. Then $cost_1 = -2$. When $i = 2$, our best choice is to swap one of $a$ and $b$ with one of $d$ and $e$; say we swap $a$ with $d$. The cut goes down from eight to six, so $cost_2 = 2$. Finally, the third choice pairs $u_3 = b$ with $v_3 = e$, with $cost_3 = 6$.

The fourth through eighth pairs involve only the unnamed nodes in Fig. 8.22, and we can see that these will not lead to a better partition, since the

(1)  pick $S$ and $T$ to be random disjoint subsets of the nodes,
     with $n/2$ nodes each;
     **repeat**
(2)      **for** $i := 1$ **to** $n/2$ **do begin**
(3)          find that pair of unselected nodes $u_i$ in $S$ and $v_i$ in $T$
             whose exchange makes the largest decrease or
             smallest increase in the cut;
(4)          $cost_i :=$ that change in cost;
(5)          make $u_i$ and $v_i$ selected
         **end**;
(6)      find $\ell$, $0 \le \ell < n/2$, such that $\sum_{j=1}^{\ell} cost_j$ is maximized;
(7)      move $u_1, \ldots, u_\ell$ from $S$ to $T$ and $v_1, \ldots, v_\ell$ from $T$ to $S$
         { note nothing happens if $\ell = 0$ }
(8)  **until** $\ell = 0$

**Fig. 8.21.** Kernighan-Lin algorithm.

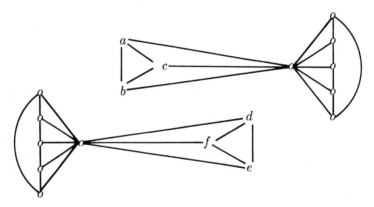

**Fig. 8.22.** Initial partition for application of Kernighan-Lin algorithm.

result of the first three swaps leaves a cut of zero. Thus, at line (6) of Fig. 8.22,
$\ell = 3$ is selected, because the running sum $cost_1 + cost_2 + cost_3$ is the largest,
namely 6. If we repeat the loop of lines (2–8) on the new partition, we of course
get no change, since all running sums will be negative. □

In order to use Algorithm 8.5 in a recursive ordering algorithm for columns
of a Weinberger array, we need to make two modifications. First, there will be
certain nodes that are not permitted to switch sides; these correspond at the
initial partition to the inputs that are required to enter at the left or required
to enter at the right. We can handle this situation by forbidding the selection
of those nodes in line (3) of Algorithm 8.5.

As the recursive partition proceeds, we shall partition subsets of the nodes, representing a group of columns whose order has not yet been decided, yet which collectively occupy a position in the order such that there is a known group of columns to the left and the remaining columns are to the right. There are wires connecting the columns in our current group to columns on the left and on the right. The easiest way to simulate this fact is to create a node in the left half of the partition representing all columns to the left. That is, each node $w$ is connected to this dummy node with an edge whose weight equals the number of columns to the left to which the column represented by $w$ is connected. A similar dummy node is placed in the right half of the partition.

The second modification concerns the fact that there is no reason to insist on a partition of each group exactly in half; we might be willing to accept a partition as long as neither side were more than twice the size of the other. We can adapt Algorithm 8.5 easily, if we introduce some dummy nodes connected to no other nodes. For example, if we add $n/2$ dummies to a set of $n$ nodes, then we shall obtain a split of the real nodes that is no worse than $1/3$ to $2/3$.

### Track Assignments

Once we have ordered the columns, we must consider how the wires connecting those columns are to be packed into tracks. Our hope is that if a good ordering has been chosen, the wires connecting columns will tend to be short, and several can share a track.

**Example 8.11:** Let us reconsider the full adder, and suppose that the lgen specifications call for the inputs $x$, $y$, and $c$, and the output $d$, all to appear on the left, while the output $s$ appears on the right. There might naturally be a requirement that $d$ appear below $c$, so it could easily become the carry-in to the next stage, located below, but we do not deal with such constraints here.

Let us assume that $t_1$ and $t_4$ are the columns chosen for the pullups that produce outputs $d$ and $s$, respectively. One possible result of Algorithm 8.5 is

$$\bar{x},\ \bar{y},\ \bar{c},\ t_1,\ t_2,\ t_3,\ t_5,\ t_6,\ t_7,\ t_4$$

The $t_i$'s represent the columns for the terms, and the correspondence was indicated in Fig. 8.20.

The ten wires needed for the full adder are shown in Fig. 8.23. To help follow the diagram, we mark the source of the signal in the wire with an asterisk, and we show taps, i.e., uses of the signal, by circles. Thus, a wire with an asterisk in a column must appear above any wire with a circle in that column. Wires (1)–(3) are the input wires, while (4)–(6) are the outputs of the inverters. Wire 7 connects the terms $t_2$ and $t_3$ to the column for $d$ and $t_1$. We use asterisks for the columns that wire 7 connects, to indicate that this wire must appear above wires with circles in the connected columns. Wire 8 is similarly related to the output $s$, while wires 9 and 10 are used to carry the outputs $d$ and $s$ to

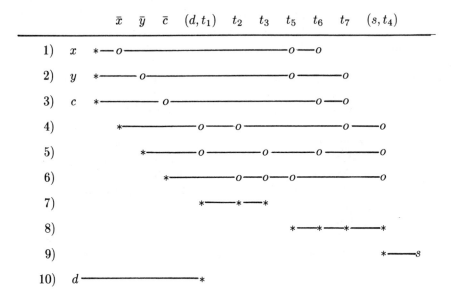

**Fig. 8.23.** A column ordering and its effect on wire lengths.

their respective sides. □

The next stage in design of the Weinberger array from the logic is to assign each wire to a track. We wish to group as many wires as we can into a single track, since by so doing we can minimize the number of tracks. In general, we cannot put two wires on the same track if they overlap, although there is an exception; the output and connector wires for the same pullup can overlap. For example, we could not put wires 1 and 8 of Fig. 8.23 on one track, because they overlap at the columns $t_5$ and $t_6$, but wires 8 and 9 can overlap because they are, respectively, the connector and output wires for column $t_4$.

However, there are other constraints, since certain wires must appear on higher tracks than others. We say that $i \succ j$, meaning that wire $i$ must appear on a higher track than wire $j$, if either one of the following conditions hold.

1.  Wire $i$ is the output wire for the pullup in column $C$, and wire $j$ taps column $C$.

2.  Wire $i$ connects the columns representing the terms for the pullup at column $C$, and wire $j$ taps one of those columns, including $C$. Call wire $i$ the *connector* for column $C$ in this case.

Note the convention based on the notation of Fig. 8.23; $i \succ j$ if and only if there is a column with an asterisk in row $i$ and a circle in row $j$.

**Example 8.12:** In Fig. 8.23, we have $4 \succ 1$, because wire 4 is the output of the inverter computing $\bar{x}$, and wire 1 is the input $x$, which taps the column for

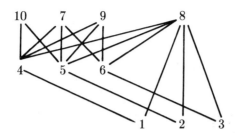

**Fig. 8.24.** The "higher than" relation for Fig. 8.23.

$\bar{x}$. By rule (2) we have $7 \succ 4$, because wire 7 is the connector for $d$, and wire 4 taps the columns for $(d, t_1)$ and $t_2$, two of the columns connected by wire 7. Similarly, $7 \succ 5$ and $7 \succ 6$. Figure 8.24 shows all the constraints introduced by the $\succ$ relation for Fig. 8.23; an edge from $i$ down to $j$ means $i \succ j$. $\square$

The relation $\succ$ must be a partial order on the wires, or else there would be a cycle in the expression itself, that is, a situation where a value computed required itself for the computation, either directly or through several intermediate values. We may therefore use one of several heuristics to select an assignment of wires to tracks in a way that respects the $\succ$ order, i.e., if $i \succ j$, then $i$ is assigned a higher track than $j$, and also respects the requirement that wires assigned to the same track not overlap.

One possible approach to the track assignment is the following "greedy" heuristic.

**Algorithm 8.6:** *Greedy Algorithm for Track Selection.*

INPUT: A set of wires, each of which is either a circuit input, the output of a pullup, or a connector among some set of columns, and each of which extends over a subinterval of the list of columns. Also given is a relation $\succ$ on the wires.

OUTPUT: An assignment of wires to tracks such that the $\succ$ ordering is respected. That is, if $i \succ j$, then $i$ is assigned to a higher track than $j$, and no two wires that share a column are assigned to the same track unless they are the connector and output wires for the same column.

METHOD: Begin by picking a topological sort of the $\succ$ order, so that $i$ appears before $j$ whenever $i \succ j$. Let the order be $i_1, \ldots, i_n$. Initially, all wires are "available." We repeatedly form a new track $T$ from wires that are still available, by beginning with $T$ empty, and considering $i_1, \ldots, i_n$ in order, inserting $i_j$ into $T$ if and only if all of the following conditions hold.

1.  $i_j$ is available.
2.  $i_j$ does not overlap any member of $T$, except that a connector and output wire for the same column may overlap.

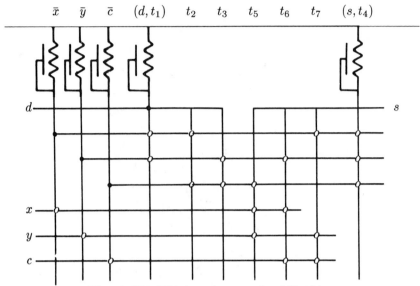

**Fig. 8.25.** Weinberger array for full adder.

3.    There is no wire $k$, either available or selected for $T$, such that $k \succ i_j$.

Any wire so selected becomes unavailable. Rule (2) protects against overlaps in one track, while rule (3) prevents a wire from being given a higher track than another wire that must precede it according to the $\succ$ relation. $\square$

**Example 8.13:** One possible topological sort of Fig. 8.24 is

10, 7, 9, 8, 4, 5, 6, 1, 2, 3

If we apply Algorithm 8.5 to this list, our first track will include 10. We add 7, because although it overlaps 10, these two wires are the output and connector for column $t_1$. Then 9 can be added, since it has no predecessors in the graph of Fig. 8.24 and does not overlap 7 or 10. Finally, 8 can also be added, because it only overlaps 9, which is another instance of the special case in rule (2). Thus, the first (highest) track will have wires 7, 8, 9, and 10.

Unfortunately, all the remaining wires overlap one another, and each must be given a track of its own. The resulting layout is shown in Fig. 8.25. As before, we use circles to represent taps. $\square$

**Wires That Change Track**

We shall not go into the detail necessary to state an algorithm, but we should remind the reader that Example 8.13 is deceptive in its simplicity. In practice, packing tracks in such a way that the $\succ$ order is preserved will lead to poor utilization of tracks. We should consider, as an alternative to Algorithm 8.5,

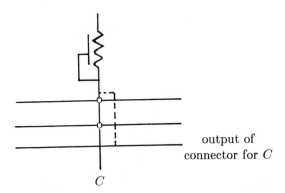

**Fig. 8.26.** Extra column used to compensate for wires in the wrong order.

a similar approach that does not refrain from packing wires into a track just because rule (3) is violated, that is, the $\succ$ order is not respected. Rather, pack wires as best we can, obeying rule (2), i.e., without packing overlapping wires into the same track.

Suppose a wire that is the output or connector for column $C$ is assigned a track that is lower than one or more taps on column $C$. Then we run a polysilicon wire up to the base of the pullup on column $C$, as illustrated in Fig. 8.26, or to the top of the column if there is no pullup. This arrangement requires extra space in the horizontal direction, of course. However, if we follow the strategy of Algorithm 8.6, first topologically sorting the wires in accord with the $\succ$ relation, we shall tend to meet relatively few wires of this sort, since we shall tend to pick wires that must be high in the array for the earlier (higher) tracks.

### Alternative Organizations for Logic Implementation

There are numerous organizations besides PLA's and Weinberger arrays that we could use as the target of a logic compilation. At least two of these have been tried in variants of the lgen compiler. In one, shown in Fig. 8.27, there is a central ground and two arrays of pullups; in effect there are two Weinberger arrays back-to-back. Horizontal wires run on either side of ground, and there will normally be some vertical wires that allow a horizontal wire not only to change tracks, but to cross to the opposite side of ground. For without such a facility, the circuit of Fig. 8.27 would really be two independent Weinberger arrays, sharing a ground.

Intuitively, the reason Fig. 8.27 may lead to logic implementations with smaller area than Weinberger arrays is that with twice as many pullups in a given horizontal space, horizontal wires can be made shorter on the average.

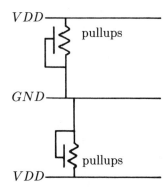

**Fig. 8.27.** An alternative form of Weinberger array.

That, in turn, may make the total area devoted to horizontal wires smaller than with Weinberger arrays. Another potential advantage of this arrangement is that it produces a circuit that tends not to be as wide as the Weinberger arrays typically constructed from logic. In compensation, the circuits of Fig. 8.27 will be about twice as high, but the improvement in the aspect ratio of the circuit is frequently beneficial.

In Fig. 8.28 we see another variation of the Weinberger array that has been tried in lgen. Here, there are two arrays of pullups, with wires from a pair of pullups, one from each side, sharing each column. Horizontal wires can tap either or both wires in a column. Figure 8.29 shows a detail of how one via, placed between paired wires, can tap either adjacent wire, and can even tap both wires in a pair, as shown to the left of the via. Note that as long as vias on one horizontal track are attached to the same wire, there is no need to worry about possible design rule violations between taps that are horizontally adjacent; they are electrically the same anyway. However, if a wire along a track ends, and another one immediately begins, there might not be room for two taps, in the geometry of Fig. 8.29. Thus, a subtle constraint on ordering of pullups and track assignments is introduced.

The reader should also note that in the geometry of Fig. 8.29, columns, consisting of two wires, require 13λ horizontal space. That compares with 8λ, the minimum space required for one of a row of pullups. Thus, the scheme of Fig. 8.28 will not actually halve the horizontal space required for a Weinberger array, but will decrease it somewhat. The advantages of Fig. 8.27 also apply to Fig. 8.28.

## SLAP

Another, rather different approach to logic implementation has been used in a system called SLAP (Silicon Layout Program), developed by S. P. Reiss and

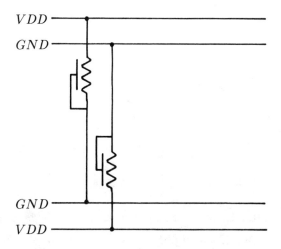

**Fig. 8.28.** Another Weinberger array variation.

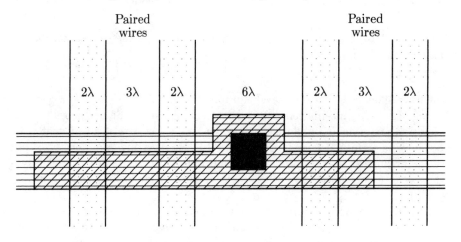

**Fig. 8.29.** Detail of Fig. 8.28.

J. E. Savage at Brown. Instead of trying to put the logic into one or a small number of arrays, SLAP builds as many levels of logic elements as there are levels in the expressions given as input.

To begin, SLAP converts logic expressions into a directed acyclic graph that is a *level graph*, meaning that all paths from any node to any of its descendant leaves have the same length. The conversion takes place by first calculating the *level* of each node, where leaves have level 0, and interior nodes are assigned, from the bottom up, a level that is one more than the largest level possessed

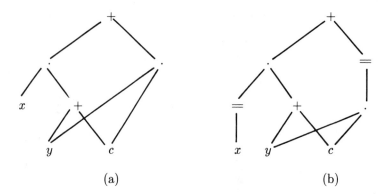

(a)                                         (b)

**Fig. 8.30.** Conversion of an expression to a level graph.

by any of its children. Then, suppose node $v$ is a child of node $u$, and these
nodes have levels $\ell$ and $m$, respectively, where $\ell < m - 1$. Then we introduce
$m - \ell - 1$ dummy nodes with an identity operator, denoted $=$, between $u$ and
$v$.

**Example 8.14:** Let us consider an alternative form of the expression for the
carry-out of the full adder

$$d = x(y + c) + yc$$

which we show as a directed acyclic graph in Fig. 8.30(a). The lower or-node,
having only leaves as children, is at level 1, as is the and-node on the right. The
other and-node, having children at levels 0 and 1, has level 2 itself, while the
root, having a child at level 2, is given level 3. By inserting two identity nodes,
we can arrange that each interior node has only paths to its descendant leaves
whose length equals the level of the node. The resulting graph is shown in Fig.
8.30(b). □

Having produced a level graph, we must select an order for the nodes
at each level, and replace them by circuit elements that perform the desired
logical operation. All connections between nodes are now wires running in the
channel between two rows of logic elements. We shall not discuss heuristics
for selecting the ordering of the logic elements. However, Sections 9.5 and 9.6
discuss channel routing and give some heuristics for minimizing the width of
the channels between the rows of logic elements.

Figure 8.31 shows a possible routing of wires between the logic elements for
the graph of Fig. 8.30(b). We have assumed that the AND and OR functions
are implemented directly by logical elements and that the left-to-right order
of nodes at each level is preserved. In practice, we would probably prefer to
redesign the expression of Fig. 8.30(a) so the operations would be NAND, NOR,
inversion, and perhaps the more general form of operator that we showed in

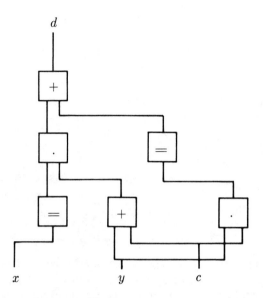

**Fig. 8.31.** SLAP implementation of logic.

Fig. 8.19. By so doing, each row of logic elements could be implemented by a Weinberger array.

## 8.4 COMPILATION OF REGULAR EXPRESSIONS

We shall now discuss how one translates a language of regular expressions, specifically the language defined in Section 7.9, into PLA's or another logic implementation. While regular expressions are no richer in their descriptive capability than finite-state languages like SLIM from Section 7.8, they are interesting in that they define sequential processes in a way that is much further removed from a particular implementation than is a finite-state language. They appear useful for processes, like controllers or communication protocols, that are expressible as actions to be taken when specified patterns of inputs are seen. Thus, we see techniques for compiling regular expressions as suggestive of what can be done in the future with languages that are at a much higher level of abstraction than those typically implemented in the early 1980's.

The particular method chosen to translate regular expressions is first to construct a nondeterministic finite automaton that is closely related to the expression. We could use standard techniques to convert the NFA into a deterministic finite automaton (*DFA*). The DFA can be implemented by choosing a coding for the states; each code is a distinct vector of $k$ bits, where $2^k$ is at least the number of states. Then, we may design a PLA that feeds back the bits representing the current state, and computes the bits of the next state by a set of terms, each of which involves the input bits and the state bits fed back.

However, we chose not to do so for two reasons. First, the DFA might have a number of states that is too large to handle. More importantly, we might lose a great deal of information about the structure of the problem, information that might be extracted from the DFA using general-purpose tools for finite automaton implementation, but only with difficulty.

Instead, we build PLA's directly from the NFA's, by picking a coding for the nondeterministic states so that we can always recognize that we are in a particular state, regardless of what other states the NFA is simultaneously in. Moreover, we find that there are really two different ways to implement the NFA for a regular expression, which we dub the "before" and "after" methods. These methods produce PLA's with different characteristics, and it is hard to predict when one method will be superior to the other.

### Construction of NFA's from Regular Expressions

We begin our discussion by giving the basic NFA construction. The language we use as input is the language of Section 7.9, but for simplicity we do not consider state symbols. The reader can easily modify our construction to take those symbols into account.

We also make a convention that when control reaches an output symbol, the control is not passed anywhere from that output position at the next cycle (although control may continue at other positions of the expression, as usual). This convention poses no hardship, since the compiler we have designed replaces an expression $E_1 U E_2$ by $E_1(U + E_2)$, where $U$ is an output symbol. Then, the control flow we expected intuitively for the first expression will be achieved formally by the second. That is, when $E_1$ is recognized on the input, the output signal $U$ will be raised, and we are also ready to recognize $E_2$.

The algorithm for translating regular expressions to NFA's must be explained in stages, each of which is a recursive definition of a function of regular expressions. As a global environment, we identify operands of the regular expression with the NFA states, as was discussed in Example 7.24, e.g. We, therefore, assume that each operand $a$ of the expression is subscripted with a unique integer $i$, and that the state corresponding to this occurrence of $a$ is $i$.

The first definition we need is the predicate $\epsilon(R)$, which is true if and only if the regular expression $R$ generates the empty string. This predicate can be defined recursively, by induction on the number of operators in the expression, as follows.

E1) (Basis—zero operators) $\epsilon(a_i)$ = false.
E2) $\epsilon(R + S) = [\epsilon(R)$ or $\epsilon(S)]$.
E3) $\epsilon(RS) = [\epsilon(R)$ and $\epsilon(S)]$.
E4) $\epsilon(R^*) = \epsilon(R?)$ = true.
E5) $\epsilon(R++) = \epsilon(R)$.

Next, we need to know those states (positions of the expression) that are initial, in the sense that they could match the first input symbol seen, if that input symbol were the right one. We need this information not about the expression as a whole, but about many of its subexpressions; the purpose of that information is to determine which states can be the successor of which states. We, therefore, define a function $init(R)$ that gives the states in $R$ that are initial. The recursive rules followed are:

I1) (Basis) $init(a_i) = \{i\}$. That is, the state corresponding to an operand is initial in the expression consisting of that operand alone.

I2) $init(R+S) = init(R) \cup init(S)$. The initial states of expression $R+S$ are those of $R$ that are initial (in $R$) and those of $S$ that are initial (in $S$).

I3) If $\epsilon(R)$ is true, then $init(RS) = init(R) \cup init(S)$; otherwise $init(RS) = init(R)$.

I4) $init(R^*) = init(R?) = init(R{+}{+}) = init(R)$. Applying one of the unary operators does not change the initial set.

Similarly, we need to know the set $fin(R)$ of positions that could match the last symbol of a string generated by $R$. The rules are practically the same as those for $init$, except for the fact that an output symbol is not permitted in the $fin$ set. The reason for this detail is so the states corresponding to output symbols will not be given successor states in the NFA.

F1) (Basis) $fin(a_i) = \{i\}$ if $a$ is an input symbol, and $fin(a_i)$ is empty if $a$ is an output symbol.

F2) $fin(R+S) = fin(R) \cup fin(S)$.

F3) If $\epsilon(S)$ is true, then $fin(RS) = fin(R) \cup fin(S)$; otherwise, $fin(RS) = fin(S)$.

F4) $fin(R^*) = fin(R?) = fin(R{+}{+}) = fin(R)$.

Now, we are ready to define the successor state function; $succ(s)$ is the set of states that are successors of state $s$. Note that since state $s$ is associated with a particular operand, which is a particular symbol $a$, the successors of $s$ may be activated only when $a$ is the next input. Initially, we let $succ(s)$ be empty for all $s$. Then we consider the parse tree for the given regular expression and examine each subexpression thereof. The rules for adding successors are the following.

S1) For each subexpression of the form $RS$, each $s$ in $fin(R)$, and each $t$ in $init(S)$, add $t$ to $succ(s)$.

S2) For each subexpression of the form $R^*$ or $R{+}{+}$, each $t$ in $init(R)$, and each $s$ in $fin(R)$, add $t$ to $succ(s)$.

Lastly, the initial states of the NFA so constructed are those in $init(R)$, where $R$ is the complete expression.

**Example 8.15:** Let us reconsider the bounce filter expression of Example 7.24, but for succinctness we shall use $a$ for "any" or dot, $o$ for *one*, $z$ for *zero*, and $U$ for *OUT*. The expression, with subscripts to name the states, is

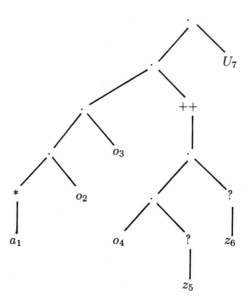

**Fig. 8.32.** Parse tree for regular expression.

$$a_1{}^*o_2o_3(o_4z_5?z_6?)++U_7$$

The parse tree for the expression is shown in Fig. 8.32.

For example, $\epsilon(a_1)$ is false, but $\epsilon(a_1{}^*)$ is true, since the empty string matches any starred expression. As another example, $init(z_5) = fin(z_5) = \{5\}$, because any one-operand expression is both $init$ and $fin$ of itself. By rules (I4) and (F4), $\{5\}$ is also $init(z_5?)$ and $fin(z_5?)$. These rules make sense here, since an expression that is an optional symbol will match that symbol as both the first and last symbol of a matching string. For a more complex example,

$$fin(o_4z_5?) = fin(o_4) \cup fin(z_5?) = fin(o_4) \cup fin(z_5) = \{4,5\}$$

The first of these steps is by rule (F3). Note the "if" condition of that rule is satisfied because $\epsilon(z_5?)$ is true. The second step is by rule (F4).

As an example of the computation of successors, note that $init(o_4z_5?z_6?) = \{4\}$ and $fin(o_4z_5?z_6?) = \{4,5,6\}$. Therefore, when we process the ++ operator, we discover by rule (S2) that $succ(4)$, $succ(5)$, and $succ(6)$ all contain 4. The complete diagram of transitions for the NFA of this expression was already shown in Fig. 7.38. There, states $N_1, \ldots, N_6$ correspond to what we have called states 1 through 6, and $OUT$ corresponds to 7. $\square$

### Before and After Interpretation of NFA's

There are two distinct ways that we could convert the successor function of an NFA into logic; we shall use PLA logic as an example, but the ideas for translation of NFA's into, say, lgen is quite similar. The first style, called the *before method*, sees a state as saying "we are ready to recognize my input symbol." This was the interpretation we put on states in Example 7.24.

The before method has the nice property that for every noninitial state, there is at most one term of the PLA needed. The term for a state checks for the input corresponding to the state, and checks that we are currently in that state. If both these conditions are met, it turns on the bits for all the successor states. If a state is initial, then two terms may be needed. The first is the same as for noninitial states, while the second checks for the start signal (rather than the state itself) and the proper input. However, the first of these two states is unnecessary if the state is not a successor of any other state, since then the start wire suffices to represent the state the first and only time that state is entered.

Fig. 7.39 is a PLA constructed from the NFA of the bounce filter using the before method. That PLA also was designed using the simple policy that for each NFA state there would be a single feedback wire. Shortly, we shall discuss other ways to code the states, but whatever method we use for state encoding, the before method gives us a number of terms that does not exceed the number of states, counting twice for initial states.

The other interpretation of states is called the *after method*. Here, a state says "we have just recognized my input symbol, and we are ready to recognize the symbol belonging to any of my successors." Thus, the graph of the NFA does not change from what it was in the before method, but the labels of the arcs change. That is, the same transitions are now made on different input symbols.

**Example 8.16:** Figure 8.33 is the same NFA as Fig. 7.38; we have even changed back to the notation of that section regarding states, with states $1, 2, \ldots, 7$ replaced by $N_1, \ldots, N_6$, and $OUT$. However, now each transition arc is labeled not by the symbol associated with its tail, as in Fig. 7.38, but with the symbol of its head. As a special case, we have labeled arcs entering an output state with $\epsilon$, indicating that the output must be made immediately, without waiting for an input. For example, since there is an $\epsilon$-arc from $N_4$ to $OUT$, we emit the output $U$ whenever we enter state $N_4$. That arrangement makes sense in the after interpretation, since when we enter state $N_4$ we have already seen the third 1. In contrast, in the before interpretation, $N_4$ said we were ready to see the third 1, but had not yet seen it. $\Box$

We design a PLA from the after interpretation of an NFA almost as we do in the before method. The significant difference is that now we may need more than one term per state of the NFA. In particular, a state that has transitions on $k$ different input symbols will, except by luck, require $k$ terms. We shall similarly need a term for each transition from the start state to a state of the

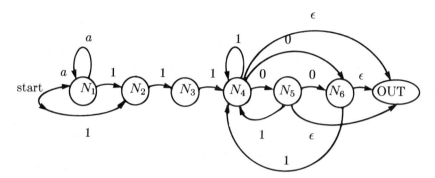

**Fig. 8.33.** "After" method of interpreting bounce filter NFA.

| 1 | 2 | 2 | 2 | 2 | 2 | 2 | 2 | | 1 | 0 | 0 | 0 | 0 | 0 | 0 |
|---|---|---|---|---|---|---|---|---|---|---|---|---|---|---|---|
| 1 | 1 | 2 | 2 | 2 | 2 | 2 | 2 | | 0 | 1 | 0 | 0 | 0 | 0 | 0 |
| 2 | 2 | 1 | 2 | 2 | 2 | 2 | 2 | | 1 | 0 | 0 | 0 | 0 | 0 | 0 |
| 2 | 1 | 1 | 2 | 2 | 2 | 2 | 2 | | 0 | 1 | 0 | 0 | 0 | 0 | 0 |
| 2 | 1 | 2 | 1 | 2 | 2 | 2 | 2 | | 0 | 0 | 1 | 0 | 0 | 0 | 0 |
| 2 | 1 | 2 | 2 | 1 | 2 | 2 | 2 | | 0 | 0 | 0 | 1 | 0 | 0 | 1 |
| 2 | 1 | 2 | 2 | 2 | 1 | 2 | 2 | | 0 | 0 | 0 | 1 | 0 | 0 | 1 |
| 2 | 0 | 2 | 2 | 2 | 1 | 2 | 2 | | 0 | 0 | 0 | 0 | 1 | 1 | 1 |
| 2 | 1 | 2 | 2 | 2 | 2 | 1 | 2 | | 0 | 0 | 0 | 1 | 0 | 0 | 1 |
| 2 | 0 | 2 | 2 | 2 | 2 | 1 | 2 | | 0 | 0 | 0 | 0 | 0 | 1 | 1 |
| 2 | 1 | 2 | 2 | 2 | 2 | 2 | 1 | | 0 | 0 | 0 | 1 | 0 | 0 | 1 |

$start\ input\ N_1\ N_2\ N_3\ N_4\ N_5\ N_6$         $N_1\ N_2\ N_3\ N_4\ N_5\ N_6\ OUT$

**Fig. 8.34.** PLA constructed by the after method.

NFA; in effect the start signal now represents a state by itself.

Each term for state $s$ checks in the and-plane that the proper symbol $a$ is seen on the input, and it checks that the NFA is in the state $s$, perhaps among others. Then, in the or-plane, this term wire turns on the columns that represent each successor of state $s$ on input $a$. If $s$ has $\epsilon$-arcs to any output symbols, the wires for those outputs are also turned on.

**Example 8.17:** The PLA constructed by the above strategy, using a single wire for each state, is shown in Fig. 8.34. The first two rows are transitions from *start*, on inputs 1 and $a$. The latter input is the dot, or "any" symbol, and it appears in the PLA by having 2, or "don't care," in the *input* column. The next two rows represent transitions out of $N_1$ on the same two inputs. There are also a pair of transitions for states $N_4$ and $N_5$. □

Note that the PLA of Fig. 8.34 has three more rows than that of Fig.

7.39. That makes sense, because three of the NFA states have two different input symbols on which they make transitions, under the after interpretation. The reader may wonder whether the after method can ever yield a better PLA than the before method. We shall see that it can for two reasons. First, the PLA's designed in the straightforward way may be reducible by the techniques of Section 8.1. Second, we shall see when we discuss NFA state codings, that there are other ways to encode NFA's besides the straightforward one used in Figs. 7.39 and 8.34, where there is one feedback bit for each state. In practice, there is frequently a better way to code NFA states, and the after method facilitates these better methods. The result is that while before-method PLA's tend to have fewer terms, after-method PLA's will typically have fewer columns, because the states can be coded in fewer bits.

## Compatible Symbols and Conflicting States

We shall now discuss how one goes about finding more succinct codings for nondeterministic states than the obvious one-wire-per-state code. To begin, we must determine which pairs of states can be on at the same time, i.e., after some input sequence. These pairs, which we call *conflicting states*, prevent our using certain succinct codes for NFA states, so it is important to know about them. In the extreme case where all pairs conflict, we cannot do anything better than the one-wire-per-state code, taking $n$ feedback wires for $n$ states. At the other extreme, if the NFA is really deterministic, there will be no conflicts, and we can code state $i$, $0 \leq i < n$, by $i$ in binary, thereby taking $\log n$ bits to code $n$ states. Random pairs of conflicts will give us something between $\log n$ and $n$ feedback wires.

Part of the difficulty in computing conflicting states concerns the fact that in the regular expression language of Section 7.9, more than one input symbol can be present at a time, since they are defined in terms of input wires, and two symbols could be defined in ways that do not require any one wire to be both 0 and 1. We say symbols $a$ and $b$ are *compatible* if for no wire $x$ does the definition for $a$ require $x$ to be on while $b$ requires $x$ to be off, or vice versa.

A sketch of an algorithm to compute all pairs of conflicting states for an NFA is given in Fig. 8.35. It computes a Boolean predicate or matrix $conflict(s, t)$ that is set true if and only if states $s$ and $t$ conflict. However, we assume that $s$ and $t$ are states that correspond to an input symbol, since there is no reason to care about conflicts involving output states; they can never be coded because their signal is wanted outside.

The initialization depends on whether the before or after method NFA is wanted, and the iteration phase, when we build more conflicting pairs by seeing where we can go with compatible input symbols, also depends on the method implicitly. That is, since the addition of new conflicting pairs depends on the symbols under which transitions are made, the test for whether $u$ and $v$ conflict

```
{ initialize }
for all states s and t do
    if s = t then conflict(s, t) := true
    else conflict(s, t) := false;
if before method NFA then
    for all initial states s and t do
        conflict(s, t) := true
else { after method NFA }
    for all initial states s and t do
        if s and t are associated with symbols that are compatible then
            conflict(s, t) := true;

{ iteration }
repeat
    for all (possibly equal) states s and t such that conflict(s, t) do
        for each transition out of s, say to u on symbol a, and each
            transition out of t, say to v on symbol b do
            if a and b are compatible then conflict(u, v) := true
until no further changes to the set of conflicting pairs
```

**Fig. 8.35.** Algorithm to compute conflicting states.

will be different under the two different NFA interpretations.

**Example 8.18:** The bounce filter is not a good example of conflicts, because in the before interpretation, every state conflicts with every other, and even in the after interpretation, that is almost true. We, therefore, consider the meaningless but instructive expression

$$ab^{++}acU + acbcU$$

where the symbols involved are declared by

line $x, y$
symbol $a(-x, -y)$, $b(x)$, $c(y)$
output $U$

We first note that symbols $b$ and $c$ are compatible, because they depend on no wire being on and off simultaneously. Specifically, if $x = y = 1$, then both $b$ and $c$ will be seen. However, $a$ is compatible with neither $b$ nor $c$. It is not compatible with $b$ because those two symbols disagree on wire $x$, and $a$ is not compatible with $c$ because of wire $y$.

Let us thus compute the conflicts for the states of the regular expression above. First, while we have not discussed the matter, the regular expression compiler does certain algebraic manipulations of expressions, such as the re-

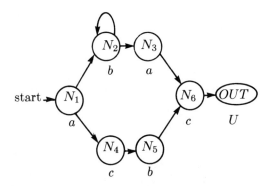

**Fig. 8.36.** NFA transitions.

placement of $E_1 U E_2$ by $E_1(U + E_2)$, when $U$ is an output symbol, discussed earlier. In the present case, the expression would be factored, both left and right, yielding the expression

$$a_1(b_2{}^{++}a_3 + c_4b_5)c_6U_7$$

We have introduced subscripts for the positions to indicate the states; we again refer to the state for position $i$ as $N_i$, but the output position 7 is called $OUT$. The transitions of this NFA are shown in Fig. 8.36. We have indicated the input symbol under each state, but we have not said whether we want a before or after interpretation. Of course, in the before case, transitions would be labeled by the symbol of the tail, and in the after case by the symbol of the head.

In either the before or after interpretation, the initialization does not produce any pairs of conflicting states, because $N_1$ is the only initial state. In the before interpretation, we find that $N_2$ and $N_4$ are conflicting, because they are successors of conflicting "states" (both $N_1$) on compatible symbols (both $a$). Similarly, $conflict(N_2, N_3)$ is true because they are each successors of $N_2$ on $b$.

The pair $(N_2, N_4)$ gives us $conflict(N_2, N_5) =$ true and $conflict(N_3, N_5)$ $=$ true, because symbols $b$ and $c$ are compatible, $N_5$ is a successor of $N_4$ on $c$, and $N_2$ and $N_3$ are successors of $N_2$ on $b$. The pair $(N_2, N_3)$ gives us no new conflicts, because these two states have all their transitions on noncompatible symbols. The same applies to the conflicting pair $(N_3, N_5)$. However, the pair $(N_2, N_5)$ yields $conflict(N_2, N_6) =$ true and $conflict(N_3, N_6) =$ true, since $N_6$ is a successor of $N_5$ on $b$ and $N_2$ and $N_3$ are successors of $N_2$ on $b$; surely $b$ is compatible with itself. No other conflicting pairs can be found; recall that we do not consider conflicts involving $OUT$. The graph of the conflict matrix is shown in Fig. 8.37(a).

Now, let us consider the conflicts in the after method NFA. Again we find that $conflict(N_2, N_4) =$ true, this time because they are successors of

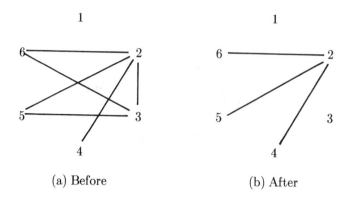

(a) Before  (b) After

**Fig. 8.37.** Conflict graphs.

conflicting "states" (both $N_1$) on compatible symbols $b$ and $c$. However, $N_2$ and $N_3$ do not conflict, because although they are successors of the same state, the symbols on which the transitions are made in the after interpretation are $a$ and $b$, which are not compatible.

The pair $(N_2, N_4)$ implies that its successor pair on symbol $b$, $(N_2, N_5)$ is also a conflicting pair, but the other successor pair $(N_3, N_5)$ is not a conflict, because the symbols causing the transitions are not compatible. Similarly, we find $conflict(N_2, N_6)$, but there are no more conflicting pairs. The graph of the conflict matrix for the after interpretation is shown in Fig. 8.37(b). $\square$

## State Coding

Now that we know the conflict matrix among the states, we can decide on a state coding. Evidently, the conflicts and, therefore, the coding will depend on whether we have chosen the before or after method, but we assume that choice is fixed at this stage.

In what follows, we make the simplifying assumption that whenever a set of states forms a clique in the graph of the conflict matrix, then it is possible for all those states to be on simultaneously. A moment's thought will convince the reader that that is not necessarily true, but the converse is true; i.e., if there is some input that makes the NFA enter some set of states $S$, then each pair of states in $S$ conflicts. Thus, by making our assumption, we may miss the fact that certain combinations of states are impossible, even though they are a clique in the conflict graph; but we shall never make the fatal error of assuming a set of states cannot be on simultaneously, when in fact they can. Errors of the latter kind would be fatal because the state coding we chose might allow us to be fooled into thinking a certain state was on, when in fact an unexpected combination of other states was on.

A state coding consists of an integer $k$, the number of bits used for the encoding, and an assignment of a vector of 0's, 1's, and 2's to each state. The 2's will function as "don't care's," as in the and-plane of PLA personalities. We shall use $C[s, i]$ to stand for the $i^{th}$ position in the code for state $s$.

Let us think of the NFA implemented by a PLA, with $k$ feedback wires corresponding to the $k$ code bits. If we enter a state $s$, then in the or-plane, we shall turn on all those wires $i$ for which $C[s, i] = 1$. Wires $i$ for which $C[s, i]$ is 0 or 2 are not turned on by state $s$, although of course they may be turned on if we are simultaneously in another state.

In the and-plane, we must detect whether state $s$ is on or not, and we must be able to do so regardless of what other states are on also, provided the set of on states forms a clique of the conflict graph. The intent of the code for state $s$ is that we shall deem state $s$ on whenever the 0's and 1's of its code match the actual situation. That is, in the and-plane, terms that require state $s$ to be on check that bit $i$ is 0 if $C[s, i] = 0$; they check that bit $i$ is 1 if $C[s, i] = 1$, and they do not check bit $i$ if $C[s, i] = 2$. In order that this check be accurate, two conditions must hold.

1.  *If $s$ is on, then $s$ is detected.* If $s$ is turned on in the or-plane, then those $i$ for which $C[s, i] = 1$ will surely be fed back as 1, so no error could occur from these bits. However, a bit $i$ where $C[s, i] = 0$ could be changed to 1 because some other state was on. To assure that does not happen, we need to know that

    if $C[s, i] = 0$ and $t$ conflicts with $s$, then $C[t, i] \neq 1$

2.  *If $s$ is detected, then $s$ is on.* The only way $s$ could be detected without being on is if there were some set of states $T$, excluding $s$, such that there is some input that makes $T$ be exactly the set of states that are on,† and $T$ simulates the code for $s$. That is, there must never be a set $T$ such that

    i)   If $C[s, i] = 0$, then for all $t$ in $T$, $C[t, i] \neq 1$.

    ii)  If $C[s, i] = 1$, then there is some $t$ in $T$ such that $T[t, i] = 1$.

**Example 8.19:** The straightforward coding scheme of devoting a unique bit to each state is easily seen to satisfy the above conditions. Here, the code for state $i$ has 1 in position $i$ and 2 elsewhere. Then condition (1) holds vacuously, as does condition (2i). Condition (2ii) is trivial, because if $C[s, i] = 1$, then there are no states $t \neq s$ with $C[t, i] = 1$, and, therefore, any clique $T$ satisfying (2ii) would have to include $s$. □

There are many ways to find better state encodings than the obvious one. The following is a simple but effective method.

**Algorithm 8.7:** *Greedy Nondeterministic State Assignment Method.*

† Recall that we approximate this condition by requiring only that $T$ be a clique (not necessarily maximal) of the conflict graph.

| $N_1$ | 0 | 1 | 2 | 2 |
| $N_2$ | 1 | 0 | 2 | 2 |
| $N_3$ | 1 | 1 | 2 | 2 |
| $N_4$ | 2 | 2 | 0 | 1 |
| $N_5$ | 2 | 2 | 1 | 0 |
| $N_6$ | 2 | 2 | 1 | 1 |

**Fig. 8.38.** State code for conflict diagram of Fig. 8.37(b).

INPUT: An NFA with conflicting pairs of states indicated.

OUTPUT: A code for each state that allows any state to be detected, as long as whatever set of states is on consists only of mutually conflicting states.

METHOD: The algorithm builds *groups* of mutually nonconflicting states in a "greedy" way. To form a single group from the remaining states, we order them highest degree (in the conflict graph) first. Then we visit each state in turn, selecting it for the group if it does not conflict with any of the previously selected states of that group. After building a group, we delete its states from the conflict graph, and repeat the above process until all states have been assigned to a group.

Next, for each group of $m$ states, we dedicate $\lceil \log_2(m + 1) \rceil$ bits. Each member of the group has a distinct, nonzero code in the bits for that group and has 2's (don't care) in the bits for all the other groups. $\square$

**Example 8.20:** Consider the conflict graph of Fig. 8.37(b) for the after method NFA of Fig. 8.36. Algorithm 8.7 will select $N_2$ for the first group, since it is the highest degree node. Then, it will add to the group the states $N_1$ and $N_3$, which do not conflict with $N_2$ or with each other. The remaining states are mutually nonconflicting, so they are placed in the second and last group.

Groups of three states require two bits for their codes. We show in Fig. 8.38 one possible state code, in which the first two bits are for the first group $\{N_1, N_2, N_3\}$, and the third and fourth bits are for the group $\{N_4, N_5, N_6\}$. Then, in Fig. 8.39 we show the PLA personality derived from the code of Fig. 8.38 and the transitions of the NFA in the after interpretation. The column $s$ represents the start signal, and the four state code bits are $c_1, \ldots, c_4$.

For example, the first line says that if the start signal is on and the input $a$ is seen ($x = y = 0$), then enter state $N_1$, represented by the state code $c_1 = 0$ and $c_2 = 1$. Note that 2's in the state code become 0's in the or-plane. The last line says that if the input $c$ is seen (wire $y$ is 1; wire $x$ is arbitrary), and we are in state $N_5$, detected by the fact that $c_3 = 1$ and $c_4 = 0$, then we enter state $N_6$ by turning on $c_3$ and $c_4$ for the next cycle. Note that we also turn on the output signal $U$, since $OUT$ is a successor of $N_6$.

For comparison, we could apply Algorithm 8.7 to the before method NFA

| 1 | 0 0 | 2 2 2 2 | 0 1 0 0 0 |
| 2 | 1 2 | 0 1 2 2 | 1 0 0 0 0 |
| 2 | 2 1 | 0 1 2 2 | 0 0 0 1 0 |
| 2 | 1 2 | 1 0 2 2 | 1 0 0 0 0 |
| 2 | 0 0 | 1 0 2 2 | 1 1 0 0 0 |
| 2 | 2 1 | 1 1 2 2 | 0 0 1 1 1 |
| 2 | 1 2 | 2 2 0 1 | 0 0 1 0 0 |
| 2 | 2 1 | 2 2 1 0 | 0 0 1 1 1 |
| $s$ | $x\ y$ | $c_1\ c_2\ c_3\ c_4$ | $c_1\ c_2\ c_3\ c_4\ U$ |

**Fig. 8.39.** PLA derived by after method using state code of Fig. 8.38.

| $N_1$ | 0 | 1 | 2 | 2 | 2 |
| $N_2$ | 1 | 0 | 2 | 2 | 2 |
| $N_3$ | 2 | 2 | 1 | 2 | 2 |
| $N_4$ | 2 | 2 | 2 | 0 | 1 |
| $N_5$ | 2 | 2 | 2 | 1 | 0 |
| $N_6$ | 2 | 2 | 2 | 1 | 1 |

**Fig. 8.40.** State coding for before NFA.

with the conflict graph of Fig. 8.37(a). There, the greedy algorithm does not perform well, selecting groups $\{\,N_1, N_2\,\}$, $\{\,N_3, N_4\,\}$ and $\{\,N_5, N_6\,\}$. The reason this coding is bad has to do with properties of group sizes that the greedy approach cannot predict. Specifically, a group of size two requires

$$\lceil\log_2(2+1)\rceil = 2$$

bits; thus this code requires six bits, just as the straightforward code does. If we take advantage of the fact that one less than a power of two is the best possible size for a group, we would prefer the grouping $\{\,N_1, N_2\,\}$, $\{\,N_3\,\}$, and $\{\,N_4, N_5, N_6\,\}$, which only requires five bits.

The resulting state code, with the first two bits for $\{\,N_1, N_2\,\}$, the third for $\{\,N_3\,\}$, and the last two for $\{\,N_4, N_5, N_6\,\}$, is shown in Fig. 8.40; the resulting PLA is shown in Fig. 8.41. For example, the first line says that when the start signal is on, and the input $a$ ($x = y = 0$) is seen, enter states $N_2$ and $N_4$. The combination of bits $c_1 c_2 = 10$ in the or-plane will cause $N_2$ to be detected at the next cycle, since we know that $c_2$ cannot be set to 1 by another state; the only state that does so, $N_1$, does not conflict with $N_2$. Similarly, the bits $c_3 c_4 = 01$ in the or-plane will cause $N_4$ to be recognized at the next cycle.

Note that a term that detects the state $N_1$ and input $a$ is unnecessary, since $N_1$ is not a successor of any state. Also observe that the predicted comparison between the before and after methods holds true here; The after PLA has fewer

| $s$ | $x$ | $y$ | $c_1$ | $c_2$ | $c_3$ | $c_4$ | $c_5$ | $c_1$ | $c_2$ | $c_3$ | $c_4$ | $c_5$ | $U$ |
|-----|-----|-----|-------|-------|-------|-------|-------|-------|-------|-------|-------|-------|-----|
| 1   | 0   | 0   | 2     | 2     | 2     | 2     | 2     | 1     | 0     | 0     | 0     | 1     | 0   |
| 2   | 1   | 2   | 1     | 0     | 2     | 2     | 2     | 1     | 0     | 1     | 0     | 0     | 0   |
| 2   | 0   | 0   | 2     | 2     | 1     | 2     | 2     | 0     | 0     | 0     | 1     | 1     | 0   |
| 2   | 2   | 1   | 2     | 2     | 2     | 0     | 1     | 0     | 0     | 0     | 1     | 0     | 0   |
| 2   | 1   | 2   | 2     | 2     | 2     | 1     | 0     | 0     | 0     | 0     | 1     | 1     | 0   |
| 2   | 2   | 1   | 2     | 2     | 2     | 1     | 1     | 0     | 0     | 0     | 0     | 0     | 1   |

**Fig. 8.41.** Before method PLA.

columns and the before PLA has fewer rows. $\square$

**Theorem 8.1:** The state code of Algorithm 8.7 yields a correctly working PLA.

**Proof:** We must check the two conditions that assure us the state $s$ is detected if and only if it is actually on. If $C[s, i] = 0$, and $C[t, i] = 1$, then $s$ and $t$ must be in the same group. Therefore, they do not conflict, and condition (1) is satisfied.

For condition (2), the only way there could be a set $T$ of states simulating the 0's and 1's of $s$'s code would be if the members of $T$ were in $s$'s group. But then, since members of a group are nonconflicting, $T$ has only one member, say $t$. As members of a group are given distinct codes, there must be some bit $i$ where $C[s, i] = 0$ and $C[t, i] = 1$, or vice versa, so $(2i)$ and $(2ii)$ cannot both be satisfied by $T$. $\square$

### Partitioning of Regular Expressions

We shall briefly mention one other optimization concerning the compilation of regular expressions; we may break an expression into several that have very limited interaction. The pieces may each be implemented, say by a PLA, and the network of resulting PLA's may take up considerably less area than a single PLA constructed by the methods we have described. Recall our discussion at the end of Section 8.1 regarding why the splitting of one PLA into two may save as much as half the area.

One important fact that makes regular expression partitioning easy is that given any regular expression, with at least two operands, we can find a subexpression that has between 1/3 and 2/3 of the operands. The expression can thus be split into two of roughly equal size, one of which is the subexpression, and the other of which is the original expression, with a new *dummy* symbol replacing the excised subexpression.

**Example 8.21:** The bounce filter expression has seven operands. We can select a subexpression like $(o_4 z_5 ? z_6 ?) + +$, which has three operands, and split the expression into

$$main = a_1 o_2 o_3 D U_7$$
$$D = (o_4 z_5 ? z_6 ?)++$$

Here, $D$ is the new dummy symbol. $\square$

Each of the two expressions resulting from a split can be split recursively, since the dummy symbol may be regarded as if it were an ordinary input symbol. In practice, one would not split expressions as small as the bounce filter; experience shows that the smallest expressions that can be split profitably are in the 25–50 operand range.

When we implement the main expression, the state corresponding to the dummy symbol $D$ must be given a wire of its own; we cannot code it in the manner discussed above. This output wire of the main PLA is connected to the "subroutine" PLA, where it becomes the start signal. Note that the start signal of the subroutine PLA may thus be raised many times, not just at the beginning of operation. A simple modification of the conflict calculation of Fig. 8.35, to take into account that the start signal may be raised at any time, is left as an exercise.

So that the "subroutine" PLA may return a signal to the calling PLA, every final state (those in $fin(R)$, where $R$ is the subexpression represented by $D$) turns on a special output wire of the subroutine PLA. This wire is fed to the calling PLA, where it becomes the signal indicating that the state corresponding to symbol $D$ is on.

There are a number of issues that are raised by the decomposition of regular expressions, but we do not go into the details here. For example, we have ignored the role of state symbols. It is possible that when we split an expression, the "label" corresponding to a state is in one part and one or more of the "goto's" to that label are in the other. These connections require additional wires running between the two PLA's.

Another issue we shall elide is the clocking involved in the wires that implement "call" and "return" for "subroutines." Depending on whether the before or after method is in use, different wires must leave or arrive at a PLA without passing through a clock gate. Moreover, if several recursive decompositions have been made, and dummy symbols are final in expressions where they appear, then we can face a situation where signals must propagate, on one clock phase, through many PLA's and over long wires. The methods of Section 5.2, converting nonsystolic networks into systolic ones can be of use here, provided the subexpressions involved do not generate the empty string. The effect will be that "subroutine" PLA's operate delayed one input cycle behind the PLA's that call them, and long propagations are avoided.

## 8.5 TOWARD SILICON COMPILATION

The goal of work in compilation and optimization algorithms for VLSI design systems is an environment similar to that in which the programmer now finds himself, one in which algorithms can be stated in one of several high-level languages, such as Pascal, or even APL, and compilers can produce a reasonably good implementation of his algorithm. However, in the VLSI world, the implementation is not a machine language program, but rather a chip that implements the algorithm. Although in the literature the term "silicon compilation" has been used to refer to almost any kind of design system, we would like to reserve the words for compilation of languages that are essentially general-purpose programming languages into layout or into a sufficiently low-level language that a layout can be obtained with relatively little extra effort.

In the early 1980's, compilers that take ordinary programming languages and produce layouts are actively researched, although existing facilities tend to produce circuits that are an order of magnitude larger than what one could produce from a lower level design. However, we believe that a more careful study of optimization techniques is in the offing, with some ideas borrowed from optimization in conventional compilers, with ideas from the design of parallel algorithms, and most likely, with some completely new concepts to be discovered. We shall therefore in this section survey the principal ideas and problems in the area of true silicon compilation.

A frequently used idea in silicon compilation is that the chip whose layout is output of the compiler will look and behave very much like a microprocessor, with a control unit and data path, as illustrated in Fig. 8.42. The data path consists of a collection of registers and arithmetic units connected by a bus or busses. The registers represent the variables of the program, and the arithmetic units perform on variables those operations that are indicated by the program. The control unit determines the sequence of events in the data path, e.g., when the contents of registers must be sent to arithmetic units, when those units perform their operations, and when the results of operations are delivered to registers.†

In the simplest case, the control can be thought of as a deterministic finite automaton, whose states correspond to the statements of the program in an obvious way. To make matters simpler, the source program should be translated into a "three-address" form of intermediate code.‡ In the intermediate code, statements have a single operator at most, e.g., $A := B + C$. Some are simple copy statements of the form $A := B$. Branching is controlled by three-address statements of the form **if** $A < B$ **goto** $C$, where "$<$" can be any comparison between $A$ and $B$, and $C$ is a statement designator, that is, a state of the

---

† An interesting exception to this generalization is the language Xi (Feldman [1982]), which generates a less specific intermediate form of output. For example, it can generate lgen code, and the lgen compiler can then produce an organization that looks nothing like Fig. 8.42.

‡ See Aho and Ullman [1977] for a discussion of intermediate languages and compilation in general.

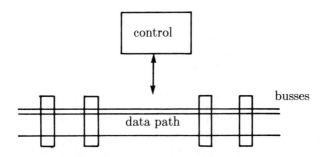

**Fig. 8.42.** Organization of chip generated by a silicon compiler.

controller DFA.

With code of the three-address form, the states corresponding to assignments or copy statements will issue a limited sequence of instructions to the data path, and then transfer control to the next state in sequence. States corresponding to branching statements will also issue a limited sequence of commands to the data path, after which the result of the comparison will be available to the data path. If the condition is met, the state indicated by the goto portion will be the next state, and otherwise, the next state in sequence will be taken. The implementation of a control of this type is quite like the implementation of SLIM programs as PLA's discussed in Section 7.8.

One might naturally compare the process of silicon compilation into a microprocessor-like organization with the programming of a microprocessor, since the same high-level source program could, in principle, serve for either. Why would one bother with silicon compilation at all, given that one could just buy a microprocessor chip for far less than one could design and manufacture the special-purpose chip by the process of silicon compilation? The answer is that the design generated by the silicon compiler can be many times faster than the microprocessor loaded with the source program compiled into machine language. Indeed, one would rarely consider designing a special-purpose chip by any means whatsoever, unless the intended purpose required such speed that a microprocessor with loaded program could not operate fast enough.

Another natural question is why the special-purpose chip should run faster than the microprocessor. One reason is that the special-purpose chip saves one level of instruction decoding and the time it takes. For example, the control in Fig. 8.42 might be a microprogram, and each step of the microcode will be decoded into a sequence of operations on the data path. In comparison, the microprocessor must read and decode each individual instruction, and once the instruction is determined, most likely go into a microprogrammed control to generate the proper sequence of events for the data path to perform.

However there are, at least potentially, much greater speedups possible

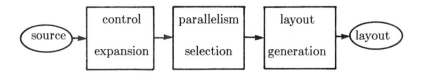

**Fig. 8.43.** Stages of silicon compilation.

from silicon compilation than are obtainable by embedding a program in a microprogram control. Two primary sources of additional speedup are

1.  *Expansion of control primitives.* A primary example is the expansion in-line of procedure calls. A procedure call is very time consuming, since a microprocessor must usually store the registers and other information on a run-time stack and restore matters when the procedure returns.

2.  *Parallelism.* When using silicon compilation, we can select the hardware we desire for the data path, and we can tailor it to the problem at hand. For example, if the source program is such that we can do many additions in parallel, then the chip may have many adders. Moreover, the bus structure can be such that there are many local busses, each connecting an adder to just a few registers, from which the adder will obtain its operands and store its results.

We shall elaborate on each of these points in what follows, through a running example. However, the reader should bear in mind the global picture of silicon compilation as a three stage process shown in Fig. 8.43.

### The Control/Data Dichotomy

One aspect of source languages for silicon compilers that is not normally dealt with in ordinary programming language compilation is the distinction between which variables represent control, and which represent data. As an example, let us consider the program in Fig. 8.44, which is written in a Pascal-like language. This language could be the source language for a hypothetical silicon compiler, although, as we shall see, there are certain syntactic changes made to the language that would help the compilation process along.

The program evidently reads sixteen numbers, stores them in an array, adds them and prints the sum. The variables used in this computation are declared in lines (1) and (2) to be arrays of 32 bits, not integers. If we are to use this program as input to a silicon compiler, the issue of how many bits long the numbers are takes on increased importance. That is, few programming languages concern themselves with the exact number of bits an integer requires; perhaps short (16-bit), long (32-bit), and double precision (64-bit) integers are defined. However, the silicon compiler can, and must, take advantage of knowing the exact width of registers required to implement variables.

(1)  $A$: **array** [1..16] **of array** [0..31] **of** Boolean;
(2)  $sum$: **array** [0..31] **of** Boolean;
(3)  $i$: **integer**;
(4)  $sum := 0$;
(5)  **for** $i := 1$ **to** 16 **do**
        **read** $A[i]$;
(6)  **for** $i := 1$ **to** 16 **do**
        $sum := sum + A[i]$;
(7)  **write** $sum$;

**Fig. 8.44.** Source language program.

Note also that the **array** [1..16] clause in line (1) is in a sense different from the other two **array** clauses in lines (1) and (2), because the former defines sixteen different registers, while the latter tells how long certain registers are. It would be reasonable for the source language of a silicon compiler to distinguish these two uses of "array" and provide another keyword for one of the purposes.

An even more important distinction is the one between the variable $i$ and the other variables of the program. The former is used only for *control*, to indicate which register in an array of registers is operated upon by an assignment statement, line (6). In comparison, the variables $sum, A[1], \ldots, A[16]$ are used as *data*. Formally, the distinction between control and data variables is that the value possessed by a control variable at any step of the program can be determined by simulating the program, and does not depend on any information read in; other variables are data variables.

### Elimination of Control Variables

While we could implement $i$ in Fig. 8.44 as a register, and even have an array $A$ of sixteen registers, any one of which could be accessed according to the value in register $i$, that may not be the optimal way to implement a chip from Fig. 8.44. A way that will lead to a faster chip, and one that probably requires less space as well, is for the compiler to do as much of the control aspects of the program as it can, leaving only the computations involving data variables to the chip.

For example, we have already mentioned the possibility that the compiler should expand procedure calls, since we shall save both space and time if the chip itself does not have to perform the calls. Another opportunity for improving the chip's performance is the expansion of loops, a process called *loop unrolling* in the code optimization literature. There may also be some opportunities for determining the direction of a branch whose condition involves only control variables, as we unroll the loop.

We see three ways that a silicon compiler could detect control variables.

1. The source language could require control variables to be declared as such.

2. We could use data-flow analysis to discover which variables were never assigned values that depended, however indirectly, on values that were read in.

3. We could perform a symbolic execution of the program. That is, we simulate the program step by step, as best we can. If the value of a variable is read, that value becomes "undefined." All operations, one or more of whose arguments are undefined, produce an undefined value. However, steps that involve only operands whose values are known as we simulate the program can be eliminated and only their effects reflected in the code. For example, an assignment $A := B + C$ that is executed at a time when $B$ is known to be 3 and $C$ is 2 can be replaced by $A := 5$. More importantly, uses of this computation of $A$ can be replaced by uses of 5 in other computations, loops, and so on. Thus, we may be able to skip the assignment $A := 5$ altogether when we revise the code. Other effects of symbolic execution include the in-line expansion of procedure calls, and the unrolling of loops.

**Example 8.22:** Data flow analysis can tell us that the value of $i$ in the loops of lines (5) and (6) of Fig. 8.44 is set to 1 the first time, and incremented once each time around.† Thus, we know that each loop is executed exactly sixteen times, and we can deduce what happens each time around the loop. Therefore, we may unroll each of these loops, replacing them by the sequence of thirty-two steps indicated in Fig. 8.45. There, we have replaced the names of registers $A[1], \ldots, A[16]$ by $A1, \ldots, A16$. Note that there are far more than thirty-two steps of intermediate code involved in executing lines (5)–(6) as written, since $i$ must be incremented and tested sixteen times in each loop. On the other hand, the total number of distinct intermediate code statements increases when we unroll loops, and it is possible that loop unrolling and similar transformations due to symbolic execution cannot be used if the resulting code is too long.

The effect of loop unrolling can also be obtained by symbolic execution. When we enter the loop of line (6), for example, we discover that $i$ has the value 1, and we do not need to generate an instruction to assign it that value. Then, the first time we encounter the statement $sum := sum + A[i]$, we know that $i$ is 1 and $sum$ is 0, so we can generate the statement $sum := A[1]$. Then, when we meet the loop incrementation statement $i := i + 1$, which is present in the intermediate code, we know that this statement can be performed at compile time and $i$ is assigned the value 2. The loop test for $i > 16$ is found false; therefore we execute $sum := sum + A[i]$ again, which we can now deduce is

---

† The syntax of the for-statement tells us that as well; we do not need data-flow analysis in this simple case. However, the reader should be advised that there is more to data-flow analysis than the most trivial examples, and a study of the subject is suggested if one is to implement a silicon compiler.

**read** $A1$;
**read** $A2$;
$\qquad$ ...
**read** $A16$;
$sum := sum + A1$;
$sum := sum + A2$;
$\qquad$ ...
$sum := sum + A16$;

**Fig. 8.45.** Unrolled loops.

$sum := sum + A[2]$

That statement is generated during the symbolic execution phase. Symbolic execution on the program of Fig. 8.44 yields essentially the statements of Fig. 8.45, followed by a write statement for *sum*. The major difference is that the first assignment to *sum* does not need *sum* on the right, because we can deduce that *sum* is 0 there. Notice how the output of the symbolic execution phase is exactly those parts of the program that involve one or more data variables. □

**Other Optimizations**

There are several other classical code optimizations that can be applied to Fig. 8.44, either before or after loops are unrolled. We shall mention a few of them here.

1. *Register allocation.* An important problem of code optimization in the ordinary sense is the allocation of the machine's registers to different variables at different times, since data in a register can generally be accessed much faster than the same data in memory. In silicon compilation, the potential for using one register to represent different registers at different times is also present, and is at least as important as in ordinary compilation, because of the large savings in area (and speed as well) that can be had if we shorten the data path.

2. *Loop jamming.* Sometimes, two loops can be combined into one, because the loop controls are the same. For example, the loops of lines (5)–(6) in Fig. 8.44 can be replaced by the single loop

   **for** $i := 1$ **to** 16 **do begin**
   $\quad$ **read** $A[i]$; $sum := sum + A[i]$ **end**

   This change is beneficial if we do not unroll the loops, as we did in Fig. 8.45, since the number of steps involved in control is cut in half.

3. *Reordering code.* Sometimes, great savings can be had if we change the order of statements. For example, the loop jamming mentioned above lets

us discover that the variable $A[i]$ is used only in the $i^{th}$ iteration of the loop. That, in turn, lets us discover that all sixteen registers of the $A$ array can share one register in the data path; that is an excellent example of the advantage of careful register allocation. Even if we do not do loop jamming before we unroll the loops, we can reorder the steps of Fig. 8.45 by shuffling the first sixteen and last sixteen steps so that

$$sum := sum + Ai$$

immediately follows **read** $Ai$. Data-flow analysis, the process of discovering where values computed at one statement are eventually used, will tell us that the shuffling of statements is legal, in the sense that it does not change the effect of the statements as a group. The heuristic for register allocation that says we should try to get the computation of a value and its uses as close together as possible, tells us that shuffling is the right permutation of the steps to try.

### Parallelism Detection and Utilization

Recall that the time we are most interested in compiling a program into a chip, rather than into machine code for a microprocessor, is when we care greatly about the speed of the program. If that is the case, then we would probably be willing to trade space for time by putting on the chip more than the minimum number of arithmetic elements and trying to design the chip so that many operations take place in parallel. The exact degree of parallelism, and therefore the degree to which we are willing to trade area for time, should probably be determined by the user of the silicon compiler, and based on the size limitations of the chip. We shall give two examples of possible methods for detecting parallelism and redesigning a chip to take advantage thereof.

**Example 8.23:** We discovered in Example 8.22 that we can implement the program of Fig. 8.44 in very little space if we unroll the loops and reorder code. In that case, all we do is read 32-bit words and sum them in a register.

However, if we want the chip to be faster, we can do much of this activity in parallel. Suppose we have unrolled the loops, by data-flow analysis or symbolic execution, to get a sequence of intermediate steps like those of Fig. 8.45. It is clear that the sixteen read-statements can be done in parallel, since none depends on values produced by any other. However, it may well not be practical to read more than one or two 32-bit words onto a chip at once, and unfortunately, we may have to forego the speedup due to parallelism here.

The sixteen addition statements present more opportunity for exploiting parallelism. We can construct from the straight-line sequence of statements an expression tree that tells how the output value $sum$ depends on the variables $A1 \cdots A16$. Although the tree we construct is skewed to the left, i.e., it represents the expression $sum = (\cdots((A1 + A2) + A3) + \cdots + A15) + A16$, we can easily apply the algebraic laws for addition to rewrite it in balanced form

$$sum = (A1 + \cdots + A8) + (A9 + \cdots + A16)$$

where each sum of eight variables is split into two sums of four variables, and so on. In this form, we can compute the sum of the first eight at the same time we compute the sum of the second eight. Recursively, we compute each sum of eight by computing two sums of four in parallel and so on.

The consequence of these observations is that if we have eight adders, sixteen registers to hold $A1, \ldots, A16$, and an appropriate bus structure, which is really a complete binary tree organization as we discussed in Section 4.3, then we can perform the addition in four stages instead of sixteen. Thus, by making use of algebraic transformations on the steps generated by a symbolic execution of code, we can make optimal use of parallelism in the execution of the program of Fig. 8.44. Note that it would be much less likely that a microprocessor running compiled code could figure out "on the fly" that there was such an opportunity for parallel computation, even if the microprocessor had an adequate supply of adders so it could even consider the issue.† □

While in Example 8.23 we assumed that the maximum possible degree of parallelism was desired, the user might prefer to get some speedup, but not at the expense of increasing the area beyond reasonable limits. If that were the case, we could build more than one but fewer than eight adders in the data path. For example, we could use four adders, and at the first stage, add four groups of four numbers in parallel, taking three additions, in sequence, for each group. The computation would then proceed in a treelike fashion as in Example 8.24.

Our next example shows how we could use data-flow analysis to detect another kind of parallelism, one that is considerably more subtle than the straightforward use of algebraic laws.

**Example 8.24:** Let us consider a program, shown in Fig. 8.46, that adds two 32-bit numbers, $x$ and $y$. Note that the array declaration in line (1) is akin to the declaration $A$: **array** [1..16] in line (1) of Fig. 8.44, since our intent is not just to tell how long $x$, $y$, and $z$ are as entities, but to allow us to address each bit of these arrays. Thus, variables in this example are at a lower level of abstraction than in the previous example, since there addition was regarded as a primitive, and here we regard only Boolean operations as primitive. The xor operation in line (6) stands for the Boolean expression that is true if an odd number of $x[i]$, $y[i]$, and $carry$ are true (the exclusive or), and the majority operator on line (7) is true if two or more of the arguments are true. Thus, this program implements a ripple-carry addition of 32-bit words.

We could implement the program of Fig. 8.46 as a ripple-carry adder on a

---

† However, an optimizing compiler could generate code calling for parallelism if the microprocessor could execute it.

(1)  $x$, $y$, $z$: **array** $[0..31]$ **of** Boolean;
(2)  $carry$: Boolean;
(3)  $i$: integer;
(4)  $carry := 0$;
(5)  **for** $i := 0$ **to** 31 **do begin**
(6)      $z[i] := \text{xor}(x[i], y[i], carry)$;
(7)      $carry := \text{majority}(x[i], y[i], carry)$
     **end**

**Fig. 8.46.** A ripple-carry addition program.

chip, but that would not be the fastest possible implementation of the program. Rather, we could use data-flow analysis to discover that the only value that is computed in one iteration of the loop of lines (5)–(7) and is needed in the next iteration is $carry$. We introduce parallelism by a "divide-and-conquer" approach to the loop. To begin, we execute the first sixteen iterations in parallel with the last sixteen iterations.

The problem is that we do not know the value of $carry$ produced by the $16^{th}$ iteration, and we need it to implement the last sixteen iterations. Fortunately, $carry$, being Boolean, has only two possible values, so we may implement the second sixteen iterations by two circuits; these two circuits are designed to run in parallel with each other and with the circuit implementing the first sixteen iterations. After all three circuits compute their values, the carry-out of the first sixteen is available, so in one more stage we may decide which of the two circuits for the last sixteen iterations produced the correct answer. The value of $z$ produced by this circuit has its first sixteen bits produced by the circuit for the first sixteen iterations; its second sixteen bits are produced by whichever of the circuits for the last sixteen iterations produced the correct answer this time.

Recursively, each of the three adders for 16-bit numbers can be implemented as three adders of 8-bit numbers working in parallel, and so on. The result is that we generate a circuit with $3^5 = 243$ adders of two one-bit variables and a known carry-in, plus a similar amount of circuitry that selects which of two partial answers is correct. We have, in effect, invented a form of carry-lookahead adder; such adders take $O(\log n)$ time to add $n$-bit numbers, making the best possible use of parallelism for addition. Note, however, that our design uses much more than the minimum possible number of gates for an adder of that speed. $\square$

## EXERCISES

8.1: Use the method of Section 8.1 to show that the expression

$$
\begin{array}{ccc@{\qquad}ccc}
0 & 0 & 0 & 0 & 1 & 0 \\
0 & 0 & 1 & 0 & 1 & 0 \\
0 & 1 & 0 & 0 & 0 & 1 \\
0 & 1 & 1 & 0 & 0 & 1 \\
1 & 0 & 0 & 1 & 1 & 1 \\
1 & 0 & 1 & 1 & 1 & 1 \\
1 & 1 & 0 & 1 & 0 & 0 \\
1 & 1 & 1 & 1 & 0 & 0 \\
\end{array}
$$

**Fig. 8.47.** PLA to be minimized.

$$xy + xz + yz + \bar{x}\bar{y} + \bar{x}\bar{z} + \bar{y}\bar{z}$$

is a tautology.

8.2: Minimize the PLA of Fig. 8.47 by raising terms and eliminating redundancy. Note that the first three columns are the and-plane and the last three are the or-plane.

8.3: Fold the and-planes of the PLA's in (a) Fig. 7.39 (b) Fig. 8.34 (c) Fig. 8.39, and (d) Fig. 8.41.

∗ 8.4: Modify Algorithm 8.3, the PLA folding algorithm, to take into account that some inputs may not need to be inverted, or may only need to be inverted, and thus, some columns may require only one wire instead of two. Apply your algorithm to the PLA's of Exercise 8.3.

∗ 8.5: Modify Algorithm 8.3 to produce PLA's in the form of Fig. 8.8.

8.6: Generate constraints for the following LAVA cells, making the same assumptions about the geometry of a pullup as in Example 8.6. Assume all wires are of minimum width, and enhancement mode transistors have channel ratio 1:1.

a–f) The circuits designed in Exercise 7.3.

g)    The cell of Fig. 7.22.

8.7: Modify the inequality generation algorithm of Example 8.6 to take electrical connectivity of wires into account. Prove that a fringe that remembers the last two nonconnected objects in each color at each point suffices.

∗ 8.8: Show that only five values (rather than three abstract coordinates and three offsets) suffice to represent fringes at a point in the method of Example 8.6.

∗ 8.9: Give an algorithm for sticks compaction in he second dimension, e.g., horizontal compaction after the abstract vertical coordinates have been made concrete.

8.10: Order the columns in Example 8.11, using Algorithm 8.5 to partition the columns recursively, with 50% dummy nodes at each stage.

8.11: Perform algebraic manipulations, order columns, and assign tracks to the wires for the following lgen programs.

a–e) Your answers to Exercise 7.5.

line x
symbol zero(x), one(−x), any()
output MISMATCH

;

((one any* zero + zero any* one) any
    + any (one any* zero + zero any* one)) MISMATCH

**Fig. 8.48.** Pattern matching regular expression program.

    f)   The bounce filter of Example 7.11.

8.12: Convert the logic expressions mentioned in Exercise 8.11 into level graphs, and design SLAP networks for each.

8.13: Modify the NFA construction of Section 8.4 to include the case where there are state symbols in the regular expression.

8.14: Translate your regular expression programs from Exercise 7.8 into (a) after, and (b) before NFA's. Compute the conflicting pairs of states, and find compact codes for the states of each NFA.

8.15: In Fig. 8.48 is a regular expression program that recognizes pattern mismatches, where one or both of the first two symbols read fail to match the last two symbols read. Note that there is one wire, denoted $x$, from which the abstract symbols *one*, meaning $x = 1$, and *zero*, meaning $x = 0$, are defined. Also note that the abstract symbol *any* represents a "don't care" symbol; *any* is always seen on the input.

    a)   Construct before and after NFA's from Fig. 8.48.

    b)   Construct the conflict graphs for these NFA's.

    c)   Code the NFA states as efficiently as possible.

   * d)   Show that any PLA to perform the function of Fig. 8.48 must use at least five feedback wires.

* 8.16: Find all the maximal conflict graphs with four nodes that allow the nodes (states) to be coded with only three bits.

* 8.17: Give an example of an NFA with three states that are mutually conflicting, yet all three cannot be on at the same time.

8.18: Modify the conflict calculation of Fig. 8.35 to handle the case where the start signal can recur at any time.

* 8.19: When constructing the PLA to implement an NFA, as in Algorithm 8.7, we can sometimes choose the particular codes for the states to allow transitions from two or more states to be implemented by one row of the PLA. Give an algorithm that searches for such combinations and selects codes accordingly.

## BIBLIOGRAPHIC NOTES

The PLA reduction methods given in Section 8.1 are from Brayton et al.

[1982]; see also Hemachandra [1982]. The folding method is similar to Hachtel, Newton, and Sangiovanni-Vincentelli [1982], although the method presented there is more efficient but less powerful. The PLA decomposition technique is from Kang and vanCleemput [1981]. Other works on PLA folding include Chuquillanqui and Perez-Segovia [1982], Egan and Liu [1982], and Paillotin [1981].

The compaction of LAVA is discussed in Matthews, Newkirk, and Eichenberger [1982]. Many other systems doing compaction of sticks languages, or similar languages where the basic compiling step is the solution of linear inequalities, have been developed. Some of these are described by Cho, Korenjak, and Stockton [1977], Dunlop [1978, 1980], Hsueh and Pederson [1979], Kedem and Watanabe [1982], Lipton et al. [1982], Lipton, Sedgewick, and Valdes [1982], and Liao and Wong [1983].

There are several theoretical papers worth noting. Lengauer [1982] and Aspvall and Shiloach [1979] discuss the compexity of compaction and the solution of linear inequalities, respectively. Maier [1979] presents an efficient algorithm to detect cycles in constraint graphs.

Compilation of lgen is discussed by Johnson [1983]. Weinberger [1967] originates the "Weinberger array." Algorithm 8.5, the partitioning algorithm used in lgen, is from Kernighan and Lin [1970]. For information on SLAP, see Reiss and Savage [1982] and Savage [1982].

The algorithms of Section 8.4, used to compile regular expressions, are taken from Trickey [1981, 1982], Trickey and Ullman [1982], and Ullman [1982]. Some better coding techniques not covered here are found in Karlin, Trickey, and Ullman [1983]. Other approaches to implementation of regular expressions appear in Floyd and Ullman [1982] and Foster and Kung [1981]. Also see Hennessy [1981] and Brown [1981] concerning implementation of deterministic finite automata.

Silicon compilation of data paths from high level languages is discussed by Shrobe [1982], Siskind, Southard, and Crouch [1982], and Tseng and Siewiorek [1983]. Fox [1983] discusses experience with the MacPitts compiler of Siskind, Southard, and Crouch [1982]; apparently the area of the resulting circuit can be ten times that of a hand design, but there is a promising decrease in the designer's time required when a silicon compiler is used. Feldman [1982] describes compilation of a high-level language into a logic form that does not imply an organization consisting of a control and a data path.

A variety of different languages have been compiled into lower-level designs. Some representative efforts are described in Ayers [1979], Johanssen [1979], Beke and Sansen [1979], Lengauer and Mehlhorn [1982], Bilgory [1982], Bergmann [1983], Wolf, Newkirk, Matthews, and Dutton [1983], and Lieberherr and Knudsen [1983].

# 9

# ALGORITHMS FOR VLSI
# DESIGN TOOLS

In the last chapter of the book we shall examine some of the problems that must be solved when we implement some of the design tools, other than compilers, that we discussed in Section 7.1. We begin with a discussion of finding intersections of rectangles, and how efficient algorithms for that problem lead to efficient algorithms for circuit extraction and design rule checking. Then, we consider algorithms for switch-level simulation of circuits and algorithms for automatic routing of wires.

## 9.1 REPORTING INTERSECTIONS OF RECTANGLES

The problem of finding, given a collection of rectangles, which pairs intersect, is a fundamental one in VLSI design tools. An example of the problem is shown in Fig. 9.1, where there are five rectangles with coordinates indicated by their lower left and upper right corners. We assume here and throughout the section that the rectangles have sides running along the axes and have vertices at integer coordinates. We shall refer to rectangles satisfying these restrictions as *rectilinear* rectangles. Algorithms of similar efficiency but greater intellectual complexity are available to handle intersections of more general shapes, and we mention some of these in the bibliographic notes. The relative simplicity of the algorithm we describe here, and the existence of even simpler, but less efficient ones, help justify the development of systems that assume all designs are composed of rectilinear rectangles.

There are several applications of the problem of reporting intersections of rectangles. Perhaps the most obvious concerns circuit extraction from CIF. The nodes of a switch language each represent a set of electrically connected rectangles. One node can include many rectangles, even on different layers. It is easy to show that two rectangles $R_1$ and $R_n$ are electrically connected if and only if we can find a sequence of rectangles $R_1, R_2, \ldots, R_n$ such that for all $i$, $R_i$ and $R_{i+1}$ intersect (abutment of rectangles counts as intersection), and are either of the same layer or one is a contact cut. There are some details that we shall defer for later, such as the question of what to do if a contact cut

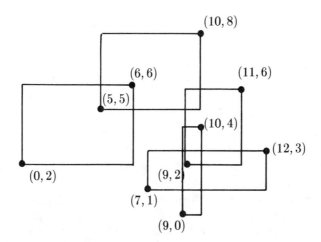

**Fig. 9.1.** Collection of rectilinear rectangles.

and another rectangle intersect or abut, but a design rule violation occurs, say because the contact cut is too close to the edge of the other rectangle. More importantly, we have to treat transistors specially. If we do not remove the portions of a rectangle where poly and diffusion overlap, then we shall wind up assuming that the two ends of a transistor channel are part of the same node of the switch diagram.

We shall cover these details shortly, after giving the basic algorithm. We shall also discuss how design rule checking can use a routine that reports the intersection of rectangles; the basic idea is that we enlarge various rectangles, and violations of the form that two rectangles are too close show up as actual intersections of the enlarged rectangles.

Suppose that there are $n$ rectangles. Given the coordinates of each, we could consider each pair of rectangles, and in $O(1)$ time decide whether that pair intersected. That is, rectangles $R$ and $S$ intersect if and only if one of the four corners of $S$ lies within $R$, and testing whether a point lies in a rectangle is easy. We generally wish to count abutting rectangles as intersecting, and if so we allow a corner of $S$ to lie on the border of $R$, as well as in the interior of $R$. Thus, finding all pairs of intersecting rectangles in $O(n^2)$ time is straightforward. However, designs of a million rectangles are not unheard of, so we might be talking about $10^{12}$ steps to do circuit extraction or design rule checking on a circuit of that size.

Fortunately, there is a better method, due to McCreight [1980], that takes time $O(n \log n + i)$, where $i$ is the actual number of intersections reported. While $i$ could be $\Omega(n^2)$, it is very unlikely that that would be the case in a circuit design. Technically, this method assumes that the coordinates of rectangles are integers and lie in a square that is $O(n)$ on a side. However, if in fact the dimensions of the circuit far exceed the number of rectangles, which is an unlikely occurrence, then we can translate coordinates into integers between 0 and $4n - 1$ as follows.

1. Form a list consisting of all the coordinates, horizontal or vertical, along which any rectangle border runs; there are at most $4n$ of these.
2. Sort the list in $O(n \log n)$ time.
3. Assign the first element of the list the number 0, the next the number 1, and so on.
4. Build a hash table† whose keys are the actual coordinates and whose values are the numbers assigned in (3).

On the average, $O(1)$ time suffices to translate each actual coordinate into the new, compressed coordinate scheme by hash table lookup; thus we shall henceforth assume that the coordinates are integers in the range 0 to $n - 1$, where $n$ is the problem size. From what we have just said, $n$ is an upper bound on both the number of rectangles and on the maximum coordinate value.

### The One-Dimensional Dynamic Problem

The first insight we need is that the problem of reporting intersections of rectilinear rectangles can be expressed as the problem of reporting the (one-dimensional) intervals that overlap a given interval. However, the one-dimensional problem is *dynamic*, in the sense that we are not given the intervals once and for all; rather the set of intervals changes, and we may be asked at any time to report those intervals in the current set that overlap with a given interval.

To see how the static two-dimensional problem translates into a dynamic one-dimensional problem, imagine a horizontal line starting at the bottom of the region containing the rectangles, and moving upward. When it meets the bottom edge of a rectangle, we insert into our set an interval $[x_1, x_2]$, where $x_1$ and $x_2$ are the left and right $x$-coordinates of the rectangle encountered. When we encounter the top edge of the rectangle, the interval representing the rectangle is deleted from the set.

To find intersections of rectangles, each time we insert an interval, we find all the intervals in the set that the new interval overlaps. Each such overlap clearly represents an intersection between the rectangles whose intervals overlap. Moreover, two intersecting rectangles will certainly have their intersection detected when the last of the two is inserted. To make sure abutting rectangles are counted as intersecting, we delete intervals with a given top coordinate after inserting rectangles whose bottoms have that coordinate.

**Example 9.1:** When the horizontal line sweeping Fig. 9.1 moves up to $y = 2$,

---

† See Aho et al. [1983] for information about some of the details we have glossed over, e.g., sorting in $O(n \log n)$ time, or implementation of hash tables.

the intervals corresponding to all but the highest rectangle are in the current set. That is, the current set of intervals is $\{ [0,6], [7,12], [9,10], [9,11] \}$. When we move past $y = 3$ we delete the interval $[7,12]$, because that interval corresponds to a rectangle that has its top at $y = 3$, and when we move past $y = 4$ we delete the interval $[9,10]$, because that interval corresponds to a rectangle with top at $y = 4$. When we reach $y = 5$, we insert the interval $[5,10]$, first determining which of the intervals in the current set $\{ [0,6], [9,11] \}$ overlap it. In this case, both intervals overlap the new one, so we report these two intersections. $\square$

## Rectangle Groups

We must now consider how to find the intervals that overlap a given interval, and we must do so in time that is proportional to the number of overlaps that we actually find. Thus, we could not simply organize the current set of intervals by, say, sorting them by left end. Even if we included a search tree structure that allowed us to find the intervals with a given left end quickly, we still might have to do a great deal of work to find all the overlaps with a given interval $[x_1, x_2]$, and yet find only a few overlaps.

For example, we might start by finding in our list the last interval with left end $x_2$ or less, since subsequent intervals surely will not overlap $[x_1, x_2]$. However, any interval prior to that in the order could overlap $[x_1, x_2]$, and so each must be searched. In practice, $[x_1, x_2]$ could be, say, the interval $[50, 60]$, yet working backwards in the list we meet intervals like $[55, 80]$ and $[48, 52]$, which do overlap, but then we meet a lot of short intervals like $[45, 47]$, $[44, 46]$, $[43, 45]$, ..., that do not overlap $[50, 60]$, until at some point we encounter $[25, 100]$, which does overlap $[50, 60]$.

To avoid wasting time on such almost-wild-goose chases, we introduce the key concept of an interval group. Each group is defined by a pair of integers that are consecutive multiples of the same power of 2. The integers $i$ and $j$ define the group $G_{ij}$; for example, we could have $G_{10,12}$, $G_{48,64}$, or $G_{48,56}$, but not $G_{16,48}$. These observations follow because 10 and 12 are $5 \times 2$ and $6 \times 2$; 48 and 64 are $2 \times 16$ and $3 \times 16$, while 48 and 56 are $6 \times 8$ and $7 \times 8$. However, there is no power of 2, say $2^k$, such that 16 and 48 are $i2^k$ and $(i+1)2^k$ for some integer $i$.

The intent is that $G_{ij}$ will consist of all those intervals in the current set that

1.  Have left end greater than $i$,†
2.  Have right end less than $j$, and
3.  Straddle the midpoint; that is, the left end is no greater than $(i+j)/2$ and the right end is no less than $(i+j)/2$.

Further, we can build a *group tree* of the groups, where the left child of $G_{ij}$ is

---

† As a special case, if $i = 0$, then the left end may be 0.

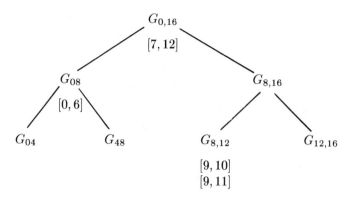

**Fig. 9.2.** Group tree.

$G_{i,(i+j)/2}$ and the right child is $G_{(i+j)/2,j}$. An important fact to observe is that any interval whose left and right ends are properly contained in the interval $[i, j]$, where $i$ and $j$ are consecutive multiples of a power of 2, will belong to $G_{ij}$ or one of its descendants. It does not necessarily belong to $G_{ij}$ itself, because it may not straddle the midpoint of $[i, j]$.

**Example 9.2:** Let us suppose $n = 16$. Thus, all coordinates are in the range 0 to 15, and we wish to represent in a group tree the intervals in the current set when the line of sweep reaches $y = 2$ in Fig. 9.1, that is, the set

$$\{ [0, 6], [7, 12], [9, 10], [9, 11] \}$$

as discussed in Example 9.1. The group tree is shown in Fig. 9.2, and the intervals are listed under the group to which they belong. For example, $[7, 12]$ is put in $G_{0,16}$, because it straddles the midpoint, 8, for that group. We do not show the tree extending down to groups indexed by consecutive multiples of 2 or 1, although we could do so. However, if all intervals are at least 1 unit long, we would not expect to find any intervals in these groups, except that the interval $[0, 1]$ would be placed in the group $G_{02}$. $\square$

**Implementing a Group**

It is much easier to find the members of a group that intersect interval $[x_1, x_2]$ than it is to find the overlaps between the given interval and a random set of intervals. In particular, let the group be $G_{ij}$, and let $m$, the midpoint, be $(i + j)/2$. Then if $x_1 \leq m \leq x_2$, every interval in $G_{ij}$ overlaps $[x_1, x_2]$; in particular, they share the point $m$.

If $x_2 < m$, then interval $[x_3, x_4]$ in $G_{ij}$ overlaps $[x_1, x_2]$ if and only if $x_3 \leq x_2$. The values of $x_1$ and $x_4$ do not matter, since we know $x_4 \geq m$ because the interval is in $G_{ij}$. Thus, it is not possible that the interval $[x_3, x_4]$ lies entirely

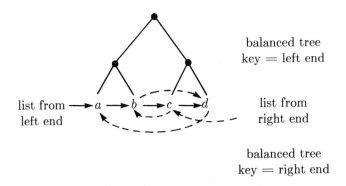

**Fig. 9.3.** Data structure for group.

to the left of $[x_1, x_2]$. We may thus examine the intervals in $G_{ij}$ in order of their left ends. At first, every interval we meet overlaps $[x_1, x_2]$, and the time spent is proportional to the number of intersections reported. Eventually, we find an interval whose left end is greater than $x_2$, whereupon we need consider no more intervals in the group. Note how the group property, the fact that all intervals straddle the midpoint, is essential here. Otherwise, scanning group members in order of their left ends might cause us to meet many intervals that were entirely to the left of $[x_1, x_2]$.

Similarly, if the interval $[x_1, x_2]$ fails to straddle $m$ because $x_1 > m$, then we must scan the intervals in order of their right ends, highest first, until we encounter one that has a right end to the left of $x_1$.

It might at first seem paradoxical that we should be able to examine the set of intervals in $G_{ij}$ both in order of left endpoint and in order of right endpoint, but in fact, all we need is two independent linked lists threading the set of intervals, one for each of the desired orders. Since we must insert and delete intervals, we shall also require two balanced tree structures, such as AVL trees or 2-3 trees,† to enable us to insert or delete at any point of the two lists in $O(\log n)$ time.

**Example 9.3:** Suppose we have four intervals: $a = [33, 40]$, $b = [35, 55]$, $c = [40, 63]$, and $d = [45, 50]$, in the group $G_{32,64}$. In order of left ends, they are $a, b, c, d$, and in order of right ends, highest first, they are $c, b, d, a$. The data structure for this group is illustrated in Fig. 9.3. □

### Searching a Group Tree

We now must consider how to find all the intervals overlapping a given interval,

---

† Again, see Aho et al. [1983] for information about these data structures.

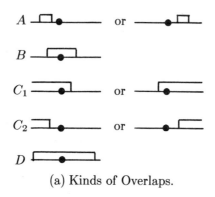

(a) Kinds of Overlaps.

| Class | Class of Left Child | Class of Right Child |
|-------|---------------------|----------------------|
| $A$   | $A$ or $B$          | No intersection      |
| $B$   | $C$                 | $C$                  |
| $C_1$ | $D$                 | $C$                  |
| $C_2$ | $C$                 | No intersection      |
| $D$   | $D$                 | $D$                  |

(b) Transition rules for interval classes.

**Fig. 9.4.** Interval classes and their propagation.

and do so in time that is $O(\log n)$ plus a term that is proportional to the number of overlaps actually found. What we do is develop a recursive procedure that searches a group and all its descendants in the group tree, and then apply the procedure to the root of the group tree. Another key idea is that we must classify the ways a given interval could relate to the interval $[i, j]$, where $G_{ij}$ is the group at the root of the group tree in question. The five possible situations are shown in Fig. 9.4(a).

For example, a class $A$ overlap is one in which the interval in question lies entirely on one side of the midpoint. Class $B$ overlaps straddle the midpoint, but do not reach either end. We know that a class $B$ interval overlaps every member of the group.

Another central observation is that the class of an interval, as far as a group $G_{ij}$ is concerned, affects the class of an interval as far as the descendants of $G_{ij}$ in the group tree is concerned. The rules for determining how the class propagates downward in the group tree are summarized in Fig. 9.4(b). In cases $A$ and $C$, where there are two symmetric forms, we assume the first form, i.e.,

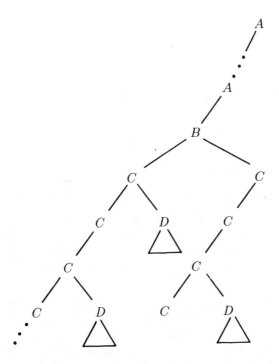

**Fig. 9.5.** Typical tree of groups to be searched for overlaps.

where the interval is biased toward the left. Also, class "$C$" means "either $C_1$ or $C_2$."

The consequence of the observations of Fig. 9.4(b) is that when we start our search for the intervals that overlap a given interval at the root of the group tree, we need to traverse a set of nodes in the manner suggested by Fig. 9.5. That is, the given interval must be in class $A$ or $B$ relative to the root group, which is $G_{0,n}$, assuming $n$ is raised to the next power of two if it is not already a power of two. A class $A$ interval requires only that we search one child and its descendants, whichever child is on the same side of the midpoint as the class $A$ interval. A class $B$ interval requires that we report intersections of the interval with all intervals in the group and that we search both children and their descendants, using as intervals one of the two pieces into which the class $B$ interval is broken by the midpoint.

The class $C$ intervals each generate one other class $C$ case, and if the class is $C_1$, it will also spawn a tree of calls on class $D$ intervals. Note that the triangles in Fig. 9.5 represent trees of class $D$ nodes. Also observe that, as the number of levels in the group tree is at most $\log_2 n$, the number of class $A$, $B$, and $C$ nodes is no greater than $2\log_2 n$.

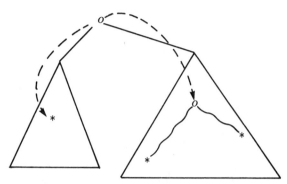

**Fig. 9.6.** Use of pointers in group tree to find nonempty groups.

We are thus almost ready to prove that searching the group tree for overlaps with a given interval takes time $O(\log n + i)$, where $i$ is the number of overlaps actually reported. We charge time spent on classes other than $D$ to the node itself if there are no overlaps to report at that node, and to the intersections reported, if there are some.

We cannot charge the cost of examining the class $D$ nodes to the nodes themselves, because there might be more than $O(\log n)$ of them. Unfortunately, it could happen that all, or almost all, of the class $D$ nodes happen to be empty, so we have no overlaps to charge their cost to. Thus, we must modify the group tree so that each node records the number of intervals it and its descendants currently contain. Further, a group node with only one nonempty descendant group in its left subtree will point directly to that group, so we do not have to follow long paths in the group tree to reach the nonempty group. For the same reason, groups with more than one nonempty descendant group in the left subtree will point directly to the lowest common ancestor of those nonempty groups. A similar pointer is used by each group to find its way quickly to nonempty groups in its right subtree.

Figure 9.6 illustrates these extra pointers at a node of the group tree. Note that the left and right child pointers are still present in the tree, and are needed as we insert and delete intervals into and from the various groups. However, it is the dotted pointers in Fig. 9.6 that are used to find groups whose members intersect a given interval of class $D$.

If, when examining a subtree of class $D$ nodes, we follow the pointers just described, the portion of the subtree searched can be viewed as a binary tree. We may use the easily proven fact that a binary tree has one more leaf than it has interior nodes, to conclude that more than half the nodes searched have nonempty groups. At a class $D$ node, we report that the interval in question overlaps all the intervals in the group of that node. Thus, we may charge the cost of searching all the class $D$ nodes to the overlaps actually reported, there being at least half as many overlaps reported as there are nodes searched. We may summarize the above observations in the following lemma.

**Lemma 9.1:** If we use the data structure of Fig. 9.3 to represent groups, and we store groups in a group tree structured as above, then we may report overlaps between a given interval and all the intervals in the current set in $O(\log n + i)$ time, where $i$ is the number of overlaps reported.

**Proof:** Fig. 9.5 indicates the nodes that must be searched. We have already remarked that the cost of searching all class $D$ nodes is proportional to the number of overlaps we report during that search. Also, except for the class $D$ nodes, only $O(\log n)$ groups are searched. We must, therefore, account for only the time spent handling nodes of classes $A$, $B$, and $C$.

*Class A:* If the interval $[x_1, x_2]$ is to the left of the midpoint for the group, then we may search the intervals in the group in order of left end, starting with the leftmost, until one with a left end greater than $x_2$ is found. As all intervals in the group extend at least to the midpoint, and therefore to the right of $x_2$, each of the intervals found do overlap the interval $[x_1, x_2]$. Thus, the time spent at this node is $O(1)$ if there are no overlaps to report, and $O(i)$ if $i > 1$ overlaps are reported. A similar argument applies if interval $[x_1, x_2]$ is to the right of the midpoint.

*Class B:* Report overlaps with all the intervals in the group. The same remark regarding time as for $A$ applies here as well.

*Class $C_1$:* Report overlaps with all intervals in the group. Again, the cost is $O(1)$ or $O(i)$, depending on whether there are intervals to report.

*Class $C_2$:* Assuming the interval $[x_1, x_2]$ extends to the left end of the interval of the group, i.e., $x_1 = k$ if the group in question is $G_{k\ell}$, examine the intervals in the group in order of left end, reporting overlaps until an interval with left end greater than $x_2$ is found. A symmetric action is taken if the interval is to the right of the midpoint. As with the other groups, this operation takes $max(O(1), O(i))$ time.

We have thus shown that each of the at most $2\log_2 n$ nodes of other than class $D$, takes an amount of time that is a constant plus a constant multiple of the number of overlaps reported. Since the searching of class $D$ nodes is charged entirely to overlaps reported, we have proved the lemma. $\square$

We summarize the ideas contained in the discussion of this section in the following algorithm and theorem.

**Algorithm 9.1:** *McCreight's Algorithm for Reporting Intersections of Rectilinear Rectangles.*

INPUT: A collection of rectilinear rectangles.

OUTPUT: A list of the pairs of rectangles that intersect.

METHOD: In what follows, we assume that coordinates of the vertices of rec-

(1)   sort the rectangles in order of the height of the bottom edge, lowest first
         and also sort them in order of the height of the top edge, lowest first;
(2)   $ACTIVE$ := empty set;
(3)   initialize the group tree, so initially all groups are empty;
              { in practice, we would initialize the group tree only as required
              by the insertions we perform }
(4)   **for** $y$ := 0 **to** $n-1$ **do begin**
(5)        **for** each rectangle with lower edge $y$, left edge $x_1$, and
              right edge $x_2$ **do begin**
(6)             report overlaps of interval $[x_1, x_2]$ with intervals in $ACTIVE$;
(7)             insert $[x_1, x_2]$ into $ACTIVE$
           **end**;
(8)        **for** each rectangle with upper edge $y$, left edge $x_1$ and
              right edge $x_2$ **do**
(9)             delete $[x_1, x_2]$ from $ACTIVE$
      **end**

**Fig. 9.7.**  Algorithm to report intersections of rectangles.

tangles are in the range 0 to $n-1$, and that $n$ is also an upper bound on the
number of rectangles. If that is not the case, translate coordinates to such a
range as we discussed at the beginning of this section. The algorithm uses a
set $ACTIVE$ of intervals, which represent rectangles. When we "report" an
overlap of intervals, we in reality list the corresponding pair of rectangles on an
output file.

The steps to be performed are shown in Fig. 9.7. Step (6), the reporting of
overlaps, uses the method described in Lemma 9.1 and the preceding discourse.
Steps (7) and (9), involving insertion and deletion of intervals consist of two
steps. First, we find the proper group in the group tree by an obvious binary
search. Since we may be inserting into an empty group or deleting the last
interval in a group, the pointers of Fig. 9.6 located at ancestors of the group
in question may need to be adjusted. We leave the details of this process as an
exercise.

Second, we insert or delete from the structure of Fig. 9.3 by finding the
place of the element by a search into one of the balanced trees. We insert or
delete it in the doubly linked lists by obvious steps and in the trees by methods
appropriate to the type of balanced tree used. The tree search and manipulation
can be done in $O(\log n)$ time by a variety of methods; we shall not discuss the
data structure techniques here. $\square$

**Theorem 9.1:** Algorithm 9.1 takes time $O(n \log n + i)$, where $i$ is the number
of intersections that it reports and $n$ is an upper bound on the number of
rectangles.

**Proof:** Translation of coordinates to the range 0 to $4n$ and sorting the rectangles in line (1) of Fig. 9.7 each take $O(n \log n)$ time at most. Line (2) takes $O(1)$ time and line (3) takes $O(n)$, although as we commented in the algorithm, we would in practice not perform line (3) but would distribute its cost over calls to the insertion operation.

As was argued in Lemma 9.1, we can perform line (6) of Fig. 9.7 in time proportional to $\log n$ plus the number of intersections reported. The insertion of line (7) and the deletion of line (9) each require $O(\log n)$ time to find the proper group in the group tree and then $O(\log n)$ more time to manipulate the group tree and the data structure of the group. Since lines (6), (7), and (9) are each executed at most $n$ times, the total time devoted to lines (7) and (9) is $O(n \log n)$, and to line (6), we devote $O(n \log n + i)$ time. The latter is the dominant term, and the running time of the algorithm is of the same order of magnitude. $\square$

## 9.2 CIRCUIT EXTRACTION ALGORITHMS

In this section we shall discuss how one performs the process of circuit extraction, that is, the translation of CIF or a similar geometry language into esim, the language described in Section 7.4, or a similar switch-level language. Recall from Section 7.1 that a switch language is essentially the lowest level of language from which we can perform a simulation of the circuit in a reasonable amount of time. To try to simulate the CIF directly requires too much time, while to extract, say, logic from CIF is a difficult problem because it is not always clear from the CIF which way signals are intended to flow in a wire, or whether in fact signals travel in both directions at times, performing logic in subtle ways.

The pivotal step in finding transistors in CIF statements is detecting where the channels are. To begin, channels may exist only where diffusion and polysilicon overlap; therefore we can use Algorithm 9.1, or a similar algorithm, to report the intersection of all poly and diffusion rectangles. Unfortunately, there are several subtleties in channel discovery.

First, not every place that poly and diffusion overlap is a transistor. It could be a butting contact, or other structure intended to serve as a via between poly and diffusion. Such situations can be detected ultimately, since we shall find that the "channel" region in a butting contact is only $\lambda$ wide and connects to only one piece of diffusion, while true transistors must have channels at least $2\lambda$ wide and normally connect to two separated pieces of diffusion that are not covered by poly. There are exceptions, such as the pullup in Fig. 7.6, which has the butting contact connecting source and gate placed at one end of the channel.

Second, a channel is not guaranteed to be contained within one piece of diffusion. The most common situation is when we wish to make good use of space by bending the channel of a long pullup into an L-shape, as suggested

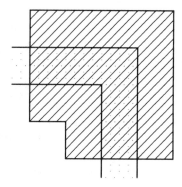

**Fig. 9.8.** Bent channel made from two rectangles of diffusion.

in Fig. 9.8. There, the channel is on two rectangles of diffusion covered by a single piece of poly. In general a channel could be made up of arbitrarily many pieces of diffusion covered by arbitrarily many pieces of poly.

To deal with the problem of detecting transistors, let us invent a new layer, *channel*. We replace certain rectangles of diffusion by doing the following steps.

1.  Use Algorithm 9.1 to find all the intersections between rectangles of diffusion and rectangles of polysilicon.

2.  For each intersection, say between diffusion rectangle $D$ and polysilicon rectangle $P$, replace $D$ by a rectangle in layer "channel," whose extent is exactly $D \cap P$, and a collection of diffusion rectangles that together cover the area $D - P$. If $D$ intersects other polysilicon rectangles, determine which of the new diffusion rectangles intersect which of the poly rectangles.

The intersection of two rectilinear rectangles is easily shown to be a rectilinear rectangle, but the difference $D - P$ may not be. Figure 9.9 shows a rare, but possible case, where a polysilicon rectangle intersects part of a diffusion rectangle. There might not be a design rule error, since another polysilicon rectangle could continue the gate of the transistor. It is an easy exercise for the reader to express the difference of rectilinear rectangles as the union of rectilinear rectangles.

### Finding Nodes and Transistors

Having broken diffusion rectangles into diffusion and channel rectangles, we next find the nodes of the switch circuit by discovering which rectangles are electrically connected. A method is summarized in the next algorithm.

**Algorithm 9.2:** *Circuit Extraction from Rectangles.*

INPUT: We are given a list of rectangles, including the channel rectangles produced from intersections of poly and diffusion by the method described

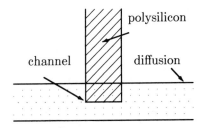

**Fig. 9.9.** Case where a diffusion rectangle becomes more than one.

above. For each channel rectangle $R$, there is a pointer to the poly rectangle that covers it; we refer to this rectangle as $POLY(R)$.

OUTPUT: A switch-level circuit with behavior equivalent to that of the given set of rectangles.

METHOD:

1. Make from the given list of rectangles a list of all intersections of rectangles.

2. We begin a process of "set merging," where initially each rectangle belongs to a set by itself, and ultimately rectangles belong to the same set if and only if they are electrically connected. That is, sets represent nodes of the switch diagram. Initialize by assigning to each rectangle a unique integer that denotes its current set, so all sets are initially singletons.

3. Consider each intersection of rectangles in turn. When we consider the fact that rectangles $R$ and $S$ intersect, we first determine whether they cause an electrical connection. Note that abutting rectangles are deemed to intersect, and will be so reported by Algorithm 9.1. Note also that rectangles on the channel layer will not be electrically connected to intersecting rectangles on other layers, such as diffusion, poly, or contact cut.†

   a) If $R$ and $S$ are electrically connected, i.e., they are both in the same layer, or one is contact cut and the other is not channel, then merge the sets containing $R$ and $S$. We shall give the details of an appropriate algorithm and data structure for merging these sets efficiently after we describe the main algorithm. However, let us here point out that the necessary steps are to ($i$) determine the set numbers for $R$ and $S$, say $i$ and $j$, then ($ii$) select one of $i$ and $j$, say $i$, and ($iii$) find every rectangle in set $i$ and change its set to $j$.

   b) If one of $R$ and $S$, say $R$, is channel, and the other is diffusion, contact, or implant, do not merge sets, but record the rectangle $S$ on a list

---

† Sometimes, a structure where a metal-poly contact occurs over diffusion is permitted. That diffusion, which forms part of a channel, is not thereby connected electrically to the metal or poly, because the contact cut is not assumed to penetrate through to the diffusion. Mead and Conway [1980] regards this structure as unreliable for sufficiently small $\lambda$. Thus, we shall assume in this Chapter that no such structure occurs legally.

$DCI(R)$ to be examined in steps (5) and (6) below.

c)   Otherwise, do nothing. We have a pair of rectangles that, although they intersect, do not interact directly, e.g., a metal and a poly rectangle, or a channel rectangle and covering poly rectangle.

After considering the entire list of intersecting rectangles, all rectangles other than channel rectangles have been grouped into *nodes*, which are maximal sets of electrically connected rectangles. The channel rectangles have been grouped into maximal sets of connected channel rectangles, i.e., each corresponds to either a transistor or a poly-diffusion overlap that is not intended to be a transistor, e.g., a butting contact. These groups of channel rectangles are the candidates for transistors of the switch diagram.

4.   Attempt to eliminate butting contacts, or other structures used to connect poly and diffusion, as candidates for transistors. We may also be able to detect structures that are design rule errors or likely errors, such as channels that are too thin or transistors with source and drain electrically connected. We cannot predict in advance the particular contact structures a particular extractor might want to eliminate, although we expect that, like the butting contact, they will be rigidly defined structures. Therefore, they will be detectable by considering each maximal set of channel rectangles and plotting a small local area on a grid, represented as a two-dimensional array. For example, we could discover channel that was part of a butting contact by the fact that the channel rectangle or rectangles form a $\lambda \times 4\lambda$ piece.

5.   Distinguish depletion mode transistors by examining the $DCI$ lists for implant. We can also catch certain design rule errors here by plotting the local area around the channel; for example, we can detect situations where the implant does not extend for $2\lambda$ around the channel.

6.   For each maximal set of channel rectangles constructed in (3) and not eliminated in (4), determine the nodes that are source, drain, and gate. We discover the gate by finding $POLY(R)$ for each channel rectangle $R$. We discover source and drain by examining the $DCI$ lists for the channel rectangle(s) and finding those rectangles of diffusion that are electrically connected to the channel. There should be two different maximal sets of rectangles that are included among the rectangles on the $DCI$ lists, and these may be called the source and drain nodes arbitrarily. However, if one of these is the same as the gate node, and the transistor was identified in (5) as a depletion mode transistor, then we have identified a pullup, and the node that is the gate is also the source.

7.   Generate a transistor list by considering every maximal set of channel rectangles and listing the mode (enhancement or depletion) found in (5) together with the set numbers for the gate, source, and drain. $\square$

## Performing the Set Merge

There are a number of efficient ways to implement the set merging aspect of Algorithm 9.2; many of them are discussed in Aho et al. [1974]. We mention just one approach, which is a good compromise between code writing effort and efficiency. For each set, keep a count of the number of members and a pointer to the first rectangle, on a linked list of rectangles that includes all and only the members of the set. In the data structure for each rectangle include the number of the set of which it is currently a member. When in step (3) of Algorithm 9.2 we are called upon to merge the sets containing rectangles $R$ and $S$, we find the sets $i$ and $j$ containing them from the data structure representing the rectangles themselves.

If $i = j$, we do nothing, because the sets are already merged. If $i \neq j$, we consult the counts for the sets $i$ and $j$ to determine which of them, say $i$, has the smaller number of members. Then we go down the list for $i$ and change the set of each rectangle on that list to $j$. Finally, we append the list for $i$ to the list for $j$, and set $i$ ceases to exist.

The reason this structure is efficient is that the time spend merging sets is proportional to the number of members of the smaller set, so we can cover the cost of the merge by charging a constant to each member of the smaller set. Each time a rectangle gets moved to a new set, that set is at least twice as big as the one it was in before, so a rectangle can be charged only $\log n$ times, if $n$ is an upper bound on the number of rectangles. Thus, the total cost of the set merges is $O(n \log n)$.

## Time Analysis of Circuit Extraction Algorithm

Assume as before that the number of rectangles is $O(n)$, but now assume that the rectangles include among them the channel rectangles discovered prior to Algorithm 9.2. Also assume that $i$ is the number of intersections reported by step (1) of Algorithm 9.2. Then step (1) takes $O(n \log n + i)$ time, since we shall use Algorithm 9.1. Step (2), the initialization of the data structure for set merging, takes $O(n)$ time if we use the method just described. Step (3) takes $O(i)$ to examine each pair of rectangles. If we use the data structure above, the set mergers in step (3a) can be performed in an additional $O(n \log n)$ time.

The search for butting contacts in step (4) is $O(n)$, since we need to consider each set of channel rectangles only once; plotting them takes only constant time, since as soon as we plot more than four different squares we know we cannot have a $\lambda \times 4\lambda$ "channel." We can make step (5) run in $O(n)$ time, since all we need to know is whether there is an implant rectangle on the list for a channel rectangle, a test that is made easy if we keep an implant rectangle at the head of the $DCI$ list. Note, however, that checking certain design rule errors, such as improperly placed implant, could take $O(n^2)$ time in the pathological case that $\Omega(n)$ implant rectangles covered $\Omega(n)$ channel rectangles. Finally, step (6), identifying gates, sources, and drains, surely takes no more than $O(i)$ time, an upper bound on the sum of the lengths of the $DCI$ lists; in practice it will be significantly faster on the average.

The result of considering each of the steps in Algorithm 9.2 is that the algorithm runs in $O(n \log n + i)$ time. Thus, extraction of switch diagrams takes on the same order of time as its central step, the reporting of rectangle intersections.

## 9.3 DESIGN RULE CHECKING

A design rule checker is a program that searches a circuit diagram for data that is certain, or even likely, to be an error. Since it is much more serious if a real error goes unreported than if a false alarm is raised, such programs need not be absolutely certain that an error has occurred, and many of them are made considerably more efficient because they are liberal in what they call an error. Another source of efficiency is restriction of the input form. We shall assume here that the circuit diagram is a collection of rectilinear rectangles, although in some cases, algorithms generalize naturally to all rectangles, or even to all polygons.

In this section we shall consider several approaches to design rule checking: raster-based, corner-based, and algebraic. We also consider some checks that can be made from the extracted switch diagram and we consider hierarchical design rule checking.

### High-Level Design Rule Checks

Let us briefly mention some of the checks that can be made from the extracted circuit, but that, unlike most error checks concerning design rules, cannot be made from the CIF conveniently. For example, the rule that says a pass transistor cannot feed the gate of another pass transistor is checkable, as is the requirement of a 4:1 or 8:1 pullup/pulldown resistance ratio. To do an accurate job, we must be able to distinguish pass transistors from pulldowns, and to attempt that we must use some heuristics.

To begin, it is helpful if we know which node is $VDD$ and which is ground. For example, we can then identify pullups, because they will be depletion mode transistors with drain connected to $VDD$. From each pullup source node, which is the opposite side of the channel from $VDD$, determine which nodes are on a path to ground, where a "path" is taken to mean a sequence of nodes $v_1, \ldots, v_n$ such that for $1 \leq i < n$, there is some enhancement mode transistor for which $v_i$ and $v_{i+1}$ are the source and drain nodes (not gate node), in some order. We shall leave the design of an algorithm for such a search through a switch diagram as an exercise. Note that it is not a simple depth-first search, since some paths may not lead to ground; typically they "dead end" at a gate.

**Example 9.4:** Consider the switch diagram of Fig. 7.17. The node 7 is easily identified as the source of a pullup. If we try to see where we can get from 7, we discover that we can cross from the source to drain (or vice versa) of a transistor to get to nodes 2 and 8. As node 8 is a gate, we can go nowhere. However, node 2 is ground, so we find that there is one path to ground from node 7, through the transistor with gate 6. We also discover that the latter transistor is a pulldown, so the transistor to its left, connected to node 5, does not violate the rule that a pass transistor may not feed the gate of a (non-pulldown) pass transistor.

We can also examine the channels of the pullup with source 7 and the pulldown between 7 and 2, by plotting them in a small array to estimate their resistances. That estimation is not easy if the channels are severely bent, but in most cases, a good estimate can be obtained easily by counting squares, and it may be determined whether the resistance of the pulldown network is no more than an eighth of the resistance of the pullup. We can also determine that an eighth, rather than a fourth, is the appropriate ratio because the pulldown gate node is not attached directly to the source node of a pullup. $\square$

### The Raster Approach to Design Rule Checking

If we restrict ourselves to rectilinear rectangles on a $\lambda$ grid, many simple errors can be caught by plotting the layout a few lines at a time. It is often too expensive to represent the entire circuit at once in a two-dimensional array, where each array element tells which layers are present in the $\lambda \times \lambda$ square that the array element represents. Since there are usually no more than eight layers, a byte per square suffices. However, since designs of many millions of square $\lambda$ are done, storage of the whole two-dimensional array in memory might present problems.

Fortunately, there is a simple adaptation of the ideas in Algorithm 9.1 that lets us avoid storing the entire design. As in that algorithm, we sort the rectangles, both by $y$-coordinate of the bottom edge and by $y$-coordinate of the top edge. Then we may run a scan line upward, maintaining the set of intervals that represent those rectangles through which we are now passing. For each $\lambda$ that we move, we compute a one-dimensional array by plotting each interval in the active set. That is, if $A$ is the array, and we have an interval from $x_1$ to $x_2$ representing a rectangle in layer $L$, we set to 1 the bits representing $L$ in $A[x_1]$ through $A[x_2 - 1]$.

We need to retain a number of scan lines; the actual number depends on the nature of the design rules used. In the Mead-Conway NMOS rules, retention of four scan lines is adequate. If there were other rules, such as a second layer of metal that needed to be at least $5\lambda$ wide, we would need to retain six scan

|      |      |      |
|------|------|------|
| 0000 | 0111 | 1111 |
| 1111 | 1111 | 0000 |
| 1111 | 1111 | 0000 |
| 1101 | 1110 | 1111 |
| (a)  | (b)  | (c)  |

**Fig. 9.10.** Three matrices.

lines. However, whatever the rules, we need to retain only a limited strip that grows with the circuit in one dimension only. We thus expect that the raster approach will remain useful as circuits get larger and larger in terms of square $\lambda$, provided designs are composed of rectangles on a square grid.

### The Baker-Terman Approach

The design rule checker of Baker and Terman [1980] relies on $4 \times 4$ matrices of 0's and 1's to do many checks, mostly wire spacing and wire width checks. It relies heavily on the fact that there are only $2^{16}$ such matrices, and this method would not be feasible if, say, we had to consider $6 \times 6$ matrices to check $5\lambda$ wire widths. For each layer, and for each $4 \times 4$ matrix, we ask ourselves what would happen if the 1's represented $\lambda \times \lambda$ squares where the layer was found, and the 0's represented squares where the layer was not found. For one layer, a vector of $2^{16}$ bits stores the answer for each matrix. That is, 1 means the matrix does not represent a design rule error; 0 means it does.

For each layer $L$, we consider each $4 \times 4$ square in the four scan lines we have retained. We determine the number of the matrix representing the pattern that this $4 \times 4$ square makes in layer $L$. We leave it to the reader to devise an appropriate ordering of the $2^{16}$ matrices so that this number may be determined efficiently. Next, we determine from the bit vector for $L$, whether or not this $4 \times 4$ square represents a design rule error.

**Example 9.5:** Figure 9.10 shows three $4 \times 4$ matrices. If $L$ is metal, all three represent design rule errors. The first is a situation where the wire is too narrow at a point. The second is a similar error, because along the diagonal the width of the metal wire is only $2\sqrt{2}$, which is less than 3. The third represents a situation where two metal wires are too close together.

If the layer in question is diffusion, only (c) represents an error, because $2\lambda$ wire width is sufficient. Finally, if the layer is poly, none of the matrices represent design rule errors. □

Actually, not every apparent spacing error is really an error, and certain nonerror situations can be "caught." The principal problem is illustrated in Fig. 9.10(c), for the metal layer, where the "error" would be a phantom if it turned out that the two metal wires we seem to see were actually electrically connected. We can avoid reporting such errors if we perform circuit extraction before computing any scan lines. Then, as we compute the scan lines, we attach to each square, information giving for each layer present the node number to which it belongs.

The above method is exact, but it can be expensive of space. Baker and Terman recommend a simple check in which we consider a local area (they suggest $10\lambda \times 10\lambda$) and try to discover whether the apparently too close rectangles are really connected. If a path connecting them is found in the local region, then the report of the error is canceled. If no connection is found, we still do not know that some circuitous connection does not exist; however it is better to report a false error than to omit a real one. Doing checks on $10\lambda \times 10\lambda$ regions requires retention of only another three scan lines, or a total of seven, since part of the $10\lambda \times 10\lambda$ region is in scan lines yet to be generated.

## Conditional Rules

Another sort of rule used by Baker and Terman [1980] does not involve a long vector of decisions for arbitrary matrices of 0's and 1's. Rather, design rules are expressed as conditional statements involving patterns in several layers. We shall extend our notation to use 2 as a "don't care" symbol in matrices that otherwise contain 0's and 1's. We also use the notation $L(M)$, where $L$ is a layer ($D$, $P$, $M$, and $C$, for diffusion, poly, metal, and contact cut), and $M$ is a matrix of some dimensions, consisting of 0's, 1's, and 2's. The expression $L(M)$ asserts about a region of the circuit, whose dimensions match those of $M$, that the region has layer $L$ wherever $M$ has 1, and does not have layer $L$ wherever $M$ has 0.

**Example 9.6:** The design rule that says the polysilicon gate must overlap the channel by $2\lambda$ can be expressed by the following rule and symmetric rules in the other three directions.

$$\text{if } D\begin{pmatrix} 22 \\ 00 \\ 11 \\ 11 \end{pmatrix} \wedge P\begin{pmatrix} 22 \\ 22 \\ 11 \\ 11 \end{pmatrix} \text{ then } P\begin{pmatrix} 11 \\ 11 \\ 22 \\ 22 \end{pmatrix}$$

That is, whenever we see diffusion covered by poly, but the diffusion ends, the poly must appear in the next two rows beyond the end of the diffusion. Note we must consider two rows and columns to avoid seeing an "error" when we have the $\lambda$-wide strip of a butting contact that contains both poly and diffusion.

Another example concerns contact cuts. Informally, it says that whenever we see a $2\lambda \times 2\lambda$ square of contact, it must be surrounded by metal and by either poly or diffusion. Moreover, the one of poly and diffusion that does not surround it must not approach it. Formally, the rule is expressed in Fig. 9.11. Note that the rule in Fig. 9.11 is not violated by a butting contact, because the

$$\text{if } C\begin{pmatrix} 0000 \\ 0110 \\ 0110 \\ 0000 \end{pmatrix} \text{ then } M\begin{pmatrix} 1111 \\ 1111 \\ 1111 \\ 1111 \end{pmatrix} \wedge$$

$$\left( \left( D\begin{pmatrix} 1111 \\ 1111 \\ 1111 \\ 1111 \end{pmatrix} \wedge P\begin{pmatrix} 0000 \\ 0000 \\ 0000 \\ 0000 \end{pmatrix} \right) \vee \left( D\begin{pmatrix} 0000 \\ 0000 \\ 0000 \\ 0000 \end{pmatrix} \wedge P\begin{pmatrix} 1111 \\ 1111 \\ 1111 \\ 1111 \end{pmatrix} \right) \right)$$

**Fig. 9.11.** Contact cut design rule.

$$\text{if } \begin{array}{c|c} M(0) & \\ \hline M(1) & M(0) \end{array} \quad \text{then} \quad \begin{array}{c|c} M\begin{pmatrix} 0 \\ 0 \\ 0 \end{pmatrix} & M\begin{pmatrix} 000 \\ 000 \\ 000 \end{pmatrix} \\ \hline & M(000) \end{array}$$

**Fig. 9.12.** Corner rule for metal separation.

butting contact does not have an isolated $2\lambda \times 2\lambda$ contact cut. $\square$

### Corner-Based Rules

Some improvement in efficiency can be had if we apply rules such as the above with some attention to where the *corners* are, that is, where edges of rectangles intersect. For this purpose, we do not care whether the intersecting edges belong to the same rectangle or to different rectangles. We do continue to assume that all rectangles are rectilinear.

**Example 9.7:** The rule that says metal must be separated by $3\lambda$ can be expressed by the conditional expression in Fig. 9.12 and three similar rules for other corners. This rule says that if we see a corner in which the lower left has metal, but this metal continues neither up nor to the right, then in all directions except for the lower left, there must not be metal for $3\lambda$.

Note that the rules in Fig. 9.12 do not check more than one column left of the corner and one row below the corner. That limited examination is sufficient to ensure that a design rule error will be detected at some corner. For example, applying the test of Fig. 9.12 to corner $A$ in Fig. 9.13 will not discover the design rule error. However, when we consider corner $B$, the error will be discovered by a rule analogous to that in Fig. 9.12.

One of the orientations of the corner rule analogous to that of Example 9.6 for saying that poly gates must overlap diffusion by $2\lambda$ is shown in Fig. 9.14. Note that while the rule of Example 9.6 is applied at each point along the edge

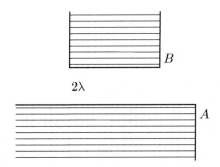

**Fig. 9.13.** Error always gets caught at one corner.

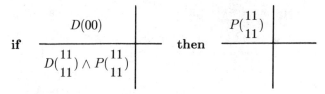

**Fig. 9.14.** Corner rule for gate overlap.

of the diffusion, the rule of Fig. 9.14 is triggered only when there is a corner. Normally, the relevant corner will be formed by the right edge of the poly, which we suppose does not extend to the right of the point shown, crossing the top edge of the diffusion. □

We can now see why corner rules may lead to efficient design rule checks. While they require us to deal with issues of where corners are found, which a pure raster approach does not, we can often simplify the rules because we know that we are at a corner. For example, the metal separation rule given above requires checking only fifteen points per corner. Moreover, we often save time because the absence of corners tells us that certain checks do not have to be made. For example, two wires running for a long way in parallel will have their separation checked when either of them begins or ends. The fact that rectangles are rectilinear assures us that if they are widely enough separated at the endpoints, then they do not violate the design rules at any point along their lengths.

### Design Rule Checks by Polygon Operations

An entirely different approach to design rule checking can be characterized as polygon manipulation. Suppose for the moment that we have found rectangle intersections and have grouped together all connected rectangles of the same layer (not all electrically connected rectangles, as heretofore). By methods of computational geometry which we shall not delve into (see Guibas and Stolfi

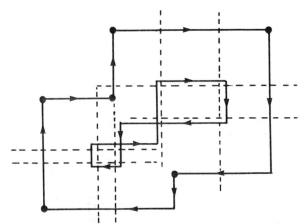

**Fig. 9.15.** Metal polygon with a design rule error.

[1983], e.g.), we can find the polygonal boundary of such groups of rectangles, representing it by its sequence of line segments, in clockwise order around the boundary. Figure 9.15 is an example of such a boundary; ignore the dashed lines for the moment.

If we wish to determine whether a spacing rule has been violated, we have only to expand the polygon by the appropriate amount. For example, if we expand all metal polygons by $1.5\lambda$, then groups of metal rectangles will intersect if and only if they are too close. In fact, we need not form the rectangles into polygons for this test. Just expand all metal rectangles by $1.5\lambda$ about their centers. Find all sets of electrically connected rectangles both before and after the expansion. If the sets are not the same, there is an error.

For another example, we can check that gates overlap diffusion by first removing "channel" regions from diffusion rectangles, that is, removing those regions where poly and diffusion overlap. Then expand all diffusion rectangles by $2\lambda$ in all directions and repeat the process of removing channels. If there are two pieces of diffusion, such as $A$ and $B$ in Fig. 9.16, that are separated by the channel in the first case, but not after expansion of the diffusion, then there is a case of a gate that does not extend far enough.

Testing for wires that are too narrow is somewhat more complex, since it is not possible for us to work with the rectangles as given; we really must combine them into polygons as in Fig. 9.15. The problem is that a wire of adequate thickness could be made up of several rectangles, none of which is by itself of adequate thickness.

Once we construct the boundary as a ring of line segments, we can manipulate the boundaries in a simple manner. Suppose we wish to check that a metal polygon is at least $3\lambda$ thick in all directions. We create a new boundary, $1.49\lambda$ inside the old one. Because we know the clockwise direction around the shape, we can construct lines parallel to each border segment, and $1.49\lambda$ to the left,

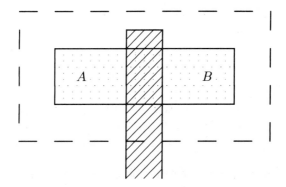

**Fig. 9.16.** Test for short gates.

traveling clockwise. We can find the intersections of the new lines, and thus form a new boundary of segments parallel to the old ones, and in the same order and direction. A necessary and sufficient condition for the polygon to be of adequate thickness (assuming that the polygon surrounds a set of rectilinear rectangles) is that the new boundary does not cross itself and is oriented in a clockwise direction.

For example, the new boundary in Fig. 9.15 does cross itself, a result of the fact that it suffers the problem alluded to in connection with Fig. 9.10(b); two corners are separated by only $2\sqrt{2}\lambda$. As another example, a simple $2\lambda \times 2\lambda$ metal rectangle will yield a new border that is (roughly) a $\lambda \times \lambda$ square. However, the new border makes a counterclockwise revolution if we follow the segments that correspond to the old border in a clockwise order.

### Hierarchical Design Rule Checking

We can often gain considerable efficiency in our design tools by taking advantage of a cell hierarchy. For example, we could have, in the previous section, done our circuit extraction hierarchically, by finding sets of electrically connected rectangles within a cell, and doing further set merging as occurrences of that cell are instantiated within larger cells. Then, if there were 1,000 copies of cell $C$ used, and that cell had two rectangles $R$ and $S$ that intersected, we would discover that fact once, instead of 1,000 times as we would if we first expanded cells to make a completely instantiated version of the circuit.

However, there may be some reasons why it is preferable to instantiate fully the circuit before doing extraction. For one, the extracted switch diagram will normally be desired in fully instantiated form, for simulation purposes, so there is not an order of magnitude to be gained in the running time if we do not fully instantiate the CIF. Moreover, the intersection-of-rectangles algorithm and the set-merging algorithm share the property that they are rather efficient on large instances, but will lose some efficiency when applied to many little instances.

On the other hand, design rule checking does not require that we fully instantiate the circuit, and in fact, we are far better off if the checker can pinpoint one occurrence of an error within a cell, rather than pointing to errors all over the chip, one at each place where the cell is instantiated. We might thus imagine a design rule checker that looks for errors within each cell only. But that is not adequate, because there could be errors involving the interaction of rectangles from two different cells. In an extreme but unlikely case, every instantiation of a cell could be overlaid with a different collection of rectangles, necessitating a completely different set of rule checks for each instantiation. Thus we are back where we started, doing work that is proportional to the number of rectangles in the fully instantiated circuit, rather than to the sum of the sizes of the cell definitions.

If hierarchical design rule checking is to be made to work, the designer of the circuit must accept some constraints. The simplest, and also most restrictive constraint, is that when a cell is instantiated, there is no rectangle from some other cell that will lie, even partially, within the boundary of that rectangle.†

**Example 9.8:** A canonical situation where the inability to lay cells over one another is a nuisance concerns the PLA built from cells in the pattern suggested by Fig. 7.29. We would most like to use an $8\lambda \times 7\lambda$ cell shown in Fig. 9.17(a), and replicate it as many times as necessary, with alternate copies mirrored about the $y$-axis, and with the shared diffusion ground wires overlapped so two cells occupy $14\lambda$ in the $x$ direction. Then, where taps are needed we would come along and place metal-diffusion vias and diffusion wires that connected to ground under the poly wires.

However, this approach violates the constraint that nothing be placed inside the border of a cell once that cell is placed, in two ways. First is the overlapping of diffusion wires. That is not so serious, because the rectangles crossing borders of cells happen to agree with what is already there. Thus, the overlap will not introduce new design rule errors within the cell. The second concerns the placement of vias and taps, and this violation is more serious. A complete reexamination of the cell would be necessary each time we placed additional rectangles that did not agree with what was already in the cell.

If we are to maintain the rule of no rectangles within the boundary of a cell that do not belong to that cell, then we need three different versions of the basic cell, shown in Fig. 9.17(b–d). Each is $7\lambda \times 7\lambda$, and includes only half of the diffusion ground wire. That in turn implies that we may have to add a $\lambda$-wide strip of diffusion at the edge of the cell array. Fig. 9.17(b) shows the

---

† Therefore, all cells must include in their design information as to what the boundary is. Languages like CHISEL allow such declarations, and even allow several different bounds, depending on whether the cell is being replicated next to itself, or being mirrored, a feature that turns out to be important for cells like those of Example 9.8, to follow.

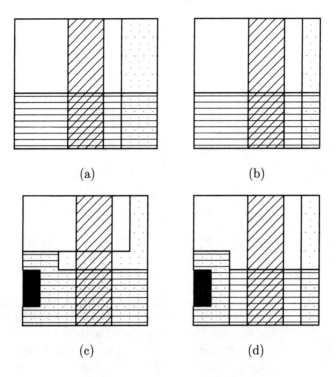

(a)  (b)

(c)  (d)

**Fig. 9.17.** Variants of PLA cell.

cell without any trace of tap. Fig. 9.17(c) shows the cell with a tap and half of the via, and Fig. 9.17(d) shows the cell with only the half via, which is needed if the adjacent cell has a tap and the present one does not. ☐

If we follow the design discipline of not allowing rectangles from outside to appear within cell boundaries, then we can use a hierarchical design rule checker that may be significantly faster than those using the instantiated circuit. The difference will be felt especially if there are many cells of reasonable size that are instantiated many times. The reason is that each instantiation of a cell can be replaced by its border, as suggested in Fig. 9.18. Information regarding which border rectangles are electrically connected could help reduce spurious errors, but is not essential. The width of the border depends on the actual design rules, but using the Mead-Conway rules, a 3λ border is sufficient because no width or distance greater than that needs to be checked. The savings comes from the fact that after doing design rule checks for the cell in isolation, we never again have to make checks within the interior, no matter how many instantiations of the cell is made.

We can use the same checking strategy if cell overlaps that agree with what

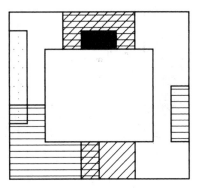

**Fig. 9.18.** A cell replaced by its border.

is already in the cell are permitted; the case of the diffusion wire in Example 9.8 is a canonical situation where that privilege is desirable. Another discipline that might be followed is to allow overlaps of the border in arbitrary ways, but only for a limited distance. For example, if in Example 9.8 we allowed the via in Fig. 9.17(c) to be drawn completely, "outside" the boundary of that cell, then it would overlap by 2λ with the adjacent cell. There would then be no need for the cell of Fig. 9.17(d). However, we can get away with this simplification only if we allow 2λ overlaps of cell borders. We could make this allowance, provided we added 2λ to the depth of the border used for cell instantiations, e.g., we used borders of 5λ for the Mead-Conway rules.

## 9.4 AN ALGORITHM FOR SIMULATION OF SWITCH CIRCUITS

We shall now consider the simulation of switch diagrams, such as those specified by esim, from Section 7.4. A language like esim includes not only a specification of transistors and their interconnection, but certain useful hints about the role played by various transistors and nodes. Most important is an indication of the size of the capacitor represented by each node; for example, the nodes that are precharged in a precharging circuit are often extremely large capacitors. Further, we are given as part of a complete esim program, a sequence of input signal changes to be made. The task of the simulator is to make each of those changes, in turn, letting the signals at the various nodes settle before each new change of input is made.

If we think about the matter, we should see that there are more ways that a circuit could compute than we can hope to plan for in advance, and therefore, any switch simulator must be imperfect. For example, even the assumption that circuits settle down between input changes is not necessarily met. Surely a circuit could oscillate, although it is not clear how useful most such circuits would be. As another example, there could be a *critical race* present in the circuit, where the final signals on certain nodes depend on the speed with which particular events, such as the switching of transistors, occur.

We shall have more to say about such problems later. However, let us remark here that more general sorts of simulators exist, ranging from

1.  Differential-equation-based simulators that handle only a few transistors, but simulate them with great precision, including the effect of nonlinearities in the resistance and capacitance as they switch, through

2.  *Timing simulators*, which attempt to estimate the times at which transistors switch by relatively simple transistor models, to the

3.  Switch simulators that we shall discuss here, which do not deal with time directly, and must assume that the relative speed of events is not relevant.

In addition to the existence of simulators that deal with switches at different levels of detail, there are also simulators for different levels of abstraction. Most notable are logic level simulators. However, these present us with problems because it is hard to extract logic from CIF or another geometry language. The difficulties with logic extraction in turn stem from the variety of ways that logic can be implemented in VLSI technologies, including but not limited to pass transistor logic, pullup/pulldown logic, and precharged logic.

In this section we shall explore one approach to switch simulation, based on the work of Bryant [1981b] and couched in his lattice-theoretic terms. We begin by discussing how signals propagate in a switch network, on the assumption that the logical state of each transistor gate is known and does not change. We then show how the solution to this problem may be iterated to simulate networks where transistor gates do switch, provided only that the circuit does not have critical races. We begin with a brief introduction to lattice theory, in particular what is called an *upper semilattice*.

### Lattice-Theoretic Notation

Let us suppose there is a set $S$ of elements (which in our application will be signal strengths) and a binary operation $\vee$ called *join* on these elements, satisfying the following laws.

1.  *Idempotence.* $a \vee a = a$ for all $a$ in $S$.

2.  *Commutativity.* $a \vee b = b \vee a$ for all $a$ and $b$ in $S$.

3.  *Associativity.* $a \vee (b \vee c) = (a \vee b) \vee c$, for $a$, $b$, and $c$ in $S$.

4.  *Existence of bottom.* There is an element $\perp$ in $S$ such that for all $a$, $a \vee \perp = a$.

Then $(S, \vee)$ is said to be an *upper semilattice*, which we shall call simply a *semilattice*.

**Example 9.9:** A common example of a semilattice is where $S$ is the set of subsets of some other set $X$, and $\vee$ is union. Clearly $Y \cup Y = Y$ for any set $Y$, so (1) is satisfied. Also, it is easy to show that union is commutative and associative, and the empty set serves as $\perp$. $\square$

Another important notational concept is the definition of $\leq$ on semilattices by: $a \leq b$ if and only if $a \vee b = b$. In our example of union as $\vee$ above, the relationship $Y \leq Z$ on sets means $Y \cup Z = Z$, i.e., $Y \subseteq Z$, so for that example, $\leq$ can be interpreted as set inclusion.

In general, $\leq$ defined as above has the properties one would expect from a partial order, and several others besides. We shall prove certain of these properties in the next lemma.

**Lemma 9.2:**
1.  $a \leq b$ and $b \leq a$ both hold if and only if $a = b$.
2.  *Transitivity.* $a \leq b$ and $b \leq c$ imply $a \leq c$.
3.  $a \leq a \vee b$.
4.  If $b \leq a$ and $c \leq a$, then $b \vee c \leq a$.

Note that (1) and (2) say that $\leq$ is a partial order on $S$, but (3) and (4) are properties that do not necessarily hold for a partial order.

**Proof:** For (1), $a \leq b$ means $a \vee b = b$, and $b \leq a$ means $b \vee a = a$. Since $a \vee b = b \vee a$ by commutativity, it follows that $a = b$. For (2), we are given $a \vee b = b$ and $b \vee c = c$. Therefore,

$$a \vee c = a \vee (b \vee c) = (a \vee b) \vee c = b \vee c = c$$

the middle step being an application of associativity. We leave (3) and (4) as exercises. $\square$

## The Signal Lattice

A node of a switch diagram can be expected to have a *state*, which tells whether that node is currently 1 (high voltage) or 0 (low voltage). In some cases, it is useful to have a state $X$, or unknown, which applies to a node that either has a voltage that is intermediate between high and low, e.g.,
1.  A node connected by two transistors, both of which are on, and which are connected, respectively, to power and ground, or
2.  A node that has a voltage we do not yet know.

The status of a node, as far as simulation is concerned, cannot be characterized solely by the state. We must also consider the *strength* of the signal. As a first cut, the strength of a signal is determined by whether the signal is the result of a current flow into or out of that node, or whether the signal is the result of trapped charge, or the absence of charge in a region of the network that is isolated from all current sources.

However, not all current sources are equally strong. The most common example is what happens at the output of an inverter when the pulldown is on. There is a source of current through the pullup resistor, and there is a way for current to flow away (i.e., a source of negative current), through the pulldown.

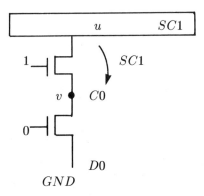

**Fig. 9.19.** A precharged circuit.

Because the pulldown has smaller resistance, it is "stronger," and the output is pulled down to state 0.

For an example involving charge, consider a precharged circuit, where there is a node with a large capacitance, and one or more paths to ground, which may be partially open. Such a situation is shown in Fig. 9.19; ignore the notation $SC1$, $C0$, and $D0$ for the moment. If the path from $u$, the precharged node, to $v$, a node part of the way to ground, is available, but the path does not reach ground, some charge will travel from $u$ to $v$. We do not want the simulator to imagine that $u$ has lost enough charge to affect its voltage. Thus, we must have the ability to say that the charge at $u$ is stronger than the (absence of) charge at $v$.

So that a simulator would be able to handle progressively more complex situations involving the competition among different strengths of current and charge, we could define a variety of lattices of signals. However, we consider here only the simplest set of signals that work for the most common kinds of logic implemented in NMOS. We must handle the distinction between currents coming through a pullup or other transistor intended to be a "large" resistance, and a current coming directly from an input pad or "small" resistance such as a network of pulldowns. We call the former strength *weak* and the latter *driven*, to conform with the notation of Baker and Terman [1980]. Note that a series connection of pulldowns actually could produce more resistance than a pullup, but we shall not recognize that possibility. In practice, the circuit extractor or a high-level design rule checker should verify that the necessary pullup/pulldown resistance ratio is met, thereby allowing us to treat all networks of pass transistors as if they had much less resistance than any pullup.

The other distinction that it seems essential to handle is between the amount of charge on a node used for precharging and the charge on any other

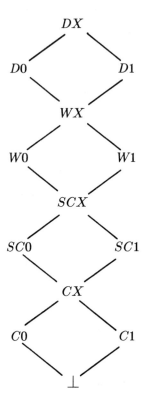

**Fig. 9.20.** The lattice of signals.

node. We shall refer to the strength of an ordinary node that holds charge (or its absence) but that does not have an unusually large capacitance as *charged*, while nodes of unusually large capacitance will be given strength *supercharged*, when they are not attached to a source of current.

In Fig. 9.20 we see the lattice of signals, including both the strength and the state. We use $D$, $W$, $SC$, and $C$ for driven, weak, supercharged, and charged, which are the four strengths, and we use 0, 1, and $X$ for the three states. The reader should observe that this same pattern could be used to handle any set of strengths based on current, like $D$ and $W$, together with any set of strengths based on charge, like $SC$ and $C$.

The lines in the diagram relate to the $\vee$ operation on signals. If $a$ and $b$ are signals, then we define $a \vee b$ to be the lowest signal that has a downward path to both $a$ and $b$. It is easy to check that the axioms of a semilattice are satisfied for this definition of $\vee$. An alternative way to look at the definition is that $a \leq b$ holds if and only if there is a path downward from $b$ to $a$. The physical significance of the $\vee$ operation is that when two signals $a$ and $b$ come together at a node, the resulting signal is $a \vee b$. The reader must verify that this point holds for any two signals, but we shall give some examples, from which the general rules should be clear.

**Example 9.10:**

1. $W0 \vee W1 = WX$. Physically, this says that a node fed by a positive current through a large resistance and drained through a similar resistance will assume a voltage that is neither 0 nor 1.

2. $C0 \vee W1 = W1$. In general, any signal of higher strength dominates a signal of lower strength, regardless of state. In the particular case, what we are saying is that if we have a node holding no charge, and current is allowed to flow into that node through a large resistor, the node will eventually have high voltage, and the strength of that node's signal will be $W$, indicating that the state is enforced by a "weak" current supply. $\square$

**Transistor Functions**

Our next task is to associate with each transistor a function from signals to signals that tells how the transistor passes signals from one end to the other. Note that the function does not depend on the direction in which the signal passes. First, let us observe that a transistor which is on will allow charge to pass from one side to the other. Thus, if the left end of a channel is a node with signal of strength $C$, then after a negligible amount of time, the right end will also be of strength $C$ and of the same state as the left end, unless there is a stronger signal than $C$ also reaching the right end. Similarly, a charge of strength $SC$ will migrate from one end of any transistor that is on, to the other.

Suppose now that one end of a transistor is a current source of strength $D$. Then if the transistor is a small resistance, i.e., a pass transistor, the current will reach the other side, in the same strength and state. However, if the transistor is a large resistance, e.g., a pullup, then the current will be weakened by passing through the large resistance and will only appear as strength $W$, with the same state, on the other side. On the other hand, a current of strength $W$ passes through any transistor that is on.

Of course, at the same time, many currents and charges may be impinging on the same node, and the lattice of Fig. 9.20 must be used to tell how they blend. That lattice has been defined to conform to the physical reality that was illustrated in Example 9.10; currents dominate charges, large charges dominate small charges, large currents dominate small, and when two charges or currents of equal strength but different states meet, the result has the same strength but state $X$. Thus, when currents and charges compete at a node, the result is the join of all those signals.

In fact, signals reach a node not only from its neighbors, but from all over the circuit, following all possible *paths*, which are sequences of transistors that are switched on. A signal starting at node $v$ to reach node $u$ is modified only by the effect of passing through the transistors along the path, which means that the signal is reduced to strength $W$ if it ever passes through a large resistor, and it is blocked completely (turned to $\perp$) if any transistor is off. Formally, we can define a function $f$, from signals to signals, to associate with each transistor in a circuit, as follows.

1.  If the transistor is off, then its associated function is $f_{off}$, where for all signals $a$,

    $$f_{off}(a) = \perp$$

2.  If the transistor is a low resistance transistor, i.e., a pass transistor, and it is on, then its associated function is $f_{strong}$, where for all signals $a$,

    $$f_{strong}(a) = a$$

3.  If the transistor is a large resistance and is on, then its associated function is $f_{weak}$, where for all signals $a$ and states $b = 0$, 1, or $X$,

    $$f_{weak}(a) = \begin{cases} a \text{ if } a \leq WX \\ Wb \text{ if } a = Db \end{cases}$$

If $P$ is a path consisting of transistors $t_1, \ldots, t_n$ with associated functions $f_1, \ldots, f_n$, then the function associated with the path $P$ is $f_n f_{n-1} \cdots f_1$, which we denote by $f_P$. As a special case, if $n = 0$ the path $P$ is empty, and we take $f_P$ to be the identity function (which also happens to be $f_{strong}$). That is, if $P$ runs from node $u$ to node $v$, then the contribution of the signal at $u$ to the signal at $v$ is $f_P(u)$. Thus, if $val_0(v)$ denotes the initial signal on the node $v$, e.g., $C1$ if $v$ starts with a small positive charge or $D0$ if $v$ is the ground node, then we can express the signals $val(v)$ obtained after each signal $val_0$ has had the opportunity to propagate wherever it can, by

$$val(v) = val_0(v) \vee \bigvee_{\substack{\text{paths } P \\ \text{from } u \text{ to } v}} f_P(val_0(u)) \qquad (9.1)$$

In practice, it is not necessary to consider all paths in (9.1); acyclic paths suffice because in a resistor network, signals cannot strengthen by going around a loop. Further, we do not have to consider paths from arbitrary nodes $u$, only those that have a signal initially, such as input pads and nodes where charge was initially trapped.

**Example 9.11:** Let us consider Fig. 9.19 again. There are three nodes, $u$, $v$, and $GND$. We have supposed that $u$ is a large capacitor and charged to 1, so initially $val_0(u) = SC1$. Node $v$, we suppose, initially has signal $C0$, perhaps because the transistor between $u$ and $v$ has just turned on, and most recently the transistor between $v$ and ground was on, leaving $v$ charged to 0, i.e., absent of charge, when that transistor turned off. Thus, $val_0(v) = C0$. Of course, the ground node is a strong source of (negative) current, so $val_0(GND) = D0$, as it always must be.

There are, in principle, two acyclic paths from sources to $v$, the path from $u$ and the path from $GND$. The latter is an off transistor, so the contribution of this path is $f_{off}(D0) = \perp$. That makes sense; it says that the ground node has no effect on $v$, because $v$ is disconnected from ground. The other path, from $u$, is through one low-resistance transistor. Thus, its contribution is

$$f_{strong}(SC1) = SC1$$

It follows that the new signal at $v$ is $val(v) = C0 \vee \perp \vee SC1 = SC1$. That is, the node $v$ becomes "supercharged" to 1. In a sense, this is unrealistic since $v$ is only "supercharged" by temporarily being part of the large capacitance at $u$. However, if $v$ were cut off from $u$, then before we used the signal at $v$, or any other node of small capacitance, to influence other nodes, we would reduce the signal to $C1$, indicating that that state is retained, but the strength reduced.

The paths to $u$ are from $v$ and $GND$. The latter contributes

$$f_{strong}(f_{off}(D0)) = f_{strong}(\perp) = \perp$$

and the former contributes $f_{strong}(C0) = C0$. Thus, the new value at $u$ is $val(u) = SC1 \vee C0 \vee \perp = SC1$. That is, the charge at $v$ and the disconnected ground do not influence the charge at $u$. It is easy to check that no signal influences ground in Fig. 9.19; thus $val(GND) = D0$, as it always must, unless there is a short circuit involving the ground node.

Let us consider another example where currents blend; the circuit of Fig. 9.21 is a NAND gate feeding its output through a pass transistor, perhaps a clock gate. We suppose that initially the nodes other than power and ground have unknown charge, represented by signal $CX$.

The acyclic paths to $u$ are

1. From $VDD$ to $v$ to $u$, with contribution

$$f_{strong}(f_{weak}(D1)) = f_{strong}(W1) = W1$$

2. From $v$ to $u$ with contribution $f_{strong}(CX) = CX$.

3. From $w$ to $v$ to $u$, with contribution $f_{strong}(f_{strong}(CX)) = CX$.

4. From $GND$ to $w$ to $v$ to $u$ with contribution

$$f_{strong}(f_{strong}(f_{strong}(D0))) = D0$$

Thus the new value at $u$ is $CX \vee W1 \vee CX \vee CX \vee D0 = D0$, the five terms corresponding to $val_0(u)$ and the four paths mentioned above. The conclusion, which is entirely expected, is that when both inputs of a NAND gate are on, the output is driven to 0. $\square$

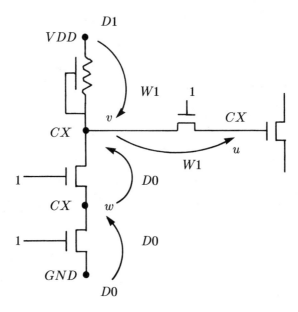

**Fig. 9.21.** Example of propagation through paths.

**Solution of Transfer Equations**

There is a natural way to calculate the values defined by (9.1). We start with $val(v) = val_0(v)$ for all $v$, and repeatedly *visit* the nodes. When we visit $v$, we do

$$val(v) := val(v) \vee \bigvee_{u \text{ adj. to } v} f_{uv}(val(u)) \tag{9.2}$$

That is, we consider all neighbors $u$ of $v$, apply to their present value the function $f_{uv}$ associated with the transistor between $u$ and $v$, and take the join of all these values and the previous value of $v$.

It seems intuitive that if we repeat the above steps until no changes occur, the result will be (9.1). Moreover, in (9.2) the new value of $val(v)$ must be $\geq$ the old value, where $\geq$, the opposite of $\leq$, is taken in the sense defined for lattices; that is, $val(v)$ can only move upward in the lattice of Fig. 9.20. Thus, convergence must occur sometime. Unfortunately, our intuition is not necessarily correct; we need a condition on circuits that is often satisfied, but need not be. We shall discuss what constraints are needed after we give an example of where the algorithm works correctly and one where it does not.

**Example 9.12:** Consider the circuit of Fig. 9.21 again, and let us use $val_i$ to refer to values after $i$ passes have been made. Then Fig. 9.22 shows the first four iterations of (9.2). For $val_1$ in the first pass, we note that $v$ is passed

|         | $u$  | $v$  | $w$  |
|---------|------|------|------|
| $val_0$ | $CX$ | $CX$ | $CX$ |
| $val_1$ | $CX$ | $W1$ | $D0$ |
| $val_2$ | $W1$ | $D0$ | $D0$ |
| $val_3$ | $D0$ | $D0$ | $D0$ |
| $val_4$ | $D0$ | $D0$ | $D0$ |

**Fig. 9.22.** Propagation of signals until convergence.

the value $CX$ by neighbors $u$ and $w$, and value $W1$ by $VDD$. It had value $CX$, so $val_1(v)$ is set to $CX \vee CX \vee CX \vee W1 = W1$. Node $w$ is passed $CX$ by $v$,† $CX$ by $u$, and $D0$ by $GND$, so it gets the new value $D0$. Also observe that nodes $v$ and $u$ are each first set to $W1$ and then to $D0$, because the signal from ground takes a larger number of steps to get to these nodes than does the signal from $VDD$. The temporary value with the wrong state is of no significance here, since we are not trying to model timing exactly, just ultimate results. However, it does make a profound difference if we change the circuit just slightly, as we shall do next.

Now, let us see what happens if we replace the pass transistor between $u$ and $v$ by a large resistance, i.e., a transistor with a long channel and a gate that is on. (Such a circuit appears in Fig. 9.24.) Nothing changes from Fig. 9.22 until we try to pass $D0$ to $u$ at the third round. That is, with $val(v) = D0$, (9.2) says

$$val(u) = W1 \vee f_{weak}(D0) = W1 \vee W0 = WX$$

Convergence occurs at this step, so the final values given by iterative application of (9.2) is $val(w) = val(v) = D0$, and $val(u) = WX$.

That result is unintuitive; it says that node $u$ is weakly supplied with current, which is true; but it also says we cannot deduce what the direction of the current is, which is false, because with $v$ at low voltage, any charge at $u$ will be drained away. Thus, $W0$ is the correct value for $u$. Note also that the value $W0$ is $val(u)$ as given by (9.1). □

**Distributive Functions**

It turns out that a sufficient condition for the iteration of (9.2) to converge to (9.1) is that the functions associated with transistors be distributive over $\vee$. To assure that only acyclic paths need to be followed, we also require that the functions be nonincreasing. That is, if $f$ is the function associated with any

---

† Note that we are not making new values available immediately, but are basing each round of new values on the complete set of old values. This decision was made so we could explain more succinctly what the algorithm does. In practice, we would use new values as soon as they were available, thus speeding up convergence somewhat.

transistor, and $a$ and $b$ are any signals, we say $f$ is *nonincreasing* if $f(a) \leq a$, and we say $f$ is *distributive* if $f(a \vee b) = f(a) \vee f(b)$. It is easy to show that the functions $f_{off}$, $f_{weak}$, and $f_{strong}$ are all nonincreasing.

The problem with the second part of Example 9.12, where the transistor between $u$ and $v$ was made to have large resistance, is that $f_{weak}$ is not distributive, because $f_{weak}(D0 \vee W1) = f_{weak}(D0) = W0$, but

$$f_{weak}(D0) \vee f_{weak}(W1) = W0 \vee W1 = WX \neq W0$$

On the other hand, $f_{strong}$ is really the identity function, and $f_{off}$ maps everything to $\perp$, so it is easy to see that these two functions are distributive. Thus, the way out of our dilemma is to show that for most "reasonable" circuits, we do not really need $f_{weak}$. First, we show that if all transistor functions are distributive, (9.2), iterated, converges to (9.1). We begin with the fact that we need not consider cyclic paths in our calculation, and then we show that after $i$ iterations of (9.2) $val(v)$ is the join over all paths of length $i$ or less, of the signal propagated to $v$ along that path.

**Lemma 9.3:** If $P$ is a cyclic path of transistors with distributive, nonincreasing functions, then there is some shorter path $Q$, connecting the same endpoints, such that for any signal $a$, $P(a) \leq Q(a)$, or equivalently, $P(a) \vee Q(a) = Q(a)$.

**Proof:** To begin, it is an easy exercise to show that the composition of distributive functions is distributive, and therefore, $f_R$ is distributive for any path $R$ composed of transistors with distributive functions. Now, if $P$ has a cycle, we can write $P$ as $P_1 P_2 P_3$, where $P_2$ is nonempty, and begins and ends at the same node. Let $Q$ be $P_1 P_3$; note that $Q$ is a legal path through the chip. Next, define $f$, $g$, and $h$ to be the functions $f_{P_1}$, $f_{P_2}$, and $f_{P_3}$, respectively. Then, $f_P = hgf$ and $f_Q = hf$. We must show that for any signal $a$, $hgf(a) \leq hf(a)$, i.e., $hgf(a) \vee hf(a) = hf(a)$.

By distributivity of $h$, $hgf(a) \vee hf(a) = h(gf(a) \vee f(a))$. By the nonincreasing property for $g$, $gf(a) \leq f(a)$, so $gf(a) \vee f(a) = f(a)$. Thus

$$hgf(a) \vee hf(a) = hf(a)$$

as we wished to prove. $\square$

The consequence of Lemma 9.3 is that if we have a join of the functions associated with a collection of paths, applied to some signal, and both $P$ and the shorter path $Q$ are members of that collection, then we can drop $f_P$ out of the join, because its effect is lost in favor of $f_Q$. Note the general rule that if $a$ and $b$ are two terms in a join of many terms, and $a \leq b$, then we may delete $a$ from the join without changing the value.

**Theorem 9.2:** If all transistor functions are distributive and nonincreasing, and the circuit has $n$ nodes, then the iteration of (9.2) will converge after at most $n$ passes, and it will converge to the same solution as (9.1).

**Proof:** Let $\mathcal{P}_i(v, u)$ be the set of paths of length at most $i$ that originate at $u$ and terminate at $v$. We shall show, by induction on $i$, that after $i$ iterations of (9.2),

$$val_i(v) = \bigvee_u \bigvee_{\substack{P \text{ in} \\ \mathcal{P}_i(v,u)}} f_P(val_0(u)) \tag{9.3}$$

The basis, $i = 0$, is trivial, since the only path $P$ in $\mathcal{P}_0(v, u)$ is the empty path, in the case $u = v$. As $f_P$ is then the identity, Equation (9.3) simply claims that $val_0(v) = val_0(v)$.

Now suppose (9.3) is true for $i - 1$. Applying (9.2) sets

$$val_i(v) = \bigvee_{w \text{ adj. } v} f_{wv}(val_{i-1}(w)) \tag{9.4}$$

where $f_{wv}$ is the function associated with the transistor between nodes $w$ and $v$. By the inductive hypothesis,

$$val_{i-1}(w) = \bigvee_u \bigvee_{\substack{P \text{ in} \\ \mathcal{P}_{i-1}(w,u)}} f_P(val_0(u)) \tag{9.5}$$

Since each $f_{uv}$ is assumed distributive, and distributive functions are easily shown to distribute not only over joins of two values, but over joins of any finite number of elements, we may substitute (9.5) into (9.4) and distribute each $f_{wv}$ over the join of terms given by (9.5) to obtain a join over all paths of the form $u, \ldots, w, v$ for some neighbor $w$ of $v$, where the path $u, \ldots, w$ is of length $i - 1$ or less. This condition describes all and only the paths from $u$ to $v$ of length $i$ or less and proves (9.3).

We must prove that the iteration of (9.2) converges after at most $n$ stages, where $n$ is the number of nodes. We have Lemma 9.3 to tell us that the joins over paths need not include any cyclic paths, since if there is any path of length at most $i$ from $u$ to $v$, then there is an acyclic one of length at most $i$. Therefore, $val_{n+1}(v) = val_n(v)$. In practice, convergence may come before $n$ stages, and if ever we find that $val_{i+1}(v) = val_i(v)$ for all $v$, then repeated applications of (9.2) would continue to produce the same results, and we may stop. $\square$

We may summarize what has been said by giving an algorithm that uses the proposed solution to (9.2) to perform simulation. When we simulate a circuit, we wish not only to solve the equations that propagate signals on the assumption that transistor gates do not change. Rather, we must allow inputs to change in some designated sequence. Furthermore, after an input change we may require several rounds of signal propagation, because one round may cause gates to change, necessitating another round.

The simulation algorithm given below requires that the circuit obey certain restrictions on the use of large resistors. A second algorithm will remove the restraint and provide better chances for simulating complex circuits in the bargain.

**Algorithm 9.3:** *Simple Simulation Algorithm for Distributive Transistor Functions.*

INPUT: We are given a circuit with transistors classified as large or small resistances, and nodes classified as large or small capacitances. We assume, however, that if a transistor is a large resistance, e.g., a pullup, then it is attached at one end to an input of strength $D0$ or $D1$, i.e., an input pad, $VDD$, or $GND$.† We are also given a sequence of circuit input changes to simulate.

OUTPUT: The response of each node of the circuit to the input sequence.

METHOD: Given our input assumptions, we may change the circuit so all large resistors have one end at a node that is attached nowhere else, and bears the signal $W0$ or $W1$ whenever in the original circuit the corresponding signal would be $D0$ or $D1$, respectively. To do so, we simply split each input node attached to more than one large resistor into one node for each such resistor. Now, we may associate the function $f_{strong}$ with large resistance transistors. Even though that function is a lie, we know the transistor will never be asked to pass a driven signal, and so the circuit as a whole will not notice the change.

Now all transistors have distributive functions associated with them, and we are ready to simulate the given sequence of input changes. Note that when an input to the original circuit changes, we must change all nodes into which that input has been split by the above alteration. Frequently, only $VDD$ will be split, and this issue does not arise.

To initialize, we set all input nodes to their initial value and all other nodes to $CX$ or $SCX$, depending on whether they are small or large capacitances. The main loop of the algorithm is

> *converge*;
> **while** input changes remain to be done **do**
> > **begin** make changes; *converge* **end**

The routine *converge* consists of the following three steps, repeated until after two consecutive iterations, the resulting signals at each node are the same.‡ It will be evident from the nature of the steps that further iterations would continue to produce the same set of signals for the nodes.

1.   Set each node except the inputs to charged strength, with the same states that they had before. The strength is either $C$ or $SC$, depending on whether the node is a small or large capacitor. The purpose of this step is seen only after the first iteration, when we may have a node that has a signal representing current flow, say $W1$, but a gate controlling access

---

† In practice, this assumption will be met, since pullups will normally have one end attached to $VDD$. Note how the problem in Example 9.12 was caused by a large resistor violating this condition.

‡ Note, however, that there is no guarantee that convergence will ever occur.

to that node has just been turned off, leaving the charge but removing the promise that if the charge were drained it could be replenished by a current. Thus, we set the signal to $C1$, and trust that if the node is in fact not cut off it will be restored to strength $W$. If we do not take this step for every node, then two adjacent nodes that have a signal like $W1$, but have just been isolated from the sources of current, could appear to keep each other supplied with current.

2.   Set all enhancement mode transistors to on if their gates are in state 1, and off if their gates are in states 0 or $X$. All depletion mode transistors are assumed to be on. The effect of this decision is that signals will propagate along only a subset of the paths that are actually open, since some of the unknown gates may in fact be in state 1.

3.   Iterate Equation (9.2) until of the solution to that equation is reached. □

The effect of running Algorithm 9.3 on a circuit will depend on the circuit in subtle ways. For example, if their are critical races, that fact will not necessarily be detected, and the simulation will force a particular resolution of the race. Worse, since for a single execution of Algorithm 9.3 we treat transistor gates in state $X$ as if they were 0, the solution $val(v)$ that we get when we iterate (9.2) accounts for only a subset of the available paths to $v$. As we mentioned, the result of joining a subset of the terms is always $\leq$ the result of joining the full set, but that is a small consolation. Fortunately, in practice the unknown states usually get resolved rapidly, unless the circuit really has a bug that causes an oscillation in voltage, or a voltage that actually is intermediate between $VDD$ and ground.

**Example 9.13:** Figure 9.23 shows a cascade of two inverters. We have split the $VDD$ node feeding both pullups into two distinct nodes, each a source of signal $W1$, as required by Algorithm 9.3. We show one step of the simulation, where the input to the first inverter has been set to 0, and its output has signal $CX$; perhaps this is the first step of the simulation. The input to the second inverter is thus $CX$ as well, which we take to indicate that the transistor is off.

When we iterate (9.2) we find that $val(v)$ is set to $W1$, since the ground signal $D0$ cannot reach it, but the signal $W1$ from the pullup can. As the pulldown of the second inverter is off, $val(u)$ is also set to $W1$.

The second time through the loop of $converge$, we set $v$ and $u$ to signal $C1$ at step (1). We then discover that the second pulldown is on, because its gate has state 1. Thus, the second time we do step (3), we converge to the proper signals $val(v) = W1$ and $val(u) = D0$. The third pass will produce the same values, and $converge$ is finished. We would then make the next of any remaining input changes, e.g., turning on the input to the first inverter, and we would repeat $converge$. □

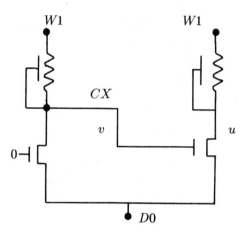

**Fig. 9.23.** Pair of inverters.

### An Improved Simulation Algorithm

There is a more effective, but more time consuming way to simulate circuits; the method allows us not only to use the nondistributive $f_{weak}$ in its proper role, but it resolves the state of certain gates that Algorithm 9.3 would leave in state $X$. The proof of validity for the algorithm is omitted, but we present the computation details here.

**Algorithm 9.4:** *Bryant's Simulation Algorithm.*

INPUT: A switch circuit with transistors classified as large or small resistances and nodes classified as large or small capacitances. A sequence of input changes to simulate is also provided.

OUTPUT: The response of each node of the circuit to the input sequence.

METHOD: The algorithm is the same as Algorithm 9.3 except in the procedure *converge*. There are two major changes. First, the application of (9.2) is a two-phase process, where first we deduce the proper strength without reference to the state of signals, and then we propagate states, ignoring all states that are associated with the wrong strength. The second change is that when we propagate states, we do so twice. The first time we allow only 0's to pass through transistors with gate in state $X$, and the second time we allow only 1's to pass through these transistors. Gates that get the same state in either case are assigned that state, while nodes that get 0 in one case and 1 in the other are assigned state $X$. The steps of *converge* are thus:

1.   Set all noninput nodes to strength $C$ or $SC$ as appropriate, without changing the state. This step is the same as in Algorithm 9.3.
2.   Set all signals to state $X$ without changing the strength.

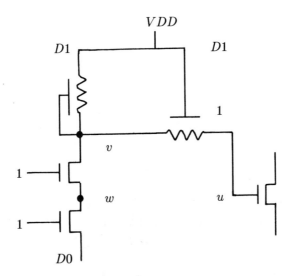

**Fig. 9.24.** Circuit with large resistor not attached to an input.

3.  Assuming all transistors with gates in state $X$ are on, iterate (9.2) until it converges. The result will be that each node has its proper strength, but has state $X$ regardless of what is proper.

4.  Solve (9.2) again, with the initial conditions from step (1). However, before taking the join over all nodes $u$ adjacent to $v$, delete terms $f_{uv}(val(u))$ in either of two situations.

    a)  The transistor between $u$ and $v$ has its gate in state $X$, and the state of $val(u)$ is 1; i.e., do not propagate 1's through ambiguous transistors.

    b)  The strength of $f_{uv}(val(u))$ is not equal to the strength of the solution from step (3).

5.  Repeat step (4), but with 1 replaced by 0 in (a). That is, now inhibit the passage of 0's through ambiguous transistors.

6.  Modify the solution of step (3) by changing the state, but not the strength, of certain nodes. Specifically, if the state of node $v$ is the same in the solutions of steps (4) and (5), then change $val(v)$ to have that state; otherwise, leave the state at $X$. □

**Example 9.14:** Let us consider how Algorithm 9.4 behaves on the problem circuit from Example 9.12, which we illustrate in Fig. 9.24. In step (3), we start with nodes $u$, $v$, and $w$ all having signal $CX$, and with $VDD$ and ground given signals $DX$, indicating that they are strong sources of current, but omitting the direction of the current. The iteration of step (3) is shown in Fig. 9.25. Note how the signal $DX$ at $v$ becomes $WX$ at $u$, because of the large resistor between those nodes.

|        | $u$  | $v$  | $w$  |
|--------|------|------|------|
| $val_0$ | $CX$ | $CX$ | $CX$ |
| $val_1$ | $CX$ | $WX$ | $DX$ |
| $val_2$ | $WX$ | $DX$ | $DX$ |
| $val_3$ | $WX$ | $DX$ | $DX$ |

**Fig. 9.25.** Iteration for computation of strengths.

|        | $u$  | $v$  | $w$  |
|--------|------|------|------|
| $val_0$ | $\perp$ | $\perp$ | $\perp$ |
| $val_1$ | $\perp$ | $\perp$ | $D0$ |
| $val_2$ | $\perp$ | $D0$ | $D0$ |
| $val_3$ | $W0$ | $D0$ | $D0$ |
| $val_4$ | $W0$ | $D0$ | $D0$ |

**Fig. 9.26.** Iteration to compute states.

Next, we propagate the states. Since there are no transistors with gate $X$, steps (4) and (5) will produce the same result. The iteration for computation of states is shown in Fig. 9.26. The key point to notice is that at the first iteration, the signal $W1$ does not reach $v$, because its strength is not the same as the final strength from Fig. 9.25. $\square$

### Some Speedups to the Algorithms

The above algorithms can profit by several modifications that make them run faster. The first is the use of an *active list* of nodes. These are the nodes whose signals we currently need to recompute. When we iterate (9.2), we initially place all nodes on the active list. After considering a node $v$ by applying (9.2) to it, we remove it from the active list, but if $val(v)$ has actually changed, we put on the active list those of its neighbors that are connected to $v$ by on transistors.

Similarly, when during the iteration of (9.2) we discover that a gate has changed state, we may place the source and drain nodes for that transistor on the active list immediately. However, if the transistor changes from on to off, we have to worry that we have isolated a large region of the chip from all current sources, and must change all these nodes to strength $C$ or $SC$.†

One way to avoid setting all nodes to charged strength and iterating over all of them is to partition nodes into *classes*, where two nodes are in the same class if and only if they are connected by a path that follows transistors from source to drain only (not through the gate). These paths and classes are computed

---

† Recall this phenomenon motivated step (1) in *converge*, where we set all noninput nodes to a charged level of strength.

only once, before the simulation begins, and the paths may jump from source to drain or drain to source of any transistor, without regard to the value on the gate (which is not defined before simulation begins anyway).

Typically, classes are very small, although giant classes, such as those containing a bus wire and its associated pulldowns, can have hundreds of nodes. The advantage of partitioning nodes into classes once and for all, before starting the simulation, is that if during the simulation we turn off a transistor, the only nodes that could be affected by having their strength downgraded to $C$ or $SC$ are those in the same class as either the source or drain node. Thus, we need only to change the signal for these nodes and place them on the active list.

## 9.5 THE PI PLACEMENT AND ROUTING SYSTEM

The system called PI (Placement and Interconnect) is, at the time of this writing, under development at MIT by R. L. Rivest and his associates. The system tackles one of the most difficult and important combinatorial problems in VLSI systems: given a collection of cells, with ports on the boundaries, and a collection of *nets*, which are sets of ports that need to be wired together, find a good way to place the cells and run the wires so that the wires are short and the whole layout takes a small amount of area.

Many systems rely on the user's judgment more than is theoretically necessary. For example, a sticks system like LAVA relies, for placement of cells, on the user selecting an order in which cells are abutted. LAVA then designs the cells so that the wires running through or between them abut correctly. Life would be simpler, and perhaps better layouts would result, if we designed only the basic cells in LAVA, and a global placer and router worried about putting them where they best fit, then ran the wires as necessary to effect the proper interconnections.

Like most combinatorial problems, it is not feasible to solve the placement and routing problem exactly for instances consisting of more than a few cells and a few nets. Thus, for years people have looked for heuristic solutions that would run in a relatively small amount of time, be relatively compact, and minimize the delay due to signals propagating through long wires. The key contribution of PI is that it partitions the problem into a number of independent subproblems, each of which must be solved heuristically, but which, because of their comparative simplicity, can be studied and possibly solved with far better results than can the placement and routing problem as a whole.

We should note the similarity of the PI approach to that of lgen, discussed in Section 8.3. In both cases, a hard problem is broken into a cascade of simpler ones, and once a solution to an early problem has been selected, it is not changed for the sake of solving later problems in the cascade. In fact, the same idea of minimal-cut partitioning figures heavily in both systems.

An overview of the steps taken by the PI system is shown in Fig. 9.27.

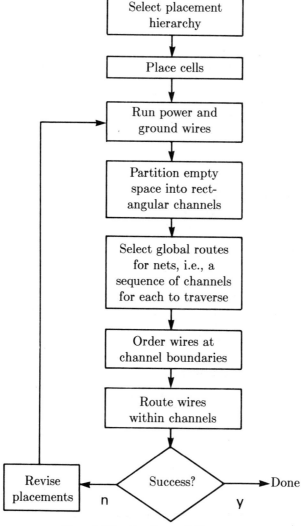

**Fig. 9.27.** Outline of PI system.

It should be remembered that the system is in a state of development. While we shall mention heuristics for performing each phase, the heuristics are not necessarily the best possible, and they may not be used in PI several years hence. The central message of this section is not the heuristics; it is the way PI partitions the problem.

## The Placement Hierarchy

The first stage is for the cells to be grouped into a hierarchy, that is, a binary tree. We can think of each node of this tree as representing a collection of cells that will be placed inside a rectangle. The two children of a node represent two subrectangles that compose the larger rectangle, either horizontally or vertically, with a gap between them to run wires. The selected hierarchy will thus represent the approximate relative placements of all the cells; the placement is refined at later stages until it becomes exact.

A hierarchy is selected by starting with a (not necessarily binary) tree consisting of a root, representing all cells, and a child of the root for each cell. We then convert the tree into a binary tree by a sequence of applications of two types of moves. The particular move to make at any time is determined by applying a priority function to the various alternatives, in an attempt to estimate how good any move is. The two types of moves are:

1. *Bottom-up combination.* Take two nodes of the tree, say $u$ and $v$, that have the same parent, $p$. Create a new node $w$, with children $u$ and $v$ and with parent $p$. This move is a good one if the cell or combination of cells represented by each of $u$ and $v$ have a good reason to be placed close together. For example, they may have ports on many of the same nets, and/or their estimated dimensions may make it likely that they fit together well. That is, if they have roughly the same height, then they should be grouped and placed horizontally, i.e., side-by-side, while if they have roughly the same width, then it would be advantageous to pair them vertically, i.e., one above the other. This choice is especially good if the adjacent sides of $u$ and $v$ have many nets in common.

2. *Min-cut partition.* Take some node $v$ with more than two children, and attempt to partition the children into two groups of roughly the same size (say, at least one-third in each group), so the number of nets with ends in both groups is minimized. Create two new nodes to be the parents of the nodes in each group. These nodes become the only children of the node $v$, although if one of the groups consists of a single member, that member and the node for the group are identified. The intent is that the children of $v$ will be laid out in two rectangles, adjacent either horizontally or vertically, and that by selecting the two groups to minimize the number of wires between them, we are likely to minimize the area needed for wires.†

**Example 9.15:** Suppose we have three cells, $A$, $B$, and $C$. $B$ has ports

---

† Note that the partition of nodes to minimize interconnections between groups is the same strategy recommended for ordering the pullups of a Weinberger array in lgen's optimization algorithm of Section 8.5. Although the latter is a one-dimensional problem, and the placement problem is two-dimensional, both are essentially the same graph-theoretic problem of recursively partitioning a graph to minimize interconnections (the *min-cut problem*), and the Kernighan-Lin algorithm is appropriate in both cases.

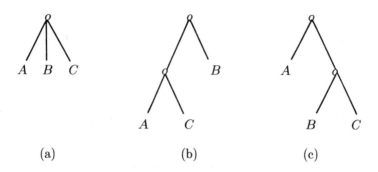

(a)                           (b)                           (c)

**Fig. 9.28.** Creation of placement hierarchy.

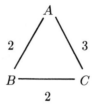

**Fig. 9.29.** Weighted graph representing nets between cells.

attached to nets 1 and 2, while $A$ and $C$ have ports attached to all three nets 1, 2, and 3. Note that this problem is essentially trivial; we introduce it as a running example only to illustrate certain heuristics used by PI. The reader must understand that all problems we discuss become complex at an exponential rate as the problem size increases.

Initially, we have a tree with a root representing $\{A, B, C\}$, and leaves representing the three cells individually, as in Fig. 9.28(a). To apply the min-cut heuristic, we must draw a graph whose nodes are the cells and whose edges are labeled by the number of nets that the cells connected have in common. This graph is shown for our example in Fig. 9.29. For example, $A$ and $C$ share all three nets, while $A$ and $B$ share only nets 1 and 2. The best way to partition the nodes of this graph into groups of roughly equal size is to group $A$ and $C$ together, since the sum of the weights of the edges cut is 4. If we instead grouped $B$ with either $A$ or $C$, the cost of the cut would be 5. The choice of partition $\{\{A, C\}, \{B\}\}$ yields the tree of Fig. 9.28(b), which is binary, so we are done in this simple example.

Before making the best min-cut partition, PI will consider whether it might be better off first pairing two nodes that are children of the same node. For example, it might be that $B$ and $C$ have the same height and that their ports for net 2 are at the same height. Further, suppose that each port of $B$ is on

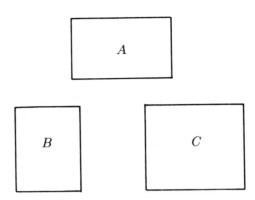

**Fig. 9.30.** Placement of cells.

the right border and the corresponding port of $C$ is on the left border. Each of these factors suggests that aligning $B$ and $C$, with $B$ to the left of $C$, could be an excellent choice. Thus, we might instead pair $B$ and $C$, forming the tree of Fig. 9.28(c). We shall suppose that this is the choice actually made, and the resulting relative placement of the cells is as shown in Fig. 9.30. Note that the rectangle consisting of the whole layout is composed of two, placed vertically and representing $\{A\}$ and $\{B, C\}$, respectively. The latter rectangle is in turn composed of two rectangles, for $B$ and $C$. $\square$

**Power and Ground Wires**

The PI system next selects routes for the power and ground wires to take. These wires are subject to the constraint that they must always run in metal and, therefore, must never cross. Further, their width may vary, depending on the power consumption of the various cells, information which is made available when the cells are input to PI.

The heuristic that PI uses to route the power and ground wires is based on heuristics for the traveling salesman problem.† If we consider a graph whose nodes are the cells and whose edge weights are estimates of the distance between cells, e.g., the distance between their centers, then a tour is an ordering of the cells. It is known that an optimal tour of points in the plane never crosses itself, so the tour defines an inside and an outside. We may run power on the inside and ground on the outside, or vice versa, and be sure that our power and ground wires never cross. We may have to be a bit careful where the wires go

† This problem is to find a minimum-cost tour of a graph with weighted edges. A *tour* is a cycle that visits each node exactly once, and the cost of the tour is the sum of the weights of the edges traversed. See Aho et al. [1983] for a discussion of the problem and heuristics for its solution.

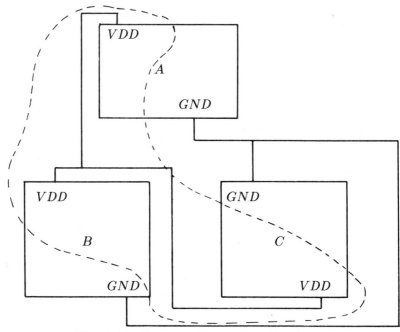

**Fig. 9.31.** Power and ground routing.

when they come near a cell, since the power and ground ports for a cell may
not be on the most convenient side. However, if we agree to run wires clockwise
around the cells, we can always make the wires reach where they must go.

**Example 9.16:** Suppose that $A$ and $B$ have their $VDD$ port at the upper left
and their ground ports at the lower right, while the situation is reversed for $C$.
We would find the optimal tour to be $A$, $B$, $C$, and back to $A$, so we must draw
a path connecting the nodes in that order, but distorted so that all power ports
are inside and all ground ports are outside. Such a path is shown by a dotted
line in Fig. 9.31, and resulting connections of power and ground are shown by
a solid line. Note that the paths are not actually cycles; they can be broken
at one point around the tour. Of course the power and ground pads must be
placed on these wires, and we must determine the width of these wires in their
various segments based on the current that each must actually carry. □

**Channel Creation**

At this stage, the cells have been placed, with gaps between them suitable for
running whatever power and ground wires there must be in those gaps, and
some space for the wires of the nets. We need not be too careful about how
wide the gaps are, because if we are unsuccessful in finding a wiring for the nets
in the space allotted, we shall try again, with a better idea of how big each gap
really must be.

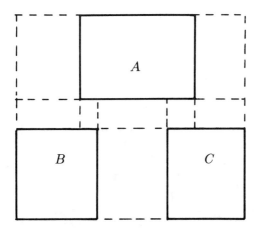

**Fig. 9.32.** Border extensions.

The PI system next divides all empty space around cells into *channels*, which are defined here to be arbitrary rectangles of empty space. The goal of the system is to make the partition as simple as possible; "simplicity" is formally defined by the condition that the total length of the edges separating the channel rectangles be minimized. As usual, the minimization problem cannot be solved exactly, for large problems, so we must resort to a heuristic. A simple example of an appropriate heuristic is to:

1.  Extend all cells' borders until they meet another cell border or the boundary of the circuit.
2.  Consider each line segment of the extended lines, in the order largest first. Delete the segment, in its turn, if we can do so without violating the condition that all channels be rectangular. That is, we cannot delete a border segment, and we cannot delete a segment if so doing would create an *L*- or *T*-shaped region.
3.  If necessary, add channels around the borders of the circuit.

**Example 9.17:** Figure 9.32 shows Fig. 9.30 with cell borders extended. Considering the extended segments in size order, we might first eliminate the two extensions of the bottom of *A*. The next longest nonborder segments are those that cross the gap above *B* and *C*, and below *A*. However, we cannot delete the extensions of *A*'s sides, because that would create an *L*-shape. The extensions of the interior sides of *B* and *C* can be eliminated. We are now done, except for the addition of border channels. The result is shown in Fig. 9.33; ignore the numbers and Greek letters for the moment. □

## Global Routing

The next task is the global routing of the nets, that is, the selection, for each net, of a tree of channels that it will occupy. The algorithms used here by PI are

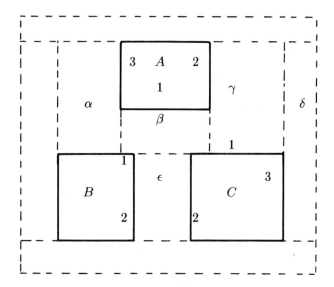

**Fig. 9.33.** Selected channels.

essentially shortest-path algorithms. Our paths will hop among the points that are the midpoints of channel edges, beginning and ending at ports belonging to the same net. The length of any edge in a path is the ordinary Euclidean distance between the points in question. Edges are available to take us between any two points in the same channel. A point on the border between channels is deemed to belong to both channels, a fact that allows paths to pass from one channel to the next.

When a net consists of two ports, we simply use Dijkstra's single-source shortest path algorithm† to expand out from one point until the other point is reached. When a net consists of more than two points, PI uses a generalization of Dijkstra's algorithm, where we push out from all ports on the net simultaneously, extending the set of points (midpoints of channel boundaries) reached from each port in such a way that at all times the points below a certain distance have been reached, and the points above that distance have not been reached.

To perform this activity, we keep a priority queue of the distance from the nearest port on the net to all the unreached neighbors of points that have been reached. As the wavefronts of reached points about each port expand, at times there will be a point just reached from one port that borders a channel previously reached from another. If that is the case, we now know the channel in which these two paths will join, and we permanently join them, allowing only the wavefront from one of the two ports to continue expanding.

† See Aho et al. [1983].

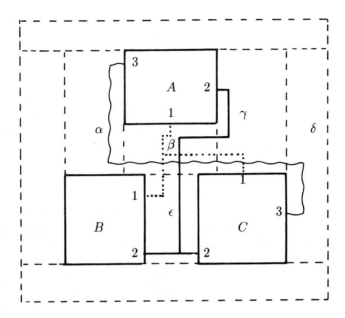

**Fig. 9.34.** Global routing of nets.

**Example 9.18:** The net 3 in Fig. 9.33 has only two ports. We might start at port $3_C$,† and next move to the middle of the wall $\gamma\delta$. Of course Dijkstra's algorithm is systematically exploring all paths from point $3_C$, but we shall indicate only the winning path. The shortest path then clearly continues by crossing channel $\beta$ to wall $\alpha\beta$, and from there to the point $3_A$.

The net 1 consists of three points, but the wavefronts from them converge very quickly. Based on the dimensions in Fig. 9.33, the shortest hop from any of the three ports is from $1_C$ to wall $\beta\gamma$. Then, the next shortest is from $1_B$ to wall $\beta\epsilon$. Thus, all three ports meet in channel $\beta$.

Net 2 requires more effort. We again expand from all three ports on the net; immediately we find that $2_B$ and $2_C$ meet in channel $\epsilon$. We must continue to expand from one of these, say $2_B$, and from the remaining port, $2_A$. The shortest hop available is from $2_A$ to wall $\beta\gamma$, and the next choice is from $2_B$ to $\beta\epsilon$. At this point, we have converged in channel $\beta$, and we have a tree connecting the three ports of net 2. All three nets are shown in Fig. 9.34. ☐

---

† For succinctness, we use the notation $i_X$ for the port on net $i$ on cell $X$. Technically, PI allows more than one port from the same net to appear on a cell, but in our running example, no ambiguity is caused. We also use the notation $ab$ to stand for the wall between channels $a$ and $b$.

## Ordering the Crossings Between Channels

One of the key ideas of PI is that it is feasible to pick the points at which wires cross the walls between channels, prior to selecting the positions of the wires within channels. This point of view is in keeping with the PI philosophy of separating hard problems into simpler ones. By applying a heuristic to pick the order and position of wires across each wall, the channels may then be wired independently of one another. Since channels tend to be small, powerful wiring algorithms can be brought to bear on the problem of wiring difficult channels.

Suppose we are considering a particular vertical wall, and we want to select the height where a certain wire $w$, belonging to net $n$, should cross. There are conflicting considerations, but two of the most important are the following.

1.  If the net $n$ leaves one of the channels on either side of the vertical wall by the other vertical wall, we would like the wire $w$ to have the same height at both walls, so it will not have to jog within the channel.

2.  If, say, there is a port of net $n$ on the bottom of one of the channels separated by the wall, then we would like $w$ to cross low on the wall.

Furthermore, not any set of heights is acceptable for the wires crossing a wall. For example, we would normally like wires to be in metal, so that the delay in propagating signals is minimized.† If wires run in metal, which is $3\lambda$ wide and requires $3\lambda$ separation, we cannot possibly allow two wires to cross with their centerlines less than $6\lambda$ apart. However, since we shall normally have to add some vias to the metal wires, so polysilicon or diffusion wires can connect them to ports on the top or bottom of the channel, we often need $7\lambda$ between the centerlines of wires. In fact, since we may have to introduce more horizontal wires inside the channel, it may be necessary to separate the wires at the walls by more than the theoretical minimum.

PI then makes repeated passes over the walls. Initially, it assumes that all points where wires cross walls occur at the midpoint of that wall. At each wall, it considers each net $n$ that crosses the wall, and all net $n$'s points on the other borders of the two channels separated by the wall. Assuming the wall in question is vertical, it takes a weighted average of the heights of the other points on net $n$, first moving (conceptually) those below the bottom of the wall to the bottom of the wall, and moving those above to the top of the wall. The weights can be adjusted so that points on walls opposite to the wall in question, i.e., the other vertical walls of the surrounding channels, are given greatest weight. Points on the top or bottom walls of a channel are given more weight if they are close to the wall in question.

---

† There are several exceptions to this rule. For example, we have ignored the fact that metal power and ground wires have been laid down at a previous step, and if such a wire crosses the channel, we must use another layer, probably polysilicon. Also, wires across a very narrow channel may not have the room to change layers and, therefore, must run in whatever layer their ports can accept.

Having calculated a desired height for each point on the wall, we adjust these values so that no design rules are violated. For example, if there are $k$ wires crossing the wall, we might divide the wall into $k + 1$ equal parts, and put the centerline of each wire on one of the $k$ dividing lines, preserving order. Presumably the wall is high enough that such a spacing will provide at least $6\lambda$ between centerlines, and in most cases, more than that will be both necessary and available.

This process is repeated for each wall, and as long as there are changes in the position of one or more crossing points of any wall, we shall repeat the process for all walls. In practice, convergence occurs quickly, and, in particular, the weight given to aligning crossings of a net on opposite walls of a channel tends to cause wires that are traveling through several channels to be straight lines, as they intuitively should be.

**Example 9.19:** Let us consider the wall $\beta\gamma$, which in Fig. 9.34 is crossed by three wires, one from each of the three nets. We shall measure heights by letting the bottom of this wall be height 0, and the top of the wall be height 1. Initially, the vertical position of wires crossing another wall will be assumed to be at the midpoint of that wall. For the purposes of this example, the weighting factor given to the various points will be assumed uniform; in practice, weights favoring points on opposite walls and nearby points on adjacent walls would be used.

Thus, consider net 1. This net leaves channel $\beta$ both at top and bottom. Therefore, the two connection points contribute 0 and 1, to the average. The point on net 1 in channel $\gamma$ is on the bottom, so it contributes 0. We now find that the average height is .33, suggesting that the optimal place for net 1 to cross the wall $\beta\gamma$ is a third of the way up.

Now consider net 2. There are two connecting points; the one in channel $\beta$ contributes 0, since it is on the bottom of that channel, while the one in channel $\gamma$ contributes 1 since it is above the top of the wall $\beta\gamma$. The average of these heights suggests that .5 is the preferred height for net 2.

Finally, net 3 also has two other points on channels $\beta$ and $\gamma$. The one across channel $\beta$ is at height .5, because of the initial conditions. The other point has an estimated height equal to the midpoint of wall $\gamma\delta$.† Let us suppose that this midpoint is higher than the top of wall $\beta\gamma$, so this point is given height 1, and the average for net 3 is .75.

Now we have a preferred order for the crossings of wall $\beta\gamma$; it is net 1 lowest, then 2, then 3. If .33 is too close to .5, we now revise the values to separate them adequately. Suppose for sake of the example that we assign these heights .25, .5, and .75, respectively. When we next consider wall $\alpha\beta$, we shall discover that the preferred height for the crossing of that wall by net 3 is

---

† Note the fact that point $3_C$ is below the level of the wall $\beta\gamma$ is not taken into account by the algorithm, as we consider only the channels $\beta$ and $\gamma$ now.

.75 also, since there is a strong attraction for that value, due to the crossing of $\beta\gamma$ at that height.

Unfortunately, when we consider wall $\gamma\delta$, we shall also discover that it is desirable to have net 3 cross that wall low. When we next consider the crossing of $\beta\gamma$, depending on the actual weighting function used, we may find it advantageous to lower the crossing there for net 3. If we lower it enough so its estimated height is below that of net 2, then we might revise the height of the crossing point for net 3 to .5, which in turn makes the crossing of wall $\alpha\beta$ follow, and even makes it possible that at the next round, the estimated height for net 3 at wall $\beta\gamma$ will be below that of net 1. If that is the case, then we shall converge to the arrangement actually shown in Fig. 9.34. $\square$

### Channel Routing

Having selected positions for all wall crossing points, PI can route the wires within each channel, one at a time, knowing that what it does in one channel does not affect any other. Moreover, it no longer is necessary to distinguish between points on the borders of a channel that are due to ports of a cell, and points representing wall crossings. Of course, since the widths of channels were initially selected somewhat arbitrarily, there is no guarantee that a given channel can be routed by even the most clever or exhaustive algorithm. That is not terribly serious, since when even one channel cannot be routed, almost the whole process of wiring starts over, as was indicated in Fig. 9.27. As long as no ports are so close together that they can never be wired legally, e.g., poly ports with centerlines $3\lambda$ apart, eventually each channel will become wide enough that it can be routed.

PI uses a set of channel routing algorithms, starting with a trivial one that in essence runs wires for each net independently and then checks for design rule errors. If one fails, the next in line, being more powerful but slower, is called. If all fail, the channels have their widths adjusted.

Here, we shall describe one simple heuristic for wiring in a channel; it is called the *greedy router* because it scans the channel from one edge, always doing what seems most sensible at the time, without looking ahead to problems that may be in the offing. In what follows, we assume that the scan begins at the left edge and proceeds rightward.

We must organize the channel into horizontal *tracks*, which are bands that are sufficient to run a wire of the appropriate layer, which we here assume is metal. Horizontal tracks initially correspond to the points on the left edge, but we may wind up creating additional tracks between these initial tracks. If, say, horizontal tracks run in metal, then we shall need $7\lambda$ per track, although sometimes we can do with less, if the positions of vias are fortunate and we can jog wires around them.

We scan rightward, in steps (*columns*) just sufficient to reach the next port on the top or bottom, or less if we meet no ports but have enough room to run wires vertically. For example, we could move in steps of $6\lambda$ if poly wires are run (so they can cross metal) and arbitrary metal-poly vias need to be placed. However, we look ahead just enough that we do not decide a column has no ports and, therefore, run certain vertical wires in the column, only to find that a few $\lambda$ to the right is a port for which no room remains to connect it to a wire. Thus, in effect, each port belongs to one vertical grid line or track, the one closest to its actual position.

In each column, zero or more vertical wires may be run. The greediness of the algorithm is embodied in the following local strategy for considering which wires to run.

**Algorithm 9.5:** *Rivest-Fidducia Greedy Channel Router.*

INPUT: A rectangular channel with ports belonging to particular nets at designated positions on the border.

OUTPUT: A wiring that connects the ports on the same nets, or an indication of failure. The failure of the algorithm to wire the nets does not imply that no such wiring exists.

METHOD: In each column, we do all of item (1) below that we can do; then we do all of (2) that is yet possible, then consider (3), and then (4).

1. Make vertical connections to points on the top or bottom of the channel. Connections to a track bearing the net to which the point belongs are made in the appropriate layer, which we here assumed was poly. If the point must be connected to in another layer, a via must change layers at a place above or below any horizontal wires.† If one or more points belong to nets not represented among the tracks at this column, then we connect the points to the nearest empty tracks. If there are no empty tracks, then we create one, as close to the point as possible. The new track is conceptual only, and when we actually lay out the wire we shall have to make room, if possible, by adjusting the heights of the wires along the other horizontal tracks. Again we may fail, if room for an extra track cannot be made.

2. If there is one net that currently occupies more than one track, and two or more of such tracks can be connected by a vertical wire in the column without conflicting with the wires laid down already in (1), then we run this vertical wire. To select which of the tracks will continue to bear the net (the others become empty), prefer the highest if the next point on the net is on the top, and the lowest if the next point is on the bottom. If the next point is at the right, prefer the track closest to its height.

---

† Note that at this point the algorithm may fail, since there may not be room for the vias. Or, there may be two ports to be wired, one on top and one on bottom, but they must be connected to tracks on the opposite sides, and there is not room for two vertical wires in this column.

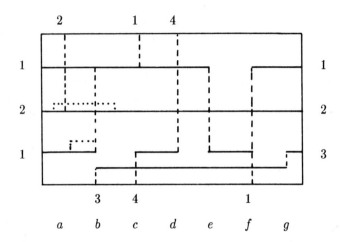

**Fig. 9.35.** Result of greedy channel routing.

3.  If nets borne by two or more tracks still exist after (2), and we cannot connect them, jog one or more of the wires to an empty track so the difference between the highest and lowest tracks bearing the net is reduced.

4.  As much as possible, move wires to empty tracks that are closer in height to the leftmost point, among those to the right, on the same net as that wire. □

**Example 9.20:** Figure 9.35 shows a channel wired using Algorithm 9.5. Initially, we are given only the locations of the points on the boundary. We assume that the scale is such that we can fit seven columns, named $a$ through $g$, vertically. We also assume for convenience that all ports on the top and bottom can be connected in poly, and that the ports are located exactly at some column. In practice, we might have to introduce vias, and/or jog the vertical wires a small distance left or right, so they would meet the ports where connections could permissibly be made.

In column $a$ we must respond to rule (1) of Algorithm 9.5, which says we should connect the wire in the middle track to the port on net 2, at the top. Since all the tracks are occupied in column $a$, there is no room to do anything else, so we go on to column $b$.

In column $b$, we must find a track for net 3, since there is a port belonging to that net, at the bottom. Since no tracks are empty, we create a new track at the bottom. Whether or not there is room depends on the vertical scale. Even if there technically is not room, we may be able to make room by moving the tracks above it upward slightly, in an expanding set of detours. We have shown by dotted lines (which do not represent diffusion here) the two tracks above the new one being detoured upward in decreasing amounts. This picture would be consistent with the view that there was slightly more than $7\lambda$ between tracks at the left edge, so detours do not propagate indefinitely.

Also in column $b$ we have room for an instance of rule (2), which tells us to give priority to collapsing wires from the same net. We continue the upper track as net 1, because the next port on net 1 appears at the top edge. Note that we could not, in column $b$, move the wire for net 3 into the track vacated by net 1, because we assume that there is not room in one column to run two vertical wires next to one another.

In column $c$, we connect to the port of net 1 on the top and introduce net 4 into the empty track; both actions are by rule (1). There is now no room to make any other moves in column $c$. Column $d$ requires a connection to the port of net 4, and no room for other moves remains.

Since net four ends at column $d$, and column $e$ has no ports, there is room to make a move by rules (2), (3), or (4). We first consider (2), but no net occupies more than one track, so rule (3) is considered next. We cannot use rule (3) for the same reason, and consider (4). The best opportunity to move a wire closer to its next point of connection is to move the wire in the top track down to the empty track. This we do in column $e$.

Column $f$ is used to connect to the point on net 1, and there is then room to bring net 1 up to the height of the next point on that net, the one on the right edge. Column $g$ is used to move net 3 up to the desired track, which was vacated by net 1 at the previous column. It is interesting to note that if column $g$ were not there, we would fail, as net three could not connect to its point at the right. Probably, the greedy algorithm should, in rule (4), give priority to moving nets to the height of one of their points on the right edge, as these jogs are essential to make. In comparison, a detour like that made by net 1 in columns $e$ and $f$ can sometimes help but appears less essential in practice. $\square$

## 9.6 OPTIMAL ROUTING

There are some limited cases where we can give algorithms that run in polynomial time, yet produce minimal-area wiring for a special case of channel routing. There is even one such case, "river" routing in a channel with ports on top and bottom, that comes up frequently in practice. However, before proceeding to this case we shall introduce a concept that yields lower bounds on the size of a channel in which particular sets of ports can be routed.

### Channel Density

Let us assume a channel is a rectangle with ports on the borders, as in the previous section. Suppose we draw any vertical line across the channel. We may count the number of nets with ports on both sides of the line; ports on the line count for neither side. Define the *channel density* to be the maximum, over all vertical lines, of the number of nets with ports on both sides of the line. Call a line *critical* if it yields this maximum. We could similarly define a channel density based on horizontal lines, but the ideas are similar to what we discuss for vertical lines, and we shall not discuss a horizontal version of channel density further.

The central fact about channel density that makes it useful in proving lower bounds on channel height is that the number of wires that must cross a critical line equals the channel density. Thus, for example, if we expect all horizontal wires to be carried in metal in tracks at some minimum separation, then the number of tracks can be no less than the channel density. As another example, if we cross a critical line with metal and poly wires, there are $m$ metal wires and $p$ poly wires, and the channel density is $d$, then $m + p = d$. If the height of the channel is $h$, measured in $\lambda$, then $h \geq 6m - 3$ and $h \geq 4p - 2$, because of the need for wire separations.† Combining these two inequalities in the proportion 2:3, we get $5h \geq 12m + 12p - 12 = 12d - 12$. That is, $h \geq \frac{12}{5}(d - 1)$.

**Example 9.21:** If we draw a vertical line in Fig. 9.35 between columns $c$ and $d$, we see that all four of the nets have points on both sides. Thus, since Algorithm 9.5 normally runs all horizontal wires in metal tracks, we know that four tracks are required, and thus the result of Algorithm 9.5 is optimal, as far as height of the channel is concerned. That is, as long as horizontal wires run in tracks, one to a track, no algorithm, however complex, could wire the problem of Example 9.20 in fewer tracks. □

Calculating the channel density is straightforward. The key observation is that as we slide the vertical line to the right, the number of nets it separates cannot change until we reach a port. Thus, it suffices to visit the ports in order, from the left, keeping a running total for each net, of the number of ports on either side of the current position. When we visit the next port, we see if there is another port on the opposite side with an identical horizontal position. We then update the counts for the net or nets containing this port or ports, and we recompute the local channel density by seeing if the status of any of these nets has changed, i.e., are all the ports for the net now to the left of the line, or were they formerly all to the right, and now they are split. In this manner, we can keep track of the maximum number of nets that are ever split, and this number is the channel density.

If there are $n$ nets and $m$ ports, we can sort the ports in $O(m \log m)$ time. We initialize the counts for the nets in $O(n)$ time, which is $O(m)$, since it makes no sense for there to be more nets than ports. Finally, the work described above is proportional to the number of ports, so the time to sort dominates, and we can compute the channel density in $O(m \log m)$ time. Algorithms such as the

---

† We here suppose that it is permissible to run wires for long distances on different layers, one on top of the other. Because of capacitance between the wires, this practice is frowned upon in some situations.

greedy one mentioned in Section 9.5 frequently attain or come close to the channel density. The bibliographic notes mention the existence of algorithms that are guaranteed to come within a constant factor of the channel density.

## River Routing

*River routing*† is channel routing restricted in the following way.

1.    Wires run in one layer only; thus they do not cross.
2.    Each net consists of two points, one on top and one on the bottom. There are no ports on the sides.
3.    The order of the nets is the same on top and bottom, as it must be if we are to run wires in one layer.

Despite the stringent conditions of the river routing problem, it comes up quite frequently in practice. Often, one designs a chip in several modules, with the intent of connecting them together to make the final circuit. When designing the modules we are able, frequently, to control the order of ports so they are the same on pairs of modules to be connected. It is also common that we have a strong preference to run the wires in one layer only. For example, if the modules being connected are far apart, we wish to connect in metal for speed. If the modules are close, but the ports are closely packed to save space, then we may be forced to run wires in poly. We could even connect ports that were $3\lambda$ apart by connecting them alternately in poly and diffusion, if slow diffusion wires were acceptable and we could arrange that the ports accept connections in these alternating layers. In these applications, it is common that the model of a rectangular channel does not hold precisely; the borders of the modules may be irregularly shaped. However, the principles we show here are easily seen to carry over to the more general case.

## The River Routing Model

The model we shall discuss is one in which there is a sequence of ports on two horizontal lines. On the upper line are ports $Q_1, \ldots, Q_n$ and on the bottom are ports $P_1, \ldots, P_n$. We must run a wire between $P_i$ and $Q_i$ for each $i$, subject to certain constraints. Initially, we shall assume that wires are lines of zero width, and lines may not come closer that one unit of distance. As a consequence, we must assume that ports on a line are separated by at least one unit; they may be separated by more. This abstraction models wires that have finite thickness and finite separation, and are allowed to run in arbitrary curves. For example, if our wires are really metal wires, then our unit of distance is $6\lambda$. Later, we shall modify the model to handle wires that are allowed to run horizontally and vertically only.

We make the assumption that it is possible to connect wires to ports either

---

† So called because the wires look like the current in a river.

vertically or horizontally. Frequently in practice we shall wish to make all connections to ports vertically, because we cannot be sure that running a wire horizontally, next to the channel border, will not cause a design rule violation. If that is the case, we can simply pretend that the ports are one unit closer to the center of the channel than they are. The only time this assumption will yield the wrong answer is if the nets all align exactly. In that case, we would get the answer that the channel top and bottom had to be separated by two units, when in fact, one would do. Zero will not do, because technically the material behind one wall cannot come into contact with the material behind the other.

We shall identify the name of a point with its horizontal position, measured from some arbitrary origin. A fundamental partition of the problem is into *blocks*, which are either *left blocks* or *right blocks*. A left block is a maximal consecutive sequence of nets $i, \ldots, j$ such that each net moves left as we go from bottom to top. That is, for all $k$, $i \leq k \leq j$, $Q_k < P_k$. Right blocks are defined analogously. It will become clear that all the problems of wiring occur within a single block, and there is no interaction between them.

A net $k$ with $Q_k = P_k$ is a special case belonging to neither a left nor a right block. Evidently, any such net can be wired straight across without interfering with the wiring of any other net. Thus, we shall ignore such nets and concentrate on wiring of a single block, which we assume to be a left block.

### Free Wiring of a Left Block

We now consider how to wire a left block of nets using wires that run in arbitrary curves, in such a way that the separation between the top and bottom of the channel is minimized. We show how to calculate a lower bound on the separation, and then we show how to run the wires in a way that meets the bound.

Consider a port $P_i$ on the channel bottom and the net consisting of $P_j$ and $Q_j$, for some $j > i$. If at any point, the wire for net $j$ comes within less than $j - i$ units from $P_i$, it will not be possible for all of the nets $i + 1$ through $j - 1$ to fit through the gap between $P_i$ and wire $j$ and still maintain their unit distance. As wires cannot cross each other or the channel border, there is no way for these nets to "go around" $P_i$ or net $j$. The situation is shown in Fig. 9.36.

Further, if to the left of $P_i$, net $j$ falls lower than $j - i$ units above the channel bottom, the same argument tells us that too many wires will run between wire $j$ and the bottom for all to be separated by a unit. The result of this observation is that about each port on the channel bottom we can draw a collection of forbidden regions; about $P_i$ we draw a quarter circle of radius $j - i$ in the upper right quadrant, to indicate the region from which net $j$ must be excluded, for $j > i$. This quarter circle extends horizontally leftward from

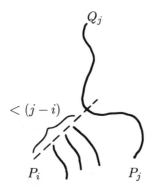

**Fig. 9.36.** Problem with running nets too close to ports.

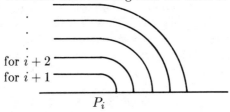

**Fig. 9.37.** Forbidden regions.

its maximum point, as shown in Fig. 9.37.

Let us define $f_i(x)$ to be $i$ for $x \leq 0$, to be 0 for $x \geq i$, and to be the height of the circle of radius $i$ about the origin, for $0 < x < i$; that is

$$f_i(x) = \sqrt{i^2 - x^2}$$

in that region. Thus, $f_i$ is the equation of the $i^{th}$ barrier about the point $P_i$ in Fig. 9.37, but with $P_i$ translated to the origin. Thus, what we have said so far implies that if we translate all ports so $P_i$ is the origin, then for $j > i$, $Q_j$ must lie above the curve $f_{j-i}$. That constraint could mean nothing, if $Q_j - P_i \geq j - i$, because then $f_{j-i}(Q_j - P_i) = 0$. However, if $Q_j$ is at a position where $f_{j-i}$ is positive, then it must be that the channel separation is at least great enough to keep the top wall of the channel, and $Q_j$ with it, at height $f_{j-i}(Q_j - P_i)$ or more. We may thus show the following theorem.

**Theorem 9.3:** In order that a left block with ports $P_1, \ldots, P_n$ on the bottom and $Q_1, \ldots, Q_n$ on the top be wirable by river routing, it is necessary and sufficient that the separation be at least the maximum, over all $i$ and $j$, with $i < j$, of $f_{j-i}(Q_j - P_i)$.

**Proof:** The necessity was argued above. For sufficiency, we must show how to wire the left block within this separation. What we do is run each net in

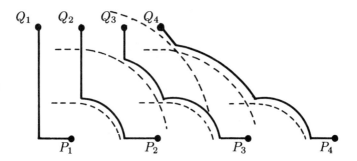

**Fig. 9.38.** Wiring a left block.

turn, from the left, running it on top of any barriers that pertain to it. When we get far enough left to be under the port on the top, we move straight up. Since barriers never decrease as we go left, when we get under the destination port, we shall be as high as ever we were. Thus, if we are wiring net $j$ and traveling along the $(j-i)^{th}$ barrier about $P_i$, we know we never got higher than $f_{j-i}(Q_j - P_i)$, and therefore, we have not gotten higher than the separation of the channel.

It is easy to show that this wiring does not allow two wires to come too close. For if nets $j$ and $j+1$ come too close, it cannot be while either is on its vertical segment, because $Q_{j+1} - Q_j \geq 1$. Thus, let the $j^{th}$ net be traveling along the barrier $j - i$ about $P_i$ at some point $p$ where net $j + 1$ comes too close. Then one of the barriers that net $j + 1$ had to avoid was barrier $j - i + 1$ about $P_i$, which is always at least one unit away from the $(j - i)^{th}$ barrier, as long as the height of point $p$ exceeds 0. Should the height of $p$ be 0, then since $p$ must be to the left of $P_j$ (because wires only move left in our wiring scheme), the first barrier about $P_j$ would keep net $j + 1$ safely away from $p$. $\square$

**Example 9.22:** Consider the left block shown in Fig. 9.38. We assume that the horizontal positions of $Q_1, \ldots, Q_4$ are 0, 1, 2, and 3, respectively, while those of $P_1, \ldots, P_4$ are 1, 3, 5, and 7. Trying all combinations, we discover that the channel separation is determined by the pair $i = 1$ and $j = 4$. There, we get a lower bound on the separation of

$$f_{4-1}(Q_4 - P_1) = f_3(2) = \sqrt{3^2 - 2^2} = \sqrt{5}$$

The first net has no relevant barriers, so it runs horizontally until it is under $Q_1$, whereupon it moves directly up. The second net runs along the bottom until it meets the first barrier about $P_1$, which it follows, until it gets under $Q_2$ and goes straight up. The third net moves horizontally until it meets the circle of radius 1 about $P_2$, which it follows until the point at which the circle of radius 2 about $P_1$ becomes higher. The latter circle is followed until the wire is under $Q_3$.

The fourth net repeats the same sort of pattern. It begins traveling horizontally, meets the circle of radius 1 about $P_3$, then the circle of radius 2 about $P_2$, and finally the circle of radius 3 about $P_1$. When it reaches the $x$-coordinate of $Q_4$ on that barrier, it is exactly at $Q_4$, since that barrier determined the separation necessary for wiring to be successful. $\square$

Using the technique implied by Theorem 9.3, we can find the separation for an arbitrary river routing problem and run the wires in $O(n^2)$ time if there are $n$ nets. First, we break the problem into left and right blocks. For each block we determine the necessary separation; right blocks can first be converted to left blocks by taking the mirror image. Then, we take the maximum separation implied by any of the blocks and wire each block by the algorithm given in Theorem 9.3.

## Rectilinear River Routing

We have discussed several reasons why we often restrict layouts to be composed of rectilinear rectangles. These reasons range from the restrictions of certain fabrication processes to the relative simplicity of certain design tool algorithms when dealing only with rectilinear rectangles. Thus, we should consider what happens if we restrict the wires in a river routing problem to be composed of rectilinear rectangles; when we do, we discover that sometimes the necessary channel separation goes up considerably, although at other times it is similar to that given by Theorem 9.3.

Our first consideration is how exactly we should model rectilinear wiring. If we allow arbitrary collections of rectangles to make up a wire, then we are really back to the case of unrestricted wires, since we could make the rectangles infinitesimally short, or even make them long, but at infinitesimally different positions as we moved along an arbitrary curve. This view of rectilinear wiring is not realistic, since the number of rectangles needed to comprise a wire grows without bound, and no fabrication process can be expected to follow arbitrary curves to the precision implied by this view.

We thus assume that wires are not only composed of rectilinear rectangles, but the rectangles themselves have sides on a discrete grid. In fact, to make things tractable, we shall assume that the grid equals the minimum separation between ports, and further, that all ports are placed on grid lines. We continue to assume that wires have zero width, and therefore, we model a rectilinear wiring by selecting for each wire a path along the grid.

Using this model, we show how to determine the minimum channel separation, and we give an algorithm that meets this separation. The algorithm is described on the assumption that wires run only along grid lines. If, however, ports are allowed to be off grid lines, and/or the grid on which wires allowed to run is smaller than the minimum distance between wires, the algorithm for wiring that we give generalizes easily, but the result may not be optimal.

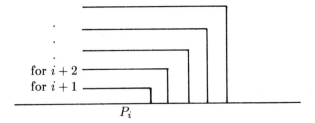

**Fig. 9.39.** Forbidden regions for rectilinear wiring.

Algorithms for optimal wiring under a variety of restrictions such as the above are known, and references are provided in the bibliographic notes.

Different wiring restrictions frequently are characterized by a different barrier shape, while the ideas behind Theorem 9.3 continue to hold. Our chosen model is no exception, and here the barriers are square, as suggested by Fig. 9.39. The proof that barriers are square is given partially by the next lemma, which shows only that wires cannot go within the barriers of Fig. 9.39. The proof that no larger barriers are needed will come when we give an algorithm for optimal wiring based on those barriers.

**Lemma 9.4:** If ports and wires are constrained to run along grid lines, then the wire for net $j$ cannot pass through a grid point $p$ whose $y$-coordinate is less than $j - i$ and whose $x$ coordinate is less than $P_i + j - i$, for any $i < j$.

**Proof:** Figure 9.40 shows such a point $p$. Draw a line diagonally downward and to the left. Then $j - i - 1$ nets, those numbered $i + 1$ through $j - 1$, must cross that line. If the diagonal goes through $P_i$, e.g., because $p$ is one unit below and to the left from the corner of the barrier, then only $j - i - 2$ points are available on the line for the crossing of the $j - i - 1$ nets numbered $i + 1$ through $j - 1$. If the diagonal meets the channel bottom to the left of $P_i$, then $j - i - 1$ points are available on the line, but it must allow for the crossing of the $j - i$ nets numbered $i$ through $j - 1$.

Finally, suppose that the diagonal hits the channel bottom $d$ units to the right of $P_i$. Then the $j - i - d$ ports $P_{i+d}$ through $P_{j-1}$ are all on or to the right of the line, because ports are separated by at least one unit. However, the number of available points on the line is only $j - i - d - 1$, one for each of the vertical grid lines from $P_i + d$ up to, but not including, the grid line of $p$, which is $P_i + j - 1$ or less. Thus, the number of available grid lines is again less than the number of lines that must cross, and we conclude that the presence of point $p$ on net $j$ makes legal wiring impossible. $\square$

The algorithms for computing minimum channel separation and finding the wiring are essentially the same as those implied by Theorem 9.3, although we shall later give a faster way of computing the separation. First, we formalize the barriers of Fig. 9.39 as the family of functions $g_i(x)$, where $g_i(x) = 0$ if

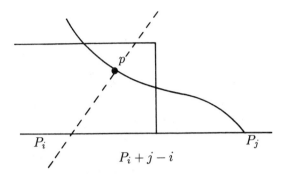

**Fig. 9.40.** Diagram for Lemma 9.4.

$x \geq i$ and $g_i(x) = i$ if $x < i$.

**Theorem 9.4:** If wires and ports are constrained to be on grid lines, then the minimal separation for a left block with ports $P_1, \ldots, P_n$ on the bottom and $Q_1, \ldots, Q_n$ on top is the maximum over all $i < j$, of $g_{j-i}(Q_j - P_i)$.

**Proof:** Lemma 9.4 states that it is not possible for net $j$ to run below $g_{j-i}$ anywhere, and so certainly $Q_j$ could not be below that barrier. Thus the separation could not be less than stated by the theorem.

To show this separation is sufficient, we modify the wiring strategy of Theorem 9.3 by requiring wires to follow the square barriers given by the $g_i$'s, rather than the round ones given by the $f_i$'s. The proof that the resulting wiring is legal follows the lines of Theorem 9.3 and Lemma 9.4, and we leave it as an exercise. $\square$

**Example 9.23:** Let us reconsider the left block of Fig. 9.38, discussed in Example 9.22. The separation is again determined by the pair $Q_4$ and $P_1$, but here the separation is 3, not $\sqrt{5}$. The reason is that

$$g_{4-1}(Q_4 - P_1) = g_3(2) = 3$$

Under the rules for rectilinear wiring, the nets follow the paths illustrated in Fig. 9.41. For example, net 4 goes horizontally until it meets $g_1$ with origin at $P_3$, follows that barrier until it meets $g_2$ with origin at $P_2$, then meets $g_3$ with origin at $P_1$, on which curve it finds $Q_4$. $\square$

### A Linear Time Algorithm for Computing Separation

Sometimes it is advantageous to compute the channel separation quickly. Performing the wiring could take $\Omega(n^2)$ time, because there may be that many rectangles to lay down. However, in a situation like that of PI discussed in the previous section, we may need to compute required channel widths many times before the cells are adjusted to their desired positions. There could

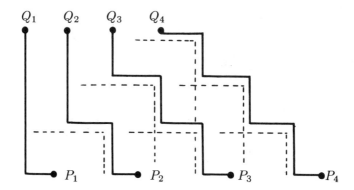

**Fig. 9.41.** Wiring a left block under rectilinear restrictions.

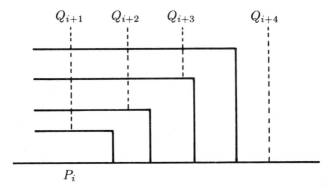

**Fig. 9.42.** Property of rectangular barriers.

be considerable speedup if we did not actually wire each channel, but could calculate the required separation quickly, wiring only when all channels were found adequate.

Thus, it is significant that there is an $O(n)$ time algorithm to compute the separation for a left block of $n$ nets, and therefore computing channel separation for arbitrary rectilinear river routing problems can be done at that speed. In $O(n)$ time we cannot hope to look at all pairs $i < j$ to compute the separation required by that pair. Rather, we shall show that a single pass through the top and bottom ports allows us to find all the candidate pairs $(i, j)$ that could possibly determine the maximum.

The key idea is illustrated in Fig. 9.42, where we see the separations determined by $P_i$ when compared with $Q_{i+1}$, $Q_{i+2}$, and so on. For a while, each $Q_j$ determines a separation one larger than the previous $Q_{j-1}$. Then, at some point, $Q_{i+4}$ in Fig. 9.42, the port "falls off" the barrier, because the difference

$i := 1; j := 2; separation := 0; \{ \text{initialize} \}$
**while** $j \le n$ **do**
    **if** $Q_j - P_i < j - i$ **then begin**
        $\{ Q_j$ is on the nonzero portion of the $j - i^{th}$ barrier about $P_i \}$
        $separation := max(separation, j - i);$
        $j := j + 1$
    **end**
    **else** $\{ Q_j$ is not affected by a barrier about $P_i \}$
        $i := i + 1$

**Fig. 9.43.** Calculation of separation.

$Q_j - P_i$ is at least $j - i$. The sequence of ports can never "get back on" their respective barriers about $P_i$, because the next barrier's nonzero portion ends only one unit to the right of where $Q_j$'s did, yet $Q_{j+1}$ must be at least one unit right of $Q_j$. The conclusion is that once we find a $j$ for which $g_{j-i}(Q_j - P_i) = 0$, that is, $Q_j - P_i \ge j - i$, we no longer need to consider $P_i$ and can move on to $P_{i+1}$. In fact, we do not need to backtrack on the upper row; we can begin by comparing $P_{i+1}$ with $Q_j$. These ideas are formalized in the following algorithm and theorem.

**Algorithm 9.6:** *Karplus' Channel Separation Algorithm.*

INPUT: The positions of ports $P_1, \ldots, P_n$ on the bottom of a channel and $Q_1, \ldots, Q_n$ on the top. All ports are assumed to lie on integer grid points.

OUTPUT: The minimum separation needed to wire $P_i$ to $Q_i$ for all $i$, using one-layer, rectilinear wires running along grid lines.

METHOD: We break up the river routing problem into blocks, calculate the maximum separation required by each block, and return the largest of these. For a block of a single net, the required separation is 1 if the ports are not aligned vertically, and 0 if they are. For left blocks of $n > 1$ nets, we compute *separation* by the algorithm given in Fig. 9.43. For right blocks, we mirror them and treat them as left blocks in the same way. □

**Theorem 9.5:** Algorithm 9.6 requires $O(n)$ time and correctly computes the largest separation of any pair $P_i$ and $Q_j$.

**Proof:** The linearity of the time is obvious, since at each iteration of the loop, either $i$ or $j$ increases. We can never get $j < i$, since in a left block the test of the if-statement, $Q_j - P_i < j - i$, always holds when $i = j$. When $j$ exceeds $n$, the algorithm ends, and by what we just observed, this must occur within $2n$ iterations of the loop.

To prove the correctness of the algorithm, we must show that the right combinations of $i$ and $j$ are always considered. Suppose that the maximum separation is obtained when we compare $P_{i_0}$ with $Q_{j_0}$. Then it is certainly the case that

$$Q_{j_0} - P_{i_0} < j_0 - i_0 \qquad\qquad (9.6)$$

or else $Q_{j_0}$ would have "fallen off" the relevant barrier.

There must be some value of $j$ for which $i$ becomes equal to $i_0$. If $j \leq j_0$ there is no problem; we shall set $j$ to $j_0$ before increasing $i$. In proof, note that for all $j \leq j_0$, we have $Q_j \leq Q_{j_0} - j_0 + j$, since each port is at least one unit to the left of the port to its right. If we subtract $P_{i_0}$ from each side of this inequality, then by (9.6), we have

$$Q_j - P_{i_0} \leq Q_{j_0} - j_0 + j - P_{i_0} < j - i_0$$

which means that the test of the if-statement in Fig. 9.43 will be satisfied from the time $i$ first becomes equal to $i_0$ until the time $j$ becomes equal to $j_0$, at which time *separation* will be set to its maximum value.

The only other possibility is that when $i$ first becomes equal to $i_0$, $j$ is already greater than $j_0$. Then consider the time at which $j$ reached its current value, $j_1$. We only increment $j$ in connection with an assignment to *separation*, where that variable becomes equal to at least $j - i$. Here, $j$ is $j_1 - 1$. The value of $i$ at that time was no greater than $i_0 - 1$. As $j_1 > j_0$, *separation* was set to a value at least as great as $j_1 - 1 - (i_0 - 1) = j_1 - i_0 > j_0 - i_0$. The conclusion in this case is that the pair $P_i$ and $Q_j$ did not determine the maximum separation, as supposed. $\square$

## EXERCISES

9.1: Consider the set of rectangles forming the inverter in Fig. 7.6, whose CIF definition is in Fig. 7.7.
   a)   Convert this set of rectangles to a one-dimensional dynamic problem.
   b)   Run Algorithm 9.1 on these rectangles to find the intersecting rectangles, including those rectangles of the same layer that abut.
   c)   Use Algorithm 9.2 to extract a switch circuit from this set of rectangles.

* 9.2: Show how to maintain the pointer structure on the group tree, developed in Section 9.1 so trees of class $D$ nodes can be searched efficiently, when there are insertions and deletions to groups that make groups empty or make empty groups nonempty.

9.3: Find the design rule errors in Fig. 1.24 using:
   a)   The Baker-Terman approach described in Section 9.3.
   b)   Corner-based rules.
   c)   Polygon-based rules.

9.4: Show that the difference of rectilinear rectangles is a union of rectilinear rectangles.

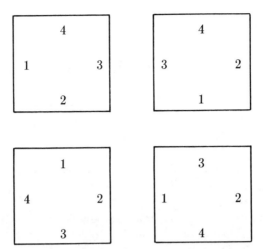

**Fig. 9.44.** Cell layout for Exercise 9.12.

*9.5: Give an algorithm to find the resistance (in squares of channel) between a point and ground, through a network of pass transistors.

9.6: In Algorithm 9.1, we considered two rectangles to be connected if they touched only at a corner but had the same layer. What effect does this decision have on the process of circuit extraction and design rule checking?

9.7: In Fig. 7.17 is a switch diagram representing a shift register. Suppose that the input, node 5, is held at 1, while the clock inputs, nodes 3 and 4, alternate, in the sequence of four steps: 00, 01, 00, 10, where the first bit represents the state of node 3, and the second bit the state of node 4. Simulate this circuit and inputs:

a)   Using Algorithm 9.3.

b)   Using Algorithm 9.4.

*9.8: What does the existence of many node states, e.g., charged, supercharged, weak, and driven, tell us about the validity of the model of Section 9.4, where we assumed that a wire could carry only a 0 or 1 value?

9.9: Prove parts (3) and (4) of Lemma 9.2.

9.10: Show that the composition of nonincreasing, distributive functions is nonincreasing and distributive.

*9.11: Give an example of an NMOS circuit that cannot be simulated correctly if we restrict resistances to two sizes, as we did in Section 9.4.

9.12: Consider the four cells placed as in Fig. 9.44, with ports on nets numbered 1 through 4. The following questions refer to the PI router discussed in Section 9.5.

a)   Assuming ports numbered 1 represent power, and ports numbered 2 represent ground, how would the power and ground nets be laid out?

   b)   Partition the space between and around the cells into channels op-
        timally.
   c)   (In what follows, we assume that nets 1 and 2, like 3 and 4, are ordinary
        nets, not power or ground.) Select routes for each of the nets to take
        through the channels.
   d)   Select orders for the crossing wires at each channel boundary.
   e)   Use a simple heuristic to route wires within channels. How wide does
        each of the channels have to be?

9.13:  In this exercise we use the notation $(\alpha, \beta, \gamma, \delta)$ to represent a channel.
       Here, $\alpha$ and $\beta$ represent the left and right sides, respectively, each being a
       sequence of integers representing the net number for each track, from the
       top; 0 represents the absence of a net at a track. $\gamma$ and $\delta$ similarly represent
       the nets associated with the columns at the top and bottom of the channel,
       respectively; the lists proceed from left to right. Apply Algorithm 9.5 to
       wire each of the following channels, if possible, introducing extra tracks
       only to the extent necessary.
   a)   (310, 424, 0201000, 2301410).
   b)   (321, 050, 0004050, 1020304).
   c)   (000, 000, 12345, 54321).

* 9.14:  Can Algorithm 9.5 optimally wire any left block in a river routing problem?

** 9.15: Suppose we wish to route a channel with ports on top and bottom, to be
         connected in pairs, but not necessarily in order. Show that if wires are
         allowed to run in two layers, either of which can connect to any port, then
         a channel separation equal to four times the channel density suffices to
         wire the channel.

** 9.16: Suppose that all is as in Exercise 9.15, but the wires in the two layers
         are restricted by requiring that all horizontal wires run in one layer and
         all vertical wires in the other, with vias where corners are turned. This
         restriction has the effect of forbidding two wires to turn corners at the
         same grid point, while the model of Exercise 9.15 permits that if the wires
         are of different layers. Prove in this case that wiring $n$ pairs of points, in
         order, with a channel density of 1, can require separation $\Omega(\sqrt{n})$. Hint:
         Consider the problem of shifting over one, where each $Q_i$, the $i^{th}$ port on
         the top has position one grid line to the left of $P_i$, the corresponding port
         on the bottom.

9.17:  What separation, as a function of $n$, is required for the problem of shifting
       over one, for river routing assuming
   a)   Arbitrary wires.
   b)   Rectilinear wires.
   * c)  Two-layer wires in the model of Exercise 9.15.

9.18:  Complete the proof of Theorem 9.4 by showing that the wiring along
       rectilinear barriers is legal.

** 9.19: Suppose wires are allowed to run not only rectilinearly, but diagonally, between two opposite corners of a grid square. Give an optimal algorithm to compute the separation for a river routing problem.

** 9.20: Suppose that we are given a river routing problem, but that the opposite sides of the channel are allowed to slide horizontally, relative to one another. The optimal *offset* is that relative position in which the separation is minimized. Give an efficient algorithm (one that is polynomial in the number of ports, not in the number of possible offsets) to compute the optimal offset.

## BIBLIOGRAPHIC NOTES

Algorithm 9.1 is from McCreight [1981]. The efficiency of reporting intersections of rectangles and general shapes has been considered by Shamos and Hoey [1976], Bentley and Wood [1980], Bentley and Ottmann [1981], Mairson and Stolfi [1982], Preparata and Nievergelt [1982], Chin and Wang [1982], and Szymanski and Van Wyk [1983]. A general discussion of the subject is in Guibas and Stolfi [1983]. We should also be aware of the algorithm by Guibas and Saxe [1983] that groups rectangles into classes of connected rectangles without having to report all intersections. Such an algorithm could be more efficient than Algorithm 9.1 if the number of intersections was high.

Some of the material in Section 9.3 on design rule checking is from Horowitz [1981]. The Baker-Terman approach is from Baker and Terman [1980], and the corner-based system is from Arnold and Osterhout [1982]. The efficient use of Boolean mask operations is discussed by Lauther [1981] and Wilmore [1981]. Eustace and Mukhopadhyay [1982] discuss a finite automaton based approach to design rule checking.

The simulation algorithm of Section 9.4 is taken from Bryant [1981b]. Related material on the implementation of these ideas is found in Bryant [1980, 1981a, 1983]. The idea of grouping nodes into classes for efficiency when deciding what nodes have to be reset to "charged" strength is covered in Baker and Terman [1980]. The discussion of distributive functions over lattices is from Kam and Ullman [1977].

We find other approaches to simulation in Newton [1979] and Nham and Bose [1980]. Logic simulation is discussed in Sherwood [1981], Berman [1981], and Leinwand [1981]. The issue of timing simulation and verification of timing constraints is examined in McWilliams [1980], Hitchcock [1982], Jouppi [1983], and Ousterhout [1983]. Lelarasmee and Sangiovanni-Vincentelli [1982] represents recent progress in the simulation of the nonlinear behavior of transistors.

The PI placement and routing system is described in Rivest [1982] and Baratz [1981]. The cell partitioning is discussed in Fidducia and Mattheyes [1982], while Algorithm 9.5, the greedy channel router, is from Rivest and Fidducia [1982].

Related ideas on min-cut placement are found in Breuer [1977] and Lauther [1979], while Preas [1979] and Preas and Gwyn [1979] discuss hierarchical placement and routing. Soukup [1981] surveys the subject of automatic placement and routing.

There has been a considerable amount of work on channel routing. Some of the earlier works on the subject are Hashimoto and Stevens [1971], Kernighan, Schweikert, and Persky [1973], and Deutsch [1976], while a representative sample of more recent work is found in Kawamoto and Kajitani [1979], Hightower and Boyd [1980], and Chan [1983]. Hightower [1974] and Yoshimura and Kuh [1982] survey the subject.

A variety of problems about channel routing have been shown $\mathcal{NP}$-complete. Some of these results are found in Sahni and Bhatt [1980], Szymanski [1981], and Arnold [1982a, b].

Optimal river routing was first considered formally by Tompa [1980], who gave the optimal algorithm in Section 9.6 for general wiring. Algorithm 9.6 on rectilinear separation and Exercise 9.20 on offset calculation are from Dolev et al. [1981]. Exercise 9.19 and other results on different kinds of wiring constraints are found in Dolev and Siegel [1981]. Other results on river routing are found in Leiserson and Pinter [1981] and Siegel [1983].

Exercise 9.15, on two-layer channel routing is from Rivest, Baratz, and Miller [1981], while Exercise 9.16 on restricted two-layer wiring is by Brown and Rivest [1981]. Other results on the complexity of multilayer wiring are found in Pinter [1981, 1982a, 1982b, 1983] and Preparata and Lipski [1982].

There are several other restricted routing problems that have received attention recently. For example, LaPaugh [1980a, b] and Gonzalez and Lee [1982] discuss the routing of wires around the borders of a single rectangle. Raghavan and Sahni [1982] examine the optimal routing of ports in a single line.

# APPENDIX

## A.1 BIG-OH AND BIG-OMEGA

We use expressions like "$f(n)$ is $O(n^2)$" to mean that the growth rate of the function $f(n)$ is no greater than that of $n^2$. We do not imply by this that $f(n)$ is proportional to $n^2$; it could be less, like $n \log n$ or $\sqrt{n}$. Formally, we say $f(n)$ is $O(g(n))$ [or $f(n) = O(g(n))$], if there are positive constants $c$ and $n_0$ such that for all $n \geq n_0$, $f(n) \leq cg(n)$. Thus, $f(n)$ may be wildly bigger than $g(n)$ for small $n$, but after a while, by the time it reaches $n_0$, it settles down to be at most proportional to $g(n)$; the constant of proportionality is upper bounded by $c$.

**Example A1.1:** Generally, we shall talk about positive functions of positive integers. For example, $2n^2$ is $O(n^2)$. In proof, pick $c = 2$ and $n_0 = 1$. It is also true that $2n^2$ is $O(n^3)$. We could pick the same $c$ and $n_0$, or we could pick $c = 1$ and $n_0 = 2$, because $2n^2 \leq n^3$, provided $n \geq 2$.

On the other hand, $n^3$ is not $O(n^2)$. Suppose it were, and the definition is satisfied by $c$ and $n_0$. That is, for all $n \geq n_0$, $n^3 \leq cn^2$. Pick

$$n = max(n_0, c + 1)$$

Then we have $c + 1 \leq c$, which is obviously false. □

The notation "$f(n)$ is $\Omega(g(n))$" is used to bound the growth rate of $f$ from below. In particular, $f(n) = \Omega(g(n))$ will always be true when $g(n) = O(f(n))$. However, for technical reasons, we sometimes want to bound the growth rate of a function from below when that function is only high from time to time. For example, the running time $T(n)$ of a prime-testing program might be very low for all the even $n$'s, but would have repeated and growing peaks for the prime values of $n$. Thus, formally we define $f(n) = \Omega(g(n))$ to mean that there is a constant $c > 0$ such that for an infinite number of values of $n$ we have $f(n) \geq cg(n)$.

**Example A1.2:** $.001n^2$ is $\Omega(n^2)$, as we may pick $c = .001$. For an example where big-omega is not just the inverse of big-oh, let $f(n)$ be $n$ for even $n$, and $n^3$ for odd $n$. Then $f(n)$ is $\Omega(n^2)$, because we may pick $c = 1$ and consider only the odd $n$'s. Also, $n^2$ is $\Omega(f(n))$, since we may pick $c = 1$ and consider only the even $n$'s. Note that neither $f(n) = O(n^2)$ nor $n^2 = O(f(n))$ is true. □

```
procedure search(v);
begin
     mark v;
     for each successor w of v do
          if w is unmarked then
               search(w);
     nextnum := nextnum + 1;
     dfn[v] := nextnum
end
```

**Fig. A2.1.**  Depth-first search procedure.

## A.2 DEPTH-FIRST SEARCH

Depth-first search is a method of visiting the nodes of a graph; the method has some useful properties. It is characterized by the policy that when at a node $v$, we try to visit a successor node that has never been visited. If there is no such node, then we retreat from $v$ by returning to the node from which we visited $v$. Thus, the arcs followed as we visit new nodes form a tree, called a *depth-first spanning tree*.

While the visiting strategy does not necessarily leave any trace, it is frequently useful to establish some order of the nodes visited. One natural order is the order of first visit, which corresponds to a preorder traversal of the depth-first spanning tree. A possibly more useful ordering, which we call *depth-first ordering*, is the order of retreat from a node; this order corresponds to a post-order traversal of the depth-first spanning tree.

Figure A2.1 sketches a recursive procedure *search*, which is used to compute an array $dfn[v]$ that holds the depth-first number for each node $v$. We initially set *nextnum* to 0, set each node in the graph to "unmarked," and call *search(v)* for some node $v$. If the graph is connected, all the nodes will be visited and given a value of $dfn$. If not, we repeat the search process on one of the nodes that are still "unmarked." In this case, the result is a depth-first spanning forest, consisting of two or more trees.

**Example A2.1:** Figure A2.2(a) shows a graph whose nodes have been numbered by the order of first visit, on the assumption that we begin at $a$ and visit its successor $b$ before we visit its successor $c$, and from $b$, we visit $d$ before $c$. Note that from $f$, we do not visit $b$, because by the time we reach $f$, $b$ has been marked. Similarly, when we eventually reach $c$, from $b$, we do not revisit $d$, because that is marked, and when we reach $e$, we do not visit $a$.

We show the depth-first ordering, the order in which we finally retreat from each node after considering all its successors in the for-loop of *search*, in Fig. A2.2(b). We also illustrate the depth-first spanning tree that resulted from

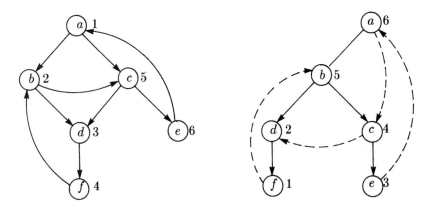

**Fig. A2.2.** Depth-first search and ordering.

the particular order of successors that we chose; tree edges are solid, and the others are dashed. Note that if, say, we had visited $c$ from $a$ before visiting $b$, a different tree would result. $\square$

There are several useful properties of depth-first search. Suppose that we draw the tree such that the children of each node $v$ are those of its successors that were first visited from $v$, and the left-to-right order of the children is the order in which they were visited. Then it can be shown that no arc of the original graph travels from left to right in the tree. All arcs travel from a node to either an ancestor in the tree, a descendant, or a node to the left. Another important property is that if the graph is acyclic, then whenever there is an arc $v \rightarrow w$, it will be the case that $dfn[w] < dfn[v]$. Thus, the depth-first numbering is the reverse of a topological ordering of the acyclic graph. Further details about depth-first search, pre- and postordering, and topological sort can be found in Aho et al. [1983].

## A.3 REGULAR EXPRESSIONS AND NONDETERMINISTIC AUTOMATA

The notation of regular expressions is designed to express succinctly patterns of events, in particular those events that can be described by finite automata and therefore can be implemented by a circuit of finite size. Regular expression notation is algebraic, consisting of operators and operands. The meaning attached to such an expression is a set of strings, those strings that fit the pattern.

Operands of regular expressions are abstract symbols; we shall use letters or digits in examples. Each symbol can be thought of as denoting the occurrence of an event, such as a wire becoming high, a wire becoming low, or a set of wires having some specific combination of values. The systems of events described are assumed to be synchronous. The state is sampled at discrete times, and the occurrence of events is determined for that time unit and expressed as an abstract symbol. The operators of regular expressions are:

1.  +, standing for union.
2.  Juxtaposition, standing for concatenation, to be defined below.
3.  *, the *Kleene closure*, which is a postfix unary operator that informally means "any number of."

We associate sets of strings with expressions by an inductive process on the number of operators. If $E$ is an expression, $L(E)$ is the set of strings it denotes. The basis, zero operators, is handled by the rule that if $a$ is an operand, $L(a)$ is $\{a\}$, that is, the set consisting of one string; that string has length one and has the symbol $a$ in its lone position. For the induction:

1.  If $E = E_1 + E_2$, then $L(E) = L(E_1) \cup L(E_2)$. In this sense, + represents union.
2.  If $E = E_1 E_2$, then $L(E) = L(E_1)L(E_2)$, that is,

    $$\{\, wx \mid w \text{ is in } E_1 \text{ and } x \text{ is in } E_2 \,\}$$

    The string $wx$ is the *concatenation* of the strings $w$ and $x$, i.e., the string formed by taking $w$ and following it by $x$.
3.  If $E = E_1{}^*$, then $L(E) = (L(E_1))^*$, that is

    $$\{\,\epsilon\,\} \cup L(E_1) \cup L(E_1)L(E_1) \cup L(E_1)L(E_1)L(E_1) \cup \cdots$$

    Here, $\epsilon$ stands for the empty string, the string of length zero, which we may think of as representing the selection of no strings from $L(E_1)$. The successive terms in $(L(E_1))^*$ represent the selection of one string from $L(E_1)$, the concatenation of two selected strings from $L(E_1)$, and so on. It is in this sense that the Kleene closure represents the selection of zero or more members of a set.

When writing regular expression, we are free to use parentheses to indicate the order of application of operators. We may also use the default precedence of operators, that * binds first, then concatenation, and + binds last.

**Example A3.1:** The regular expression 0 represents the single string $\{0\}$. The expression 001 is the concatenation of the expression 0 with the expression 01, which in turn is the concatenation of the expression 0 with the expression 1. It represents the set with the single string $\{001\}$.

The expression $0^*$ represents $\{\epsilon, 0, 00, 000, \dots\}$, that is, the set of all strings of 0's. The expression $(01 + 011)(10 + 110)$ represents the concatenation of the sets $\{01, 011\}$ and $\{10, 110\}$, which is $\{0110, 01110, 011110\}$. Note that the string 01110 is formed in two ways by the concatenation of a string from the first set and one from the second. As a final example, the expression $10^*$ represents the set of strings $\{1\}\{\epsilon, 0, 00, \dots\}$, that is, $\{1, 10, 100, \dots\}$. Note that the expression is interpreted as $1(0^*)$, because closure takes precedence

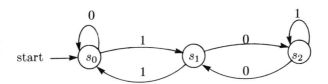

**Fig. A3.1.** DFA for computing residues modulo 3.

over concatenation in the absence of parentheses telling us otherwise.

### Nondeterministic Automata

A *nondeterministic finite automaton* (*NFA*) consists of a finite set of states and rules for jumping between states in response to inputs. The inputs are abstract symbols, and we can see an NFA as a recognizer of patterns, that is, sets of strings, just as regular expressions are generators of patterns. Technically, for each state $s$ and each input symbol $a$ there is a set of zero or more next states. If at some time, state $s$ is one of the states that is on, and $a$ is the current input symbol, then at the next time unit, each of the next states for $s$ and $a$ will be on. State $s$ will not necessarily be on; it is only on if it is in the successor set for some state $t$ and input $a$. It is important to realize that more than one state can be on at a time if the automaton is nondeterministic. If for each state and input there is a unique next state, then the automaton is *deterministic* (a *DFA*), i.e., it is the ordinary variety of automaton or "sequential machine."

Normally, one state of the NFA is designated the *start state*; initially, only the start state is on. For output, we may select one or more states to be *final states* (or *accepting states*). The output is deemed to be 1 whenever one or more of the final states are on, and is 0 otherwise. If we wish to get more than a single bit of output from the NFA, we may associate output symbols with some or all of the states. An output signal is assumed to be raised whenever one or more of the states associated with that symbol are on. Thus, the NFA can raise more than one output signal at a time.

Sometimes we find it useful to permit $\epsilon$-*transitions*. In addition to transitions on input symbols, for each state $s$ there is a set of zero or more states that are turned on whenever $s$ is turned on, without waiting for an input symbol.

The easiest way to picture NFA's is as directed graphs, where the states are the nodes. An arc from $s_1$ to $s_2$ is labeled by the set of input symbols and/or $\epsilon$, for which transitions from $s_1$ to $s_2$ are permitted. Then after reading a sequence of input symbols $w$, the set of states that are on is exactly those to which there is a path labeled $w$ from the start state. Note that any number of $\epsilon$-transitions is permitted along the path.

**Example A3.2:** Figure A3.1 shows a three-state automaton that happens to be deterministic. It reads a sequence of 0's and 1's, and indicates by its state

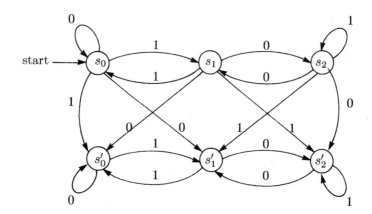

**Fig. A3.2.** NFA for computing residues of strings with errors.

the value of the number read (high-order bit first), modulo 3. That is, the automaton is in state $s_i$ if the remainder is $i$, for $0 \leq i \leq 2$. For example, the path labeled 111001 from the start state $s_0$ leads back to $s_0$; the entire sequence of states is $s_0, s_1, s_0, s_1, s_2, s_1, s_0$. That makes sense, since 111001 in decimal is 57, which is divisible evenly by 3.

Suppose we suspected that one bit of the input string might be in error. We could compute all the possible residues (remainders) if we introduced a new set of states $s_i'$, for $0 \leq i \leq 2$, where we enter $s_i'$ if there is some way a one-bit change in the input could produce an integer whose residue is $i$. For example, if we want to know whether a change of zero or one bit could lead to an input divisible by 3, we make states $s_0$ and $s_0'$ be the final states.

Suppose we start in $s_0$ only, and the input is 101. There are transitions from $s_0$ to $s_1$ and $s_0'$ labeled 1, so after reading the first input we are in set of states $\{s_1, s_0'\}$. The next input, 0, takes the first of these states to $s_2$ and $s_0'$, while the second is taken to $s_0'$ only. Thus, after the second input we are in $\{s_2, s_0'\}$. The third input, 1, takes these two states to $\{s_2, s_1'\}$. One interpretation of the fact we are in those two states is that the residue of 101, or 5 in decimal, is 2; this is what $s_2$ tells us. Also, $s_1'$ tells us that by making a single error, we get numbers that leave remainder 1 when divided by three. The three possible errors leave 001, 111, and 100, which all represent numbers with remainder 1. The fact that we are in no other states tells us there are no other possibilities, which we already discovered by considering all one-bit errors. $\square$

More information on regular expressions, DFA's, and NFA's can be found in Hopcroft and Ullman [1979].

## A.4 SKETCH OF THE FLASHSORT PARALLEL SORTING ALGORITHM

There is a complicated parallel sorting algorithm that with probability 1 (in the sense of Theorem 6.8)† finishes in time $O(\log n)$. The basic idea comes from Reif and Valiant [1983]. Whether it is in practice more efficient than the $O(\log^2 n)$ sort of Algorithm 6.1 for realistic networks is open to question, but it is instructive, and for completeness we give a sketch here. Many details are left as exercises.

We begin with a set of $n$ elements, one at each node on the top rank (rank 0) of a butterfly of width $n = 2^k$. There is a constant $c$ such that each node can store at least $c \log_2 n$ elements.‡ The larger the constant $c$, the more likely the algorithm is to succeed on any given trial, but as long as there is a probability greater than zero of success, the expected time needed for success will still be proportional to $\log n$.

The algorithm is similar to Quicksort (see Aho et al. [1983]), in the sense that we attempt to establish *splitters*, which are elements that recursively divide sets of elements into two subsets, one consisting of the elements less than the splitter, and the other consisting of the elements greater than or equal to the splitter. In particular, we construct, for each binary string $x$ of length zero through $k$, a subset $S_x$ of the elements, and a splitter $\sigma_x$ in $S_x$. Imagine we start with $S_\epsilon$ equal to the set of all elements; recall that $\epsilon$ represents the empty string. From each set $S_x$, where $x$ is of length shorter than $k$, we pick $\sigma_x$ at random from $S_x$ and divide $S_x$ into $S_{x0} = \{ a \mid a \text{ is in } S_x \text{ and } a < \sigma_x \}$ and $S_{x1} = \{ a \mid a \text{ is in } S_x \text{ and } a \geq \sigma_x \}$.

Our goal is to build the tree of splitters suggested by Fig. 6.15. There we indicate the particular node of the butterfly at which each splitter is selected. Sets of columns into which we desire the members of each set to move are also suggested. That is, the *zone of* $S_x$ is all those columns whose numbers, counting from the left, have $k$-bit binary representations that begin with $x$.

The sorting process can be thought of as one where initially each element knows it is in $S_\epsilon$. As splitters are chosen, they are broadcast to the relevant zone; $\sigma_x$ is broadcast to the zone of $S_x$. An element that knows it is in $S_x$ will find its way to the zone of $S_x$, and position itself on rank $\mid x \mid$.†† When the element is told of the value of $\sigma_x$, it compares that with itself and decides whether it is in $S_{x0}$ or $S_{x1}$. In the former case, it moves vertically to the zone of $S_{x0}$, and in the latter it uses the diagonal of the butterfly to enter the zone of $S_{x1}$. In either case it moves downward, to the rank with the next higher number.

---

† We refer to this sense as "with high probability" in what follows.

‡ By storing extra elements in nodes of the same column, we could run the algorithm with only a constant number of elements allowed at any node.

†† $\mid x \mid$ stands for the length of string $x$.

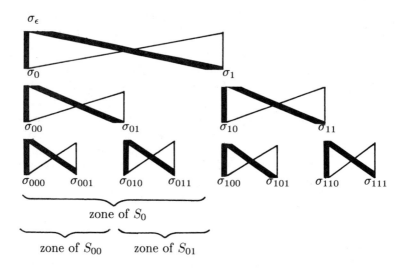

**Fig. A4.1.** Splitter tree and zones of the butterfly.

We shall initially describe the algorithm as if all $n$ elements could be sorted at once by this process. In practice, should we build the splitter tree for all $n$ elements in one pass, there would very likely be one large set $S_x$, where $\mid x \mid = k$, since whenever we split, one half is larger than the other, and often, the larger half is $3/4$ or more the size of the set split. Thus, we shall see that it is necessary to sort only a fraction of the elements, selected at random, and use these as the top portion of the splitter tree. The entire set of elements is then divided into sets, based on these splitters, and moved down to the appropriate rank $r$.

Then, the sets $S_x$, where $\mid x \mid = r$, are flashsorted in parallel, by a recursive application of the portion of the algorithm we shall describe, resulting in the splitter tree being extended additional ranks. We continue in this manner, until after $O(\log n)$ passes, the entire splitter tree is constructed, with no more than $O(\log n)$ elements in any set, with high probability. From that condition, producing the sorted order is relatively easy. Moreover, the first pass takes $O(\log n)$ time, and each subsequent pass takes a fraction less than 1 of the time required for the previous pass. Thus, the time for all the passes is $O(\log n)$. Further, although the algorithm could fail, because one set is too big, the probability of failure is less than 1, so repeating the algorithm until success only multiplies the time by a constant factor.

We shall now give the steps needed to perform one pass. Note that we begin by constructing a "splitter tree" that is not the final splitter tree of the complete algorithm, but rather is used to sort the sample of elements, all of which will become splitters in the "real" splitter tree.

1.  We begin by using Algorithm 6.2 to permute the given elements at random. This step needs to be performed only on the first pass, and it takes $O(\log n)$ time with high probability.

2.  Select splitters, and partition the elements into sets $S_x$, where $\mid x \mid = k$. Elements begin on the top rank and wind up on the bottom rank, by a method we shall describe in detail later.

3.  In $O(\log n)$ time, each element determines its position in the sorted order as follows. Initially, each set $S_x$, where $\mid x \mid = k$ is at the bottom rank. It sends a count of the number of elements upward, and these counts are summed in an appropriate way and then passed down the splitter tree, so each set can know how many elements appear to its left. With high probability, there are at most $O(\log n)$ elements in any set, so the elements themselves can know their positions in $O(\log n)$ time. The details of this step are left as an exercise.

4.  Each element now becomes a splitter of the "real" splitter tree. The element whose position is $n/2$ becomes $\sigma_\epsilon$, those at positions $n/4$ and $3n/4$ become $\sigma_0$ and $\sigma_1$, and so on. The elements are sent to the desired nodes by an application of Algorithm 6.2, so the whole process takes $O(\log n)$ time with high probability.

**Splitter Selection**

We now sketch some of the details necessary to make step (2) work in time $O(\log n)$. There are three kinds of packets that travel through the butterfly.

1.  The elements themselves.

2.  Splitter packets, giving the value of some splitter $\sigma_x$.

3.  Candidate packets. When an element $a$ meets a splitter packet $\sigma_x$, and $a$ knows it is in $S_x$ but does not know whether it is in $S_{x0}$ or $S_{x1}$, it determines which, and sends a packet containing itself as a candidate for $\sigma_{x0}$ or $\sigma_{x1}$, whichever is appropriate.

To see that packets get where they are needed quickly, we need certain rules governing them. First, a given node will send and receive only one packet at any time unit. To assure that all types of packets reach where they are needed, we devote every third time unit to packets of each type. We also kill packets when they are not needed, and follow certain priority rules. There rules are:

1.  Element packets move down the butterfly, reaching rank $r$ when it knows the set $S_x$ to which it belongs, where $\mid x \mid = r$. Conflicts, the desire of two elements to reach the same node at the same time, are decided at random.

2.  Splitter packets move down the butterfly from the node at which they are selected, moving both vertically and diagonally. When they reach the bottom, they move up the butterfly, vertically only. In this manner, they reach only the zone in which they are relevant. If a splitter packet $\sigma_x$

meets a splitter packet $\sigma_y$, where $\mid y \mid > \mid x \mid$ (note $x$ must be a prefix of $y$ for this to happen), $\sigma_x$ has priority.

3.   Candidates for splitters move vertically and diagonally down the butterfly, always moving toward the column of the node at which the splitter is to be selected. When they reach that column, they move upward. The first candidate to reach that node is selected, and it is broadcast by splitter packets. If a candidate packet for splitter $\sigma_x$ meets a candidate for splitter $\sigma_y$, where $\mid y \mid > \mid x \mid$, we kill the former, since a candidate for $\sigma_x$ has evidently been selected. If $x = y$, priority is determined at random.

## The Complete Algorithm

Now, we must describe how the four steps above fit into the entire algorithm. First, note that if we want, with high probability, no more than $O(\log n)$ elements to appear in the same set at the bottom of the tree, then we must be sure that this condition is likely to apply to the largest set. On the average, a set is divided with one fourth in one part and three fourths in the other. Thus, if we start with $m$ of the $n$ elements, and split $k = \log_2 n$ times, the set that we reach by following the larger subset at each split will have roughly $m(3/4)^k$ members.

We wish that number to be about 1, so allowing for statistical fluctuations, the probability that more than $c \log n$ elements will be in this set will decrease exponentially with $c$. Thus, we would like $\log_2 m$ to be about $k \log_2(4/3)$, i.e., $m = n^{0.41}$. That formula is only approximate, because assuming that the effect of a number of divisions into random fractions between $1/2$ and 1 is the same as all fractions being $3/4$ is conservative; the actual effect will tend to produce a slightly smaller set on the average.†

On the other hand, we also have to consider the possibility that some set formed from the smaller part at one or more splits actually winds up larger than the set that takes the larger part of all its splits. However, we claim that there is some $p > 0$ such that we can let $m = n^p$ and sort $m$ of the $n$ elements, chosen at random, with a high probability of success in steps (1–4) above. Thus, the above steps are really executed with $m$ elements, not $n$, although the number of levels we have available is still $k = \log_2 n$.

To complete the description of a pass, we add a step before step (1) in which we select the $m = n^p$ elements at random. Then, after step (4), which now installs only the $m$ elements selected as the splitters in the top $pk$ levels, we take all $n$ elements and move them through the tree of splitters by a process similar to steps (1–4) above, with splitters being broadcast to their zones and elements traveling down the butterfly. This process takes time $O(\log n)$ with high probability.

---

† To see why, compare multiplying $m$ by 1 half the time and by $1/2$ half the time with multiplying by $3/4$ all the time.

The effect is that we are left with $n^{1-p}$ sorting subproblems to be solved in parallel on the bottom $(1-p)k$ ranks of the butterfly. The problems may not involve equally sized sets of elements, but the differences will, with high probability, be negligible. When we solve the subproblems, because of their smaller size, we need only $(1-p)$ times as much time as to solve the whole problem. It is thus easy to show that repeating this process until we get down to sets of size 1 will take $1+(1-p)+(1-p)^2+\cdots$ times the $d\log n$ time needed for one pass. As that sum is $(\frac{d}{1-p})\log n$, we see that with high probability, $O(\log n)$ time suffices for the entire algorithm.

Of course, in addition to leaving out some algorithmic details, we have completely elided the probabilistic analysis. The reader is referred to Reif and Valiant [1983] for this aspect of the proof.

# BIBLIOGRAPHY

Abelson, H. and P. Andreae [1980]. "Information transfer and area-time trade-offs for VLSI multiplication," *Comm. ACM* **23**:1, pp. 20–23.

Aho, A. V., J. E. Hopcroft, and J. D. Ullman [1974]. *The Design and Analysis of Computer Algorithms*, Addison-Wesley, Reading Mass.

Aho, A. V., J. E. Hopcroft, and J. D. Ullman [1983]. *Data Structures and Algorithms*, Addison-Wesley, Reading Mass.

Aho, A. V. and J. D. Ullman [1977]. *Principles of Compiler Design*, Addison-Wesley, Reading Mass.

Aho, A. V., J. D. Ullman, and M. Yannakakis [1983] "On notions of information transfer in VLSI circuits," *Proc. Fifteenth Annual ACM Symposium on the Theory of Computing*, pp. 133–139.

Ajtai, M., J. Kolmos, and E. Szemeredi [1983]. "An $O(n \log n)$ sorting network," *Proc. Fifteenth Annual ACM Symposium on the Theory of Computing*, pp. 1–9.

Aleliunas, R. [1982]. "Randomized parallel computation," *Proc. ACM Symp. on Principles of Distributed Computing*.

Arnold, M. H. and J. K. Ousterhout [1982]. "Lyra: a new approach to geometric layout rule checking," *Proc. Nineteenth Design Automation Conference*, pp. 530–536.

Arnold, P. B. [1982a]. "Complexity results for channel routing," TR–21–82, Aiken Computation Lab., Harvard Univ., Cambridge, Mass.

Arnold, P. B. [1982b]. "Complexity results for single row routing," TR–22–82, Aiken Computation Lab., Harvard Univ., Cambridge, Mass.

Aspvall, B. and Y. Shiloach [1979]. "A polynomial time algorithm for solving systems of linear inequalities with two variables per inequality," *Proc. Twentieth Annual IEEE Symposium on Foundations of Computer Science*, pp. 205–217.

Atallah, M. J. and S. R. Kosaraju [1982]. "Graph problems on a mesh–connected processor array," *Proc. Fourteenth Annual ACM Symposium on the Theory of Computing*, pp. 345–353.

Atrubin, A. J. [1965]. "An interactive one-dimensional real time multiplier,"

*IEEE Trans. on Computers* **EC-14**:3, pp. 394–399.

Ayres, R. [1979]. "Silicon compilation—a hierarchical use of PLA's," *Proc. Sixteenth Design Automation Conference*, pp. 314–326.

Baker, C. M. and C. Terman [1980]. "Tools for verifying integrated circuit design," *Lambda* **1**:3, pp. 22–30.

Baratz, A. E. [1981]. "Algorithms for integrated circuit signal routing," Ph. D. thesis, Dept. of EECS, MIT, Cambridge, Mass.

Batali, J., N. Mayle, H. Shrobe, G. Sussman, and D. Weise [1981]. "The DPL/Daedelus design environment," in *VLSI-81* (J. P. Gray, ed.), Academic Press, New York, pp. 183–192.

Batcher, K. [1968]. "Sorting networks and their applications," *Spring Joint Computer Conference* **32**, pp. 307–314, AFIPS PRESS, Montvale, N. J.

Baudet, G. M. [1981]. "On the area required by VLSI circuits," in Kung, Sproull, and Steele [1981], pp. 100–107.

Beke, H. and W. Sansen [1979]. "CALMOS: a portable software system for the automatic interactive layout of MOS/LSI," *Proc. Sixteenth Design Automation Conference*, pp. 102–108.

Bentley, J. L. and T. A. Ottmann [1979]. "Algorithms for reporting and counting geometric intersections," *IEEE Trans. on Computers* **C-28**:9, pp. 643–647.

Bentley, J. L. and T. A. Ottmann [1981]. "The complexity of manipulating hierarchically defined sets of rectangles," unpublished memorandum, Univ. of Karslruhe, W. Germany.

Bentley, J. L. and D. Wood [1980]. "An optimal worst-case algorithm for reporting intersections of rectangles," *IEEE Trans. on Computers* **C-29**:9, pp. 571–577.

Bergmann, N. [1983]. "A case study of the F.I.R.S.T. silicon compiler," *Proc. Third Caltech Conference on VLSI*, pp. 413–430, Computer Science Press, Rockville, Md.

Berman, L. [1981]. "On logic comparison," *Proc. Eighteenth Design Automation Conference*, pp. 854–861.

Bhatt, S. N. and C. E. Leiserson [1981]. "Minimizing the longest edge in a VLSI layout," unpublished memorandum, Laboratory for Computer Science, MIT, Cambridge, Mass.

Bhatt, S. N. and C. E. Leiserson [1982]. "How to assemble tree machines,"

*Proc. Fourteenth Annual ACM Symposium on the Theory of Computing*, pp. 77–84.

Bianchini, R. and R. Bianchini Jr. [1982]. "Wireability of an ultracomputer," DOE/ER/03077-177, Courant Inst., New York.

Bilardi, G., M. Pracchi, and F. P. Preparata [1981]. "A critique and appraisal of VLSI models of computation," in Kung, Sproull, and Steele [1981], pp. 81–88.

Bilgory, A. [1982]. "Compilation of register transfer language specifications into silicon," UIUCDCS–R–82–1091, Dept. of CS, Univ. of Illinois, Urbana, Ill.

Borodin, A. and J. E. Hopcroft [1982]. "Routing, merging, and sorting on parallel models of computation," *Proc. Fourteenth Annual ACM Symposium on the Theory of Computing*, pp. 338–344.

Borodin, A., J. von zur Gathen, and J. E. Hopcroft [1982]. "Fast parallel matrix and GCD computations," *Proc. Twenty-Third Annual IEEE Symposium on Foundations of Computer Science*, pp. 65–71.

Brayton, R. K., G. D. Hachtel, L. A. Hemachandra, A. R. Newton, and A. L. M. Sangiovanni-Vincentelli [1982]. "A comparison of logic minimization strategies using EXPRESSO: an APL program package for partitioned logic minimizalization," *Proc. IEEE Intl. Conf. on Circuits and Computers*.

Brent, R. P. and L. M. Goldschlager [1982]. "Some area time tradeoffs for VLSI," *SIAM J. Computing* 11:4, pp. 737–747.

Brent, R. P. and H.-T. Kung [1980a]. "The chip complexity of binary arithmetic," *Proc. Twelfth Annual ACM Symposium on the Theory of Computing*, pp. 190–200.

Brent, R. P. and H.-T. Kung [1980b]. "On the area of binary tree layouts," *Information Processing Letters* 11:1, pp. 44–46.

Brent, R. P. and H.-T. Kung [1981]. "The area-time complexity of binary multiplication," *J. ACM* 28:3, pp. 521–534.

Breuer, M. A. [1977]. "Min-cut placement," *J. Design Automation and Fault-Tolerant Computing* 1:4, pp. 343–362.

Brown, D. J. and R. L. Rivest [1981]. "New lower bounds for channel width," in Kung, Sproull, and Steele [1981], pp. 178–185.

Brown, D. W. [1981]. "A state-machine synthesizer—SMS," *Proc. Eighteenth Design Automation Conference*, pp. 301—305.

Brown, H. and M. Stefik [1982]. "Palladio: an expert assistant for integrated circuit design," HPP–82–5, Dept. of Computer Science, Stanford Univ., Stanford,

CA.

Browning, S. A. [1980]. "The tree machine: a highly concurrent computing environment," Ph. D. thesis, Dept. of Computer Science, CIT, Pasadena, CA.

Bryant, R. E. [1980]. "An algorithm for MOS logic simulation," *LAMBDA*, Oct., 1980, pp. 46–53.

Bryant, R. E. [1981a]. "MOSSIM: a switch level simulator for MOS LSI," *Proc. Eighteenth Design Automation Conference*, pp. 786–790.

Bryant, R. E. [1981b]. "A switch-level simulation model for integrated logic circuits," MIT/LCS/TR-259, MIT, Cambridge, Mass.

Bryant, R. E. [1983]. "A switch-level model and simulator for MOS digital systems." 5065:TR:83, Californai Inst. of Technology, Pasadena, CA.

Chandra, A. K., L. J. Stockmeyer, and U. Vishkin [1982]. "A complexity theory for unbounded fan-in parallelism," *Proc. Twenty-Third Annual IEEE Symposium on Foundations of Computer Science*, pp. 1–13.

Chang, T. L. and P. D. Fisher [1982]. "A programmable systolic array," *IEEE COMPCON*, pp. 48–53.

Chan, W. S. [1983]. "A new channel routing algorithm," *Proc. Third Caltech Conference on VLSI*, pp. 117–140, Computer Science Press, Rockville, Md.

Chazelle, B. M. and D. P. Dobkin [1980]. "Detection is easier than computation," *Proc. Twelfth Annual ACM Symposium on the Theory of Computing*, pp. 146–153.

Chazelle, B. M. and L. M. Monier [1981]. "Model of computation for VLSI with related complexity results," *Proc. Thirteenth Annual ACM Symposium on the Theory of Computing*, pp. 318–325.

Chin, F. Y., J. Lam, and I.-N. Chen [1982]. "Efficient parallel algorithms for some graph problems," *Comm. ACM* **25**:9, pp. 659–665.

Chin, F. Y. and C. A. Wang [1982]. "Optimal algorithms for the intersection and the minimum distance problems between planar polygons," TR82–8, Dept. of Computer Science, Univ. of Alberta, Edmonton, Alberta.

Cho, Y. E., A. J. Korenjak, and D. E. Stockton [1977]. "FLOSS: an approach to automated layout for high-volume designs," *Proc. Fourteenth Design Automation Conference*, pp. 138–141.

Chuquillanqui, S. and T. Perez-Segovia [1982]. "PAOLA: a tool for topological optimization of large PLA's," *Proc. Nineteenth Design Automation Conference*, pp. 300–306.

Dekel, E., D. Nassimi, and S. Sahni [1979]. "Parallel matrix and graph algorithms," TR 79–10, Dept. of C. S., Univ. of Minn., Minneapolis, MI.

Deutsch, D. [1976]. "A dogleg channel router," *Proc. Thirteenth Design Automation Conference*, pp. 425–433.

Dolev, D., K. Karplus, A. Siegel, A. Strong, and J. D. Ullman [1981]. "Optimal wiring between rectangles," *Proc. Thirteenth Annual ACM Symposium on the Theory of Computing*, pp. 312–317.

Dolev, D., F. T. Leighton, and H. W. Trickey [1983]. "Planar embedding of planar graphs," unpublished memorandum, Stanford Univ., Stanford, CA.

Dunlop, A. E. [1978]. "SLIP: symbolic layout of integrated circuits with compaction," *Computer Aided Design* **10**:6, pp. 387–391.

Dunlop, A. E. [1980]. "SLIM—the translation of symbolic layouts into mask data," *Proc. Seventeenth Design Automation Conference*, pp. 595–602.

Egan, J. R. and C. L. Liu [1982]. "Optimal bipartite folding of PLA," *Proc. Nineteenth Design Automation Conference*, pp. 141–146.

Eustace, R. A. and A. Mukhopadhyay [1982]. "A deterministic finite automaton approach to design rule checking for VLSI," *Proc. Nineteenth Design Automation Conference*, pp. 712–717.

Feldman, S. I. [1982]. "The circuit design language xi," unpublished memorandum, Bell Laboratories, Murray Hill, NJ

Fidducia, C. M. and R. M. Mattheyes [1982]. "A linear-time heuristic for improving network partitions," *Proc. Nineteenth Design Automation Conference*, pp. 175–181.

Fischer, M. J. and M. S. Paterson [1980]. "Optimal tree layout," *Proc. Twelfth Annual ACM Symposium on the Theory of Computing*, pp. 177–189.

Fisher, A. L. [1981]. "Systolic algorithms for running order statistics in signal and image processing," in Kung, Sproull, and Steele [1981], pp. 265–272.

Fisher, A. L., H.-T. Kung, L. M. Monier, H. Walker, and Y. Dohi [1983]. "Design of the PSC: a progammable systolic chip," *Proc. Third Caltech Conf. on VLSI*, pp. 287–302, Computer Science Press, Rockville, Md.

Floyd, R. W. and J. D. Ullman [1982]. "The compilation of regular expressions into integrated circuits," *J. ACM* **29**:2, pp. 603–622.

Fortune, S. and J. Wyllie [1978]. "Parallelism in random access machines," *Proc. Tenth Annual ACM Symposium on the Theory of Computing*, pp. 114–118.

Foster, M. J. and H.-T. Kung [1980]. "The design of special purpose VLSI chips," *Computer* **13**:13, pp. 26–40.

Foster, M. J. and H.-T. Kung [1981]. "Recognize regular languages with programmable building blocks," in *VLSI–81* (J. P. Gray, ed.), Academic Press, New York, pp. 75–84.

Fox, J. R. [1983]. "The MacPitts silicon compiler: a view from the telecommunications industry," *VLSI Design*, May, 1983, pp. 30–37.

Galil, Z. and W. J. Paul [1981]. "An efficient general-purpose parallel computer," *Proc. Thirteenth Annual ACM Symposium on the Theory of Computing*, pp. 247–262.

Garey, M. R. and D. S. Johnson [1979]. *Computers and Intractability: a Guide to the Theory of $\mathcal{N}\mathcal{P}$-completeness*, Freeman, New York.

Goates, G. B. and S. S. Patil [1981]. "ABLE: a LISP-based layout modeling language with user-definable procedural models for storage/logic array design," *Proc. Eighteenth Design Automation Conference*, pp. 322–329.

Goldschlager, L. [1978]. "A unified approach to models of synchronous parallel machines," *Proc. Tenth Annual ACM Symposium on the Theory of Computing*, pp. 89–94.

Gonzalez, T. F. and S.-L. Lee [1982]. "An $O(n \log n)$ algorithm for optimal routing around a rectangle," TR–116, Prog. in Math. Sciences, Univ. of Texas, Dallas, TX.

Greene, J. W. and A. El Gamal [1983]. "Area and delay penalties in restructurable wafer-scale arrays," *Proc. Third Caltech Conference on VLSI*, pp. 165–184, Computer Sciecne Press, Rockville, Md.

Guibas, L. J., H.-T. Kung, and C. D. Thompson [1979]. "Direct VLSI implementation of combinatorial algorithms," *Proc. Caltech Conf. on VLSI*, pp. 509–525.

Guibas, L. J. and F. M. Liang [1980]. "Systolic stacks, queues, and counters," *Proc. MIT Conf. on Advanced Research in VLSI*.

Guibas, L. J. and J. B. Saxe [1983]. "Solution to problem 80-15: computing the connected components of a collection of rectangles," to appear in *J. Algorithms*.

Guibas, L. and J. Stolfi [1983]. *Computational Geometry*, unpublished manuscript, Xerox PARC, Palo Alto, CA.

Hachtel, G. D., A. R. Newton, and A. L. M. Sangiovanni-Vincentelli [1982]. "Techniques for programmable logic array folding," *Proc. Nineteenth Design*

*Automation Conference*, pp. 147–155.

Hambrusch, S. E. [1981]. "VLSI algorithms for the connected component problem," CS–81–9, Dept. of C. S., Penn State Univ., State College, Pa.

Hambrusch, S. E. and J. Simon [1982]. "Solving undirected graph problems in VLSI," unpublished memorandum, Dept. of Computer Science, Penn State Univ., State College, Pa.

Hardy, G. H. and E. M. Wright [1938]. *The Theory of Numbers* University Press, Oxford.

Hashimoto, A. and Stevens, J. [1971]. "Wire routing by optimizing channel assignment in large apertures," *Proc. Eighth Design Automation Conference*, pp. 155–169.

Hemachandra, L. A. [1982]. "GRY: a PLA minimizer," unpublished memorandum, Dept. of Computer Science, Stanford Univ., Stanford, CA.

Hennessy, J. L. [1981]. "SLIM: a simulation and implementation language for VLSI microcode," *LAMBDA*, April, 1981, pp. 20–28.

Hightower, D. W. [1974]. "The interconnection problem: a tutorial," *Computer* **7**:4, pp. 18–32.

Hightower, D. W. and R. L. Boyd [1980]. "A generalized channel router," *Proc. Eighteenth Design Automation Conference*, pp. 12–21.

Hirschberg, D. S., A. K. Chandra, and D. V. Sarwate [1979]. "Computing connected components on parallel computers," *Comm. ACM* **22**:8, pp. 461–464.

Hirschberg, D. S. and D. J. Wolper [1982]. "A parallel solution for the minimum spanning tree problem," unpublished memorandum, Dept. of Information and Computer Science, Univ. of Calif., Irvine, CA.

Hitchcock, R. B. [1982]. "Timing verification and the timing analysis program," *Proc. Nineteenth Design Automation Conference*, pp. 594–604.

Hoey, D. and C. E. Leiserson [1980]. "A layout for the shuffle-exchange network," *Proc. IEEE Intl. Conf. on Parallel Processing*.

Hong, J.-W. and H.-T. Kung [1981]. "I/O complexity: the red-blue pebble game," *Proc. Thirteenth Annual ACM Symposium on the Theory of Computing*, pp. 326–333.

Hong, J.-W. and A. L. Rosenberg [1981]. "Graphs that are almost binary trees," *Proc. Thirteenth Annual ACM Symposium on the Theory of Computing*, pp. 334–341.

Hopcroft, J. E. and J. D. Ullman [1979]. *Introduction to automata theory, languages, and computation*, Addison-Wesley, Reading Mass.

Horowitz, M. [1981]. "Design rule checking," unpublished memorandum, Integrated Circuits Lab., Stanford Univ., Stanford Calif.

Hsueh, M.-Y. and D. O. Pederson [1979]. "Computer-aided layout of LSI circuit building-blocks," *Proc. 1979 Intl. Symp. on Circuits and Systems*, pp. 447–477.

Johanssen, D. [1979]. "Bristle blocks: a silicon compiler," *Proc Caltech Conf. on VLSI*, pp. 303–310. See also *Proc. Sixteenth Design Automation Conference*, pp. 310–313.

Johnson, S. C. [1983]. "Code generation for silicon," *Proc. Tenth ACM Symposium on Principles of Programming Languages*.

Johnson, S. C. and S. A. Browning [1980]. "The LSI design language i," TM–1980–1273–10, Bell Laboratories, Murray Hill, NJ.

Jouppi, N. P. [1983]. "TV: an nMOS timing analyzer," *Proc. Third Caltech Conference on VLSI*, pp. 71–86, Computer Science Press, Rockville, Md.

Kam, J. B. and J. D. Ullman [1977]. "Monotone data flow analysis frameworks," *Acta Informatica* **7**:3, pp. 305–317.

Kang, S. and W. M. vanCleemput [1981]. "Automatic PLA synthesis from a DDL-P description," *Proc. Eighteenth Design Automation Conference*, pp. 391–397.

Karlin, A. R., H. W. Trickey, and J. D. Ullman [1983]. "Experience with a regular expression compiler," to appear in *Proc. IEEE Conf. on Computer Design/VLSI in Computers*.

Karplus, K. [1982]. "CHISEL, an extension to the programming language C for VLSI layout," Ph. D. thesis, Dept. of C. S., Stanford Univ., Stanford, CA.

Kawamoto, T. and Kajitani, Y. [1979]. "The minimum width routing of 2-row, 2-layer polycell layout," *Proc. Sixteenth Design Automation Conference*, pp. 290–296.

Kedem, G. and H. Watanabe [1982]. "Optimization techniques for IC layout and compaction," TR–117, Dept. of Computer Science, Univ. of Rochester, Rochester, NY.

Kedem, Z. [1982]. "Optimal allocation of computational resources in VLSI," *Proc. Twenty-Third Annual IEEE Symposium on Foundations of Computer Science*, pp. 379–386.

Kedem, Z. M. and A. Zorat [1981]. "On relations between input and communication/computation in VLSI," *Proc. Twenty-Second Annual IEEE Symposium on Foundations of Computer Science*, pp. 37–44.

Kernighan, B., D. Schweikert, and G. Persky [1973]. "An optimum channel routing algorithm for polycell layouts of integrated circuits," *Proc. Tenth Design Automation Conference*, pp. 50–59.

Kernighan, B. W. and S. Lin [1970]. "An efficient heuristic procedure for partitioning graphs," *Bell System Technical J.* **49**:2, pp. 291–307.

Kissin, G. [1982]. "Measuring energy consumption in VLSI circuits: a foundation," *Proc. Fourteenth Annual ACM Symposium on the Theory of Computing*, pp. 99-104.

Kleitman, D., F. T. Leighton, M. Lepley, and G. L. Miller [1981]. "New layouts for the shuffle-exchange graph," *Proc. Thirteenth Annual ACM Symposium on the Theory of Computing*, pp. 278–292.

Knuth, D. E. [1969]. *The Art of Computer Programming, Vol. II: Seminumerical Algorithms*, Addison-Wesley, Reading Mass.

Knuth, D. E. [1973]. *The Art of Computer Programming, Vol. III: Sorting and Searching*, Addison-Wesley, Reading Mass.

Kosaraju, S. R. [1979]. "Fast parallel processing array algorithms for some graph problems," *Proc. Eleventh Annual ACM Symposium on the Theory of Computing*, pp. 231–236.

Kung, H.-T. [1979]. "Let's design algorithms for VLSI systems," *Proc. of Caltech Conf. on VLSI: Architecture, Design, Fabrication*, pp. 65–90.

Kung, H.-T. and P. L. Lehman [1980]. "Systolic (VLSI) arrays for relational database operations," *ACM SIGMOD International Symposium on Management of Data*, pp. 105–116.

Kung, H.-T., and C. E. Leiserson [1978]. "Algorithms for VLSI processor arrays," *Symposium on Sparse Matrix Computations*, Knoxville, Tenn.

Kung, H.-T., L. M. Ruane, and D. W. L. Yen [1981]. "A two-level pipelined systolic array for convolutions," in Kung, Sproull, and Steele [1981], pp. 255–264.

Kung, H.-T. and S. W. Song [1981]. "A systolic 2–D convolution chip," CMU–CS–81–110, Dept. of C. S., Carnegie-Mellon Univ., Pittsburg, Pa.

Kung, H.-T., R. Sproull, and G. Steele (eds.) [1981]. *VSLI Systems and Computations*, Computer Science Press, Rockville, Md.

LaPaugh, A. S. [1980a]. "A polynomial time algorithm for routing around a rectangle," *Proc. Twenty-First Annual IEEE Symposium on Foundations of Computer Science*, pp. 282–293.

LaPaugh, A. S. [1980b]. "Algorithms for integrated circuit layout: an analytic approach," Ph. D. thesis, Dept. of EECS, MIT, Cambridge, Mass.

Lauther, U. [1979]. "A min-cut placement algorithm for general cell assemblies based on a graph representation," *Proc. Sixteenth Design Automation Conference*, pp. 1–10.

Lauther, U. [1981]. "An $O(n \log n)$ algorithm for boolean mask operations," *Proc. Eighteenth Design Automation Conference*, pp. 555–559.

Lehman, P. L. [1981]. "A systolic (VLSI) array for processing simple relational queries," in Kung, Sproull, and Steele [1981], pp. 285–295.

Leighton, F. T. [1981]. "New lower bound techniques for VLSI," *Proc. Twenty-Second Annual IEEE Symposium on Foundations of Computer Science*, pp. 1–12.

Leighton, F. T. [1982]. "A layout strategy for VLSI which is provably good," *Proc. Fourteenth Annual ACM Symposium on the Theory of Computing*, pp. 85–98.

Leighton, F. T. and C. E. Leiserson [1982]. "Wafer-scale integration of systolic arrays," *Proc. Twenty-Third Annual IEEE Symposium on Foundations of Computer Science*, pp. 297–311.

Leighton, F. T. and G. L. Miller [1981]. "Optimal layouts for small shuffle-exchange graphs," in *VLSI-81* (J. P. Gray, ed.), Academic Press, New York, pp. 289–299.

Leighton, F. T. and A. L. Rosenberg [1982]. "Three dimensional circuit layouts," unpublished memorandum, MIT, Cambridge, Mass.

Leinwand, S. M. [1981]. "Process oriented logic simulation," *Proc. Eighteenth Design Automation Conference*, pp. 511–517.

Leiserson, C. E. [1979]. "Systolic priority queues," *Proc. Caltech conf. on VLSI*, pp. 199–214.

Leiserson, C. E. [1980]. "Area efficient graph algorithms (for VLSI)," *Proc. Twenty-First Annual IEEE Symposium on Foundations of Computer Science*, pp. 270–281.

Leiserson, C. E. [1983]. *Area efficient VLSI computation*, MIT Press, Cambridge, Mass.

Leiserson, C. E. and R. Y. Pinter [1981]. "Optimal placement for river routing," in Kung, Sproull, and Steele [1981], pp. 126–142.

Leiserson, C. E., F. M. Rose, and J. B. Saxe [1983]. "Optimizing synchronous circuitry by retiming," *Proc. Third Caltech Conference on VLSI*, pp. 87–116, Computer Science Press, Rockville, Md.

Leiserson, C. E. and J. B. Saxe [1981]. "Optimizing synchronous systems," *Proc. Twenty-Second Annual IEEE Symposium on Foundations of Computer Science*, pp. 23–36.

Lelarasmee, E. and A. L. M. Sangiovanni-Vincentelli [1982]. "RELAX: a new circuit simulator for large scale MOS integrated circuits," *Proc. Nineteenth Design Automation Conference*, pp. 682–690.

Lengauer, T. [1982]. "The complexity of compacting hierarchically specified layouts of integrated circuits," *Proc. Twenty–Third Annual IEEE Symposium on Foundations of Computer Science*, pp. 358–368.

Lengauer, T. and K. Mehlhorn [1981]. "On the complexity of VLSI computations," in Kung, Sproull, and Steele [1981], pp. 89–99.

Lengauer, T. and K. Mehlhorn [1982]. "HILL—hierarchical layout language, a CAD system for VLSI design," TR–A82/10–FB10, Univ. des Saarlandes, Saarbrucken, West Germany.

Lev, G., N. Pippenger, and L. G. Valiant [1981]. "A fast parallel algorithm for routing in permutation networks," *IEEE Trans. on Computers* **C-30**:2, pp. 93–100.

Liao, Y.-Z. and C. K. Wong [1983]. "An algorithm to compact a VLSI symbolic layout with mixed constraints," *IEEE Trans. on Computer Aided Design of Integrated Circuits and Systems* **CAD-2**:2, pp. 62–69.

Lieberherr, K. J. and S. E. Knudsen [1983]. "Zeus: a hardware description language for VLSI," to appear in *Proc. Twentieth Design Automation Conference*.

Lipton, R. J., S. C. North, R. Sedgewick, J. Valdes, and G. Vijayan [1982]. "ALI: a procedural language to describe VLSI layouts," *Proc. Nineteenth Design Automation Conference*, pp. 467–474.

Lipton, R. J. and R. Sedgewick [1981]. "Lower bounds for VLSI," *Proc. Thirteenth Annual ACM Symposium on the Theory of Computing*, pp. 300–307.

Lipton, R. J., R. Sedgewick, and J. Valdes [1982]. "Programming aspects of VLSI," *Proc. Ninth ACM Symposium on Principles of Programming Languages*, pp. 57–65.

Lipton, R. J. and R. E. Tarjan [1979]. "A planar separator theorem," *SIAM J.*

*Applied Math.* **36**:2, pp. 177–189.

Lipton, R. J. and R. E. Tarjan [1980]. "Applications of a planar separator theorem," *SIAM J. Computing* **9**:3, pp. 513–524.

Lipton, R. J. and J. Valdes [1981]. "Census functions: an approach to VLSI upper bounds," *Proc. Twenty-First Annual IEEE Symposium on Foundations of Computer Science*, pp. 13–22.

Luk, W. K. [1981]. "A regular layout for parallel multiplier of $O(\log^2 n)$ time," in Kung, Sproull, and Steele [1981], pp. 317–326.

Lyon, R. [1981]. "Simplified design rules for VLSI layout," *Lambda* **2**:1, pp. 54–59.

Maier, D. [1979]. "An efficient method for storing ancestor information in trees," *SIAM J. Computing* **8**:4, pp. 599-618.

Mairson, H. G. and J. Stolfi [1982]. "Reporting line intersections in the plane," unpublished memorandum, Dept. of Computer Science, Stanford Univ., Stanford, CA.

Matthews, R., J. Newkirk, and P. Eichenberger [1982]. "A target language for silicon compilers," *IEEE COMPCON*, pp. 349–353.

McCreight, E. M. [1980]. "Efficient algorithms for enumerating intersecting intervals and rectangles," CSL–80–9, Xerox PARC, Palo Alto, CA.

McWilliams, T. M. [1980]. "Verification of timing constraints on large digital systems," *Proc. Eighteenth Design Automation Conference*, pp. 139–147.

Mead, C. A. and L. A. Conway [1980]. *Introduction to VLSI Systems*, Addison-Wesley, Reading Mass.

MOSIS [1983]. Design rule documentation files, MOSIS, ISI, Marina Del Rey, Ca.

Mukhopadhyay, A. [1979]. "Hardware algorithms for nonnumeric computation," *IEEE Trans. on Computers* **C-28**:6, pp. 384–394.

Nassimi, D. and S. Sahni [1980]. "Finding connected components and connected ones on a mesh-connected parallel computer," *SIAM J. Computing* **9**:4, pp. 744-757.

Newkirk, J. and R. Matthews [1983]. NMOS cell library, Information Systems Laboratory, Stanford Univ., Stanford, CA.

Newton, A. R. [1979]. "Techniques for the simulation of large-scale integrated circuits," *IEEE Trans. on Circuits and Systems* **CAS-26**:9, pp. 741–749.

Nham, H. N. and A. K. Bose [1980]. "A multiple delay simulator for MOS LSI circuits," *Proc. Eighteenth Design Automation Conference*, pp. 610–617.

Ousterhout, J. K. [1983]. "Crystal: a timing analyzer for nMOS VLSI circuits," *Proc. Third Caltech Conference on VLSI*, pp. 57–70, Computer Science Press, Rockville, Md.

Paillotin, J. F. [1981]. "Optimization of the PLA Area," *Proc. Eighteenth Design Automation Conference*, pp. 406–410.

Papadimitriou, C. H. and M. Sipser [1982]. "Communication complexity," *Proc. Fourteenth Annual ACM Symposium on the Theory of Computing*, pp. 196–200.

Parker, A., D. Thomas, D. Siewiorek, M. Barbacci, L. Hafer, G. Lieve, and J. Kim [1979]. "The CMU design automation system," *Proc. Sixteenth Design Automation Conference*, pp. 73–80.

Paterson, M. S., W. L. Ruzzo, and L. Snyder [1981]. "Bounds on minimax edge length for complete binary trees," *Proc. Thirteenth Annual ACM Symposium on the Theory of Computing*, pp. 293–299.

Pinter, R. Y. [1981]. "Optimal routing in rectilinear channels," in Kung, Sproull, and Steele [1981], pp. 160–177.

Pinter, R. Y. [1982a]. "On routing two-point nets across a channel," *Proc. Nineteenth Design Automation Conference*, pp. 894–902.

Pinter, R. Y. [1982b]. "The impact of layer assignment methods on layout algorithms for integrated circuits," Ph. D. thesis, MIT, Cambridge, Mass.

Pinter, R. Y. [1983]. "River routing: methodology and analysis," *Proc. Third Caltech Conference on VLSI*, pp. 141–164, Computer Science Press, Rockville, Md.

Preas, B. T. [1979]. "Placement and routing algorithms for hierarchical integrated circuit layout," Ph. D. thesis, Stanford Univ., Stanford, Calif.

Preas, B. T. and C. W. Gwyn [1979]. "General hierarchical automatic layout of custom VLSI circuit masks," *J. Design Automation and Fault-Tolerant Computing* **3**:1, pp. 41–58.

Preparata, F. P. [1978]. "New parallel-sorting schemes," *IEEE Trans. on Computers* **C27**:7, pp. 669–673.

Preparata, F. P. [1981]. "A mesh-connected area-time optimal VLSI integer multiplier," in Kung, Sproull, and Steele [1981], pp. 311–316.

Preparata, F. P. and W. Lipski, Jr. [1982]. "Three layers are enough," *Proc. Twenty-Third Annual IEEE Symposium on Foundations of Computer Science*, pp. 350–357.

Preparata, F. P. and J. Nievergelt [1982]. "Plane sweep algorithms for intersecting geometric figures," *Comm. ACM* **25**:10, pp. 739–747.

Preparata, F. P. and J. E. Vuillemin [1979]. "The cube-connected cycles: a versatile network for parallel computation," *Proc. Twentieth Annual IEEE Symposium on Foundations of Computer Science*, pp. 140–147.

Preparata, F. P. and J. E. Vuillemin [1980]. "Area-time optimal VLSI networks for multiplying matrices," *Information Processing Letters* **11**:2, pp. 77–80.

Preparata, F. P. and J. E. Vuillemin [1981]. "Area-time optimal VLSI networks for computing integer multiplication and discrete Fourier transform," *Proc. 8th Intl. Colloquium on Automata, Languages, and Programming*, pp. 29–40, Springer-Verlag, New York.

Raghavan, R. and S. Sahni [1982]. "Optimal single row router," *Proc. Nineteenth Design Automation Conference*, pp. 38–45.

Reif, J. H. and P. G. Spirakis [1982]. "The expected time complexity of parallel graph and digraph algorithms," TR–11–82, Aiken Computation Lab., Harvard Univ., Cambridge, Mass.

Reif, J. H. and L. G. Valiant [1983]. "A logarithmic time sort for linear size networks," *Proc. Fifteenth Annual ACM Symposium on the Theory of Computing*, pp. 10–16.

Reiss, S. P. and J. E. Savage [1982]. "SLAP—a silicon layout program," CS–82–17, Dept. of Computer Science, Brown Univ., Providence, R. I.

Rivest, R. L. [1982]. "The PI (Placement and Interconnect) System," *Proc. Nineteenth Design Automation Conference*, pp. 475–481.

Rivest, R. L., A. E. Baratz, and G. Miller [1981]. "Provably good channel routing algorithms," in Kung, Sproull, and Steele [1981], pp. 153–159.

Rivest, R. L. and C. M. Fiduccia [1982]. "A 'greedy' channel router," *Proc. Nineteenth Design Automation Conference*, pp. 418–424.

Rosenberg, A. L. [1981]. "Three-dimensional integrated circuitry," in Kung, Sproull, and Steele [1981], pp. 69–79.

Rosenberg, A. L. [1982]. "The Diogenes approach to testable fault-tolerant networks of processors," CS–1982–6.1, Dept. of Computer Science, Duke Univ., Durham, N. C.

Ruzzo, W. L. and L. Snyder [1981]. "Minimum edge length planar embeddings of trees," in Kung, Sproull, and Steele [1981], pp. 119–123.

Sahni, S. and A. Bhatt [1980]. "The complexity of design automation problems," *Proc. Seventeenth Design Automation Conference*, pp. 402–411.

Savage, C. [1977]. "Parallel algorithms for graph-theoretic problems," R–784, Dept. of C. S., Univ. of Illinois, Urbana, Ill.

Savage, C. [1981]. "A systolic data structure chip for connectivity problems," in Kung, Sproull, and Steele [1981], pp. 296–300.

Savage, J. E. [1981a]. "Planar circuit complexity and the performance of VLSI algorithms," in Kung, Sproull, and Steele [1981], pp. 61–67.

Savage, J. E. [1981b]. "Area-time tradeoffs for matrix multiplication and related problems in VLSI models," *J. Computer and System Sciences* **20**:3, pp. 230–242.

Savage, J. E. [1982]. "Three VLSI compilation techniques: PLA's, Weinberger arrays, and SLAP, a new silicon layout program," CS–82–24, Dept. of Computer Science, Brown Univ., Providence, R. I.

Schwartz, J. T. [1980]. "Ultracomputers," *ACM Trans. on Programming Languages and Systems* **2**:4, pp. 484–521

Sequin, C. H. [1982]. "Managing VLSI complexity: an outlook," unpublished memorandum, Dept. of EECS, Univ. of California, Berkeley, CA.

Shamos, M. I. and D. Hoey [1976]. "Geometric intersection problems," *Proc. Seventeenth Annual IEEE Symposium on Foundations of Computer Science*, pp. 208–215.

Shaw, D. E. [1982]. "The NON-VON supercomputer," unpublished memorandum, Dept. of Computer Science, Columbia Univ., New York, NY.

Sherwood, W. [1981]. "A MOS modeling technique for 4-state true-value hierarchical logic simulation," *Proc. Eighteenth Design Automation Conference*, pp. 775–785.

Shiloach, Y. and U. Vishkin [1981]. "Finding the maximum, merging and sorting in a parallel computation model," *J. Algorithms* **2**:1, pp. 88–102.

Shrobe, H. E. [1982]. "The data path generator," *IEEE COMPCON*, pp. 340–344.

Siegel, A. [1983]. "Fast optimal placement for river routing," unpublished memorandum, Dept. of Computer Science, Stanford Univ., Stanford, CA.

Siegel, A. and D. Dolev [1981]. "The separation required for general single-layer wiring barriers," in Kung, Sproull, and Steele [1981], pp. 143–152.

Siegel, H. J. [1977]. "Interconnection networks and masking schemes for single instruction stream—multiple data stream machines," Ph. D. thesis, Princeton Univ., Princeton, N. J.

Siegel, H. J. [1979]. "Interconnection networks for SIMD machines, *Computer* **12**:6, pp. 57–65.

Siskind, J. M., J. R. Southard, and K. W. Crouch [1982]. "Generating custom high performance VLSI designs from succinct algorithmic descriptions," *Proc. MIT Conf. on Advanced Research in VLSI.*

Soukup, J. [1981]. "Circuit layout," *Proc. IEEE* **69**:10, pp. 1281–1304.

Stefik, M., A. Bell, D. Bobrow, H. Brown, L. Conway, and C. Tong [1982]. "The partitioning of concerns in digital system design," HPP–82–2, Dept. of Computer Science, Stanford Univ., Stanford CA.

Steinberg, D. and M. Rodeh [1980]. "A layout for the shuffle-exchange network with $\Theta(N^2/\log^{3/2} N)$ area," TR 088, IBM Scientific Center, Haifa, Israel.

Stolfo, S. J. and D. E. Shaw [1982]. "DADO: a tree-structured machine architecture for production systems," unpublished memorandum, Dept. of Computer Science, Columbia Univ., New York, NY.

Stone, H. S. [1971]. "Parallel processing with the perfect shuffle," *IEEE Trans. on Computers* **C-20**:2, pp. 153–161.

Storer, J. A. [1980]. "The node cost measure for embedding graphs on a planar grid," *Proc. Twelfth Annual ACM Symposium on the Theory of Computing,* pp. 201–210.

Szymanski, T. G. [1981]. "Dogleg channel routing is NP-complete," unpublished memorandum, Bell Laboratories, Murray Hill, N. J.

Szymanski, T. G. and C. J. Van Wyk [1983]. "Space efficient algorithms for VLSI artwork analysis," to appear in *Proc. Twentieth Design Automation Conference.*

Thompson, C. D. [1979]. "Area-time complexity for VLSI," *Proc. Eleventh Annual ACM Symposium on the Theory of Computing,* pp. 81–88.

Thompson, C. D. [1980]. "A complexity theory for VLSI," Ph. D. thesis, Carnegie-Mellon Univ., Pittsburg, Pa.

Thompson, C. D. [1981]. "The VLSI complexity of sorting," in Kung, Sproull, and Steele [1981], pp. 108–118.

Thompson, C. D. and H.-T. Kung [1977]. "Sorting on a mesh connected parallel computer," *Comm. ACM* **20**:4, pp. 263–271.

Tompa, M. [1980]. "An optimal solution to a wire routing problem," *Proc. Twelfth Annual ACM Symposium on the Theory of Computing*, pp. 161–176.

Trickey, H. W. [1981]. "Good layouts for pattern recognizers," *IEEE Trans. on Computers* **C-31**:6, pp. 514–520.

Trickey, H. W. [1982]. "Using NFA's for hardware design," unpublished memorandum, Stanford Univ., Dept. of C. S.

Trickey, H. W. and J. D. Ullman [1982]. "A regular expression compiler," *IEEE COMPCON*, pp. 345–348.

Tseng, C.-J. and D. P. Siewiorek [1983]. "Facet: a procedure for the automated systhesis of digital systems," unpublished memorandum, Dept. of CS, Carnegie-Mellon Univ., Pittsburg, PA.

Ullman, J. D. [1982]. "Combining state machines with regular expressions for automatic synthesis of VLSI circuits," STAN–CS–82–927, Computer Science Dept., Stanford Univ., Stanford, CA.

Upfal, E. [1982]. "Efficient schemes for parallel communication," *Proc. ACM Symp. on Principles of Distributed Computing.*

Valiant, L. G. [1981]. "Universality considerations in VLSI circuits," *IEEE Trans. on Computers* **C-30**:2, pp. 135–140.

Valiant, L. G. [1982a]. "A scheme for fast parallel communication," *SIAM J. Computing* **11**:2, pp. 350–361.

Valiant, L. G. [1982b]. "Fast parallel computation," TR–17–82, Aiken Computation Lab., Harvard Univ., Cambridge, Mass.

Valiant, L. G. and G. J. Brebner [1981]. "Universal schemes for parallel communication," *Proc. Thirteenth Annual ACM Symposium on the Theory of Computing*, pp. 263–277.

Vuillemin, J. E. [1980]. "A combinatorial limit to the computing power of VLSI circuits," *Proc. Twenty-First Annual IEEE Symposium on Foundations of Computer Science*, pp. 294–300.

Weinberger, A. [1967]. Large scale integration of MOS complex logic: a layout method," *IEEE J. of Solid State Circuits* **SC-2**:4, pp. 182–190.

Williams, J. D. [1978]. "STICKS, a graphical compiler for high level LSI design," *Proc. 1978 National Computer Conference*, pp. 289–295, AFIPS Press, Montvale, New Jersey.

Wilmore, J. A. [1981]. "Efficient boolean operations on IC masks," *Proc. Eighteenth Design Automation Conference*, pp. 571–579.

Wise, D. S. [1981]. "Compact layouts of banyan/FFT networks," in Kung, Sproull, and Steele [1981], pp. 186–195.

Wolf, W., J. Newkirk, R. Matthews, and R. Dutton [1983]. "Dumbo, a schematic-to-layout compiler," *Proc. Third Caltech Conference on VLSI*, pp. 379–393, Computer Science Press, Rockville, Md.

Yao, A. C. [1979]. "Some complexity questions related to distributive computing," *Proc. Eleventh Annual ACM Symposium on the Theory of Computing*, pp. 209–213.

Yao, A. C. [1981]. "The entropic limitations of VLSI computations," *Proc. Thirteenth Annual ACM Symposium on the Theory of Computing*, pp. 308–311.

Yoshimura, T. and E. S. Kuh [1982]. "Efficient algorithms for channel routing," *IEEE Trans. on Computer-Aided Design of Integrated Circuits and Systems* **CAD-1**:1, pp. 25–35.

# INDEX

## T

# Z